SMALL SHIPS

— FOURTH EDITION —

A BOOK OF STUDY PLANS FOR TUGS, FREIGHTERS, FERRIES, EXCURSION BOATS, TRAWLER YACHTS, HOUSEBOATS & FISHING VESSELS

WORKING VESSELS & WORKBOAT HERITAGE YACHT DESIGNS
FROM THE BOARDS OF THE

BENFORD DESIGN GROUP
P. O. BOX 447
ST. MICHAELS, MD 21663
Voice: 410-745-3235
Fax: 410-745-9743

PUBLISHED BY

ST. MICHAELS, MARYLAND

TILLER PUBLISHING
P. O. Box 447
St. Michaels, MD 21663

Voice: 410-745-3750
Fax: 410-745-9743

Photo Credits:

Front cover photo of 38' Tug Yacht ***Loafer*** by Monica Brown, the Florida Bay Coaster yard Reuben Trane and 50' ***Florida Bay*** and ***Duchess*** by Jay Benford. Back cover photos of the Solarium 44 ***Duchess*** courtesy of Richard Brashear. Photos in the book supplied by owners and others as noted. All other photos by Jay Benford.

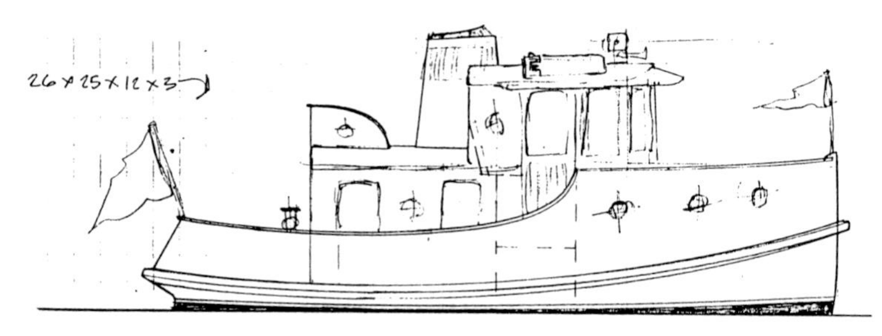

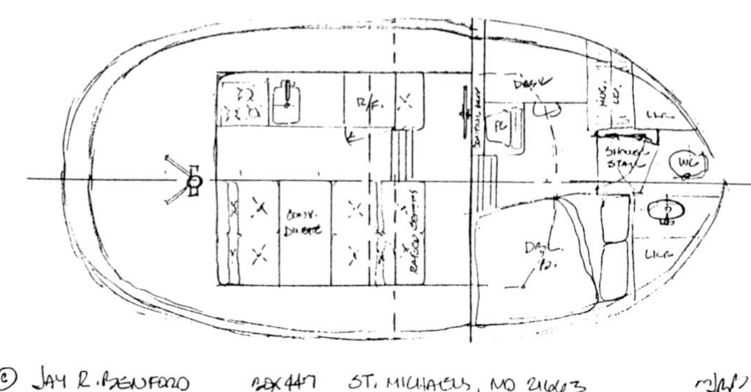

Dedication:

To my friend, the late Ron Brown.

He never forgot that these boats had to "look" right in order to be right. I miss the pleasure of working with him and his unflagging optimism.

Also, many thanks to all the clients who gave us the opportunity and challenge to create these small ships.

The variety of the work has kept us interested and burning the midnight oil at our "idea factory".

BY THE SAME AUTHOR:

CRUISING BOATS, SAIL & POWER, 4 editions in 1968, 1969, 1970 & 1971. Design catalog and article reprints. OP*

PRACTICAL FERRO-CEMENT BOATBUILDING, with Herman Husen, 3 editions in 1970, 1971 & 1972. Best-selling construction handbook, a how-to on ferro-cement. OP*

DESIGNS & SERVICES, 7 editions, 1971, 1972, 1987, 1988, 1990, 1993 & 1997. Catalog of plans and services of our design firm.

BOATBUILDING & DESIGN FORUM, 1973. A monthly newsletter with more information on ferro-cement boatbuilding & other boatbuilding information. OP*

THE BENFORD 30, 3 editions in 1975, 1976 & 1977. An exposition on the virtues of this design and general philosophy on choosing a cruising boat. OP*

CRUISING DESIGNS, 4 editions in 1975, 1976, 1993 & 1996. A catalog of plans and services and information about boats and equipment.

DESIGN DEVELOPMENT OF A 40M SAILING YACHT, 1981. A technical paper presented to the Society of Naval Architects & Marine Engineers, at the fifth Chesapeake Sailing Yacht Symposium, and in the bound transactions of that meeting.

CRUISING YACHTS, 1983. A hard cover book with a selection of Benford designs covered in detail, including several complete sets of plans, a lot of information about the boats and how they came to be. Eight pages of color photos.

THE FLORIDA BAY COASTERS, A Family of Small Ships, 1988. A book of study plans of these Benford designed freighter yachts. OP*

SMALL CRAFT PLANS, 2 editions in 1990 and 1997. A book with fifteen sets of full plans for 7'-3" to 18'-0" dinghies and tenders.

SMALL SHIPS, 3 prior editions, 1990, 1992, &1995. A book full of study plans for Benford designs for tugs, freighters (like the Florida Bay Coasters), ferries and excursion boats. Ten pages of color photos.

POCKET CRUISERS & TABLOID YACHTS, VOLUME 1, 1992, revised 1996. A book with 6 complete sets of plans (11 boats including the different versions) for boats from 14' to 25', including 14' & 20' Tug Yachts, 17' & 25' Fantail Steam Launches, a 14' Sloop, a 20' Catboat, and 20' Supply Boat & Cruiser.

* OP = Out of print

What size is it?

We've endeavored to keep most of the boats at the same scale throughout this book. This worked well with keeping them at 1/8"=1'-0" until we got past about the sixty-footers, and we switched to 3/32"=1'-0" at that point, and 1/16"=1'-0" for one of the one hundred-footers and some of the preliminaries. Many of the plans have been reproduced with scale strips on them for a handy check of the scale to use.

With the unusual situation of an author also in charge of the content and layout of the book, I must take full blame for any inconvenience in some of the boats being shown with their layouts turned sideways. Doing this allowed us to put more drawings in what became a book more than twice the size of the original volume.

A Second Opinion....

Several magazines have graciously given us their permission to reprint some interviews and articles about our designs. These will provide a second opinion about the boats and our design work. We again thank the publications where these originally appeared for this helpful addition to this book.

Acknowledgments:

Others besides myself (JRB on the drawings) whose drawing talents have contributed to these drawings are Peter Dunsford (PAD), Tom Fake (TWF), Ed Frank (EMF), Gary Grant (Grant), Brian Harris (BAH), George Hockley (GOH), Bob Perry (RHP or Perry), Tatsuaki Suzuki (TS or chop mark), Jon Stivers (JSS), and Bruce Williams (BEMW). Many thanks to them for their help, to Craig Goring for a couple of decades of engineering and design work, to Steve Worden and Dona Benford for help in cleaning up the errors and typos in the text, and to Tom Fake for his help in the lengthy process of assembling the drawings into the form you see them in the book.

Introduction:

I spent my apprentice years (1962-1969) working for others, doing a great variety of design work. A significant part of this work involved commercial vessels, and my interest and love for small ships has continued to grow in the ensuing years while running my own design office.

This book is a collection of our working vessel and workboat heritage powerboat designs. I find this type to have a refreshing frankness about them, with the absence of tricky styling and glitter and glitz. While too many of the production boats are busy selling sizzle, we've been delivering steak — to those astute enough to recognize the difference.

Some of the designs shown in this book are available as stock boats, or as stock plans for you or a professional builder to use to build your next small ship. Drop us a line or give us a call. We'll be happy to put you in touch with the current builders.

Some of the rest of the plans are ideas — pipe dreams of what might be or ought to be .... If one of these appeals to you, let us know and we'll talk about what it would take to turn it into a reality. Or if one suggests a variation, we're happy to talk about creating a new boat — this is the mainstay of our business.

— About The Fourth Edition —

Seven years ago we put together the first edition of this book of study plans. It proved to be very popular and we've since sold out that edition and the expanded second and third editions that followed. In the course of talking to people and reading their letters about the book over that time, there have been a number of constructive suggestions amongst the genuinely delighted responses. The materials added to this further expanded edition are in answer to suggestions and questions we've had over the last several years and we'll look forward to getting feedback about how we can continue to improve this book.

The new material added to this edition will be found throughout the book, along with a few pages at the end, including an index. In interests of economy we've put some of the design work of the last couple years in pages numbered 16A, 16B and so forth, so as not disrupt the page numbering already in place for the majority of the book. We hope you enjoy the book as much as we've enjoyed creating the designs.

Jay R. Benford
Motor Yacht ***Odyssea***
St. Michaels, Maryland
June 1997

TABLE OF CONTENTS

Section/Page Subject/Designs

Reuben Trane photo

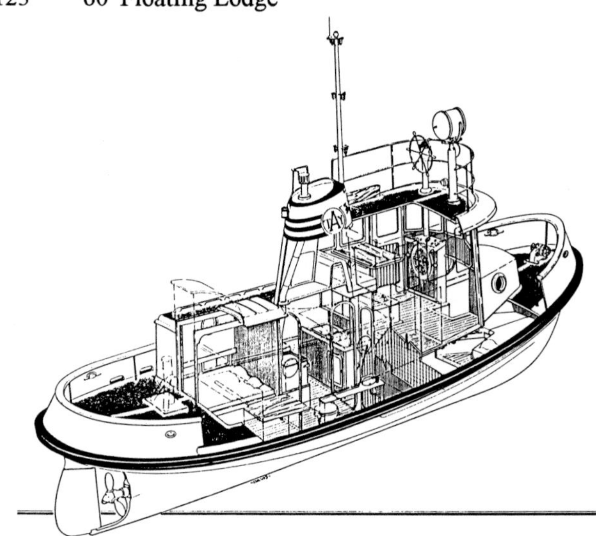

Section/Page Subject/Designs

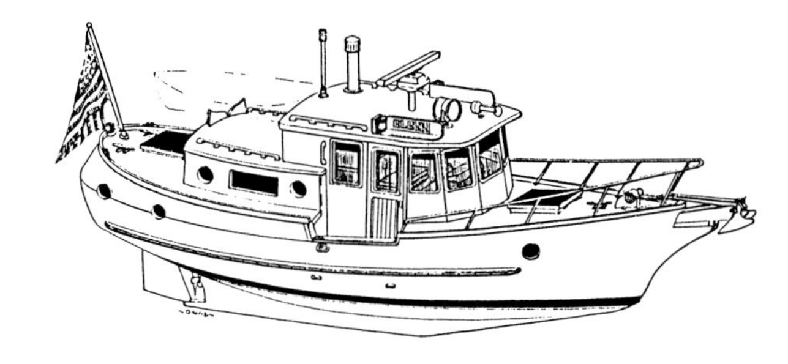

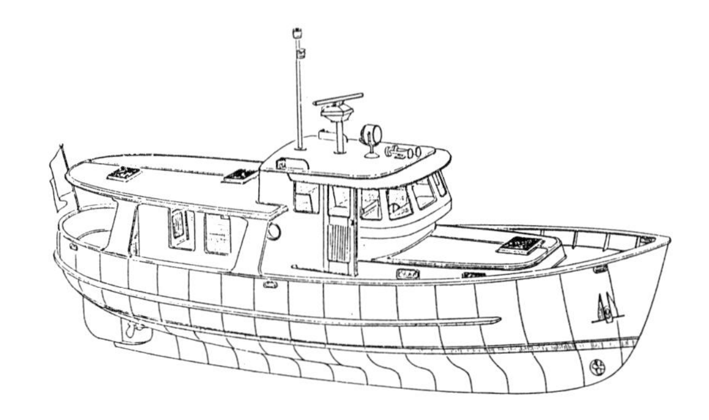

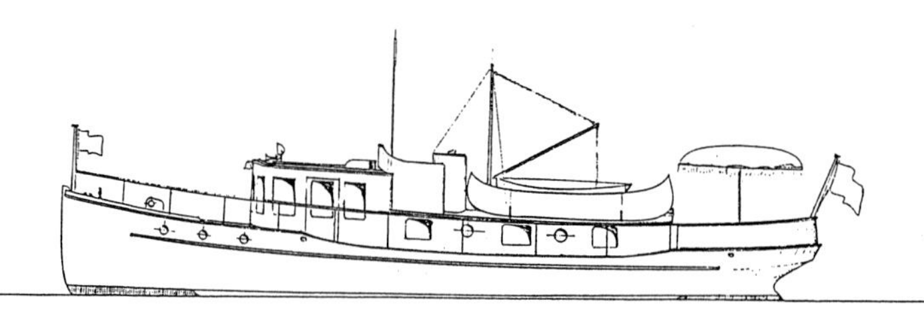

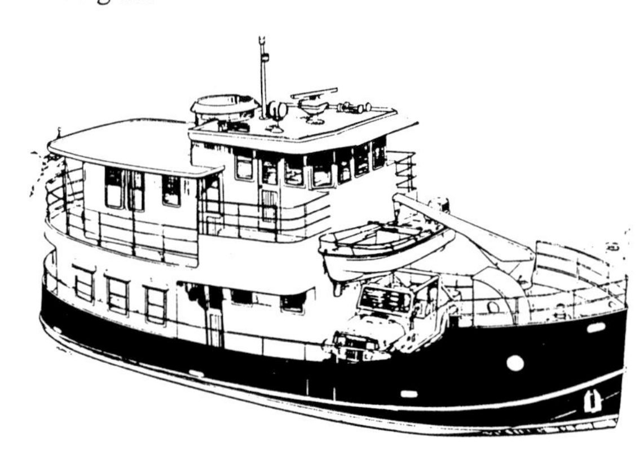

MORE COMFORTS THAN HOME

Jay Benford's stunning Florida Bay Coaster 65 is a true home away from home. It even has a space to park the car

STORY BY DOUG HUNTER

PHOTOS BY JIM WILEY

And it came to pass that cruising yachtsmen looking for something solid, proven and reliable in hull design came upon the fishing trawler, and rejoiced greatly, and stuck master staterooms down below where the cod were once piled high on ice. And this enthusiasm for the fishing trawler inspired sane men to cast their eyes about the rest of the commercial fleet. Soon they were leaving dock with their mates and mateys in luxurious comfort, in hulls inspired by lobster boats, and shrimpers, and crabbers, and tugs, and you name it. ▸

The Florida Bay Coaster 65 boasts 1,305 square feet of enclosed living space. Then there's the deck area

Jerry Trane (below) got brother Reuben's sights bent away from sailboats to powerboats, then brought in designer Jay Benford (right) to deliver the ultimate liveaboard. All the appliances (opposite, bottom) are from Sears

With time it seemed the designers had left no port unprobed for fresh inspiration. What could possibly be left for the cruising fraternity to adopt? The dredging barge? The oil rig?

Of course not. But think. Think hard. A kind of boat you have probably been aboard on numerous occasions, probably with the family, and have travelled in comfort from points A to B, the whole time without ever considering what a great yacht this thing would make.

You don't mean —

I do.

* * *

Jay Benford is one of those rare yacht designers whose work is startling and handsome at the same time. There are lots of designers who come up with wild and daring ideas, but too often gimmickry commands the foreground of their vision. Benford's work is firmly anchored in tradition. Sometimes he salutes it with faithful recreations. More often than not he seizes upon the essential elements of the past, tosses in some contemporary yearnings, and whirls the whole thing like a Rubik's cube. You're left with something that makes you think of days gone by even while wondering: why hasn't someone done that before?

On one level Benford is a renovator. Like someone who buys a handsome brick Victorian home and guts the thing, laying in banks of skylights and open-concept floor plans, Benford seizes on familiar, graceful forms and goes wild within their essential structure. On another level he is a rehabilitator, taking styles of yachts or ships that have either been consigned to the past or deemed unsuitable to the cruising life and making them relevant. More than relevant. Seductively logical.

I defy anyone who loves ships and the sea to peruse Benford's eclectic portfolio and not come across something that makes you feel warm and kind of dizzy, as if you've indulged yourself too long in a hot bath. I am particularly taken by his drawings of the Solarium 44, a plumb-bow fantail motor yacht. Its elliptical aft superstructure surrounds a saloon and galley in a parade of windows which drop open like drawbridges. If you are not overcome by the urge to take this design to the nearest remote anchorage and start lollygagging around in the solarium, then you are without a soul.

The Solarium 44 is typical of Benford's inspirations in that the concept is shot through with evidence of an eye for the essential needs of the good life. This quality is what links the Solarium 44 to the Kanter 64, a rakish pilothouse sloop built by Canada's Kanter Aluminum

OMNI
W.T
DOOR CLOSE
IN

Yachts. And it's what makes them kinfolk of the Florida Bay Coaster 65, the astonishing crowning glory of the Benford pleasure principle.

I don't know anyone who has a thing about ferries the way Benford does. In fact, I don't know anyone other than Benford who has shown a professional interest in making the ferry the cruising yacht of tomorrow. There's nothing in his background, no adolescent signpost, that points to this conviction. He grew up on the south shore of Lake Ontario in Rochester, N.Y. "My folks took me sailing before I could walk," he says. "I've always been cruising." At age 12 he started reading library books on yacht design, and he wound up in the naval architecture program at the University of Michigan.

Benford left one year before completing his degree ("I was too impatient to get to work.") and apprenticed with John Atkin in Connecticut. There followed 18 years as a designer around Puget Sound in the Pacific Northwest – eight of them in Seattle, 10 of them in Friday Harbor. He moved back east, to Maryland, five years ago.

It's out west where you'll find one of Benford's best-known boats, for which he personally is virtually unknown. If you have travelled from Vancouver's downtown waterfront to the shops and restaurants and marine businesses of Granville Island, then you have been a passenger on the Granville Island ferry.

The 20-footer whirls back and forth across False Creek like a diesel-powered waterbug, its passengers snug within the cabin that stretches the length of the boat. The pilot sits dead centre, over the engine; to give him a good view while seated, Benford raised the helm seat and with it the centre of the cabintop in a kind of cupola.

I have a set of blueprints for the ferry. The basic plans were completed Oct. 16, 1983 – by then Benford had relocated to Maryland. But somewhere between that fall day and Feb. 7, 1984, the ferry concept had wormed its way into Benford's imagination, and he had produced a *cruising* version of the False Creek ferry. He gave it a galley, an enclosed head compartment and a settee that converted into a double berth. He called it the Friday Harbor Ferry.

There was no turning back. The concept of the ferry-as-cruiser gave rise to a 34-foot and a 45-foot Friday Harbor Ferry. This served to pique the interest of brothers Reuben and Jerry Trane.

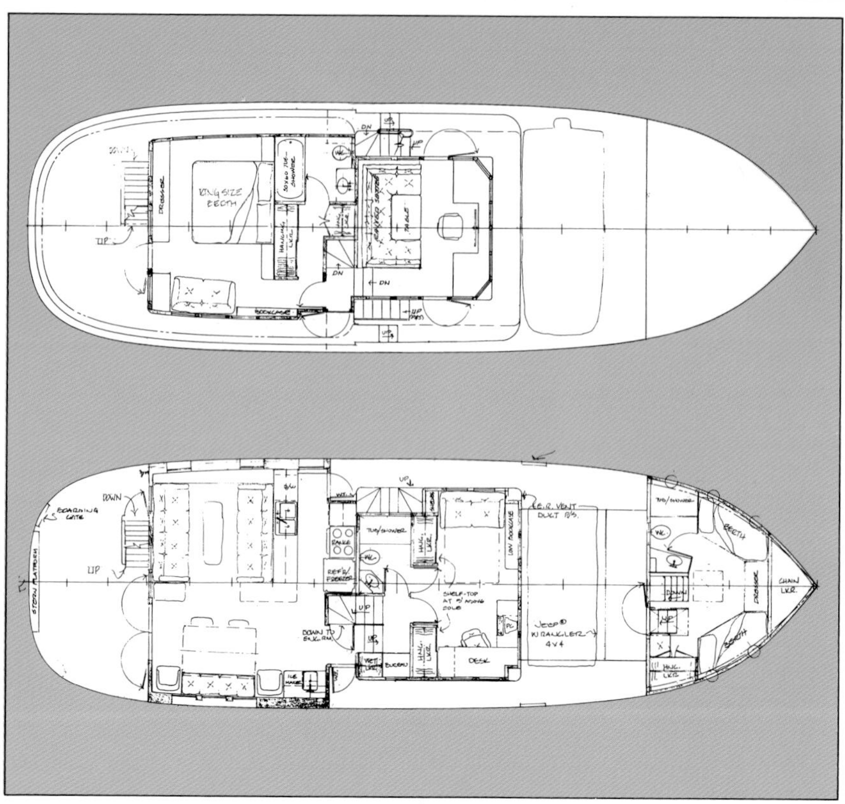

Reuben is the sailor, Jerry the powerboater. About nine years ago Reuben founded the Florida Bay Boat Company and launched a highly idiosyncratic series of coastal cruising sailboats known as Hens. The Mud Hen was the first model, which championed the Reuben Trane philosophy that cruising in any boat under 30 feet was really camping, so let's not kid ourselves. (The Hen moulds were recently sold to Mirage Fibreglass of Palatka, Fla.)

Jerry got Reuben's sights twisted over to powerboats about three years ago when they saw the design for the 45-foot Friday Harbor Ferry. This definitely wasn't camping, but they loved the concept. The aesthetics needed some work, though. What if Benford could bring a tramp-steamer flavour to his ferries-cum-cruisers?

Benford came up with two designs, a 50-footer and a 65-footer, both of which have been built, both of which were on display at the Miami International Boat Show last February. Showgoers were drawn aboard like iron filings to a magnet. There was nothing else even close to them in style or intent.

That intent is to provide luxurious, self-sufficient cruising. Where some powerboats conspire to drag along the comforts of home, Benford's coastal cruiser determines to drag along the whole home. The enclosed-floor-plan space in the 65, which is the subject of the accompanying photographs, is 1,305 square feet. That's bigger than the ground floor of my house, and the ground floor of my house has three bedrooms, a kitchen, a bathroom, a living-room and a dining room. The conclusion (for me, anyway) is obvious: the 65 has every right to call itself a home. After all, the appliances (which include a washer and dryer and a large freezer) are from Sears. There's even a place to park your car.

Your car in the case of the Coastal Cruiser 65 is a Jeep Wrangler, which rests atop a 10x12-foot cargo bay. The bay is big enough to accommodate a golf cart, and both the Jeep and the cart are moved around with a hydraulic crane mounted on the foredeck. The automobile is a key ingredient in the many Coaster studies Benford has developed. Even the smallest version, a 40-footer, can accommodate a Suzuki Samurai.

The tall superstructure and the sight of a Jeep on deck give the Coastal

The master cabin in the 65 is a masterpiece of air and light. From the outside, the 50 and 65 (right) demonstrate their tramp- steamer/ferry heritage

CONTINUED

Cruiser a top-heavy appearance, yet the design is quite stable. The hull form and generous beam give the boat more stability than a lot of conventional motor yachts. Nor is the 65 as ponderous as I had imagined. While we weren't able to leave the dock, Benford says that the 65 cruises at 10 to 10.5 knots, and achieves hull speed at only three-quarter rpm, thanks to the pair of Caterpillar 3208TAs tucked into their own walk-in engine rooms. The Cats are so up to the job that the 65 can reach its cruising speed on one engine alone. Each engine room also houses a 20-kw Kohler generator built on a Yanmar diesel block. Up top, controls for the engines and bow thruster are located outboard as well as on the bridge (which is equipped with commercial ship windows and a rotating clear-view screen), and there's an MMC remote forward.

As pretty as this vessel is inside, it means business. The 65 is built to American Bureau of Shipping standards for small steel ships up to 200 feet in length. The hull is ¼-inch thick, with 3/16ths plate used on the lower works and ⅛-inch on the upper works. Four watertight doors divide the belowdecks into six compartments; framing is both transverse and longitudinal. Beneath the aft deck is a hold area large enough to house a proper scuba diving shop. A compressor allows you to run air-powered tools anywhere on the boat.

Such attention to detail does not come cheap. The base price of the 65 is US$595,000. Fully equipped, it goes for US$750,000. I don't know if that includes the Jeep. At three-quarters of a million bucks all up, I don't think the Jeep's status as an option is that important.

Not many Canadians are in a position to acquire one of these fine craft. They're certainly in a position to appreciate one. About a year ago Wye Heritage Marina in Midland, Ont., sounded out its clientele on their interest in a possible new breed of cruiser suited to poking around Georgian Bay. The marina, which has metal fabricating capability, included a drawing in its newsletter designed to stimulate brainstorming. The drawing was of a Florida Bay Coaster.

I lounged around with Benford for a while in the main living area of the 65. There aren't many boats that I could truly live aboard. This was one of them, perhaps the only one. Living aboard invariably means compromise. Not on this boat. There were plenty of walls for paintings, a nice office area forward, hard-wearing oak floors, and no weird storage areas in the kitchen (definitely not a galley) propagated by curving topsides. I confessed to Benford that the only drawback I could find was a lack of room for my piano, a cabinet grand.

No problem. "Take out the wet bar," said Benford, pointing to a feature on the starboard side that was clearly superfluous, given its proximity to the upright fridge-freezer in the kitchen. Besides, he had a 55 in the works with a piano.

But if you own a Donzi 16 runabout, a Lamborghini 4x4, a mini-sub and a Hughes 500 helicopter, nothing less than the 100-foot Florida Bay Coaster will do. Benford has drawn only an outboard profile for that model. If you own more than a piano — say a string quartet — I'm sure he could squeeze it in. ⚓

Doug Hunter is the editor of CY

North Star shows one of these small ships with very traditional and elegant interior accommodations. Each is built to order with the owners' choices in style, decor and outfitting. Photos courtesy of Reuben Trane.

A Long Trane Coming

by Rick Friese

The "HEAD CHICKEN" turns his talent to steel motoryachts

For quite a few years Reuben Trane, president of the Florida Bay Coaster Company, was Head Chicken of the Florida Bay Boat Building Company.

"Head Chicken?" you ask. "What's that?"

Well, the Florida Bay Boat Building Company built "hens" — the Peep Hen, Mud Hen, Bay Hen, Marsh Hen and Sand Hen. These trailerable sailboats ranged from small 14-footers to medium-sized 24-footers, but all were Grade A.

The hens reflected Trane's thinking about what a sailboat should do and what a sailboat should be. For example, the Peep Hen, no larger than a good-sized rowboat, boasted a cuddy cabin with two quarter berths (large enough for the 6′5″ Trane), full sitting headroom, portable head, galley area large enough for a two-burner stove, portable cooler, counter space for food preparation and storage for canned goods and crew gear. Add a cockpit that seated four adults comfortably for day sailing, a hull that drew just nine inches with the centerboard up, and you certainly have a lot of sailboat in 14 feet.

A couple of years ago Trane's interest turned to building motoryachts (40 to 100-plus feet), and the lines of thinking he pursued with his hens carried over. The absolutely functional design and execution of his Florida Bay Coasters is consistent with his belief that boats should be designed to do a specific job properly.

Although Trane, tall, trim, with salt and pepper hair and a moustache, is only in his early 40s, he has been sailing, paddling, rowing and powering about in boats for 30 years. He's both a designer and builder. And he's won an Academy Award for *Manhattan Melody*, produced when he was a student filmmaker at Columbia.

But that's another story.

The Coaster 50, designed by Jay Benford and built in Florida by the Florida Bay Coaster Company, represents Reuben Trane's idea of the ultimate shoal-draft, coastal cruiser. She's seen here on Card Sound, Florida.

Trane's experience in small boats, which includes cruising to Bermuda through the Bahamas and along the East Coast of the United States, has if nothing else taught him to value what works in practice rather than what

Reprinted courtesy of **Cruising World** magazine from their October 1989 issue.

Accompanied by son Jimmy and friend Cheryl Nelson, Trane points out a sight during a recent cruise of South Florida.

appears to be a good idea at the boat show. "My success in sailboat building," he says, "was a result of always designing a boat *I* would like to use."

Trane has carried this philosophy over into the Florida Bay Coaster project. His first Coaster, a 50-footer, was a result of what Trane and his late wife Rosanne wanted in a boat when their kids (Flemming, a freshman at the University of Florida, and Jimmy, an 8th grader) were grown and out of the house. Ideally, they decided they would still like to maintain a home, but gunkhole and cruise six months each year.

Half a year on a boat is a lot of time and Trane didn't want to rough it. In fact, he wanted all of the comforts and conveniences of home, including oversized staterooms, saloon and galley, all with vistas of the water. Also, for convenience ashore and afloat, he wanted to carry some type of car and small runabout. His stated design goal was "...to create as homelike a space as possible aboard a functional, comfortable, easily maintained vessel."

To achieve this design goal, he began to review portfolios of various naval architects. Jay Benford's designs, which are reminiscent of proper, small ships, and his design philosophy seemed to mesh well with Trane's. Indeed, their collaboration has produced a very effective team.

Because each Coaster is custom built to meet the owner's specific cruising requirements, Benford and Trane decided to build in steel, which is easily modified and not restricted by casting molds. It also has the additional advantages of being both extremely rugged and, with today's current coatings, moderate in maintenance requirements.

Trane decided to build the Coasters in Palatka, Florida, on the St. Johns River. He says, "There was an available labor force familiar with steel boatbuilding, the wage structures were reasonable compared to many other parts of the country and it was close enough to our marketing operation in Miami that I could personally oversee the construction."

The first Florida Bay Coaster slid down the ways in late January 1988 and sailed immediately for her introduction to the yachting world at the Miami International Boat Show. Sitting on her own protective skegs with no supporting cradle in the parking lot of the Miami Beach Convention Center, the Coaster drew constant attention. With her wheelhouse equal in size to a Mississippi River pusher, she looked like a small, rugged, purposeful ship that somehow got lost and wandered into the local yacht club. She also carried a full-size Jeep on deck, with a 14-foot runabout above, just forward of the wheelhouse. Both could be launched with the deck-mounted crane in less than 10 minutes.

And Trane, in his characteristic deck shoes, Levis™ and checked lumberjack shirt, was kept busy answering questions from the curious tire kickers and the serious buyers.

As interest in the Coaster series grew, Trane convinced his brother Gerry and his sister-in-law Shirley to join the organization. With their arrival, a new 65-foot Coaster was launched, complete with an on-deck hot tub, "back porch and patio."

Said Trane: "Knowing that the boatbuilding industry has had its share of dreamers who could not produce what they conceived — or if they did, the boat frequently didn't match the original expectations — I built the first two boats as demonstrators, or, as is said in the industry, 'on spec'! But why should I worry? I've always built boats that I personally would want to use. After all, the worst thing that could have happened is that Gerry and I each would end up with a wonderful vessel that is functional, comfortable, easily maintained, and ideally suited for cruising and gunkholing.

That does make sense, doesn't it?

Photojournalist Rick Friese, of Palm Beach Gardens, Florida, operates a photography business called Yacht Shots.

Sails is the first one of the three 45' Florida Bay Coaster sisterships built to date.

The action photo (above right) shows ***Sails*** powerfully making her way through a stiff chop. Note how the pilothouse is well above the spray and has great visibility. I crossed Pamlico Sound on a sistership in worse conditions and can testify that these boats can take rough weather.

The photos above left show two views of her saloon and the built-in seating and the dining area. The lower photos show her spacious, well-equipped, modern galley and the master stateroom on the upper deck. Note the custom entrance for the "doggie door" on the weather-tight door to the upper aft deck.

Below right is the guest cabin forward with a desk or study area in it. The photo at center right shows her comfortable aft decks on both levels. For more details and the drawings of ***Sails*** and some variations on her, see pages 34 to 47.

Dennis Kelly photos, courtesy of Sherry and Bill Welch.

The 35' ***Strumpet***, above left in the Roy Montgomery Photo and on page 164, the 32' ***Ladybug***, above right and on page 160, and the 34' Friday Harbor Ferry, below in the Steve Davis rendering and on page 104, are some of the small ships whose designs are found in this book.

Choice Waterfront Homes

Available with panoramic marine views from Maine to Florida or Puget Sound to Alaska. An ideal liveaboard, retirement, or vacation home, the 34' Friday Harbor Ferry offers a majestic variety of unique marine park settings, each available without ever leaving home. These luxurious homes offer low maintenance exteriors and warm wood interiors, with all utilities installed, including bathtub with shower stall, washer, and dryer. 2 bedrooms, complete kitchen, large sundeck, and 3 viewing porches. Economical diesel power and large tankage provides for safe and low-cost operation. Neighbors too noisy? Move your house. Tired of mowing the lawn and raking leaves? Try Friday Harbor Ferry living for fast relief.

Caution: The Sturgeon Genial warns that Friday Harbor Ferry living may be addictive and habit forming.

The Right Liveaboard For You

Cutting your ties to shore is a process that happens in stages. For some, it begins with a vague dissatisfaction, a sense of uneasiness. A tenacious feeling that says, "I know there's a better way." Eventually, this disquieting sensation crosses into our awareness, and we understand that ***we were meant to live on board!*** This can be a gut-wrenching feeling, but is frequently accompanied by joyous laughter and a warm, lighthearted afterglow that lasts for many days! If this is you, ***congratulations!*** You have chosen a lifestyle that will move you closer to your mate, your world, and your own soul - a lifestyle that is truly "good for you."

So, having made this decision to live aboard, you now need to identify the ***right*** liveaboard for this experience. Given all the companionship and help in the world, and even unlimited finances, your endeavor will be hobbled unless you begin aboard the right boat.

You know that every boat is a compromise. She's the trade-off between speed and capacity, storage and roominess, stability at speed versus stability alongside. You need to begin your search for the ideal boat with a clear idea ***of what is important to you at this time***. The ideal boat is out there - let's be sure we'll know her when we see her.

But before we go further, consider this important point: the "ideal boat" is a moving target. Over time, your priorities will change, as will the requirements you have for your vessel and how much vessel you can afford. There is no ideal boat for ***all*** the seasons of your life. At some point, you will again find yourself going through this whole lifestyle evaluation process. This is normal, this is good, and this article will help at that time, too.

Needs, Wants, & Skills

To determine the right boat, you (the collective you — whoever will live on board) need to take a hard look at your needs, your wants, and your skills. This assessment is necessary to give yourself a clear picture of what you are looking for in a boat. Take the time needed to think this whole thing through. Understand exactly what living aboard means to you. Think about it ***a lot***. Define first what you **need**, what is a "must-have" in your life afloat. Talk over with your partner the importance of hot showers, clothes storage, covered deck space, and air conditioning. Talk about shiphandling, maintenance, and stability. Find out what is required by each person - find out now, before you begin looking. Make it a completely realistic, absolutely honest assessment, or you will unfairly prejudice your liveaboard experience. (A good way to do this is to have your partner "check your work." Share together for accuracy and completeness.) I guarantee you will save time in the end by doing this work up front.

Your needs are those things you feel you ***absolutely must have***. A private Master Stateroom is a common must. A full-size refrigerator might be another, as is your upper limit on cost. (I don't mean to make suggestions here, because everyone's needs and wants are different. It is important that you understand ***your own***, to make this endeavor successful.) Write down every ***requirement***, making your list as long as it needs to be. Drop those things you aren't certain you must have. (Put those items on your list of wants for now.) Have your mate do the same, then combine the two lists. Now that you have your list of needs, don't waste time by looking seriously at a boat that won't meet them. Don't be swayed by price or prettiness. Your needs are what you feel ***you must have*** to live aboard a boat. If you begin with a vessel that won't meet your needs, how well do you think you'll fare living on it?

Your wants, however, are a different matter. Rather than making an endless list of every feature you'd like to find on your dream boat, put down the Top Ten features that ***mean the most*** to you. Have your partner do this too. Try to keep your list restricted to things not easily changed -

like tank sizes, number of berths, galley design, and so on. Now prioritize this list and combine them. This is your list of ***wants***.

A Little Guidance

More than any single factor, your family will dictate the boat you buy. Do you need extra berths for children, or guests? Guests don't require a permanent space, but your kids will like a spot that's "theirs." Not only is changing from a dinette to a bed (and back!) each day a major chore, you and your mate or guests might like that dinette for a game of cards, a late night snack, or simply visiting after the kids have gone to bed. If you have children or will soon, plan on a permanent berth for each of them. The same thought applies if you have grandkids or expect regular guests aboard.

Kids change the way you use your vessel as well. If your children are in school, your weekly routine is set for, say, nine months of the year. Yes, you can home school your kids on a 'round-the-world cruise, but give careful consideration to how you will sail and maintain your vessel, care for yourself and your family, ***and*** teach your little ones reading, writing, and arithmetic. (Plan to move in this direction slowly, feeling your way as you go, so that you are not overloaded with all these responsibilities at once.) And remember, some young children are prone to seasickness, which likely will limit the kind of cruising you do.

Where Do You Want To Go?

What is your family's vision of a life afloat? Are you planning on passagemaking, or do you need to stay near a job? If your goal is voyaging, your need for storage and tankage is greater than the vessel that stays near-coastal her whole life. Being further from civilization (which may, in fact, be ***several time zones away*** from where you are) the passagemaker and her crew will need to be better outfitted, better skilled, with more spares and back-up systems on board. Make your choice based on ***what you will do the most***. You can always add temporary tanks and additional navigation equipment, but it is almost impossible to add structures, like a big galley or open deck space.

Where you go effects the kind of systems you have on board. Is yours to be a high-tech vessel? A lot of electronics? Will there be a shop on board? How is your celestial navigation? Do you need heating and air conditioning to be comfortable year-round? Or do you follow the seasons and simply do without? And will you be marina based, or mostly at anchor in some new (to you) country? Think about how you'll generate electricity, make water, and do laundry. How much room do you need to store food? Are you near markets, or do you even speak the local language?! Think about how you'll use your boat, and plan for every ***need***. For the greatest chance of success, you need to find the boat that meets all your needs.

Skills

What has turned up, hopefully, are a number of material requirements that will satisfy your vision of living aboard. What may be less obvious are the ***skills*** required to make this vision work.

A yacht is an assemblage of diverse systems, elegantly integrated in a safe, home-like environment. And with each system comes a set of needs, too. Oil has to be changed, sails have to repaired, filters need cleaning, ***everything needs maintenance***. Guess who gets to do it. Do you have the skills? Do you have the time? Do you have the tools? Find out what your maintenance effort will be like. And be honest! Don't overload yourself with problematic systems. Keep everything as simple, straightforward, and robust as possible. It's cheaper initially and much cheaper in the long run to live aboard a low-tech vessel. Particularly if you are having a yard maintain her.

And beyond the chipping, cleaning, painting, etc. needed to keep your home afloat, there are other skills that need to be considered. As a liveaboard, especially if you are a voyager, you and your partner (your spouse, crew, kids, whoever) ***need to learn each other's skills***. It's no good having one Master Electrician who happens to be ashore for the day when the electric bilge pump doesn't. Same for navigation, cooking, first-aid, and so on. Skills shouldn't be divided, there should be lot's of overlap - a Master and an Apprentice for every needed function. You can buy the boat that's your ideal, but if you can't sail and maintain her, you'll be back on land ***fast***. Think about the skills you need, and go learn the ones you lack. This is one of the blessings of living aboard, the continual stretching of one's mind, skill set, and experiences. You grow in ways undreamed of.

Make your first year aboard as simple as possible. Learn your vessel - her maintenance needs and sailing characteristics. At the same time, you will be learning how your family adapts to cruising. Make honest, realistic assessments of your capabilities and desires. And plan every step. Make sure your skills and desires match-up, with each skill meeting the need your desire produces. Study up on the ones you need to acquire and practice in times and places that are not stressful — that is, practice storm routines like you would a fire drill so you've thought through all the possibilities and consequences of what might happen and are prepared for them.

Ease into it. Don't worry if it doesn't happen all at once. Along the way every woman that I have spoken with (who has also made the happy transition from living on land to water) says basically the same thing; "Why did I ever dread leaving?"

35' Packet Preliminary

Design Number 300
1989

This preliminary sketch led into the final design for the 35' Packet. It has a lot of the features of the final version, and some alternative ideas shown here that could be used on the later one too. This one has a ladder from the lower aft deck to the upper deck instead of the door into the saloon. The saloon has an open, family room style layout. The pilothouse is a bit shorter with the bridge wrapping around the front to a center stair to the foredeck.

The hanging locker at the forward end of the master stateroom is over the tub and shower stall and has the raised pilothouse seat over it, making it a bit short for hanging longer items.

Comparing this drawing to the final one shows the evolution of the ideas, from this rough sketch form to the final version shown on the following pages. For instance, we ended up turning the layout of the head around so that the side deck outboard of the head was shortened and the shower stall was out to the side of the boat.

These 35-footers will make very nice vacation homes or full-time liveaboards.

Particulars:

Length overall	35'-0"
Length designed waterline	34'-6"
Beam	15'-0"
Draft	3'-6"

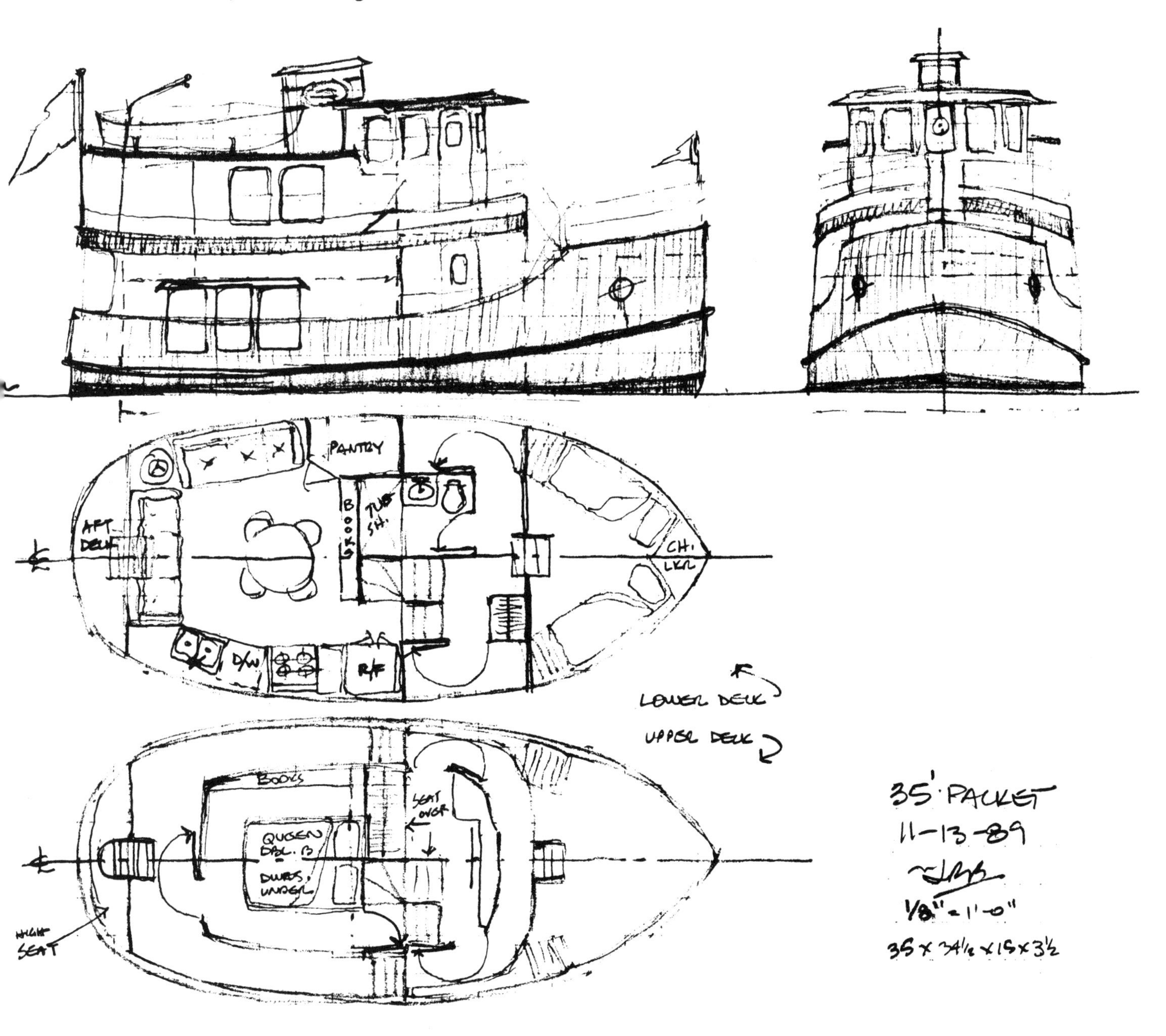

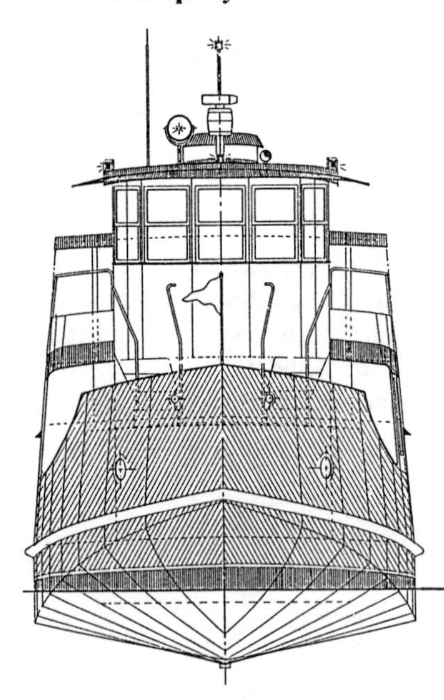

35' Packet
The Sensible Liveaboard

A Novel & Practical Design
Design Number 300
1990

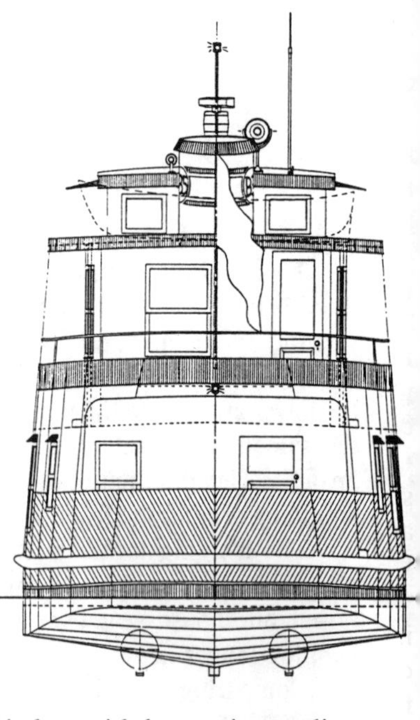

"Houses are but badly built boats so firmly aground that you cannot think of moving them. They are definitely inferior things, belonging to the vegetable not the animal world, rooted and stationary, incapable of gay transition. I admit, doubtfully, as exceptions, snail-shells and caravans. The desire to build a house is the tired wish of a man content thenceforward with a single anchorage. The desire to build a boat is the desire of youth, unwilling yet to accept the idea of a final anchorage.

It is for that reason, perhaps, that, when it comes, the desire to build a boat is one of those that cannot be resisted. It begins as a little cloud on the serene horizon. It ends by covering the whole sky, so that you can think of nothing else. You must build to regain your freedom. And always you comfort yourself with the thought that yours will be the perfect boat, the boat that you may search the harbours of the world for and not find."

Arthur Ransome
Racundra's First Cruise, 1923

For every couple that ends up living aboard there must be a thousand who dream of the possibility. What is it that holds them back? Is it giving up the creature comforts of a home ashore for "camping" on a small boat? Or is it that they think that the only boats with enough room aboard are out of sight financially and too large to be easily and affordably maintained?

If this is your concept of living on a boat, take a look at another solution. The 35' Packet is our latest liveaboard design. It's the culmination of a couple decades of relentless pursuit of one goal; the most practical and affordable home afloat. Taking our desires for a great liveaboard and combining this with over a decade's experience living aboard and about three decades of specializing in liveaboard designs, we've come up with a great boat for use either as a summer vacation home or full time liveaboard.

It's all here; two private bedrooms, a separate shower stall in the bathroom, a roomy kitchen with house size appliances, a dining table with real chairs, two full length sofas, good sized closets and dressers, a full sized washer and dryer, separate engine room, pilothouse with real glass windows at the height of most flying bridges, and plenty of porch/deck space.

Come Aboard.

Let's take a tour and look at the boat in more detail. We'll board through the rail gates just forward of amidships. This puts us on the waist decks, which lead into the foyer/bathroom area or up outside stairs to the forward deck.

From the foyer, let's go into the foc'sle, forward and down three steps. It has two standard twin (30" x 75") beds in it, with drawers under them, and a hanging closet in the bow. It's big enough to be home to a couple of kids. Or for occasional guests, on the theory that you don't want to make them too comfortable, as a way of limiting how long they want to visit. (If you expected that your guests would always be couples, there's room to have another double bed in the foc'sle as shown in the alternative Tramp layout.)

Leaving the foc'sle, up through the foyer and down a couple steps aft, leads us into the family room. The first impression is of an open, roomy and well lit space. There's a coat closet with room for coats and boots. The dining table can be expanded to seat more than four for entertaining. Opposite are two sofas that are long enough to nap on or take a temporary guest overflow. They're made with sloping seats and backs for lounging comfort, and have storage under and behind them. There are several bookshelves around them for the reference books (flora & fauna, birds, sea critters, boating stories, cruise guides, novels & biographies) and a nice big section of flat bulkhead for mounting an art piece. (Perhaps a print of the acres of lawn you no longer slave over, or maybe a sailboat slogging to weather without the comfort and shelter this boat offers?...)

All the way aft is the kitchen, with full sized house appliances like the side-by-side fridge/freezer, four burner stove with oven, microwave over the stove, and the double sink and dishwasher.

Out the back door and a couple steps up puts us onto the after deck. There's a gate on centerline for boarding if moored stern-to or getting into the dinghy. The ladder takes you up to the larger lounging porch aft of the master stateroom.

Back inside, we'll go forward and look into the engine room, by going down the stairs next to the dining table. There's comfortable sitting headroom alongside the engine, making service easy. The generator, hot water tank, pressure water set, and other small mechanical systems all live in this compartment, out in the open and readily accessible.

Returning to the saloon, we'll go up the stairs to the master stateroom. By virtue of it's location above the saloon, it's got a wonderful view of the world around us. The windows let in lots of light and air, so it's well lit and ventilated. No claustrophobic cave dwelling here. The full sized double bed is built over port and starboard drawers, and the long dresser provides more storage. Forward of the berth is full height hanging closet. Over the head of the berth are several full width shelves specially built to fit paperback books.

Forward and up a couple steps is the pilothouse. The raised settee is high enough to make for good forward visibility and long enough for stretching out for a nap. Under the settee are drawers for storing charts and navigation information. The helmsman's position, standing at the wheel, is at the usual height for flying bridges on boats this size. But, it's got real glass in the windows for easier viewing and cleaning, and it's heated and air-conditioned for four seasons comfort.

Outside of the pilothouse, the bridgewings let the helmsman look right along either side and will greatly facilitate graceful landings. (The full 360-degree wraparound rubber fender also takes the worry out of where to hang the fenders.) From the wings, steps lead down to the foredeck or aft to the walkaround deck at the stern. This afterdeck has room for some deck chairs and plenty of lounging space. Alongside the port side of the pilothouse is a ladder to the boatdeck. There's room for some lightweight small craft here, and a davit to lift them on and off. The stack can either house part of the air conditioning or be a deck storage locker.

Let's go below to the saloon and talk about her concept some more. Make yourselves comfortable on the sofas.

Did you know that despite her apparent tall height, she has higher stability than most 50-footers. This is due to her generous beam and well designed hull form. That's why when we were moving around on her and climbing to the boatdeck, she felt like a much bigger boat. With a loaded draft of three feet, she's a great gunkholer and the protective skeg under the prop and rudder means that groundings don't usually represent a trip to the prop shop. Just put her in reverse and power off.

The other important part about her height is that it still permits her to go up the Hudson River, through the Erie Barge Canal and out the Oswego River onto Lake Ontario. From

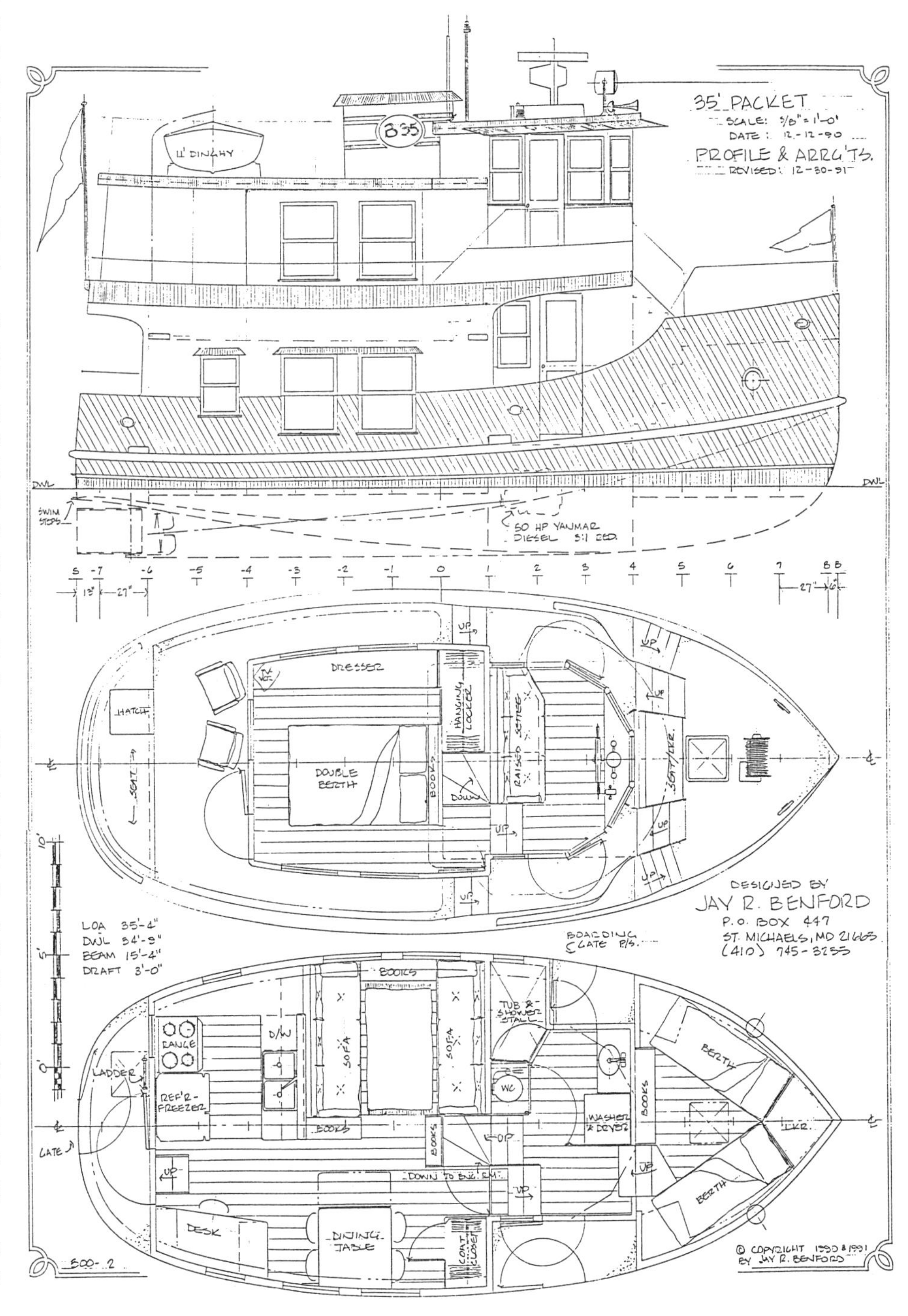

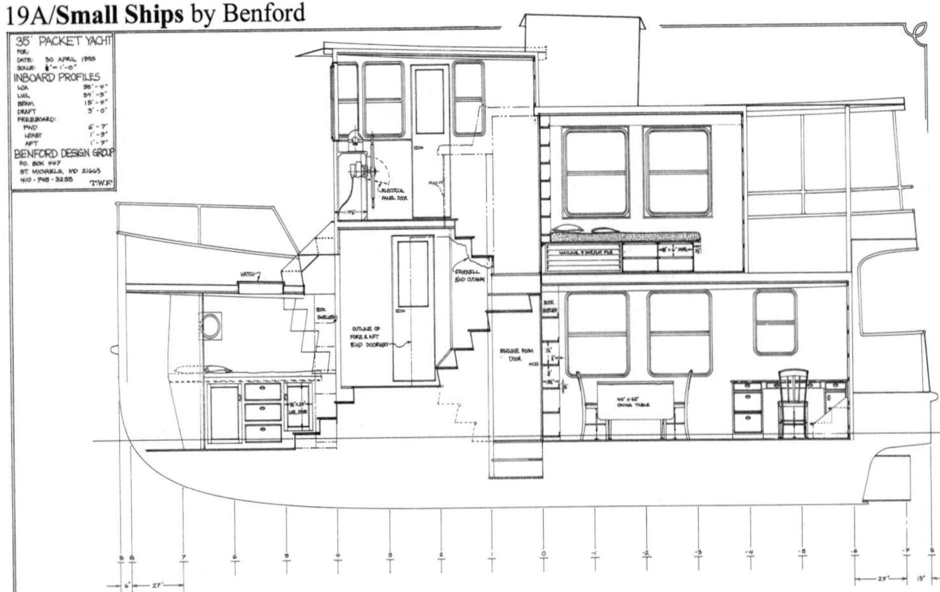

The stability of our Packet and Coaster designs has been questioned more than once, and I can pretty well just turn on a recording at this point to respond.

The bottom line, no pun intended, is that these are **very** stable boats. They have to be to carry the weight of all that tophamper and to be able to swing the vehicle overboard with the crane. One of the most common misconceptions about stability is that a deeper boat is more stable.

In fact, the opposite is usually true. Initial stability, what we think of as a boat being "stiff", is a combination of the moment of inertia of the waterplane (the outline of where the boat floats in the water) and how deep in the water is her center of buoyancy. The broader a boat is the better, for stability changes — all other things being equal — relative to the waterline beam cubed. Thus the beamier boat will have distinctly improved stability. A 20% increase in beam gives a 73% increase in stability.

The deeper hull has her center of the volume (and thus center of buoyancy) further underwater and this is a reduction of the stability. For example, think of a barge — a very broad and shallow hull that is quite stiff. The barge has a very high initial stability because she has such a wide waterplane and her center of buoyancy is quite close to the waterplane. She does not require any ballast to remain "stiff".

The alternative is one of the old wineglass section sailing yachts, which required a very large portion of her displacement as ballast to keep her upright, let along to start to feel "stiff". Our Coasters do not use ballast for stability, but rather a combination of their form and the placement of the structural and outfitting weights to give the right trim and stability.

Ultimate stability (that is when the boat takes a knockdown and rolls on her beam ends or puts the house and/or rig into the water) is more directly a function of the shape and volume of the boat above the waterline. We've designed quite a number of boats that have positive stability all the way out to 180°. (Naval architects assume that the boats are built with symmetrical port and starboard sides, and so we only need to calculate to 180° since the other 180° will be a mirror image of the first side.)

Having the luxury of using a good computer design analysis system (Fast Yacht) for looking at stability for over nine years, I've had plenty of time to sort through the aspects of what makes a boat stable and what does not. The main thing that is needed to make for a boat with a long range of stability is to have good and well placed volume above the waterline — a combination of good freeboard and deckhouse volumes placed where they will do some good for the boat as it heels over.

What I've done with the Coaster and Packet designs is to combine a broad and shallow hull form for good initial stability with large volumes of houses above the waterline for a long range of stability. The stability curve on page 89 shows the range of stability that 65' **Key Largo** has — more than enough to pass the USCG requirements to get her a 49 passenger T-Boat rating. We actually did the inclining test for this and she passed. Note that the curve dips between 60 and 90 degrees, when the cutaway of the waist decks are immersing, but the house and foc'sle volumes kick in and keep her in the positive range all the way out to 180°.

In comparison to offshore fishing vessels, the **Key Largo** has much greater initial stability which we wanted to make loading the Jeep on and off safe and we've found that this makes for greater comfort in a coastwise liveaboard. These fishing vessels, and most naval vessels too, often have a range of stability of only 65 to 70 degrees before they capsize. **Key Largo**, with her large houses, has extended this considerably....

Also see the comments on the 50' **Florida Bay** on page 48 relating more information about stability.

As a postscript to this, the **Florida Bay** and another 50' Coaster both survived Hurricane Andrew in good order, when other boats were sinking near them.

Many large cruise ships have much naval architectural work done to keep their stability moderated so that they do not have a snappy motion. The gentle, easy roll that they exhibit is done so intentionally so that they are easy on their passengers and cargo. They still have a good and safe range of stability but efforts have been made to look out for the comfort and pleasure of the paying passengers.

We're pleased that many people seem to share "Dr. Ruth levels" of interest in our Coaster designs. We'd be happy to go over any questions with them and help them get just the right version sorted out and find just the right builder to translate the dream into a reality. JRB

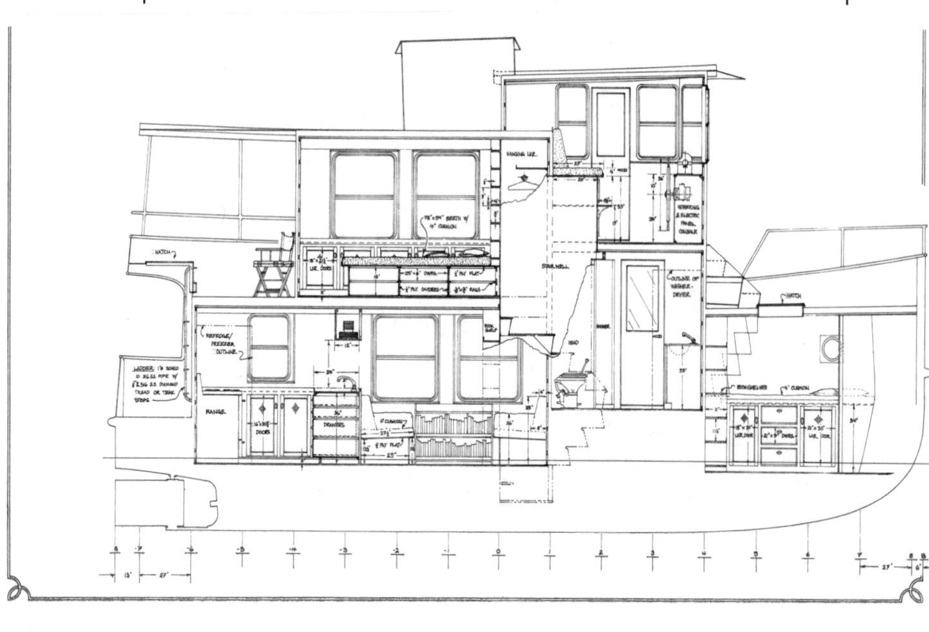

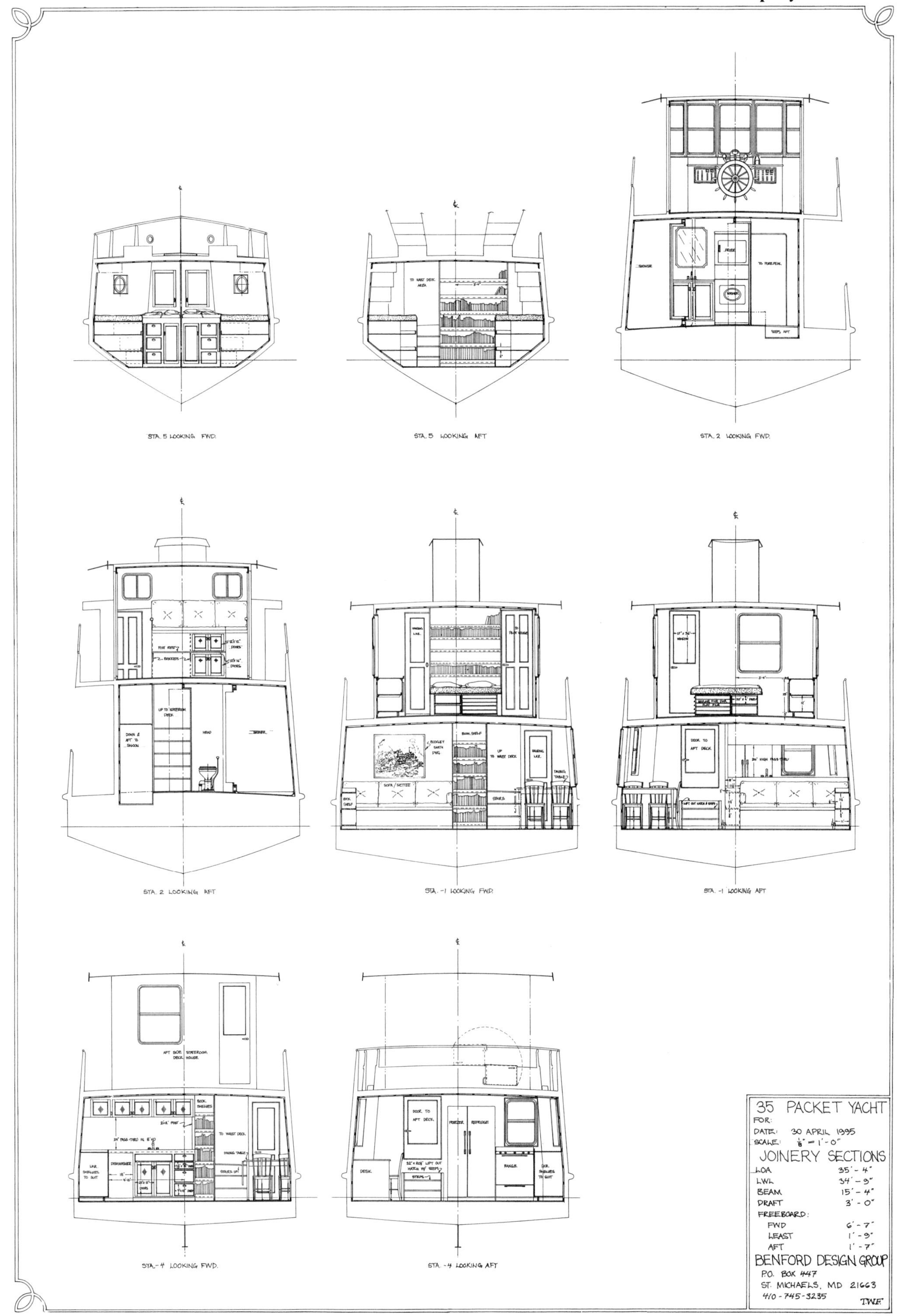
STA. 5 LOOKING FWD.
STA. 5 LOOKING AFT
STA. 2 LOOKING FWD.
STA. 2 LOOKING AFT
STA. -1 LOOKING FWD.
STA. -1 LOOKING AFT
STA. -4 LOOKING FWD.
STA. -4 LOOKING AFT
35 PACKET YACHT
FOR:
DATE: 30 APRIL 1995
SCALE: ⅛" = 1'-0"
JOINERY SECTIONS
LOA 35'-4"
LWL 34'-9"
BEAM 15'-4"
DRAFT 3'-0"
FREEBOARD:
FWD 6'-7"
LEAST 1'-9"
AFT 1'-7"
BENFORD DESIGN GROUP
P.O. BOX 447
ST. MICHAELS, MD 21663
410-745-3235
T.W.F.

there, west past the Niagara River to the Welland Canal and into Lake Erie and onward through the Great lakes to Chicago. There, it's through the bridges and connecting to the Mississippi River. This can lead you to side trips on the Illinois, the Ohio, and the Tenn-Tom. All this can be years of exploring the United States. Or, it can be a circumnavigation of the eastern half of the U.S.

Do you recall the simple structure you noticed in the engine room? The whole boat is designed to be built of materials available from a good lumberyard. Good quality plywood with fir framing is the basis for the whole boat. The sides of the houses are all double wall, with insulation built-in.

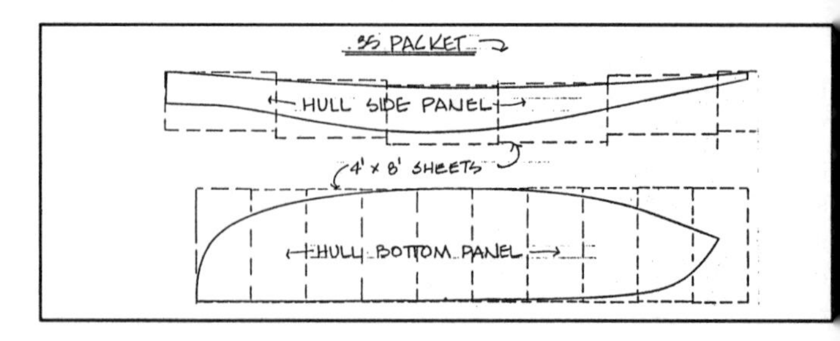

The whole structure is sealed and glued together with epoxy, and she's sheathed with a skin of Dynel or 'glass cloth set in epoxy for abrasion and impact resistance.

In designing her, we used our Fast Yacht software to refine her hull form, providing a shape that will be easily driven with modest power. We also created a developed hull surface, that can be planked up using full sheets of plywood. The computer program lets us "unwrap" the panels from the hull form, and we used this part of the program early on in the design process to be sure we could get the bottom planking out of less than 8' wide sheets. (See illustration above which shows the unwrapped panels — no attempt to "nest" pieces for optimizing use of the offcuts was made on this drawing.)

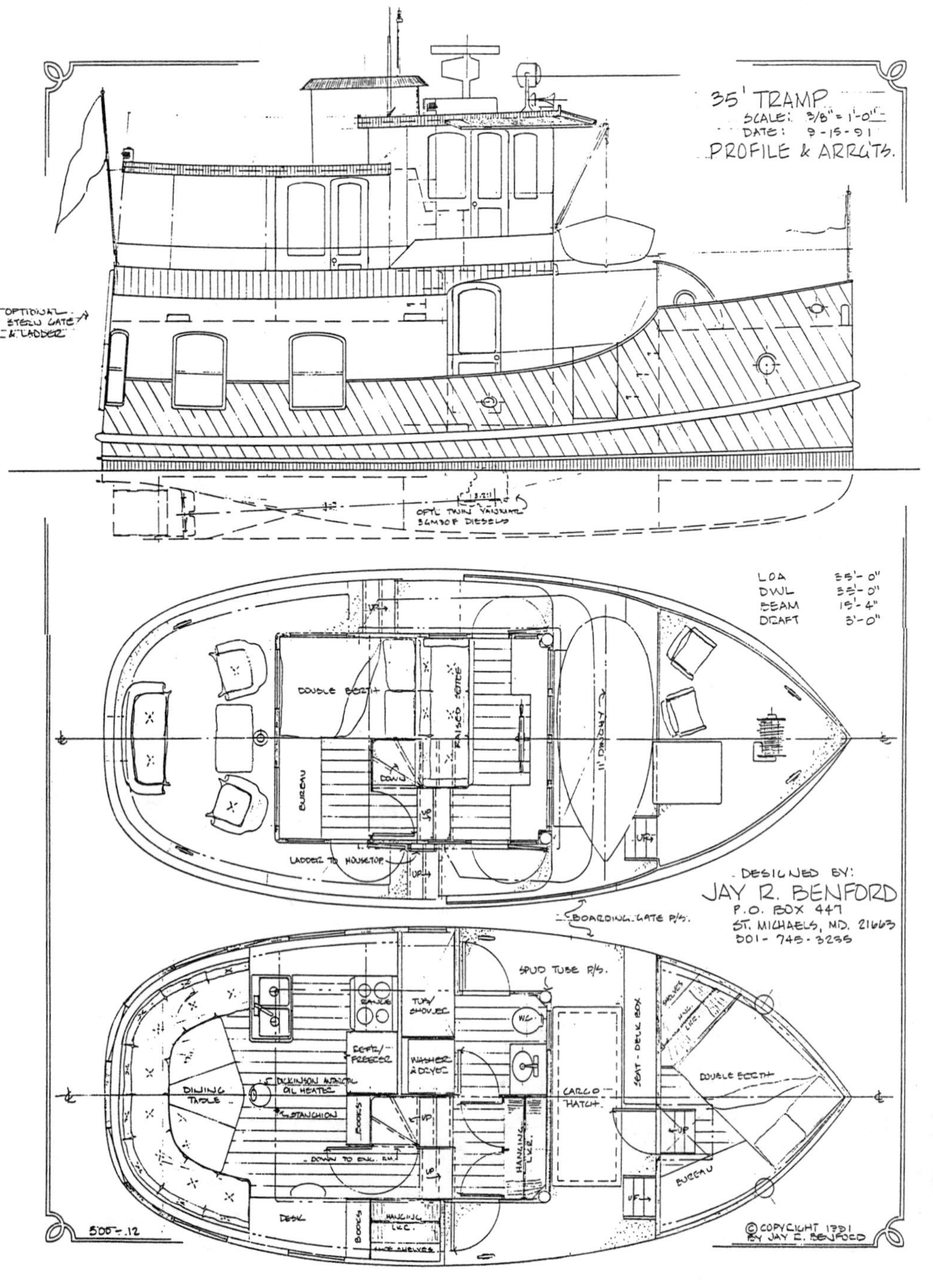

By using multiple thin layers, particularly at the bow, we are able to wrap them into the desired hull form while still staggering the butt joints so that we have good structural integrity.

We've eliminated the need to use "marine" hardware and equipment wherever possible. All the kitchen, laundry, bath, heating, and air-conditioning equipment can be good quality house equipment. Even the windows and doors can be wood framed -- just have 1/4" safety glass substituted for the standard 1/8" window glass.

As an alternative, we've designed a version with a steel hull and houses, with the upper stateroom and pilothouse optionally still built in wood. This is a couple of tons heavier than the all plywood version, and has slightly larger standard tankage by virtue of having integral tanks; 300 gallons of fuel and 500 gallons of water versus 250 and 400 in the wood version.

We always suggest larger water capacity in a liveaboard, since it permits longer periods of time between refills. Refilling in the middle of the winter is something to forestall as much as possible, and it's nice to have showers and do laundry on a regular basis.

Twin screw versions are certainly possible, though I'd think about a bow thruster as an alternative. Twins might let you end up with three of the same engine, if the right engines and generator were selected, thus simplifying keeping the right spares on board. I'd give serious consideration to a single screw if your normal moorage was reasonable to get in and out of, and you were willing to do a little practicing to get comfortable with your skill in handling her. The initial cost would be lower with only one engine to buy and have installed, and there would be slightly lowered fuel consumption when cruising. I admit that the extra nimbleness in maneuvering with twins has a lot of attraction, and this extra level of confidence might often mean she'd be used more frequently -- a goal worth pursuing....

How much weather can she stand, you ask? Well, probably more than any of us would intentionally set off into. And if she gets caught out in something nasty, the basic survival technique is to keep her speed down to the level at which the motion is reasonable for the circumstances. Thus, in a short, steep chop one would slow down more than in big ocean swells.

The structure is stout, her windows and doors are made to keep the water out, and her range of stability is greater than most offshore fishing vessels. My own experience in living aboard a stout offshore sailing vessel is that even those with that capable a boat will wait for good weather before making an open water crossing. It's nice to know your boat is capable of it, even if you don't feel like intentionally seeking it out.

The cost of materials to build the 35' Packet is less than those for building a house. And it looks like the insurance and upkeep costs will be less than a house, and if you don't like the neighbors or the neighborhood, you can always move along to another area; and still be at home.

Yes, she can be loaded on a freighter and sent to Europe to do the canals. There's only one section of the French canals that she's too big for. Or, you could send her to Seattle and spend a year or a decade cruising back and forth to Alaska and exploring the fjords of British Columbia along the way. I spent eighteen years cruising out there and didn't begin to see it all. There were still areas on the B.C. charts in the Northern areas that were just outlines and no details at all on the charts.

However, I think that for those on the East Coast, the combination of the Intra Coastal Waterway, the Great Lakes, and the Rivers systems would provide a lifetime of cruising and exploring. There are still places that are not built up and over populated, where you could settle in as a new home port, or visit on your travels. What better way to see the vastness and variety of our country? Or use this as a way to see the sites of our history and development.

Alternate Master Stateroom Layouts

We've had quite a few people interested in the idea of having a second head, ensuite in the master stateroom. After studying the problems of how to do this, we've come up with a drawing with suggestions of how to make it work. What's lost, in the addition of this head, is the three sided walk-around of the master berth. What's preserved is the full walk-around side decks and the whole afterdeck. From our experiences in prior designs of ours, these are very important spaces and we're unwilling to give them up.

Particulars:

Length over guards	36'-0"
Length-structural	35'-4"
Designed waterline	34'-9"
Beam over guards	16'-0"
Beam-structural	15'-4"
Draft, loaded	3'-0"
Freeboard:	
Forward	6'-7½"
Waist	1'-11"
Aft	1'-9"
Displacement, loaded for cruising*	40,400 pounds
Displacement-length ratio	409
Prismatic coefficient	.61
Pounds per inch immersion	2,140
Tankage: Fuel	250 or 300 Gals.
Water	400 or 500 Gals.
Stability - GM	6.7' to 4.4'

***CAUTION:** The displacement quoted here is for the boat in cruising trim. That is, with the fuel and water tanks filled, the crew on board, as well as the crews' gear and stores in the lockers. This should not be confused with the "shipping weight" often quoted as "displacement" by some manufacturers. This should be taken into account when comparing figure and ratios between this and other designs.

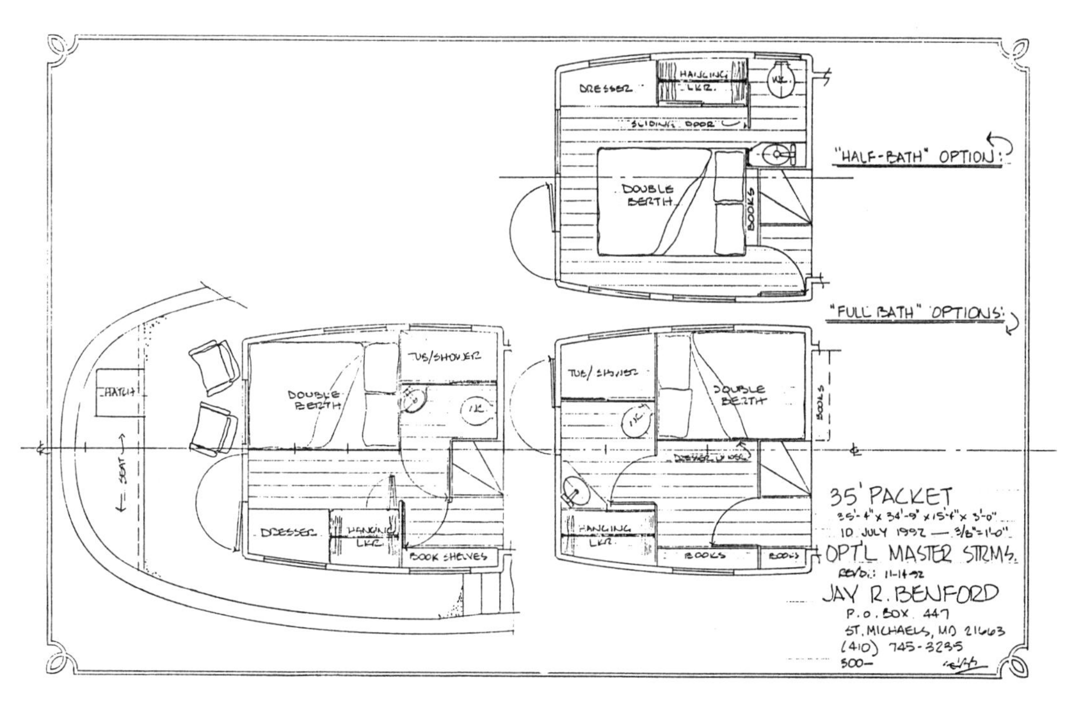

Tramp Version

An alternative, for those looking for an even more "shippy" appearance, is this Tramp version, with a well deck about five feet long. The cargo hatch opens into an area below in which to stow diving gear, salvage treasures, or bicycles. The deck space can be used for lounging, carrying a personal watercraft, motorbike, or some extra fifty-five gallon drums of fuel or water.

The forward cabin shows an alternate double berth cabin layout. It can be used with the Packet also, or the Packet's twin bed version can be used on the Tramp. A four foot headroom passage could be provided through the engine room for foul/cold weather access fore and aft.

The great cabin in the stern has the dining seating and the lounging seats combined around a large drop-leaf table. There is a ship's office desk opposite the galley.

One thing given up in this version is the extra space for walking around the berth in the upper stateroom. We kept the aft bulkhead of the stateroom in the same location, to provide good deck space for the deck chairs.

The corner posts of the pilothouse and cabin below it are tubes for the spuds. For those unfamiliar with these, spuds are weighted tubes or poles that drop through the bottom of the boat to locate the boat when mooring or serve as a pivot point for making a turn. These are a smaller version of those found on commercial vessels and can be equally handy to have. You could have either a power or manual winch to raise them back up, with controls inside or outside the pilothouse.

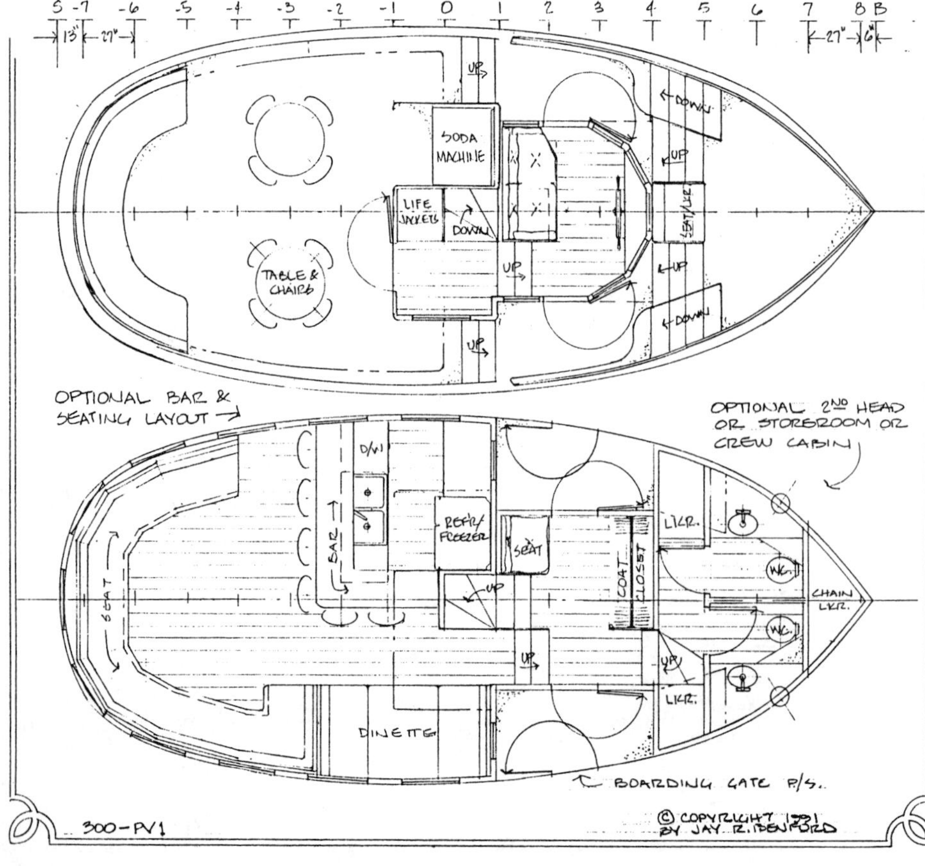

Passenger Vessel Versions

We've included two ideas for passenger carrying versions of the 35' Packet. One (PV2) has a more open layout for shorter trips. The other (PV1) has a built in galley and bar and more seating.

The pilothouses also show a couple of alternative ideas. PV2 has ladders letting the passengers have access to the upper aft deck. It also has a ladder from the starboard entry deck to the open deck above the saloon. There is a similar inside ladder to the foyer behind the pilothouse, giving a choice of inside or outside access, depending on the weather.

For more information or prices on these small ships, call or write to us at the address on the drawings or as shown in the front of this book. The pricing is being refined and updated as we go to press with the book and, of course, will change with time due to the ongoing course of inflation.

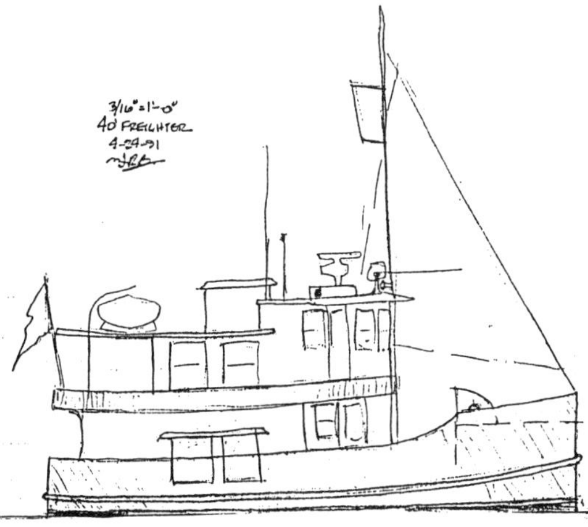

The 40' Freighter sketch at left is an idea for a modified version of the 35' Packet with a 5' well deck added. The 30' Trawler idea at right is a smaller version of the one shown on the next page. The photos below are of the first 35' Packet, now building at Custom Steel Boats.

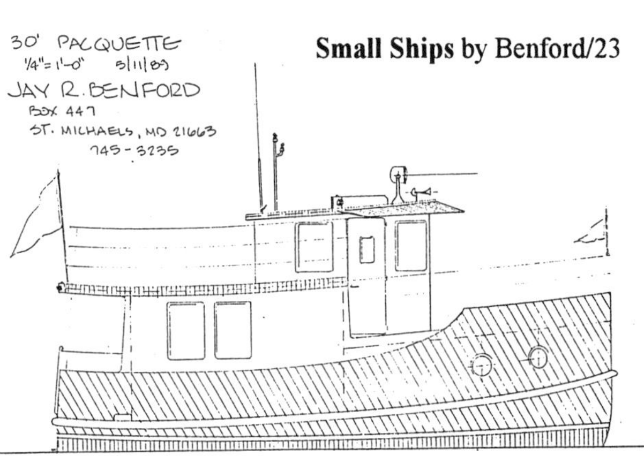

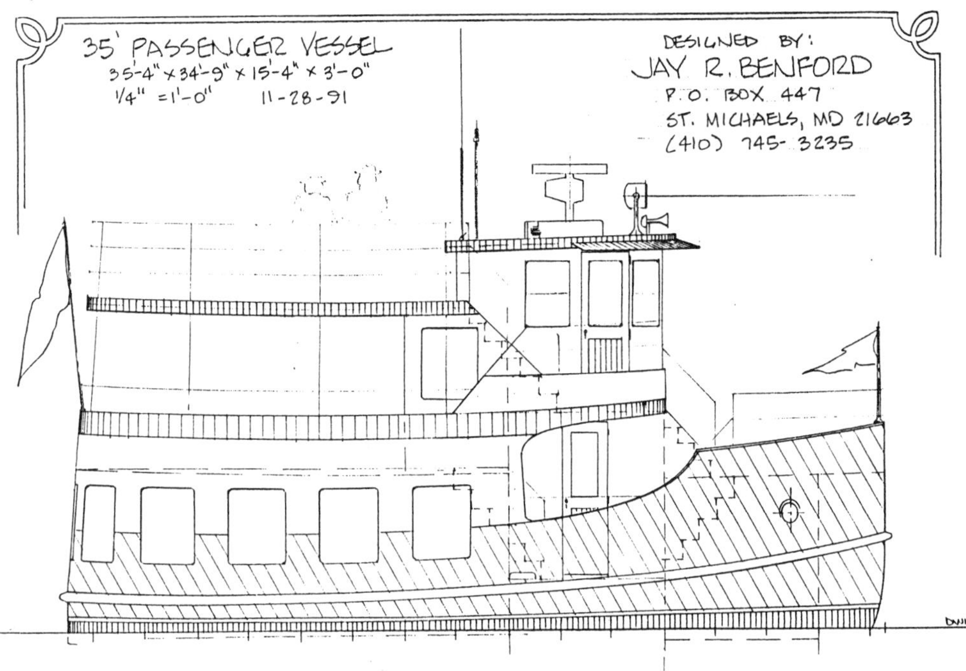

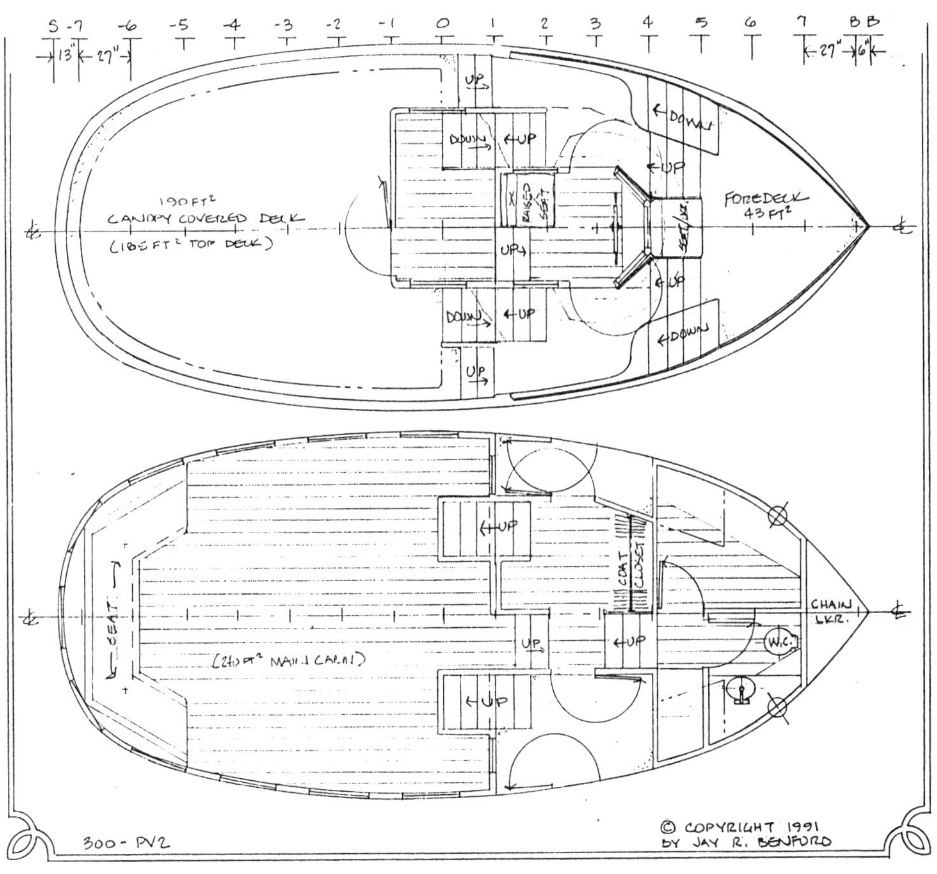

35' Trawler

Design Number 300
1994

This trawler yacht is an evolutionary development based on our 35' Packet design, with some sharing of parts. It was designed to create more of a "one-bedroom" liveaboard versus the "two-bedroom" layout on the Packet.

The accommodations are intended for use as a liveaboard for a couple, with occasional guests. The guests would berth either in the pilothouse or on the saloon settees. There is a washer and dryer plus tub and shower in the head. The galley has a full-size refrigerator-freezer, a 30" range with oven, and space for a small dish-washer, if desired. The desk would be a good place for a computer. The after deck could be screened or curtained for weather or bugs, and is often the nicest place for the dining table. It could be moved into the saloon in poor weather.

She has a ship-like Portuguese Bridge wrapping around the front of the pilothouse, giving a sheltered area for standing on deck, binoculars in place, scanning for the next mark or harbor entrance. There is plenty of boat deck space for carrying one or more small craft.

This 35-footer has many of the same features and livability found in our Friday Harbor Ferry series of designs. The difference is that this one has a pointed bow instead of the rounded bow of the ferries, which will let her go through a seaway without having to slow down quite as much. Construction could be done in either steel, fiberglass, or plywood and epoxy.

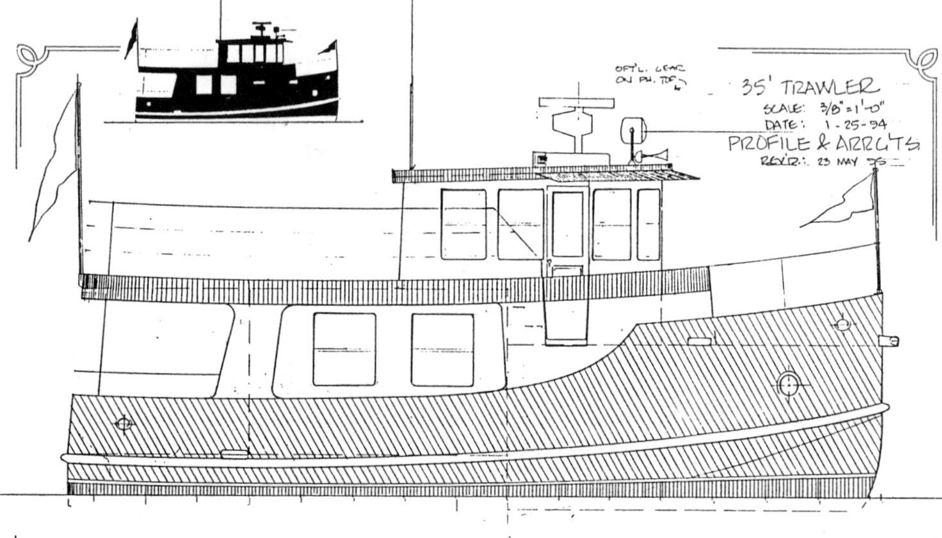

An alternative to the layout shown, for use with a family, would be to have berthing for the kids in the bow stateroom and making the berth and seat in the pilothouse into a double or queen-sized berth for the parents. However, this would take away a lot of the spacious, roomy feeling this open layout offers for full time life afloat.

Particulars:

Length over guards	36'-0"
Length-structural	35'-4"
Designed waterline	34'-9"
Beam over guards	16'-0"
Beam-structural	15'-4"
Draft, loaded	3'-0"
Freeboard:	
Forward	6'-7½"
Waist	1'-11"
Aft	1'-9"
Cruising Displ.*	40,400 lbs
Displ.-length ratio	409
Prismatic coefficient	.61
Pounds per inch imm.	2,140
Tankage: Fuel	300 Gals.
Water	500 Gals.
Stability - GM	6.7' to 4.4'

***CAUTION:** The displacement quoted here is for the boat in cruising trim. That is, with the fuel and water tanks filled, the crew on board, as well as the crews' gear and stores in the lockers. This should not be confused with the "shipping weight" often quoted as "displacement" by some manufacturers. This should be taken into account when comparing figures and ratios between this and other designs.

35' Florida Bay Coaster

Preliminary Design Idea, 1988

A predecessor to the 35' Packet and Tramp, this is an idea I sketched out just about a year after the first of the Florida Bay Coasters. It has a lot of the character, charm and feeling of the ***Florida Bay***, with the proportions condensed. The foc'sle has a cozy double berth, locker space, and a head with shower stall. I'm not sure if there would really be room to make the stall work, unless the bow were fuller than I would like to design one today. Perhaps switching the layout with the toilet forward and the shower aft would fit.

The well deck has a raised trunk/lid for access to the hold and engine room. This would make a nice workbench on deck and give standing headroom in that area in the engine room.

The great cabin aft is the saloon, with the full galley and enormous elliptical settee and dining table. There is a second head with shower stall in the port, forward corner.

The mid-level has the washer and dryer stackset and a stateroom with upper and lower berths. If the aft head were shifted a bit aft, this stateroom could be expanded to the side of the hull to port and have room to have another double berth in it or room for more lockers.

The upper, aft cabin has the living room and office space and opens onto the after deck. There's room for sitting out in several chairs there, to enjoy the view.

The pilothouse has a settee long enough to nap on, or room for several people to sit and watch where you're going. Doors open onto the wing decks on both sides. The bridge wing deck wraps around the front of the pilothouse and has an extension that spans forward to join the top of the scuttle on the foredeck, with stairs down onto the forward deck. This gives quick access to the bow for line or anchor handling. The wings also connect to the side decks leading to the aft deck, giving quick access full length for ship handling.

Particulars:

Length overall	35'-7½"
Length, hull	35'-0"
Length, DWL	34'-6"
Beam, hull	15'-0"
Draft	3'-3"

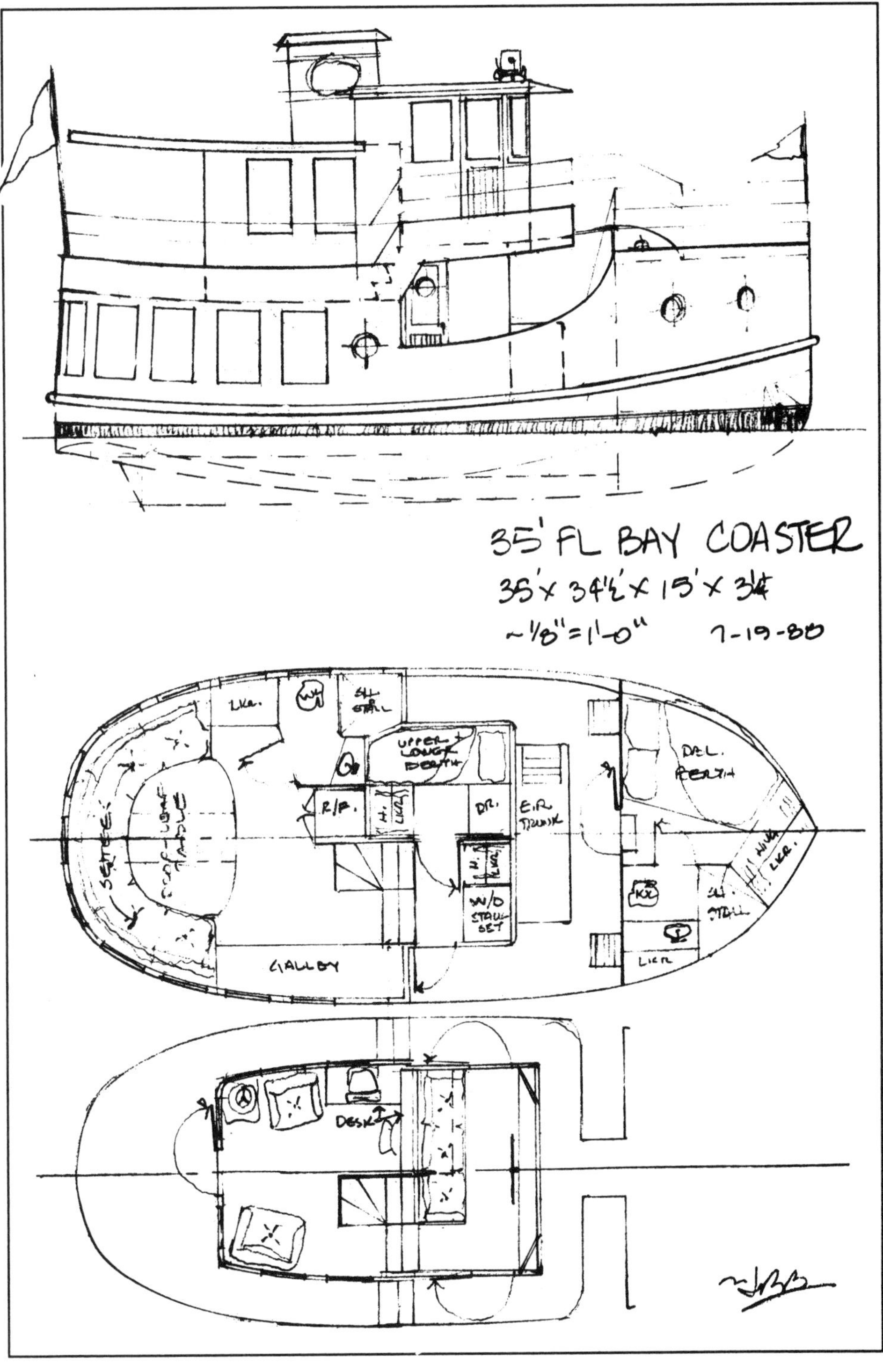

40' Coaster Ideas & Variations

Design Number 265 & 304

1985 & 1990

The following four pages have several variations on a theme of a 40-footer. They were generated over a period of a couple of years as differing thoughts and themes were pursued.

The profile below is a variation on the theme of the 40' Florida Bay Packet, with a bit different use of the spaces. This one has the saloon on the upper deck and the master stateroom down below.

The 40' Florida Bay Coaster, with the accommodations all on two levels, is an effort to make an improved version on our Ferry Yachts type layout with an enclosed bow and styling and features that have made the other Coasters so successful. My own preference would be to make the upper house a bit narrower, so that full walkaround upper decks could be fitted for better traffic fore and aft. This would involve moving some of the storage to the hold, which would be suitable for the out-of-season clothes. The stairs would have to be moved a bit inboard also and we'd move the washer and dryer down below too.

Having the accommodations all on two levels will make for more expeditious construction, and thus keep the costs lower. However, for living aboard, I would prefer the wider variety of spaces to be found on the split-level sort of layout that is found on the other versions. In cruising on four of the various different Florida Bay Coasters I've found that I very much like the separation of the spaces with the split-level approach and they work quite well with several people living and/or cruising aboard.

The version sketched with the short well deck is more like the larger Coasters, except there isn't room to carry a vehicle on deck. There is an inside passageway from the foc'sle through the hold to the after houses. The master stateroom layout could be used to substitute for the full width one on the prior design, to show what could be done in this area. The pilothouse is held a bit narrower to permit narrow stairs on each side to lead to the boat deck. The widths for these could be noticeably improved by making the pilothouse even narrower. To make this work with good width stairs and still have a pilothouse settee long enough to stretch out on requires about 20' of beam. Of course, if we do an asymmetric layout with the stairs up to the boatdeck only on one side, this can be done with only 17 to 18' of beam, as can be seen from some of the other designs in this book.

Other improvements that can be made from what's shown would be to shift the engines a bit aft, so they would be under the saloon and thus could be separated from the standing headroom area with a soundproofed bulkhead. This full headroom area can become a nice workshop, study or hobby room. Or, it can give room for more storage.

The saloon was held rather short to make for a large aft deck which can be enclosed in inclement weather, to let this become part of a larger combined living room .

Particulars:	**265**	**304**
Length overall	40'-0"	40'-0"
Length designed waterline	39'-6"	40'-0"
Beam	16'-0"	16'-0"
Draft	3'-6"	4'-0"
Freeboard:		
Forward	8'-0"	8'-6"
Least	1'-6"	1'-6"
Aft	2'-0"	1'-9"
Water tankage	600 Gals.	
Fuel tankage	600 Gals.	
Headroom	6'-7"	

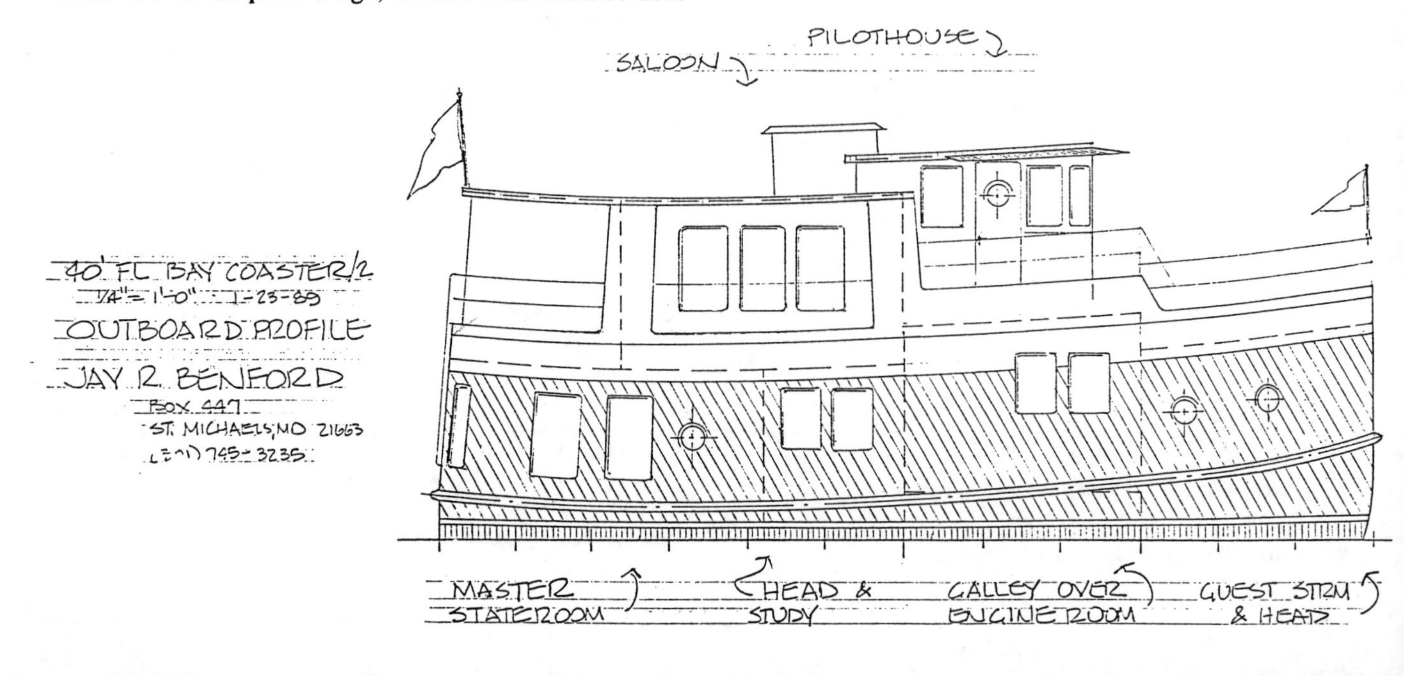

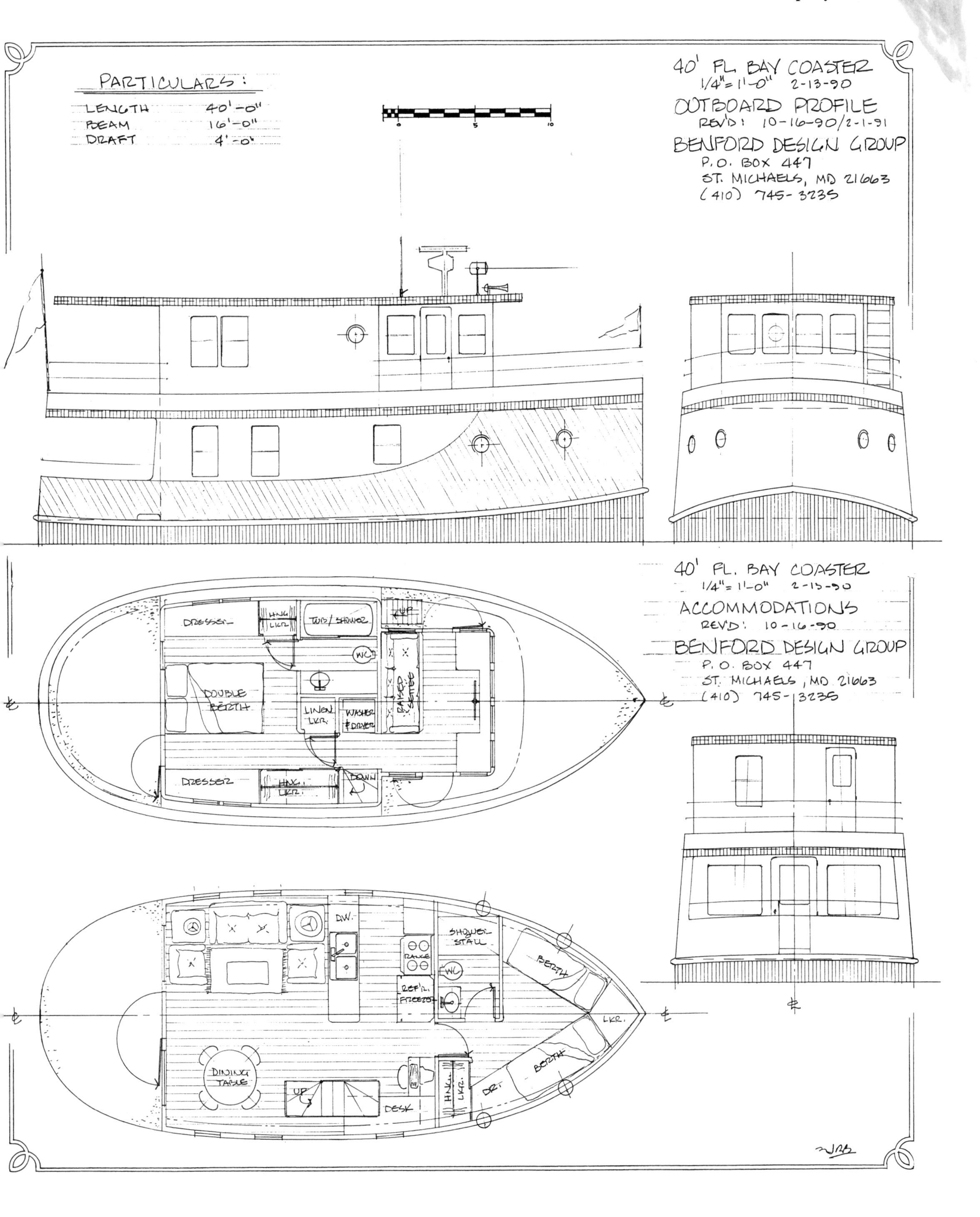
PARTICULARS:
LENGTH 40'-0"
BEAM 16'-0"
DRAFT 4'-0"
0 5 10
40' FL. BAY COASTER
1/4"=1'-0" 2-13-90
OUTBOARD PROFILE
REV'D: 10-16-90/2-1-91
BENFORD DESIGN GROUP
P.O. BOX 447
ST. MICHAELS, MD 21663
(410) 745-3235
40' FL. BAY COASTER
1/4"=1'-0" 2-13-90
ACCOMMODATIONS
REV'D: 10-16-90
BENFORD DESIGN GROUP
P.O. BOX 447
ST. MICHAELS, MD 21663
(410) 745-3235
DRESSER
HNG. LKR.
TUB/SHOWER
UP
WC
DOUBLE BERTH
LINEN LKR.
WASHER & DRYER
RAISED SETTEE
DRESSER
HNG. LKR.
DOWN
DW.
SHOWER STALL
RANGE
BERTH
WC
REF'R. FREEZER
LKR.
DINING TABLE
UP
DESK
HNG. LKR.
DRT.
BERTH
JRB

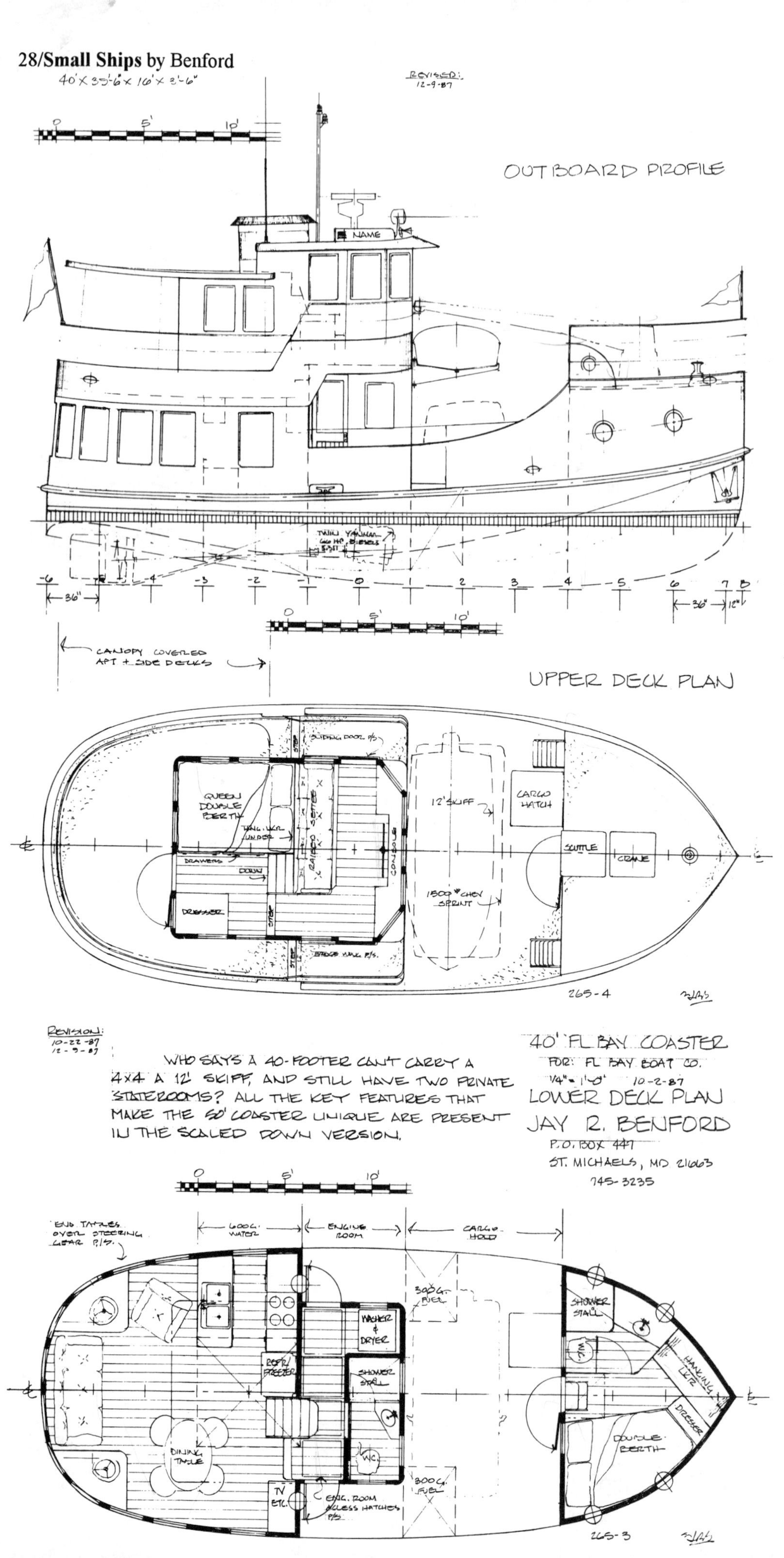

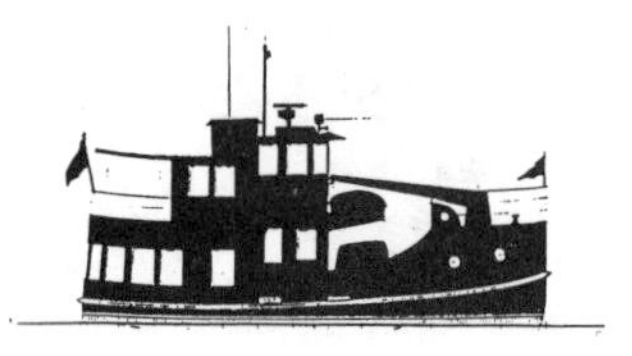

40' Florida Bay Coaster

1987

The 40' Florida Bay Coaster is the first of this small size that we drew up, as we explored what a smaller version of the ***Florida Bay*** might look like. We did some preliminary calculations and decided that we could probably have enough stability to carry a 1,500 (to perhaps as much as 2,000) pound vehicle on deck.

The accommodation plan has a cozy, private stateroom for two with its own head in the foc'sle. It has access over the deck to the after cabins.

The master stateroom is on the upper deck and the head for it is on the mid level, with the washer and dryer alongside the head. The master has been kept small so that there will be good space for sitting out on the after deck. This means that the bed only has one side access and a lot of the storage space for the clothes will be on a different level.

The great cabin style saloon has windows wrapped all around the stern and will have a lovely view of the world around it.

Tankage is called out for 600 gallons each of fuel and water. Power specified is a pair of 66 horsepower Yanmar diesels with 3.3:1 reduction gears.

This is a very shippy looking craft and would be a lot of fun to cruise and own.

40' Florida Bay Packet

1989

The 40' Florida Bay Packet has its side entry door pocketed into the side of the hull below the pilothouse. From the outside deck, the stairs lead up and aft to the walkaround side decks, instead of onto the foredeck. The entry door to the upper cabin opens onto the landing leading to the stairs to the pilothouse while the port side has outside stairs to the bridge wing from the after deck. Both sides of the bridge wings have stairs to the foredeck.

The tall exhaust stack is like the ones I did in the 1960's on some of the tugs I designed for Foss Launch & Tug. It has the benefit of getting the noise and soot well off the deck level, making it easier for both to be blown away. This means that the upper deck would be cleaner and it would be easier to hear conversations. If the engines were to be dry-exhausted, we'd have a vertical trunk at the end of the galley counter that fed up through part of the hanging locker in the master stateroom.

The accommodations layout has three bedrooms and two baths. The saloon is of the great cabin aft type, with no lower aft deck. The settee in the stern is built in so that it covers the rudder posts, stuffing boxes and steering gear.

The levels are shifted up more in the area of the pilothouse, making the pilothouse settee height such that it can overlap the head in the master stateroom. This height also means that there will be a full headroom basement space where we can have the washer and dryer.

For maximized room for living aboard, on a distinctively styled vessel, this boat has a lot to offer.

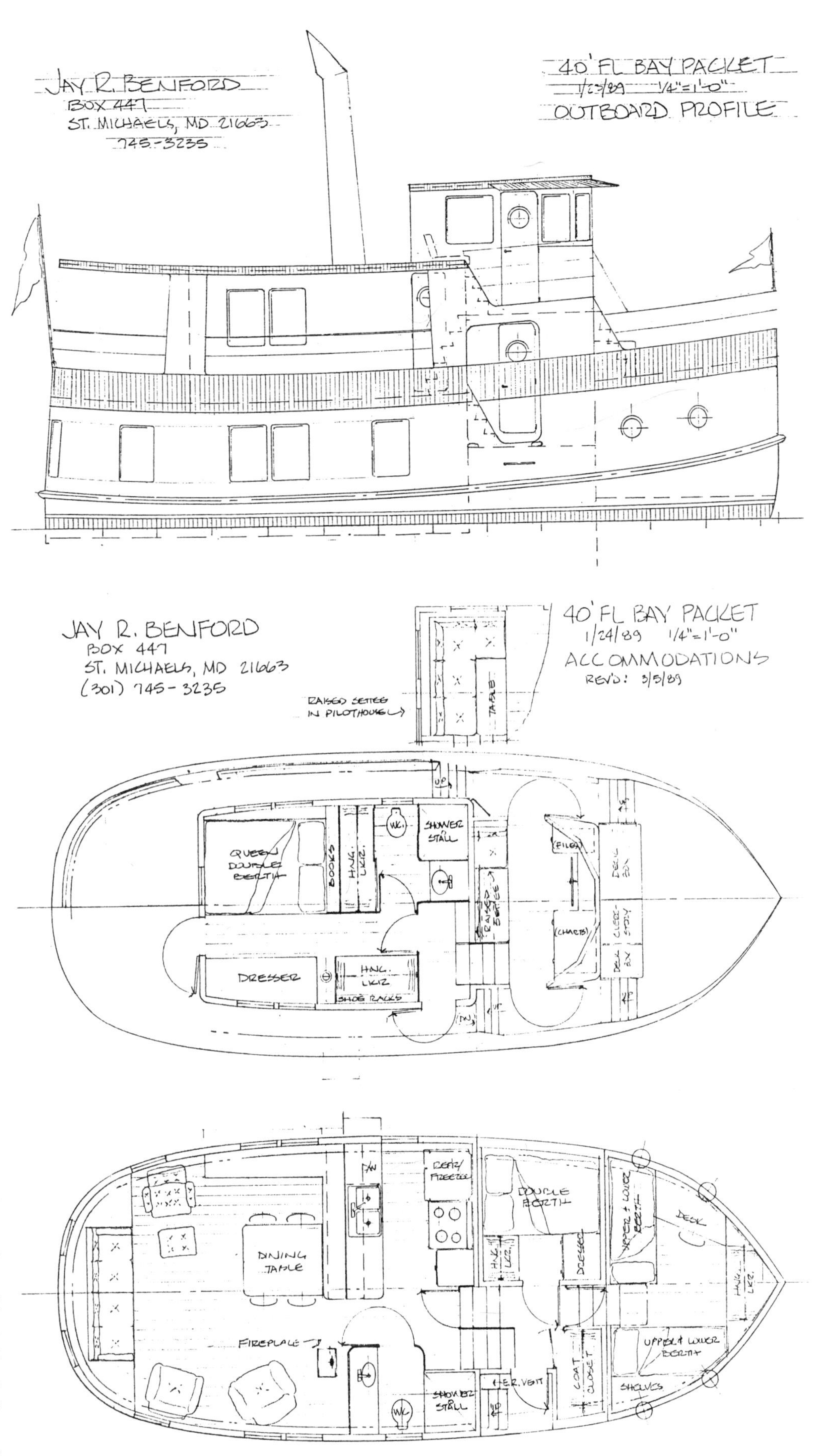

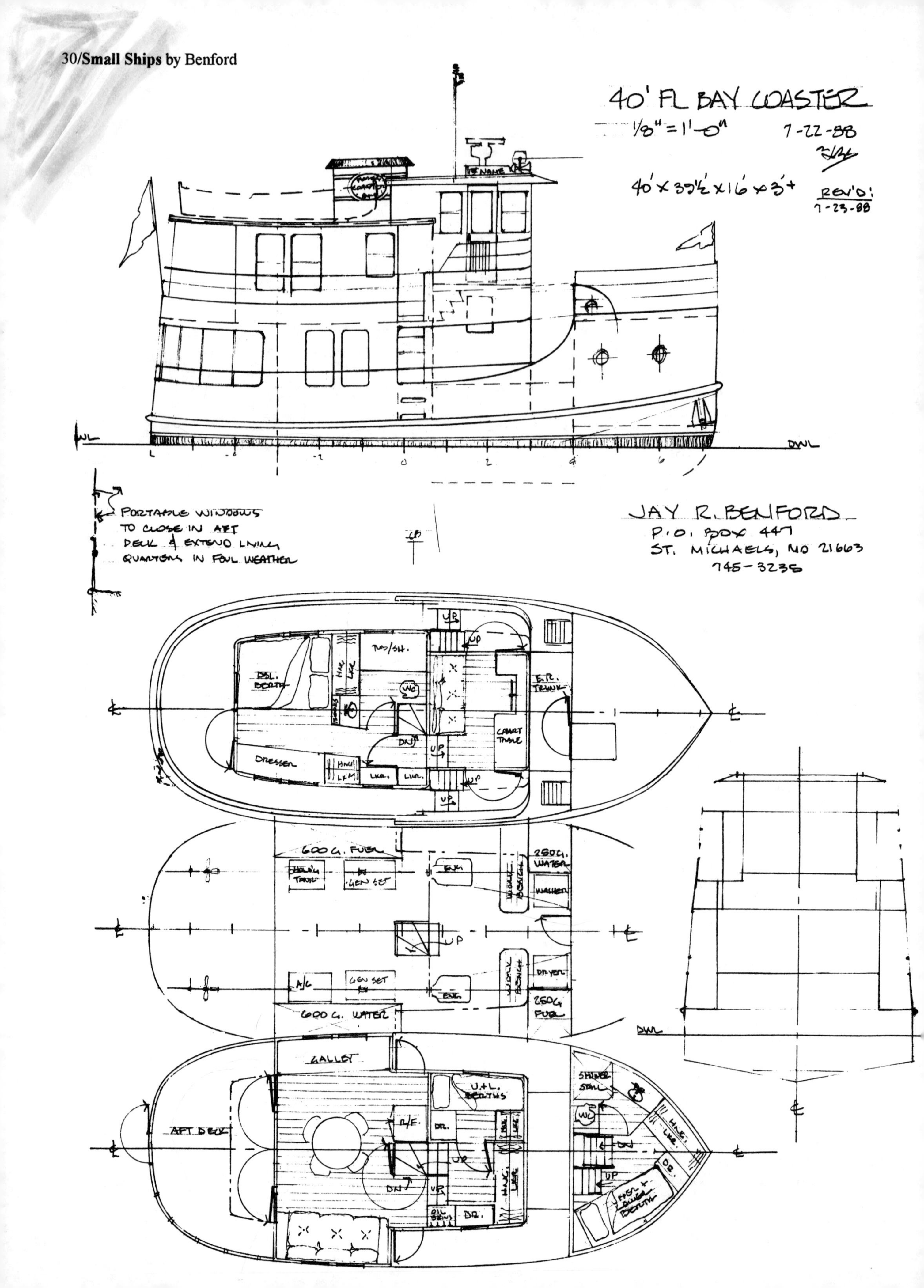
40' FL BAY COASTER
1/8" = 1'-0"
7-22-88
40' x 39½ x 16 x 3+
REV'D:
7-23-88
WL
DWL
PORTABLE WINDOWS
TO CLOSE IN AFT
DECK & EXTEND LIVING
QUARTERS IN FOUL WEATHER
JAY R. BENFORD
P.O. BOX 447
ST. MICHAELS, MD 21663
745-3235
DBL. BERTH
HNG. LKR.
TUB/SH.
WC
E.R. TRUNK
CHART TABLE
DRESSER
LKR.
UP
DN
600 G. FUEL
HOLD'G TANK
GEN SET
ENG
WORK BENCH
250G. WATER
WASHER
DRYER
250G FUEL
A/C
600 G. WATER
GALLEY
U+L BERTHS
R/F.
DR.
AFT DECK
OIL SKINS
SHOWER STALL
FOLD + DOUBLE BERTH

43' Packet & Freighter Yacht

Design Number 321
1992

The 43' Packet design evolved from our 35' Packet. Our design brief was to add a second full bathroom to the boat so that the master stateroom had its own private head, en suite. We were also to add enough length to make the lower aftdeck a good space for sitting out with deck chairs and to add a study/office space.

In doing all this, we added two feet to the beam and eight feet to the length. We were able to shift the spaces around enough to permit the engine room to shift aft and under the saloon. This weight shift let the galley portion of the saloon be switched back to the forward end. We'd had to hold it aft on the 35 to keep the boat in level trim without ballast. With the greater beam on the 43, we've gotten a little more height, enough to permit a standing headroom area in the "basement". This could be used to have a workshop, a third stateroom, and/or more storage space.

Construction calls for plywood, sealed and glued with epoxy, over sawn and/or laminated timber framing, with plywood floors and bulkheads. The exterior will be sheathed with a layer of cloth set in epoxy for abrasion and impact resistance. This provides for a simple and rugged boat with all the warmth and good insulating qualities of wood. Having lived aboard three wooden boats, this would be my first choice for materials.

Or, she could be built in steel with a bit greater structural weight. This would let her have integral tanks of a bit greater capacity, if desired. With our philosophy of designing "low-tech" steel boats, the structural weights are intentionally made a little greater in the interest of making the boat simple to build. This also makes the boat easy to maintain, which helps insure longevity.

The four inch molded rubber guard is the widest point of the boat, making for easy landings and not having to worry about having fenders hung in the right place. This makes for a lot of peace of mind when making landings single-handed.

The protective skegs under the rudders let the skipper feel there's no problem in going exploring. We particularly enjoy this peace of mind sightseeing in backwaters.

Optional fly bridge controls can be fitted on the boat deck, with the helm positioned behind the pilothouse. With the pilothouse as high as it is, though, I doubt there is much need for a fly bridge. The pilothouse can be opened up, with the doors and windows opening, and there will be lots of fresh air that way. There is room for one or two small craft to be carried on top of the master stateroom, and a davit can be installed to lift them on and off.

The Freighter version has an eight foot well deck, with room for a small (under two thousand pound) vehicle to be carried on deck. By stacking the two berths on one side in the foc'sle, there's room for a complete head en suite. The cargo hold under the vehicle can also be used for an additional berthing area, office space, workshop or more general storage.

The saloon and middle deck cabin are rearranged to suit their shorter length. The saloon has the living and dining room furniture. The mid-deck has a large galley plus an oilskin or coat closet.

The master stateroom is moved aft, keeping an en suite full head, but has a much shorter aft deck. We've kept the valuable walkaround deck space for that big ship feeling and practical ship handling.

The pilothouse has a good view over the skiff. The L-shaped settee is raised for visibility underway. The starboard aft corner is notched for the stairs to the boat deck.

The hold space shows an option for more berths if wanted, additional clothes lockers for out-of-season storage, and a fore and aft passageway from the foc'sle through the shop and laundry area to the saloon.

The tankage layout shown on the Freighter gives a range of options to keep the boat in trim. We like to design them with the center of gravity of the tankage at the center of bouyancy for the boat. This is an attempt to keep them "trim neutral" as they are filled or emptied. This variation has the fuel close to the center of the boat and water forward and aft. Thus either water or fuel could be used to keep the boat in trim athwartships. The water usage can be taken from either the forward or aft tanks to keep the boat in trim lengthwise. With sight gauges on the tanks and some attention on a daily basis to which tanks are being used, the boat can be kept in good trim.

Particulars:	**Wood**	**Steel Vers.**
Length over guards	44'-0"	44'-0"
Length-structural	43'-4"	43'-4"
Designed waterline	42'-6"	42'-6"
Beam over guards	18'-0"	18'-0"
Beam-structural	17'-4"	17'-4"
Draft, loaded	4'-0"	4'-0"
Freeboard:		
Forward	7'-6"	7'-6"
Waist	4'-1"	4'-1"
Aft	1'-6"	1'-6"
Displacement, loaded for cruising*	75,000	85,000
Displacement-length ratio	436	494
Pounds per inch immersion	3010	3010
Tankage: Fuel	500 gals.	600 gals.
Water	1,000 gals.	1,000 gals.
Stability - GM	4'	3.5'

***CAUTION:** The displacement quoted here is for the boat in cruising trim. That is, with the fuel and water tanks filled, the crew on board, as well as the crews' gear and stores in the lockers. This should not be confused with the "shipping weight" often quoted as "displacement" by some manufacturers. This should be taken into account when comparing figure and ratios between this and other designs.

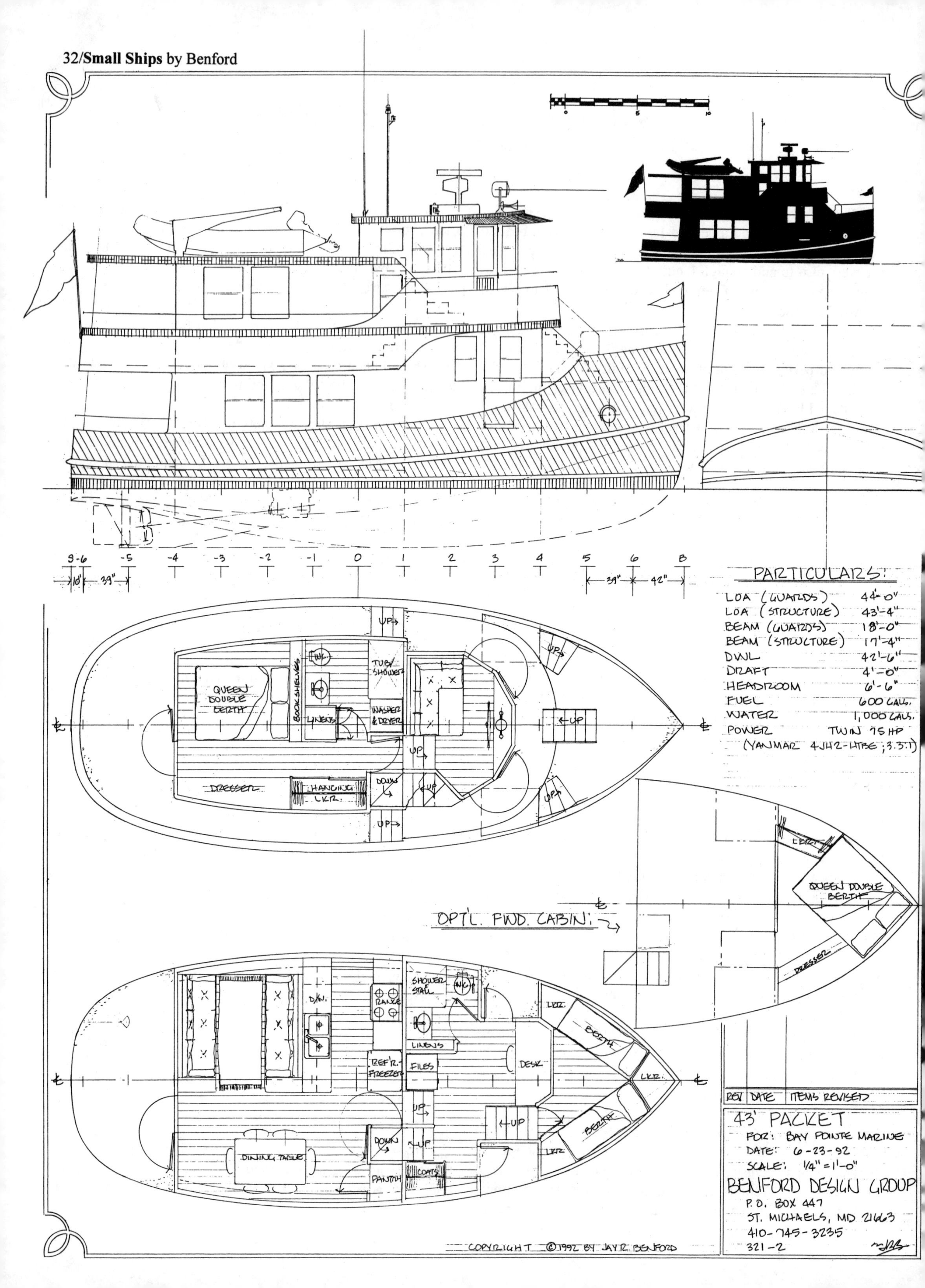
PARTICULARS:
LOA (GUARDS) 44'-0"
LOA (STRUCTURE) 43'-4"
BEAM (GUARDS) 18'-0"
BEAM (STRUCTURE) 17'-4"
DWL 42'-6"
DRAFT 4'-0"
HEADROOM 6'-6"
FUEL 600 GALS.
WATER 1,000 GALS.
POWER TWIN 75 HP
(YANMAR 4JH2-HTBE; 3.3:1)
QUEEN DOUBLE BERTH
BOOKSHELVES
TUB/SHOWER
LINENS
WASHER & DRYER
DRESSER
HANGING LKR.
UP
DOWN
OPT'L. FWD. CABIN
LKR.
DRESSER
SHOWER STALL
D/W.
RANGE
REF'R.-FREEZER
FILES
DESK
BERTH
DINING TABLE
PANTRY
COATS
REV DATE ITEMS REVISED
43' PACKET
FOR: BAY POINTE MARINE
DATE: 6-23-92
SCALE: 1/4" = 1'-0"
BENFORD DESIGN GROUP
P.O. BOX 447
ST. MICHAELS, MD 21663
410-745-3235
321-2
COPYRIGHT ©1992 BY JAY R. BENFORD

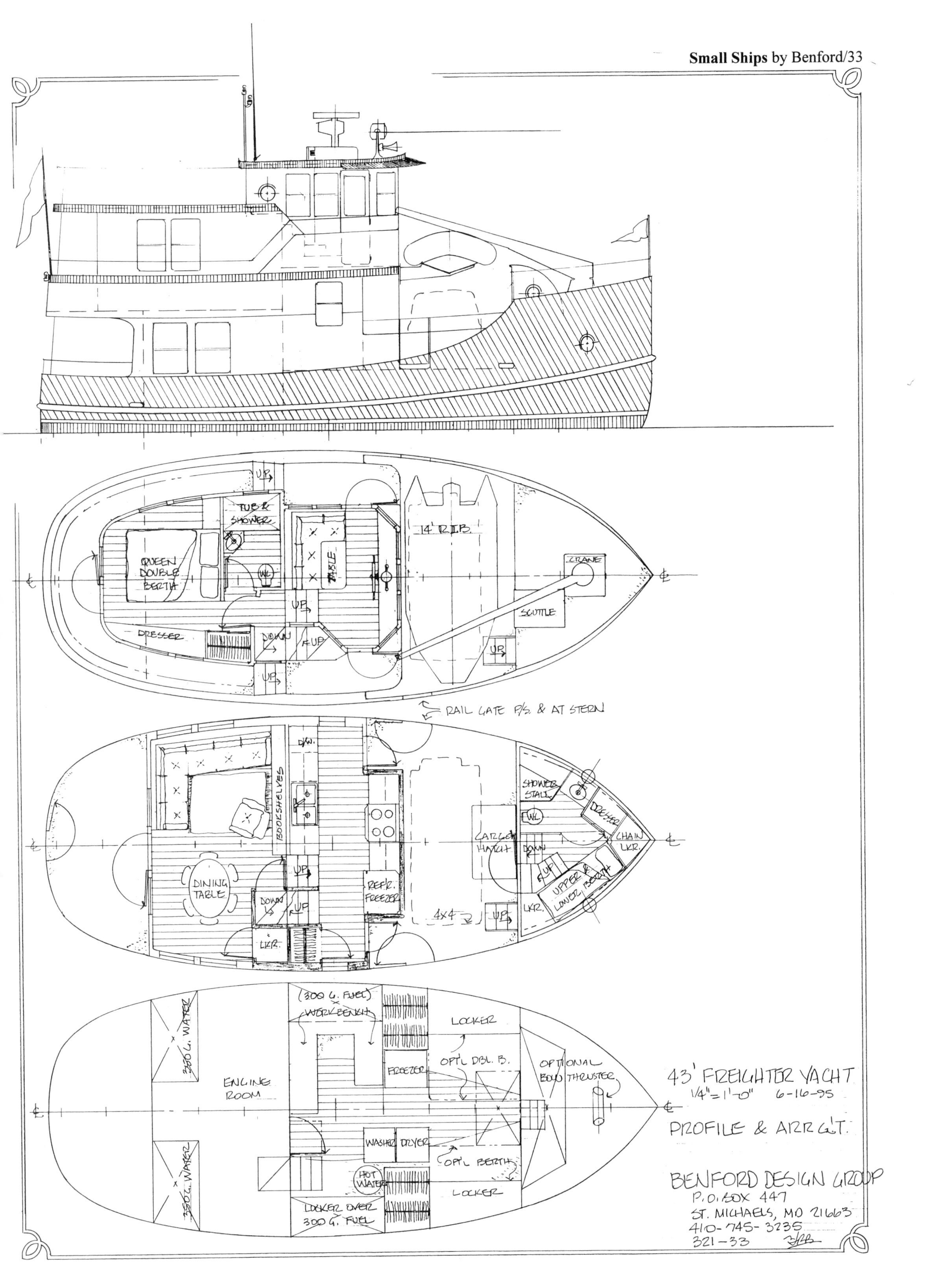
TUB & SHOWER
QUEEN DOUBLE BERTH
TABLE
14' RIB
CRANE
SCUTTLE
DRESSER
UP
DOWN
RAIL GATE P/S. & AT STERN
D/W.
BOOKSHELVES
SHOWER STALL
DRESSER
CHAIN LKR.
CARGO HATCH
UPPER & LOWER BERTH
LKR.
DINING TABLE
REFR. FREEZER
4x4
(300 G. FUEL)
WORKBENCH
LOCKER
350 G. WATER
ENGINE ROOM
FREEZER
OPT'L DBL. B.
OPTIONAL BOW THRUSTER
WASHER
DRYER
OPT'L BERTH
HOT WATER
LOCKER OVER 300 G. FUEL
43' FREIGHTER YACHT
1/4"=1'-0" 6-16-95
PROFILE & ARR'G'T.
BENFORD DESIGN GROUP
P.O. BOX 447
ST. MICHAELS, MD 21663
410-745-3235
321-33

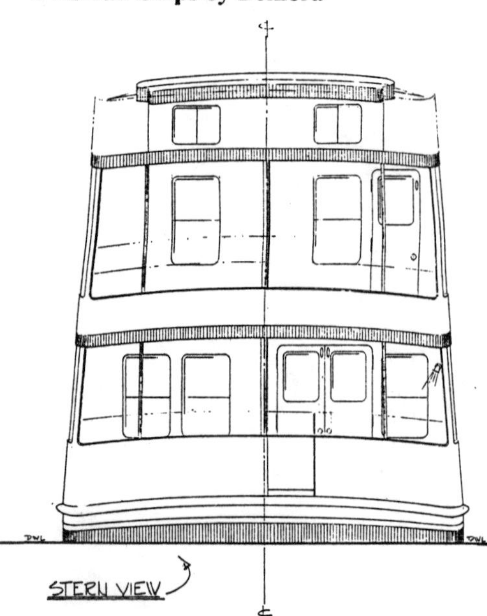

45' Florida Bay Coaster *Sails*

Design Number 270
1988

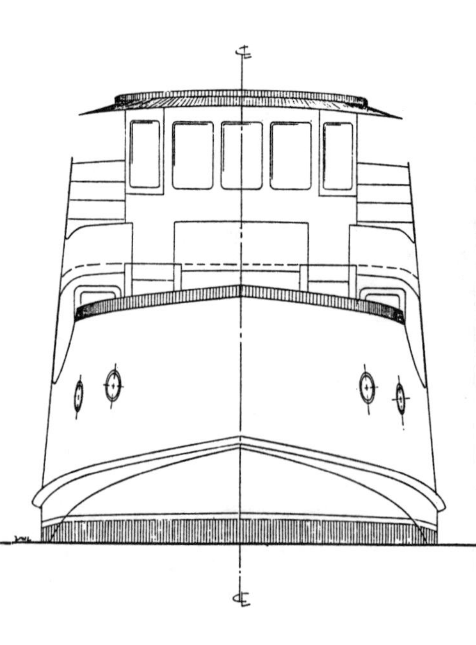

Sails was the first of the Coasters built without a well deck and crane to lift a Jeep® on and off. She was also the first of a series of three variations on this design built to date.

Closing up the well deck in evolving this design gave extra room inside, making her at least as roomy as the 50' ***Florida Bay***. We also pushed the sides of the master cabin out almost to the sides of the upper deck, leaving just a small shelf under the windows.

I had a trip aboard her, and later had a cruise on a sistership with my wife and kids aboard. The 45's have a very pleasant motion and give a feeling of great competence. We felt quite confident crossing Albemarle Sound when the larger sailboats were turning back.

The photos below show ***Sails*** on completion of her delivery trip to St. Petersburg, laying alongside ***Key Largo*** and shifted to her end moorage slip, where she has an excellent view of Tampa Bay. It's evident that she has almost the same height as ***Key Largo***, yet carries it gracefully. She has good headroom and good stability, coming through ten foot seas on one cruise. Her inclining test gave a GM of 3.5'.

The color photos on page 12 give some additional idea of how she turned out. The owners have enjoyed her and are planning a bigger one as their next liveaboard.

Particulars:

Length overall	45'-0"
Length designed waterline	44'-7"
Beam	17'-0"
Draft	4'-0"
Freeboard:	
Forward	8'-4½"
Waist	3'-10"
Aft	1'-5"
Displacement, cruising trim*	96,000 lbs.
Displacement-length ratio	484
Prismatic coefficient	.624
Pounds per inch immersion	3,408
Water tankage	1,050 Gals.
Fuel tankage	1,050 Gals.
Headroom	6'-7"

***CAUTION:** The displacement quoted here is for the boat in cruising trim. That is, with the fuel and water tanks filled, the crew on board, as well as the crews' gear and stores in the lockers. This should not be confused with the "shipping weight" often quoted as "displacement" by some manufacturers. This should be taken into account when comparing figures and ratios between this and other designs.

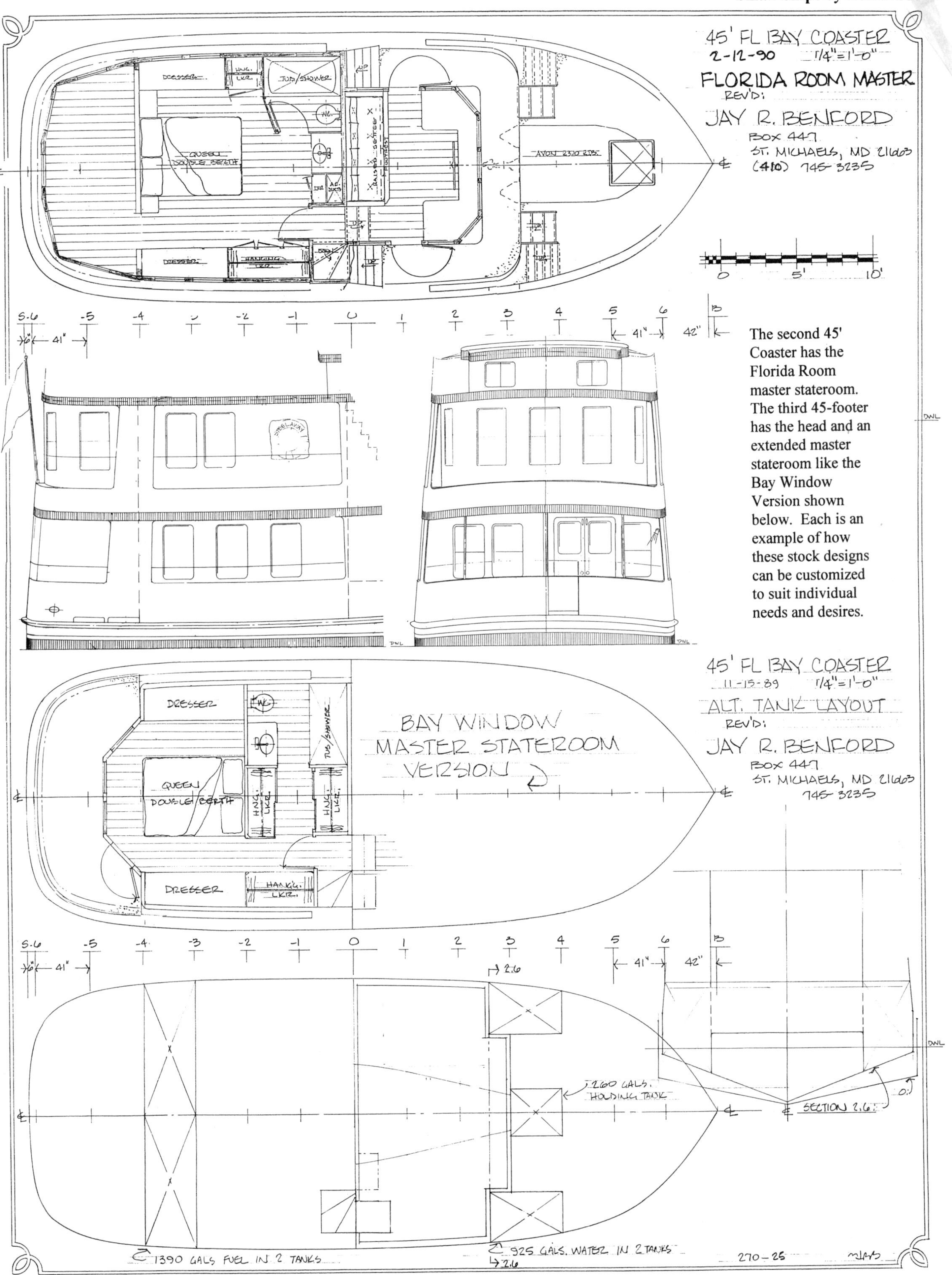

The second 45' Coaster has the Florida Room master stateroom. The third 45-footer has the head and an extended master stateroom like the Bay Window Version shown below. Each is an example of how these stock designs can be customized to suit individual needs and desires.

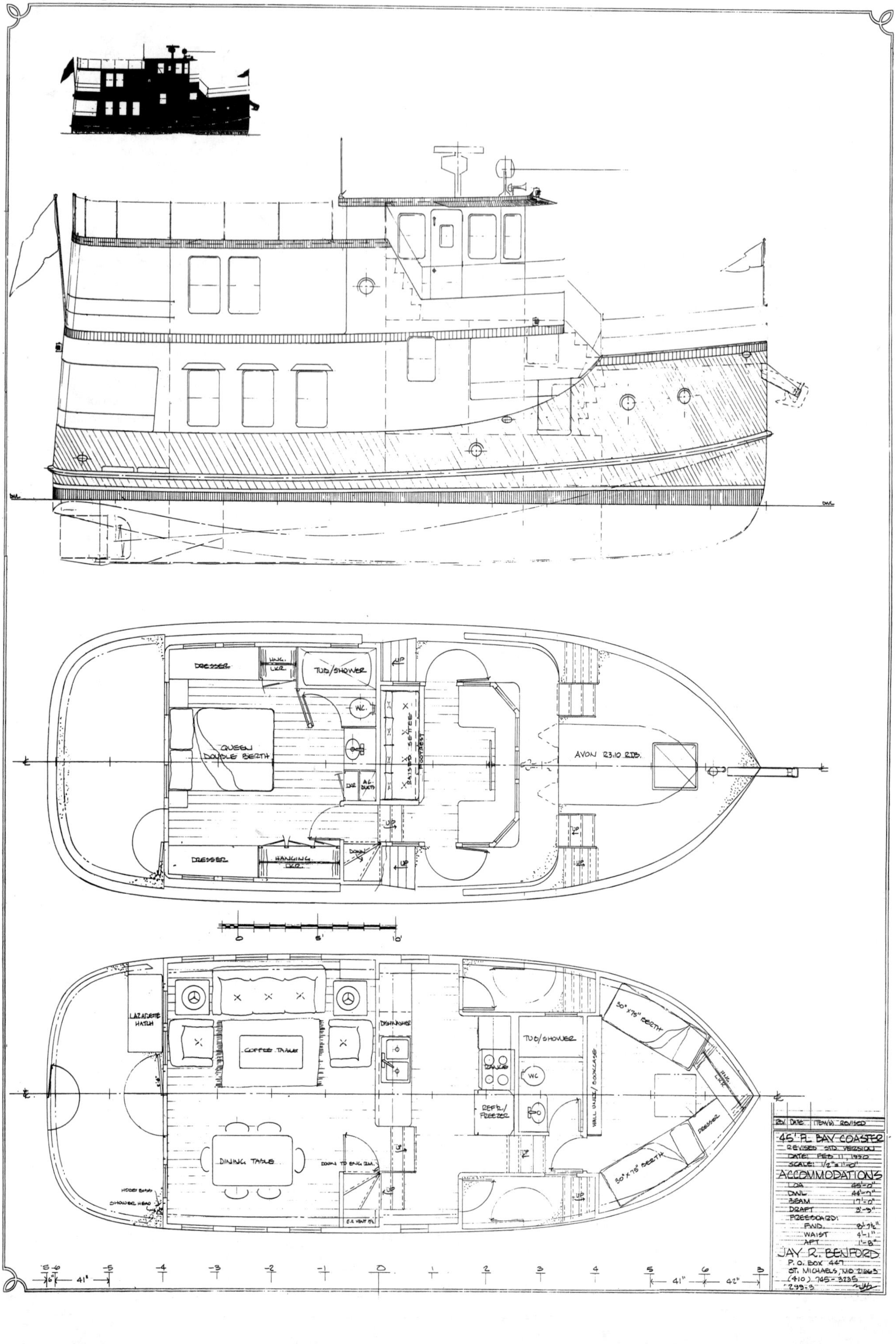

AVON R3.10 RIB.

DRESSER
HNG. LKR.
TUB/SHOWER
W.C.
QUEEN DOUBLE BERTH
RAISED SETTEE
FOOTREST
LKR
A/C DUCTS
UP
DOWN
DRESSER
HANGING LKR.

0 5' 10'

LAZARETTE HATCH
COFFEE TABLE
DISHWASHER
RANGE
REFR./FREEZER
TUB/SHOWER
WC
WALL UNIT/BOOKCASE
30" x 75" BERTH
HNG. LKR.
DRESSER
30" x 75" BERTH
DINING TABLE
DOWN TO ENG. RM.
UP
HOSE BIBS
SHOWER HEAD
E.R. VENT

REV	DATE	ITEM(S) REVISED

45' FL BAY COASTER
REVISED STD. VERSION
DATE: FEB 11, 1990
SCALE: 1/2" = 1'-0"

ACCOMMODATIONS

LOA	45'-0"
DWL	44'-7"
BEAM	17'-0"
DRAFT	3'-9"
FREEBOARD:	
FWD.	8'-7½"
WAIST	4'-1"
APT	1'-8"

JAY R. BENFORD
P.O. BOX 447
ST. MICHAELS, MD 21663
(410) 745-3235
299-3

-5-6 -5 -4 -3 -2 -1 0 1 2 3 4 5 6 B
6" 41" 41" 42"

45' FL BAY COASTER
FOR: BILL & SHERRY WELCH
DATE: DEC 16, 1988
SCALE: 1/2" = 1'-0"
SECTIONS
LOA 45'-0"
DWL 44'-7"
BEAM 17'-0"
DRAFT 3'-9"
FREEBOARD:
FWD. 8'-3"
WAIST 4'-3"
AFT 1'-8"
JAY R. BENFORD
P.O. BOX 447
ST. MICHAELS, MD 21663
(410) 745-3235
270-8
6 7-20-89 ADDED LADDER REV. STA'S +1, +4, -3
5 6-7-89 REV. STA. +1 AFT
4 5-19-89 REV. STA -1 FWD
3 4-14-89 ADD'L FRAMING, TACKWELL @ WINDOW TOWERS
2 3-31-89 FINISH VIEWS
1 3/17/89 ADD'L STATEROOM DETAILS
NO DATE ITEMS REVISED
PORT INBOARD PROFILE
MICRO & HOOD
DWL
STA. -3: LOOKING AFT
STA. +4: LOOKING FWD
STA +3: LOOKING AFT
0 5' 10'
STARBOARD INBOARD PROFILE
STA. -1 LOOKING FWD
STA +1: LOOKING AFT
STA. +1 LOOKING FWD

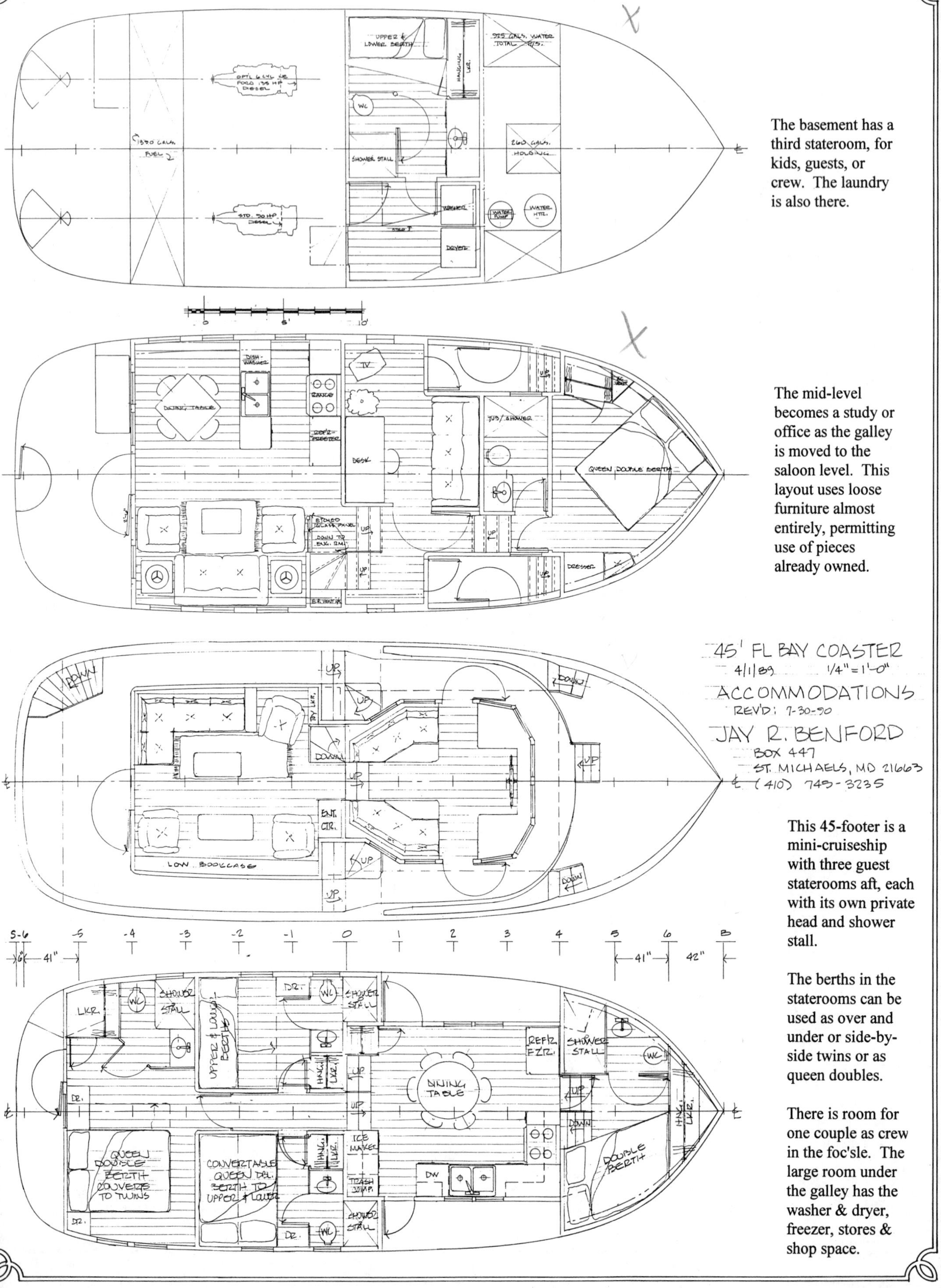

45' FL BAY COASTER
4/1/89 1/4"=1'-0"
ACCOMMODATIONS
REV'D: 7-30-90
JAY R. BENFORD
BOX 447
ST. MICHAELS, MD 21663
(410) 745-3235

The basement has a third stateroom, for kids, guests, or crew. The laundry is also there.

The mid-level becomes a study or office as the galley is moved to the saloon level. This layout uses loose furniture almost entirely, permitting use of pieces already owned.

This 45-footer is a mini-cruiseship with three guest staterooms aft, each with its own private head and shower stall.

The berths in the staterooms can be used as over and under or side-by-side twins or as queen doubles.

There is room for one couple as crew in the foc'sle. The large room under the galley has the washer & dryer, freezer, stores & shop space.

Three 45' Coaster Variations

Design Number 270

1988

The following three pages show drawings for the first three 45' Coaster variations. These ideas are drawn on a 16' beam hull versus the 17' beam on the later ones like ***Sails***. They show some more layout ideas for this smaller size freighter yacht.

The first 45' Coaster drawn has a small 4X4 on deck. She came about from being a variation of the 40' Coaster with an aft deck added and a bit more room in the master stateroom. Her after house is sunk into the hull with the sole at the waterline. Thus it is two steps up to the after deck. The upper deck is stepped up to give full headroom over the after deck. The overhang on the upper housetop is shortened so that there will be room to stand up on this deck. The upper house is asymmetrical, with a side deck along one side. The other side could have stairs alongside the pilothouse for access to the boat deck.

The head for the master stateroom is on the mid-level. A rearrangement of the upper cabin could make room for a head there by reducing the clothes storage. Then, the mid-level could be used for something else and the extra clothes storage would be in the hold.

With the reduced height from the sunken houses, the space under the mid-level and in the hold will not have standing headroom. However, there is quite a lot of storage space there.

The saloon has a variation of the layout we used on the ***Florida Bay***. I'd tried moving the furniture like this while cruising the ***Florida Bay*** and found that there was an interesting potential there.

45' Coaster-2 is a variation I drew to look at giving the spaces differing usages. It has a small after deck and the raised upper deck over it forms a seat at the stern. The asymmetrical upper house still allows outside access along one side. This would be the preferred side for making landings, since there are better deck traffic options on that side. On the main deck, the deck steps up amidships for entry into the mid-level galley. Outside this door there is a deck locker for stowing the myriad of odds and ends that make for good ship's husbandry.

Alongside the pilothouse is a set of stairs going up to the boat deck and another going down to the waist deck. This will permit quick traffic from the pilothouse down to the waist and up onto the foredeck for line and anchor handling.

The layout with the master stateroom down on the lower deck is an alternative we have had some interest in, but so far none of the Florida Bay Coasters have been built this way. After cruising with the master stateroom on the upper deck, it's hard to go back to thinking about having it anywhere else. This master stateroom does have a lot of closet space. There are dual dressers and there is room for a bank of drawers under the berth.

We developed the 45' Coaster-3 from some sketches Reuben Trane and a prospective buyer made while they were negotiating about the possibility of having it built. With the elimination of the waist deck and extension of the cabin spaces, this is becoming more like the later 45-footer ***Sails*** and her sisterships.

There are stairs at the forward end of the bridge wings, wrapped in front of the pilothouse and leading onto the foredeck. There is a notch below the window level on the front of the pilothouse for access around the stern of the 13' skiff on the foredeck. The skiff sits on a deck box with chocks at the ends to hold it. Within the deck box are four skylight hatches, giving great light and ventilation to the foc'sle, the head and the study.

The saloon is an open layout, with a long, shelf-top locker along the port side, for storage of a myriad of items. There is a breakfast bar, a dining table and sofa and chairs for eating, sitting and entertaining.

The galley is at the mid-level entry like we later used on the 45's and some of the 50's. A few steps down and forward is a study with a convertible settee/berth and the washer and dryer stacked in one corner. The foc'sle has a cozy double berth and its own head and shower stall.

The master stateroom has the berth against the after bulkhead with the hanging locker against the forward head bulkhead. This leaves room on the after bulkhead for some windows for good light and ventilation.

These three variations give some more ideas that can be used to make up your own perfect liveaboard cruiser. Let us know if they give rise to some more ideas for another version.

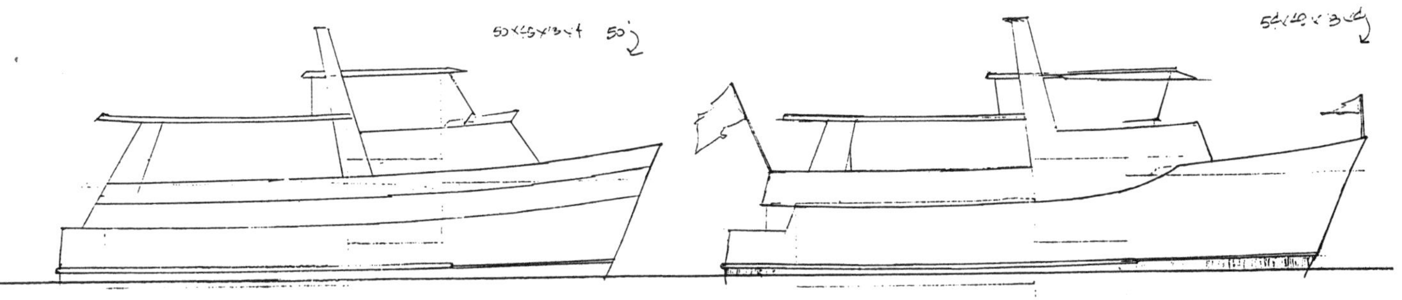

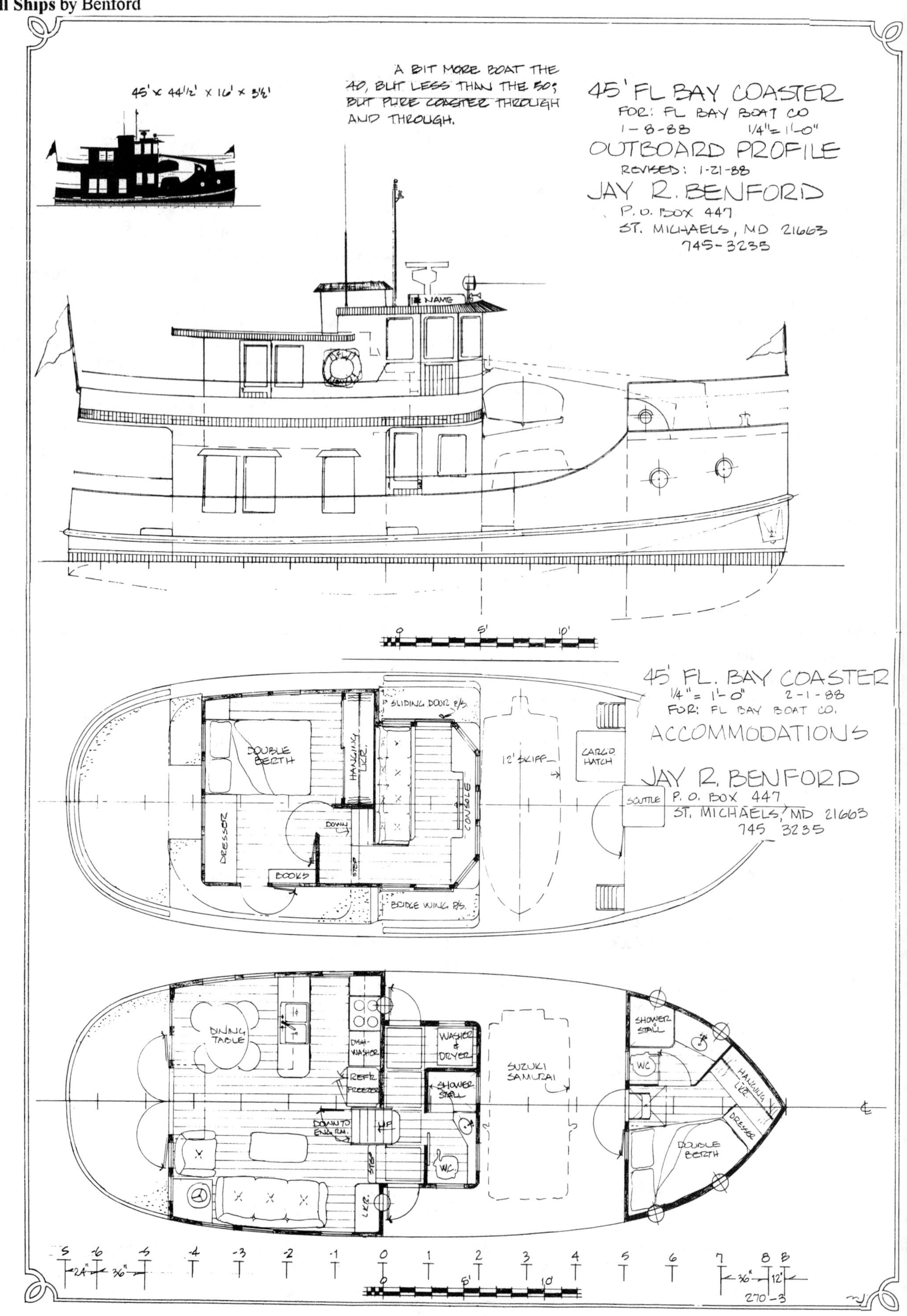
45' x 44½' x 16' x 3½'
A BIT MORE BOAT THE 40, BUT LESS THAN THE 50; BUT PURE COASTER THROUGH AND THROUGH.
45' FL BAY COASTER
FOR: FL BAY BOAT CO
1-8-88 1/4"=1'-0"
OUTBOARD PROFILE
REVISED: 1-21-88
JAY R. BENFORD
P. O. BOX 447
ST. MICHAELS, MD 21663
745-3235
NAME
0 5' 10'
45' FL. BAY COASTER
1/4"=1'-0" 2-1-88
FOR: FL BAY BOAT CO.
ACCOMMODATIONS
JAY R. BENFORD
P. O. BOX 447
ST. MICHAELS, MD 21663
745 3235
SLIDING DOOR P/S
DOUBLE BERTH
HANGING LKR.
12' SKIFF
CARGO HATCH
SCUTTLE
CONSOLE
DRESSER
DOWN
STEP
BOOKS
BRIDGE WING P/S.
DINING TABLE
DISH-WASHER
REFR. FREEZER
WASHER & DRYER
SHOWER STALL
SUZUKI SAMURAI
SHOWER STALL
WC
HANGING LKR.
DRESSER
DOUBLE BERTH
DOWN TO ENG. RM.
UP
STEP
WC.
LKR.
-5 -6 -5 -4 -3 -2 -1 0 1 2 3 4 5 6 7 8 8
24" 36" 36" 12"
0 5' 10'
270-3

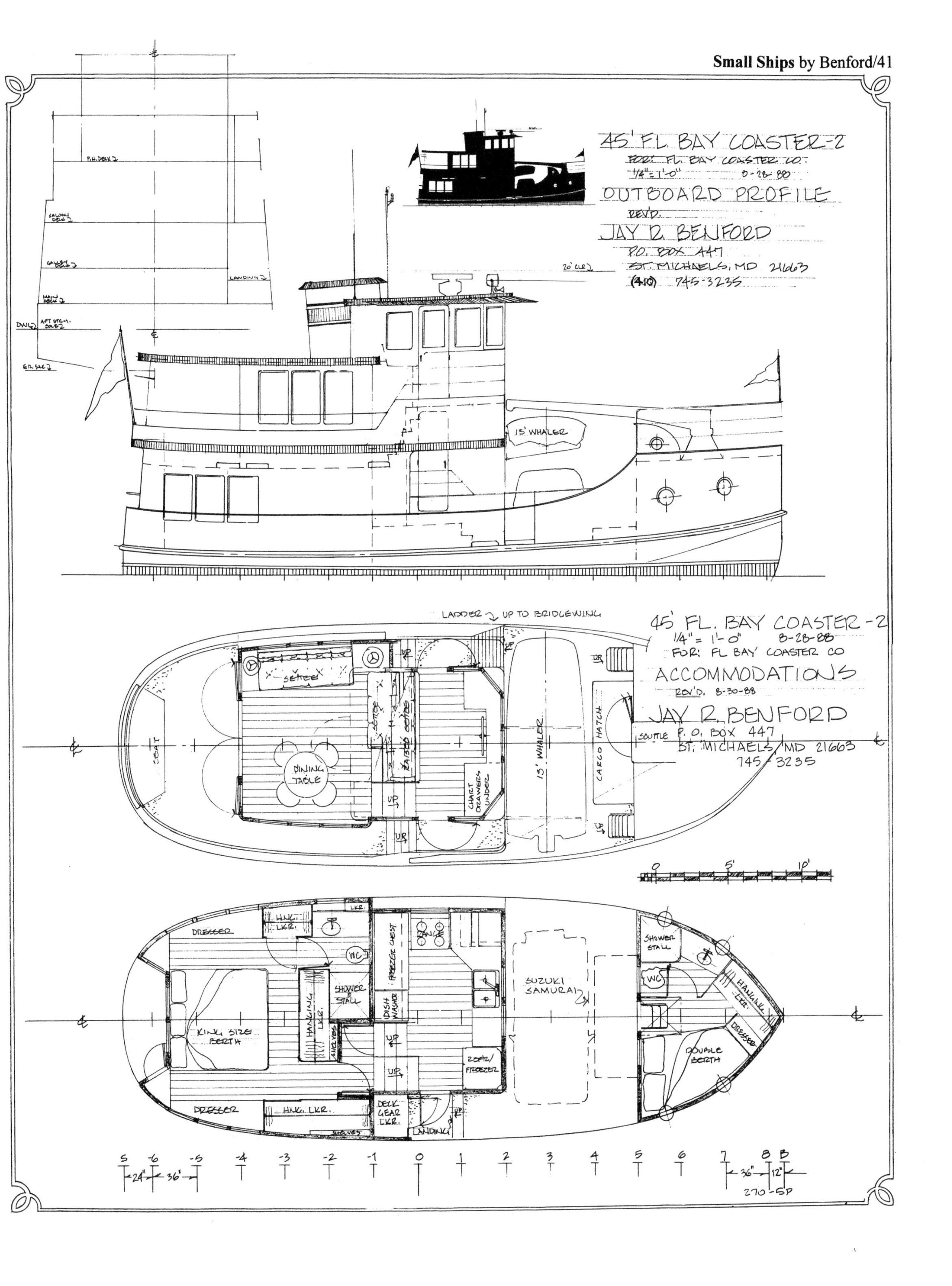

45' FL. BAY COASTER-2
FOR: FL. BAY COASTER CO.
1/4" = 1'-0"
OUTBOARD PROFILE
REV'D.
JAY R. BENFORD
P.O. BOX 447
ST. MICHAELS, MD 21663
(410) 745-3235
15' WHALER
LADDER UP TO BRIDGEWING
45' FL. BAY COASTER-2
1/4" = 1'-0" 8-28-88
FOR: FL BAY COASTER CO
ACCOMMODATIONS
REV'D. 8-30-88
JAY R. BENFORD
P. O. BOX 447
ST. MICHAELS, MD 21663
745-3235
SETTEE
SEAT
DINING TABLE
RAISED SETTEE
CHART DRAWERS UNDER
15' WHALER
CARGO HATCH
SCUTTLE
UP
DRESSER
HNG. LKR.
LKR.
WC
KING SIZE BERTH
HANGING LKR.
SHOWER STALL
SHELVES
FREEZE CHEST
DISH WASHER
RANGE
REFR/FREEZER
SUZUKI SAMURAI
DECK GEAR LKR.
LANDING
SHOWER STALL
HANGING LKR.
DRESSER
DOUBLE BERTH
24" 36"
36" 12"
270-SP

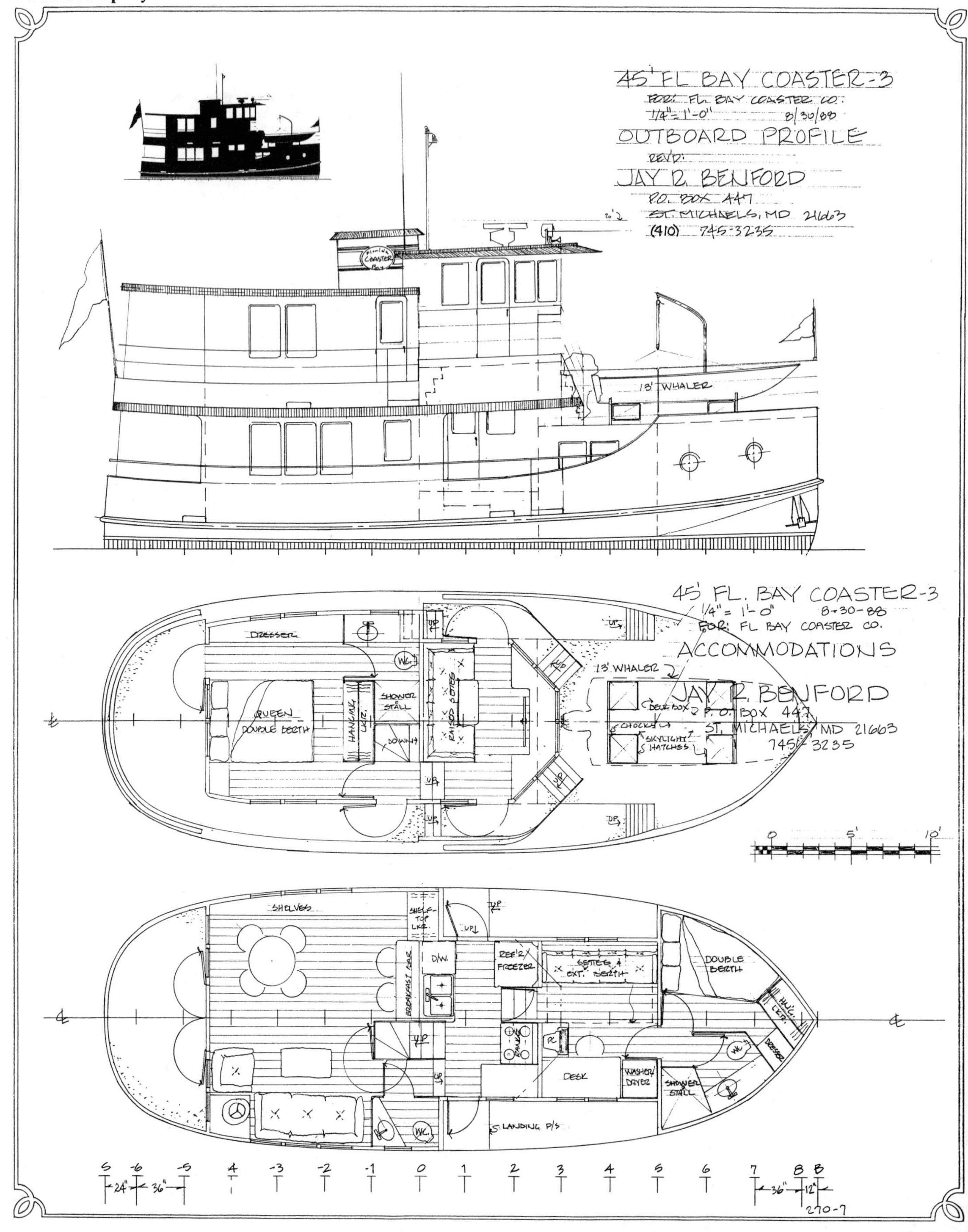

45' FL BAY COASTER-3
FOR: FL BAY COASTER CO.
1/4"=1'-0" 8/30/88
OUTBOARD PROFILE
REV'D:
JAY R. BENFORD
P.O. BOX 447
ST. MICHAELS, MD 21663
(410) 745-3235
13' WHALER
45' FL. BAY COASTER-3
1/4" = 1'-0" 8-30-88
FOR: FL BAY COASTER CO.
ACCOMMODATIONS
JAY R. BENFORD
P. O. BOX 447
ST. MICHAELS, MD 21663
745-3235
DRESSER
W.C.
SHOWER STALL
HANGING LKR.
QUEEN DOUBLE BERTH
RAISED SETEE
DOWN
UP
DECK BOX
CHOCKS
SKYLIGHT HATCHES
0
5'
10'
SHELVES
SHELF-TOP LKR.
D/W.
BREAKFAST BAR
REF'R/ FREEZER
SETTEE & EXT. BERTH
DOUBLE BERTH
H.U. LKR.
DRESSER
RANGE
PC
DESK
WASHER/ DRYER
SHOWER STALL
LANDING P/S
24"
36"
12"
270-7

45' East Coaster *Pelican*

Design Number 299
1989

The ***Pelican*** drawing and the variation on her shown on the following page are some ideas I drew up to explore what could be done in this size for a liveaboard for myself. With a couple kids at home yet and the desire for some sort of home office space, our needs are different than those that led to the creation of ***Sails***, with only a couple to be living aboard her.

The ***Pelican*** is about the same size as ***Sails***, but with considerable changes made to her. This one is a double chine hull, with a single screw in a tunnel stern to keep the draft down. She has a bow thruster shown, as an option to aid maneuvering in a breeze.

The pilothouse layout is asymmetric to permit the stairs to the boat deck to be notched into the after port corner. The head in the master could be further revised so that the traffic goes through a split hanging locker rather than through the head. The page of alternatives shows some of the variations we looked and considered in trying to come up with our ideal 45-footer.

I've shown spud tubes on her. These are large pipes that go from the upper deck level right through the bottom of the hull, being open on both ends. They have another pipe in them that is ballasted and which drops to stick in the bottom, either to anchor the boat in position or to provide a pivot point for maneuvering the boat. These are found on some commercial boats, and are quite handy for boats that work in shallow waters that don't have a lot of tidal rise and fall.

Particulars:

Length overall	45'-0"
Length designed waterline	44'-6"
Beam	17'-0"
Draft	3'-6"
Freeboard:	
Forward	9'-3"
Waist	4'-3"
Aft	2'-0"
Displacement, cruising trim*	90,000 lbs.
Displacement-length ratio	456
Water tankage	1,500 Gals.
Fuel tankage	950 Gals.
Headroom	6'-7"

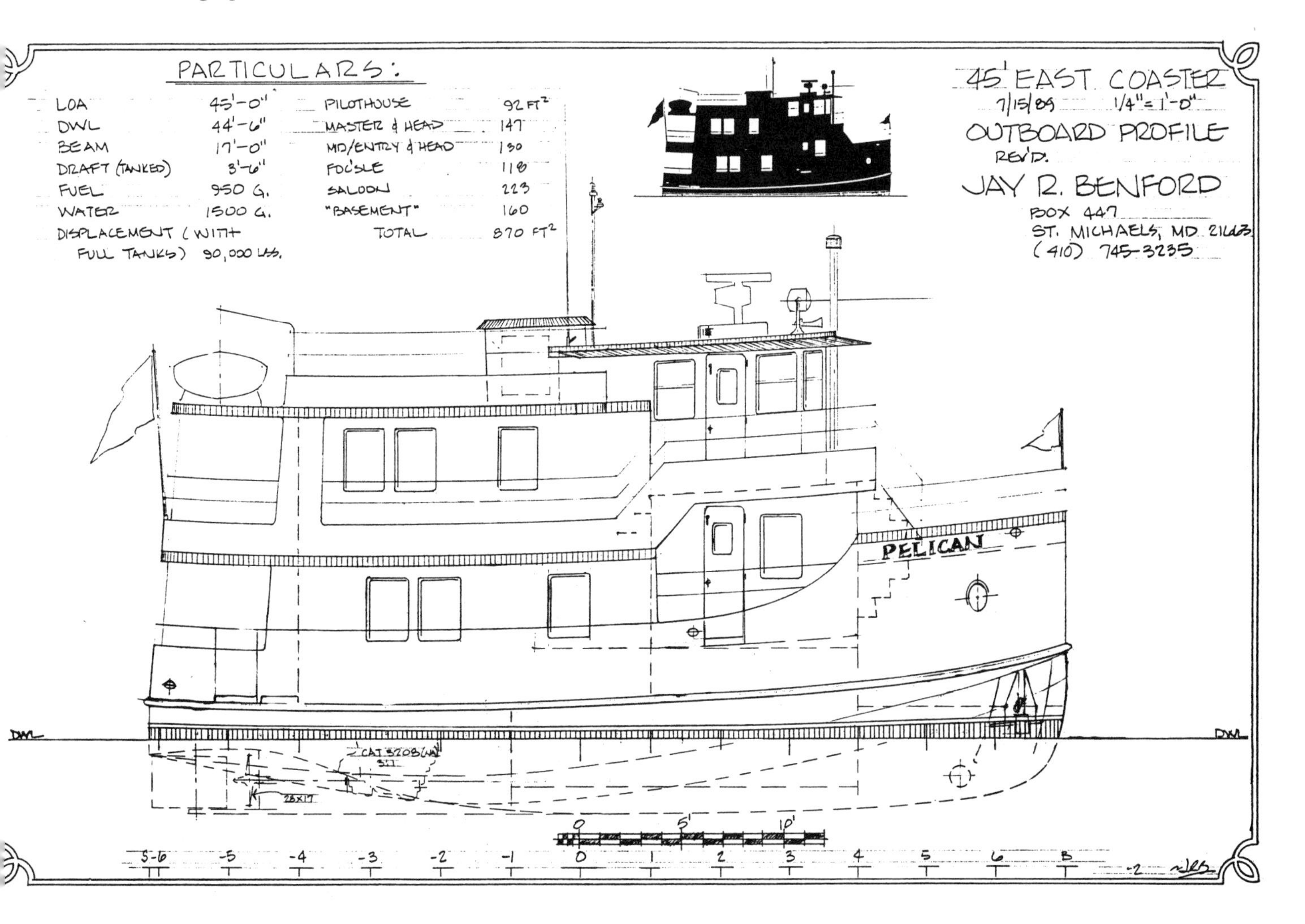

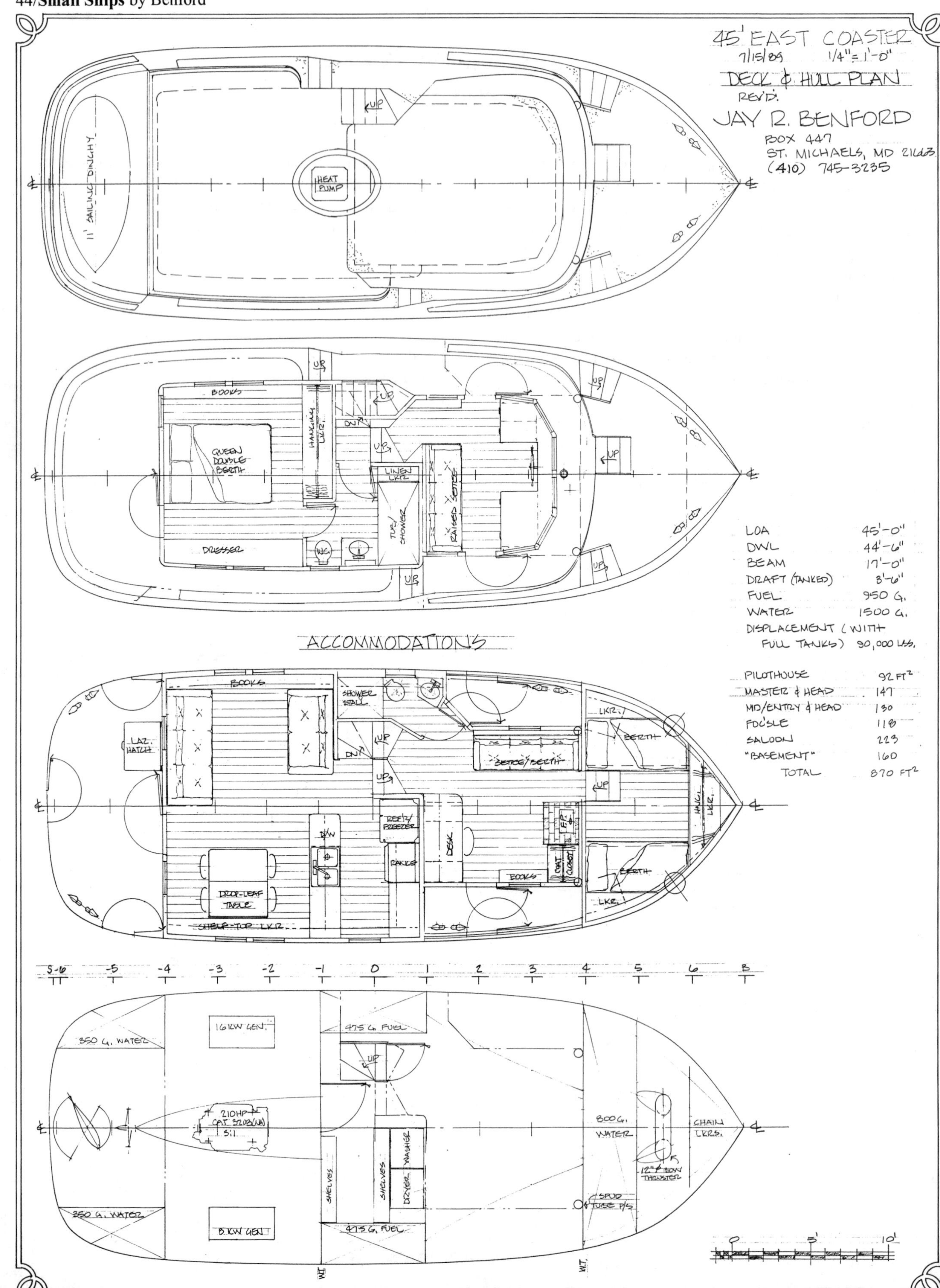

45' EAST COASTER
7/15/89 1/4"=1'-0"
DECK & HULL PLAN
REV'D:
JAY R. BENFORD
BOX 447
ST. MICHAELS, MD 21663
(410) 745-3235
11' SAILING DINGHY
HEAT PUMP
UP
BOOKS
QUEEN DOUBLE BERTH
HANGING LKR.
DN
LINEN LKR.
TUB/SHOWER
RAISED SETTEE
DRESSER
W.C.
LOA 45'-0"
DWL 44'-6"
BEAM 17'-0"
DRAFT (TANKED) 3'-6"
FUEL 950 G.
WATER 1500 G.
DISPLACEMENT (WITH FULL TANKS) 30,000 LBS.
PILOTHOUSE 92 FT²
MASTER & HEAD 147
MID/ENTRY & HEAD 130
FO'C'SLE 118
SALOON 223
"BASEMENT" 160
TOTAL 870 FT²
ACCOMMODATIONS
SHOWER STALL
LAZ. HATCH
SETTEE/BERTH
LKR.
BERTH
HNG. LKR.
REFR/FREEZER
D/W
RANGE
DESK
F.P.
COAT CLOSET
DROP-LEAF TABLE
SHELF-TOP LKR.
S-6 -5 -4 -3 -2 -1 0 1 2 3 4 5 6 B
16KW GEN.
350 G. WATER
475 G. FUEL
210HP CAT 3208(NA)
800 G. WATER
CHAIN LKRS.
SHELVES
DRYER
WASHER
12" BOW THRUSTER
SPUD TUBE P/S
5 KW GEN
W.T.
0 5' 10'

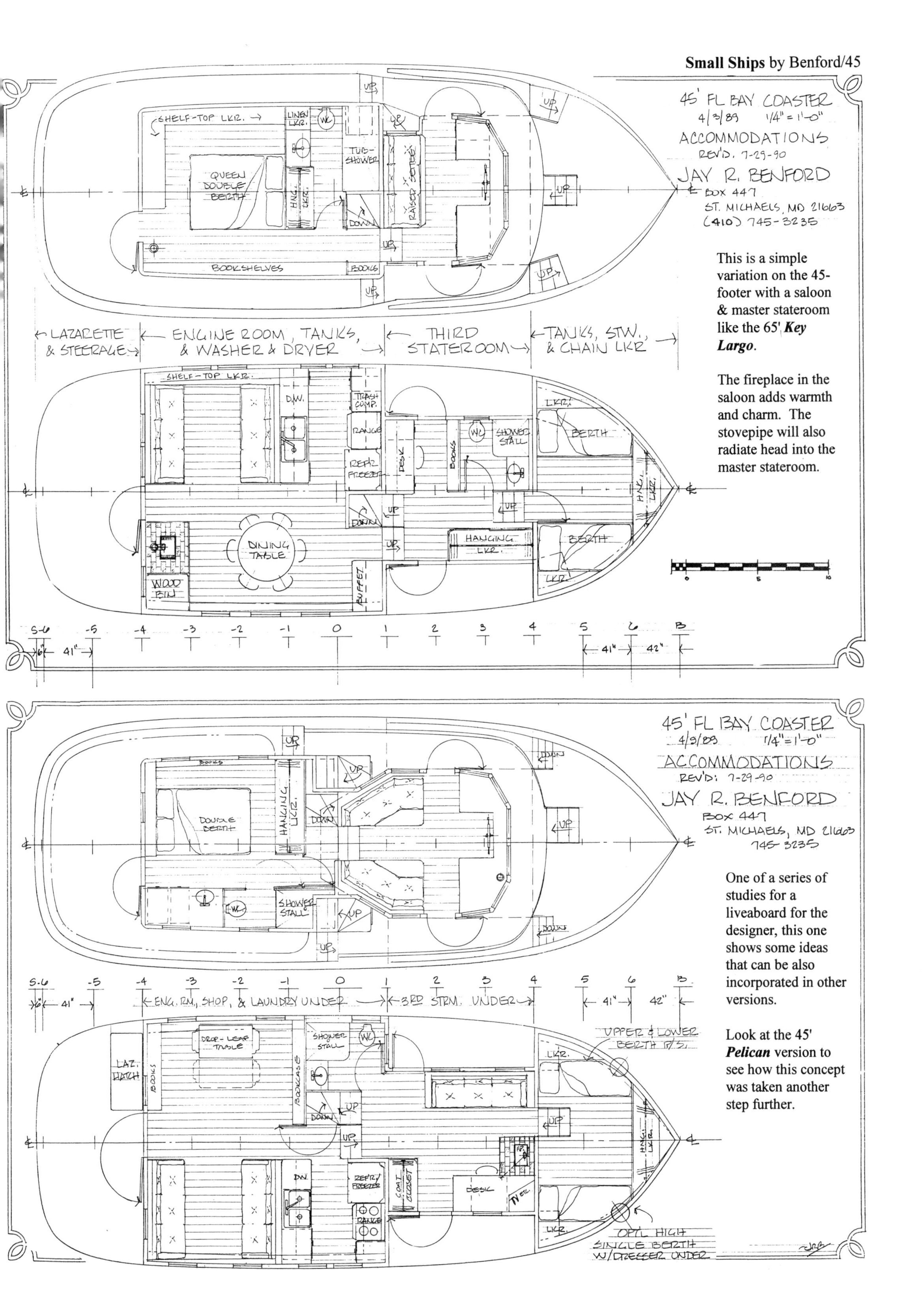

This is a simple variation on the 45-footer with a saloon & master stateroom like the 65' ***Key Largo***.

The fireplace in the saloon adds warmth and charm. The stovepipe will also radiate head into the master stateroom.

One of a series of studies for a liveaboard for the designer, this one shows some ideas that can be also incorporated in other versions.

Look at the 45' ***Pelican*** version to see how this concept was taken another step further.

45' Tramp Freighter

Design Number 270

1989

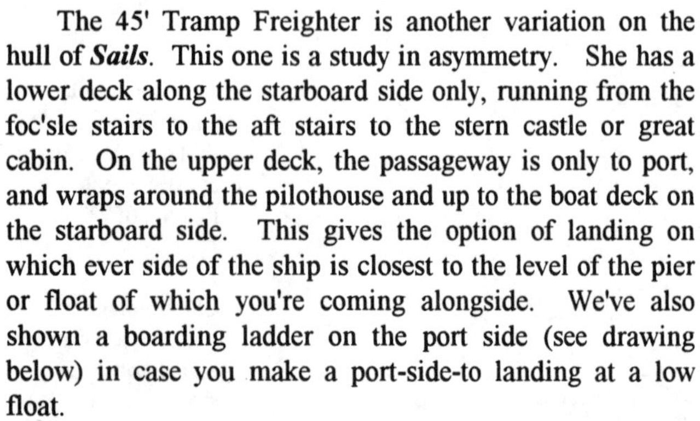

The 45' Tramp Freighter is another variation on the hull of ***Sails***. This one is a study in asymmetry. She has a lower deck along the starboard side only, running from the foc'sle stairs to the aft stairs to the stern castle or great cabin. On the upper deck, the passageway is only to port, and wraps around the pilothouse and up to the boat deck on the starboard side. This gives the option of landing on which ever side of the ship is closest to the level of the pier or float of which you're coming alongside. We've also shown a boarding ladder on the port side (see drawing below) in case you make a port-side-to landing at a low float.

The accommodations have two kids or guest staterooms forward that share the amidships head. The door from the middle stateroom through the after bulkhead closes off the passage to the head from the hall when opened from the stateroom, giving a form of private access to the head. The master stateroom on the upper deck has its own private head. The saloon is open and on one level right into the great cabin, making this one large living space. It has an open, country kitchen type of layout, instead of making the kitchen a separate area from the dining and living rooms.

The machinery would be located under the forward end of the saloon. The area under the middle stateroom and head would be a full headroom basement, with a cabin sole perhaps about half the width so that the outer areas could be lockers, berths, a desk, a workbench, a pantry, and/or space for the washer and dryer.

Particulars:

Length overall	45'-0"
Length designed waterline	44'-6"
Beam	17'-0"
Draft	3'-6"
Freeboard:	
Forward	9'-3"
Waist	4'-3"
Aft	2'-0"
Displacement, cruising trim*	90,000 lbs.
Displacement-length ratio	456
Water tankage	1,500 Gals.
Fuel tankage	950 Gals.
Headroom	6'-7"

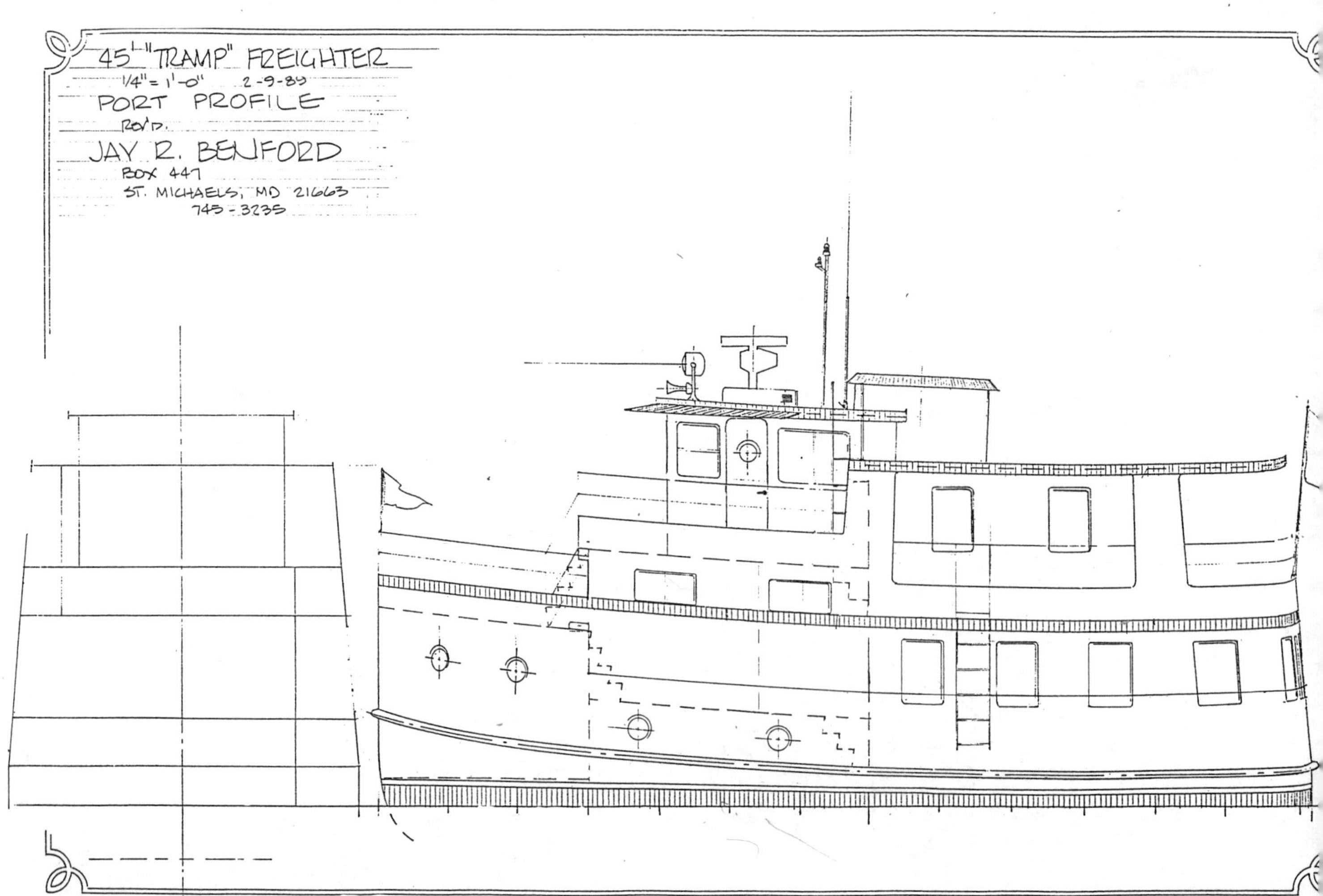

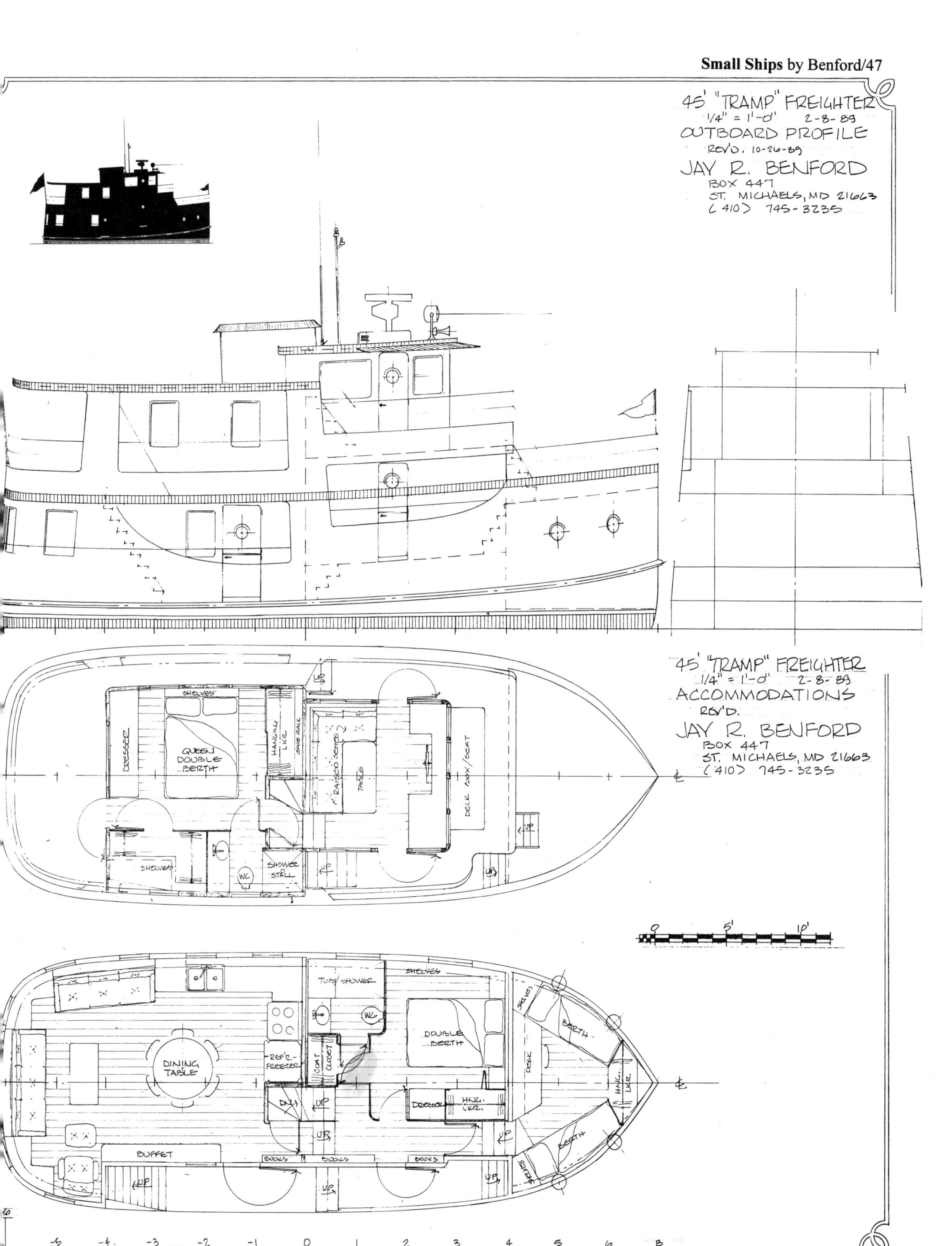

45' "TRAMP" FREIGHTER
1/4" = 1'-0" 2-8-89
OUTBOARD PROFILE
REV'D. 10-26-89
JAY R. BENFORD
BOX 447
ST. MICHAELS, MD 21663
(410) 745-3235
45' "TRAMP" FREIGHTER
1/4" = 1'-0" 2-8-89
ACCOMMODATIONS
REV'D.
JAY R. BENFORD
BOX 447
ST. MICHAELS, MD 21663
(410) 745-3235
SHELVES
DRESSER
QUEEN DOUBLE BERTH
HANGING LKR
SHOE RACK
RAISED SETTEE
TABLE
DECK BOX/SEAT
SHOWER STALL
WC
UP
TUB/SHOWER
DOUBLE BERTH
BERTH
DESK
HNG. LKR.
DINING TABLE
REF'R.-FREEZER
COAT CLOSET
DRESSER
BUFFET
BOOKS
DN
0
5'
10'
41"
42"
B

50' *Florida Bay*

Design Number 261
1987

The 50' ***Florida Bay*** was the first of the line of Florida Bay Coasters built. We had a great time designing and using her. She is still my favorite of the ones we've built so far, sentimentally and esthetically. Like all of them, she exudes a solid, big ship feel, giving you the feeling you're on a much larger vessel. You step aboard, and immediately want to cast off and go exploring. It's as if she's saying to you, "*Of course* I can take you exploring in the places you've always wanted to see."

The design commission for the 50' **Florida Bay** came about from Reuben Trane having seen my 45' Friday Harbor Ferry in his search for a new liveaboard design. He liked all the room in it, but wanted to rearrange some of the spaces. And, he wanted to carry a Jeep. We went through several rounds of sketching preliminaries, and ended up with the 60-footer shown later in this book. This design was sent out for bids.

After reviewing the bids, we decided to create a smaller version, to keep the price in a more affordable range. The 50' **Florida Bay** resulted from this, and had all the main features of the 60, with a large part of the length saved achieved by stacking the skiff over the Jeep instead of having them side by side.

I've had the pleasure of being aboard and cruising the **Florida Bay** several times, and it's always been a delight. She has an honest and forthright appearance, and her performance lives up to it, taking you surely and safely on your way.

I still continue to get questions about her stability, even with the photos of loading the car aboard testifying that she has a lot of stability. Right after the first boatshow in which she was exhibited, Reuben and I took the ***Florida Bay*** out the inlet at Fort Lauderdale and headed into the eight foot seas that were coming in from the East. The winds had dropped to about 20mph, after several days of gale force winds. We poked into them, throwing spray, until we got into the Gulf Stream and then turned South and took them on the beam for the next two hours. It was proof to us of how stiff she was in action. This bore out all our calculations and the inclining test we'd done while still at the shipyard. After watching the Jeep® rocking on its suspension but never sliding on deck, I turned to watching her action as the waves heeled her. As the eight foot seas would approach, she would heel with the approach of each wave but immediately start to right herself, heeling *into* the wave. She was effectively splitting the difference between following the wave surface and trying to stay upright. A photo I took on this outing let me measure the angle of the horizon against the known level of the railings. This indicated that she was heeling a maximum of ten degrees in these eight foot seas.

Since this time she has made a number of crossings to the Bahamas and numerous coastwise trips. My own further experiences on her and several other Coasters has given me great confidence in their ability to take rough weather. I expect that we'll be getting more cruise stories from the owners as they go far and wide over the years.

The ***Florida Bay*** is a very successful boat and I hope to see more sisterships built. I've a number of ideas on how to make some evolutionary refinements to her design, yet keeping the practicality, charm and character of the original.

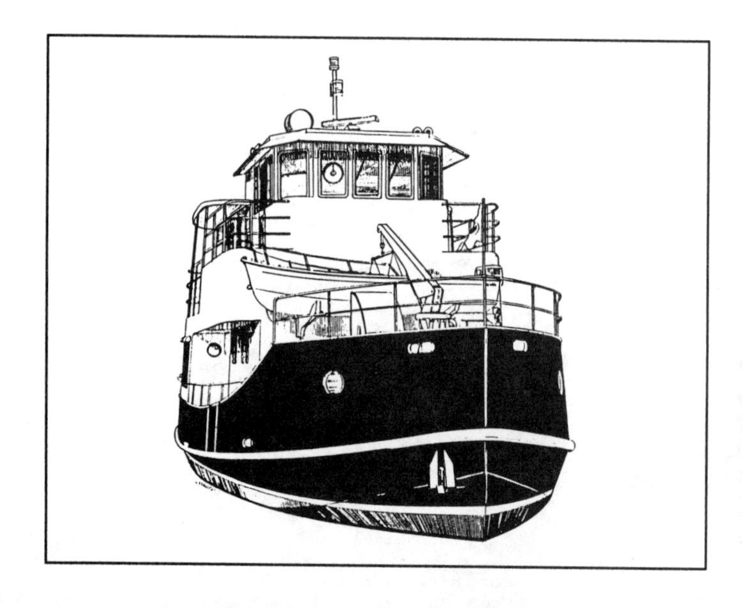

Reuben Trane photo

Florida Bay

The First Florida Bay Coaster

By Reuben Trane

The Florida Bay Coaster is a special sort of vessel: part yacht, part ship. At first glance she conjures up images of a tramp steamer, negotiating uncharted rivers, plying her trade in exotic ports of call and spending sultry nights at anchor in far away secluded coves. From her proud plumb stem to her elegant rounded transom, she evokes a feeling of purpose. Here is a vessel meant to go places and to do things, to carry her master and crew on coastwise cruising adventures, to be free and independent of the restraints of busy marinas and harbors and at home off the beaten path, along the reaches of seldom frequented waters found along America's coastline, and in her inland lakes and rivers.

Styled after the coastal freighters that plied our nation's waterways, the Coaster is designed and engineered to comfortably house and cruise the liveaboard couple or family. And, like her working sisters, she is a stout, hearty vessel. But, freed from the restraints of earning her way in the cargo trade, her interior volume is entirely given over to accommodations similar to those found in a vacation home or condominium. And left over is still room for two over-sized holds and a well deck for the Jeep®.

The Concept. The original idea for the Coaster was to create a cruising vessel with real liveaboard space, the kind of room normally associated with a shoreside home. She needed oversized staterooms, galley and main saloon along with generous storage areas, private heads and deck room. And perhaps most importantly, all of the interior spaces must have vistas of the water. Coastal cruising provides a myriad of delights for the senses; constantly changing scenery, sea birds calling to one another, and fresh breezes carrying the smell of the sea. To take full advantage of these surrounding pleasures, the Coaster provides views from all her cabins through large windows that open wide in fair weather.

The Design Elements. Intended as a liveaboard home for a man and his wife, with room for visiting guests and children, the Coaster is easily operated by a single couple. Fifty feet was decided upon as a maximum length. Twin diesels spaced wide apart provide the necessary maneuverability. A reversible electric windlass makes the setting and weighing of anchor a push-button operation. Almost everyone enjoys operating a vessel from her fly bridge, but also likes to stay out of the sun and the weather

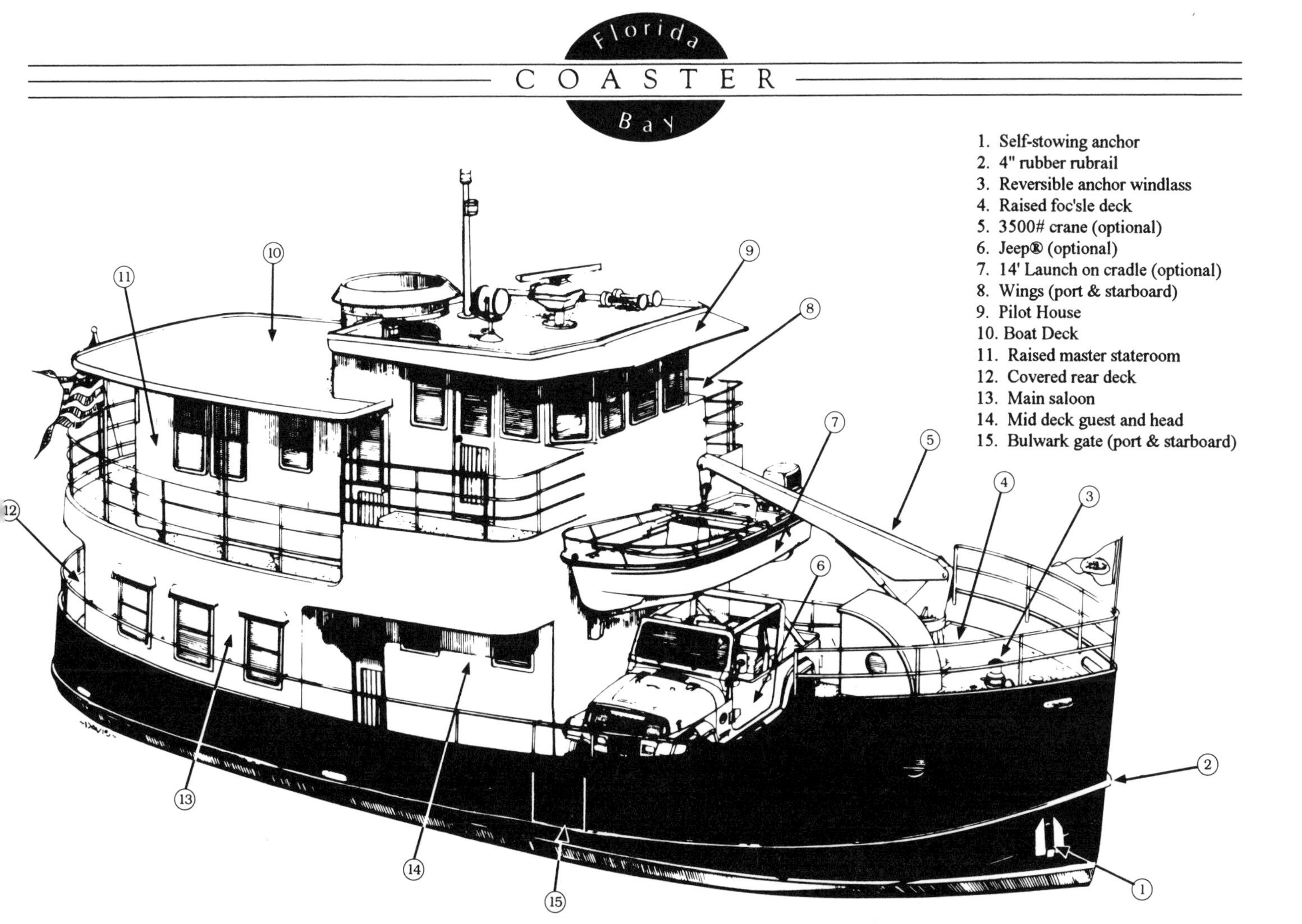

while motoring. A raised pilothouse with plenty of opening windows is the best solution for maximum visibility and protection from the elements. The Coaster's pilothouse is quite similar in both size and nature to those found aboard the Mississippi River tow boats.

It's no problem to provide a view in the main saloon of a motor vessel, but how about in the master and guest staterooms? The Coaster does this by lowering the main deck freeboard and thus the main saloon, while placing the guest stateroom on the mid-deck and the master on the sun deck over the saloon. This fresh approach allows all the living areas to be above deck, with large opening windows. Only the foc'sle, engine room and holds are below deck. An added advantage of the lower freeboard is to bring the daytime living spaces closer still to the water. Central air conditioning keeps the inside comfortable in any weather.

A major goal in the conception of the Coaster was to provide room to carry a four-wheel drive Jeep® and the means to load and off load in a variety of conditions, both alongside a wharf or nosed into a river bank. A well deck would act as a garage, while a forward mounted crane would get the Jeep® on and off. A raised foc'sle provides a good spot for the crane and makes for a pleasant double stateroom, office or workshop.

A Personal Freighter. Step into the Coaster's pilot house and take command of your own personal freighter. Squint your eyes, stretch your imagination, and a coastal steamer sits before you, awaiting orders to embark on high adventure. This is not by chance. A mini freighter has in fact a maximum number of advantages for her liveaboard owner. 110,000 pounds of displacement means that a few extra groceries for a month-long trip through the Bahamas won't spoil her trim. Neither will the addition of a davit to handle dinghies, sailboards, and Jet Skis® from her spacious boat deck. Standard is 1,000 gallons each of fuel and water with a cruising range of over 1,000 nautical miles. There's room to increase that and to add a water maker for those who are serious about their cruising.

The Coaster has two holds; the forward one is the size of the well deck and accessed through its own 3' x 5' watertight hatch, the after one is under the main saloon. Between them they provide room for bosun's supplies, canned goods, suitcases, out of season clothing, bicycles, motorcycles, water skis, snow skis, scuba gear, compressor, engine spares, accumulated treasures and a lifetime of collected memories.

Beachable. With a draft of but four feet, and a cutaway forefoot, the Coaster can ease up to the river bank, beach or sandbar and spend the night securely tied to a nearby tree. If the tide goes out, she rests firmly on the bottom supported by her full length steel grounding shoe, held upright by her load bearing propeller struts. The optional 3,500 pound capacity crane easily launches the skiff and handily lifts the Jeep® clear over the bow and sets it down on dry land. The skiff supplies local water transportation; the Jeep® overland.

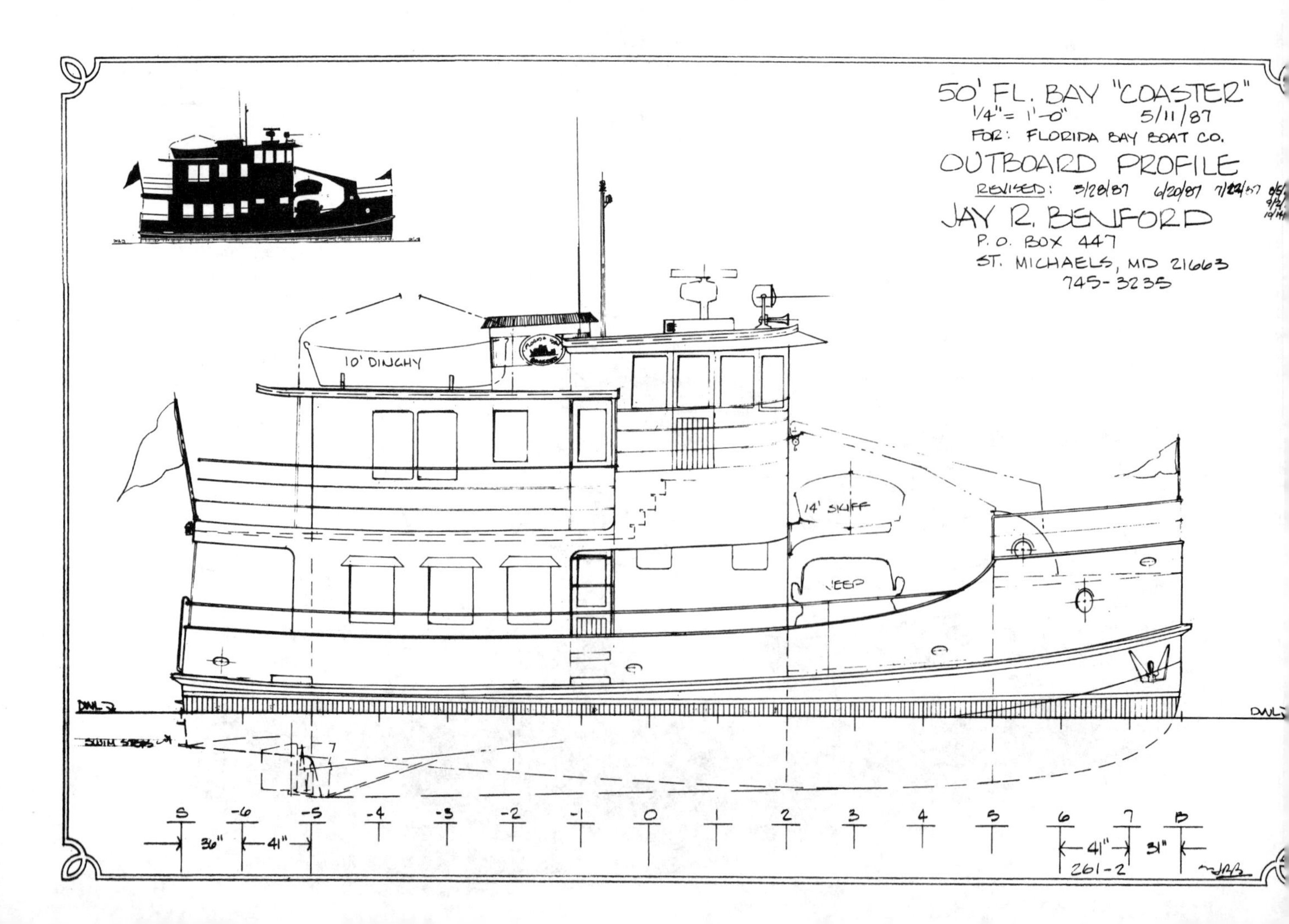

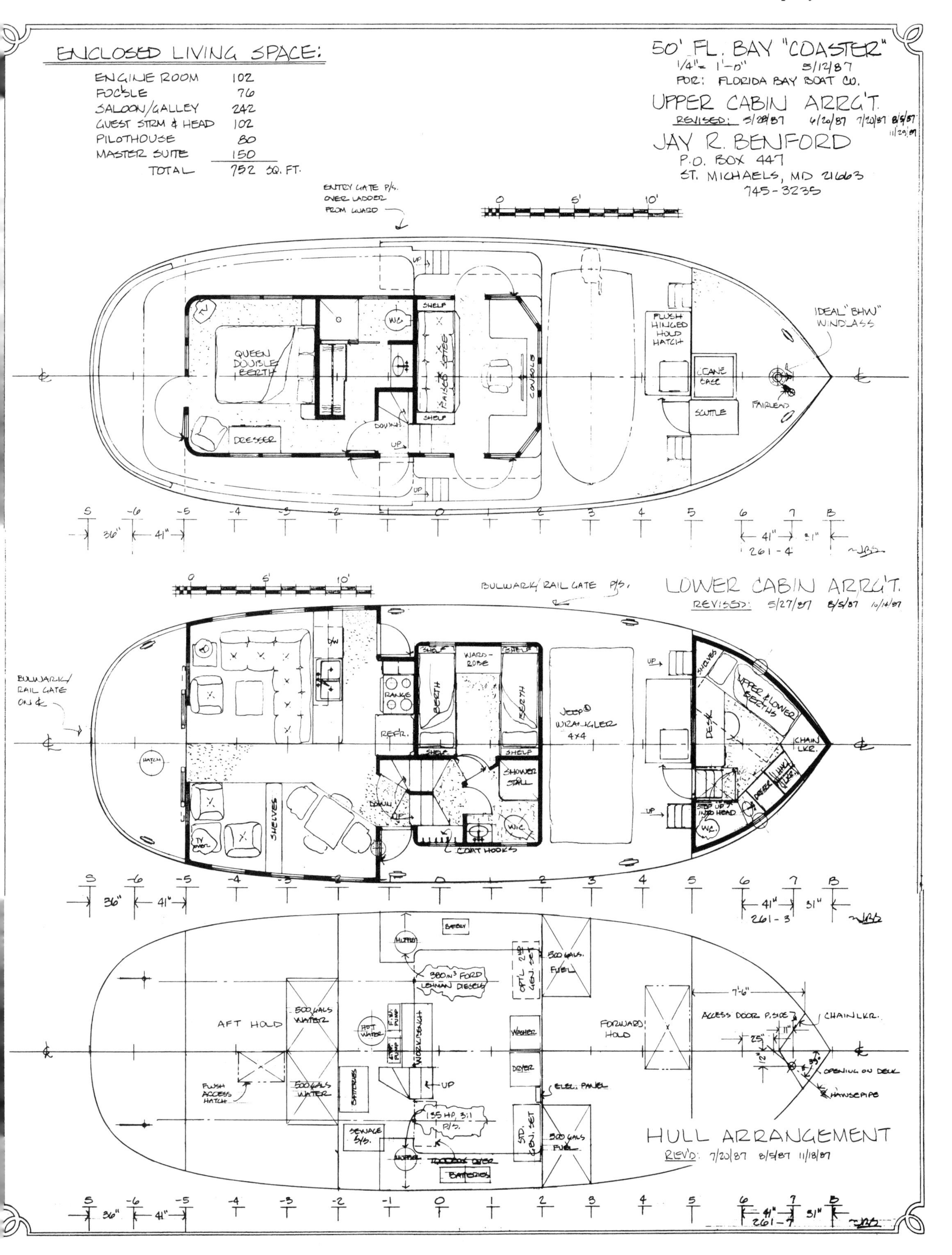
ENCLOSED LIVING SPACE:
ENGINE ROOM 102
FOC'SLE 76
SALOON/GALLEY 242
GUEST STRM & HEAD 102
PILOTHOUSE 80
MASTER SUITE 150
TOTAL 752 SQ. FT.
50' FL. BAY "COASTER"
1/4" = 1'-0" 5/12/87
FOR: FLORIDA BAY BOAT CO.
UPPER CABIN ARR'G'T.
REVISED: 5/28/87 6/20/87 7/20/87 8/5/87 11/29/87
JAY R. BENFORD
P.O. BOX 447
ST. MICHAELS, MD 21663
745-3235
ENTRY GATE P/S. OVER LADDER FROM GUARD
QUEEN DOUBLE BERTH
DRESSER
RAISED SETTEE
SHELF
CONSOLE
DOWN
UP
FLUSH HINGED HOLD HATCH
CANE CASE
SCUTTLE
IDEAL "BHW" WINDLASS
FAIRLEAD
261-4
LOWER CABIN ARR'G'T.
REVISED: 5/27/87 8/5/87 10/14/87
BULWARK/RAIL GATE P/S.
BULWARK/ RAIL GATE ON ℄
RANGE
REFR.
WARD-ROBE
BERTH
SHOWER STALL
JEEP® WRANGLER 4x4
SHELVES
UPPER & LOWER BERTHS
DESK
CHAIN LKR.
STEP UP INTO HEAD
COAT HOOKS
TV OVER
HATCH
261-3
HULL ARRANGEMENT
REV'D: 7/20/87 8/5/87 11/18/87
AFT HOLD
500 GALS WATER
HOT WATER
WORKBENCH
BATTERIES
FLUSH ACCESS HATCH
MUFFLER
BATTERY
380 in³ FORD LEHMAN DIESEL
135 HP, 3:1 P/S.
SEWAGE S/S.
TOOLBOX OVER BATTERIES
OPT'L 2ND GEN. SET
STD. GEN. SET
500 GALS. FUEL
WASHER
DRYER
ELEC. PANEL
FORWARD HOLD
ACCESS DOOR P. SIDE
CHAIN LKR.
OPENING ON DECK
HAWSEPIPE
261-7

Rugged Independence. The Florida Bay Coaster is built entirely of welded steel, the material that sets the standard for strength and durability. With the advent of epoxy primers and topcoatings, and of electronic anti-electrolysis devices, the Coaster will provide a lifetime of use with only a normal amount of maintenance. A four inch industrial rubber rubrail surrounds the entire hull at the sheerline protecting it from incidental nicks and scrapes. The five watertight compartments are each serviced by its own automatic bilge pump. Full standing headroom in the engine room facilitates access to the engine and the ship's pumps and machinery. A work bench is provided along with a clothes' washer and dryer.

The Pilot House. The raised pilothouse gives the Coaster's master a commanding view over her massive bow. At a height above the waterline equal to that of most motor yacht's fly bridges, the helm has the advantages of both superior visibility and the protection and comfort of an inside steering station. The console houses engine instrumentation, the 12vdc electrical panel, chart storage plus room for navigational instruments and plenty of working surface. A raised settee with built-in end tables gives the crew the same view as the helmsman. Opening windows provide plenty of light and fresh air. The settee is long enough for an afternoon nap, and immediately aft is room for a pilot berth. The wheel house doors open onto the port and starboard wings, extending the full beam of the vessel. They prove handy when maneuvering in close quarters and provide an excellent spot to idle away the time, watching the shoreline pass by.

The Master Suite. Down a half a flight of stairs from the pilothouse is the master stateroom. A dressing area, with plenty of closet space, opens into the shower and head. The queen-sized berth has built-in drawers and is surrounded by over-sized opening windows. The Coaster is unique in that her master stateroom is up. This results in a bright, open cabin with surrounding views of the water. Immediately aft is the owner's sundeck, with space for four deck chairs. Covered side decks allow passage forward to the wheel house and create a delightful, protected spot to linger and watch the changing scenery.

The Guest Stateroom. Half a flight down from the sun deck (half a flight up from the main deck) is the guest stateroom and adjoining head. The two extra-long berths are separated by a custom wardrobe complete with hanging locker and room to house the TV. Ample storage is found in drawers under the berths. One opening window forward and two to port bring the outside in; passing sights stream by, visible while relaxing in bed. The head is accessible from both the guest stateroom and the stair landing. Excellent cross ventilation is created in the guest cabin by leaving the connecting door open.

The Main Saloon. Life aboard the Coaster centers in her spacious main saloon. Taking full advantage of the

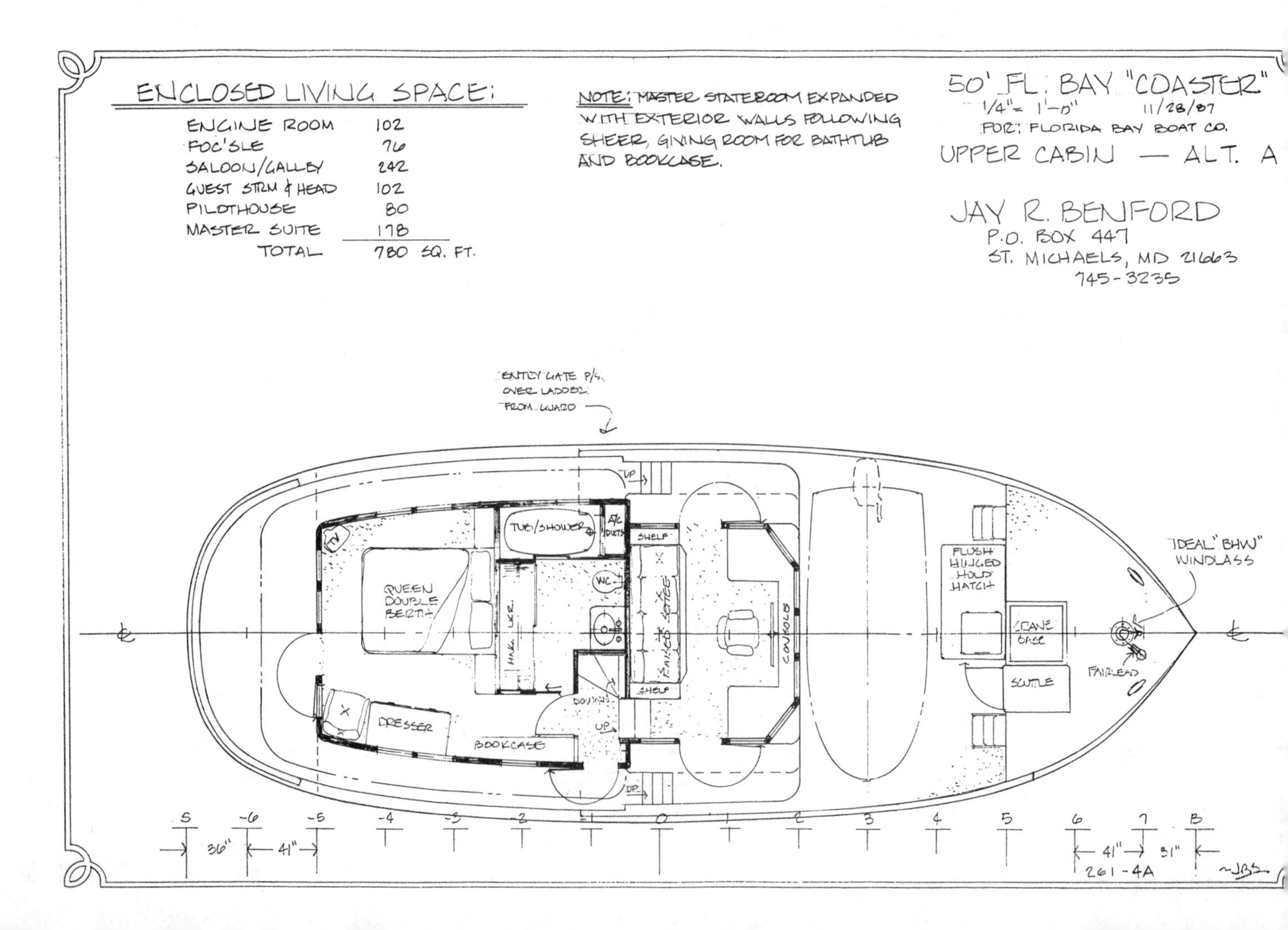

Conference on the well deck during construction.

'Tween deck being insulated and framed in.

Covered walkway alongside master stateroom leading to the starboard bridge wing.

Note hose bib and hose rack alongside ladder to foc'sle deck.

An inclining test was done right at launching, proving the COASTER has excellent stability. The results also indicated that she could be Coast Guard licensed to carry several dozen passengers for hire.

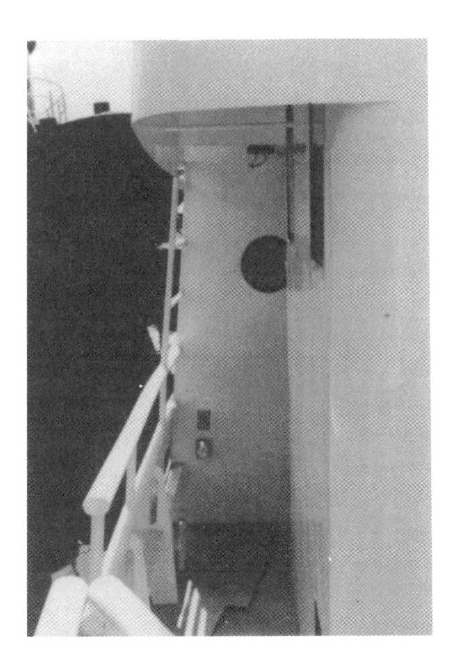

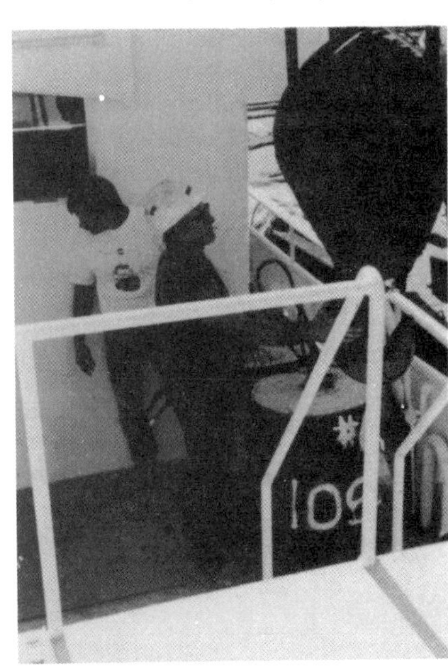

Coaster's stately beam, the saloon has room for a home style galley with all full-sized, household type appliances (electric range, side-by-side refrigerator/freezer, microwave oven, dishwasher and double stainless sink). There is plenty of cabinet and counter space. The area aft of the galley is reserved for a sectional sofa, easy chair and side tables. The dining table and chairs have their own space to starboard, opposite the galley. A bookshelf separates the dining area from a cozy aft reading corner, complete with room for two easy chairs and reading lamp. Double weathertight doors open onto the covered rear porch. Large opening windows provide light, vistas and fresh air.

The Foc'sle. Under the foredeck is a spacious foc'sle. It is simply and efficiently laid out to sleep two with good stowage, head facilities, and room to spread out. A good spot for kids, crew, the occasional guest (nice enough to be comfortable, but not so great that they'll want to extend their stay), or the owner and wife team that desire to charter their Coaster. This area could also be easily laid out as an office, workshop, storage area or sumptuous single stateroom.

The Engine Room. The heart of any motor vessel is her engine room. That of the Coaster has full standing headroom and an intelligent, efficient layout. The twin diesel engines are well outboard, with access to all four sides. Along the centerline are the genset and washer and dryer. The forward bulkhead is in fact the aft end of the fuel tanks, with sight gauges, fuel manifold and filters conveniently at hand. Aft, under the saloon deck is the 1,000 gallon water tank. Forward of it are the hot water heater, the pressure system, air conditioning pump, intake filters and the waste system. There's room left over for a functional work bench and storage for oil, filters and engine parts. Everything, including the batteries, is easily accessed or serviced in this brightly lit, well-ventilated space.

The Deck. The deck of a vessel the size of the Coaster must be safely and easily worked. All railings are 42 inches above deck level (the same height as found aboard cruise ships). The foredeck, atop the foc'sle, is easily accessed by a pair of stairs, port and starboard. Forward are a pair of cleats with fairleads, and an electric, reversible windlass with hawse pipe for the self-stowing anchor, and a chain locker. The anchor can either be handled from the foredeck or by remote control from the wheel house. The deck is heavily reinforced to carry the optional crane.

The well deck has a raised opening cargo hatch (3' x 5') with built-in access hatch. The deck is reinforced to carry the weight of the Jeep®. An optional swing away cradle is mounted over the Jeep® to carry the 14' launch (up to 15'). Two gates cut into the bulwarks swing open for entering or leaving the vessel. There are four cleats and fairleads on the main deck, two each port and starboard. The cargo hold takes up the entire space under the well deck.

The after deck provides space for several deck chairs and a coffee table. This is a delightful spot to lounge and watch

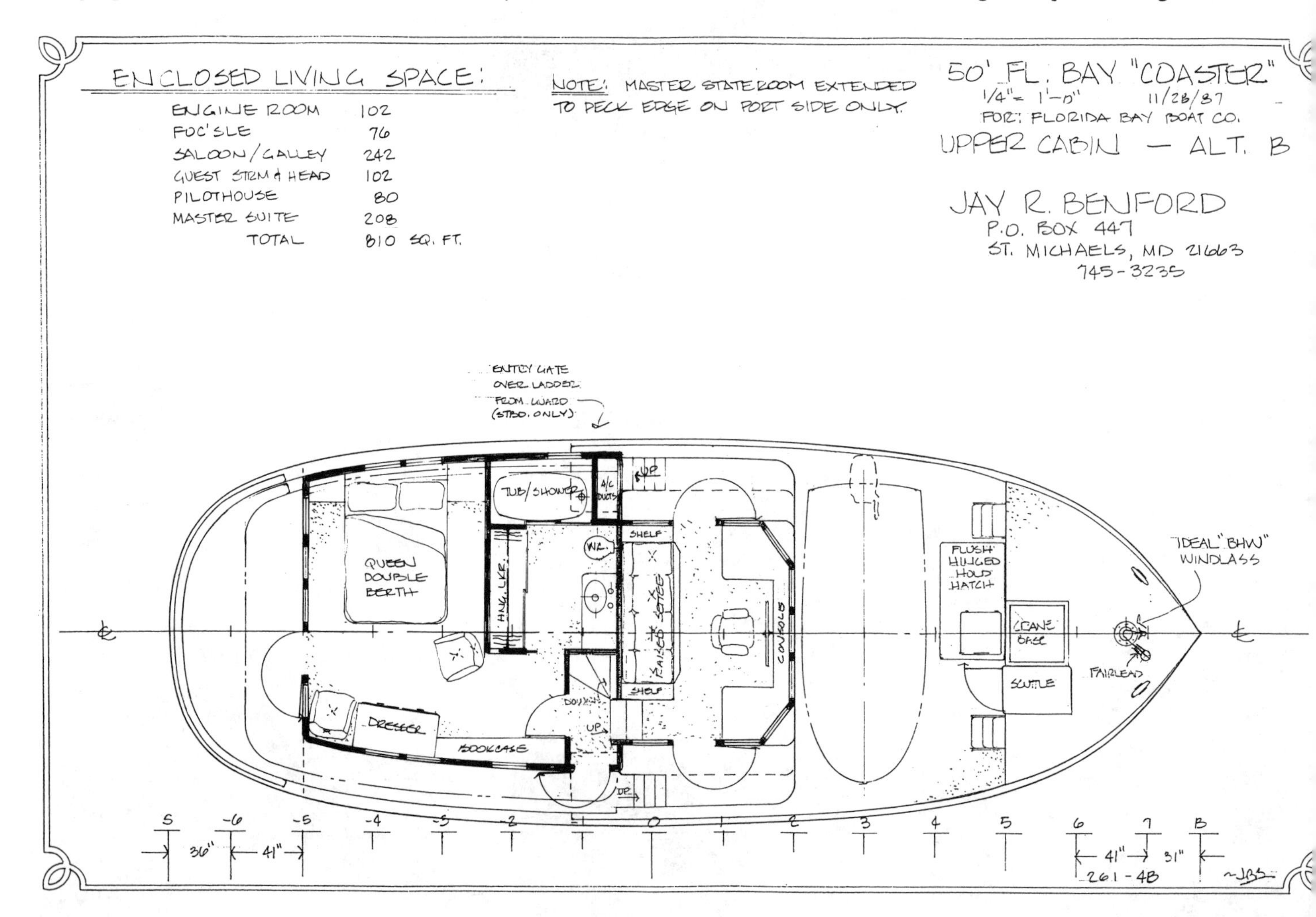

Reuben Trane photos

Launching the Jeep® onto the beach, above and right. Off loading it onto a boat launching ramp has proven to be an easy way to find a place to put it ashore. Loading the Jeep® for the first time, below right. Note the slight angle of heel, less than three degrees, indicating the great stability of these designs. Below, the Jeep® and skiff stowed on deck and ready to go cruising.

the wake trail behind. It can be entirely enclosed with weather curtains during inclement weather. A swing open gate on the centerline of the transom gives way to a built-in swim ladder. There are two mooring cleats and fairleads, one each port and starboard.

The boat deck is accessed by a ladder on the port side. It has plenty of room for dinghy-handling equipment and small boats and water craft. It also makes for a spacious sun deck, or with the addition of some folding chairs, an observation platform for races, parades and other waterborne events. Youngsters will find it a challenging dive platform. All the exterior decks are finished with non-skid epoxy paint.

Made In America. The Florida Bay Coaster is built in Florida, with the majority of her sub-components, wherever possible, being made in America. This gives her owner, along with the satisfaction of buying an American product, several tangible advantages. Delivery is made at the yard where she is built, by the men who built her, after sea trials have been conducted. In the event any problems do show up, the men and facilities are available to correct them. None of the purchase price is being spent on freight from either Europe or the Far East, none to pay duties, and none for repairs and make-ready following an ocean crossing. The price does not fluctuate with the rate of exchange. Also, an American built Coaster can be legally chartered with a licensed captain. Some of the finest American craftsmen have been assembled to fabricate the Coaster to the highest standards of boat building, equal to that found anywhere in the world.

A Final Word. The Florida Bay Coaster is truly a unique vessel. She is built for the individual who wants to spend more time aboard, cruising those special, out of the way places, with complete independence from shoreside facilities, but with a modicum of civilized comforts. Lavish she is not. Practical she is. Run her up to the beach, lift her Jeep® ashore and explore the deserted coastline of a barrier island. Find a secluded anchorage around a bend of the Mississippi; stay a month or a year before moving on. Anchor in the middle of Florida Bay in the Everglades National Park, one of this country's loveliest bodies of water, under-populated with people and overpopulated with bird and marine life. The imagination runs wild with potential cruises waiting to be taken along America's waterways. All one needs is the spirit of adventure, a willing mate, and, of course, the proper vessel. The Florida Bay Coaster is such a vessel.

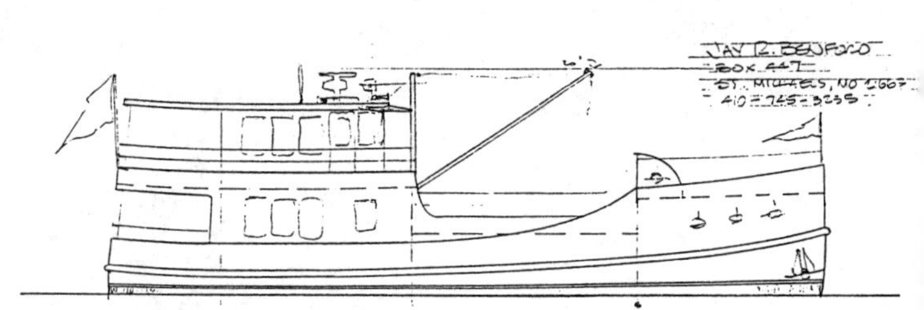

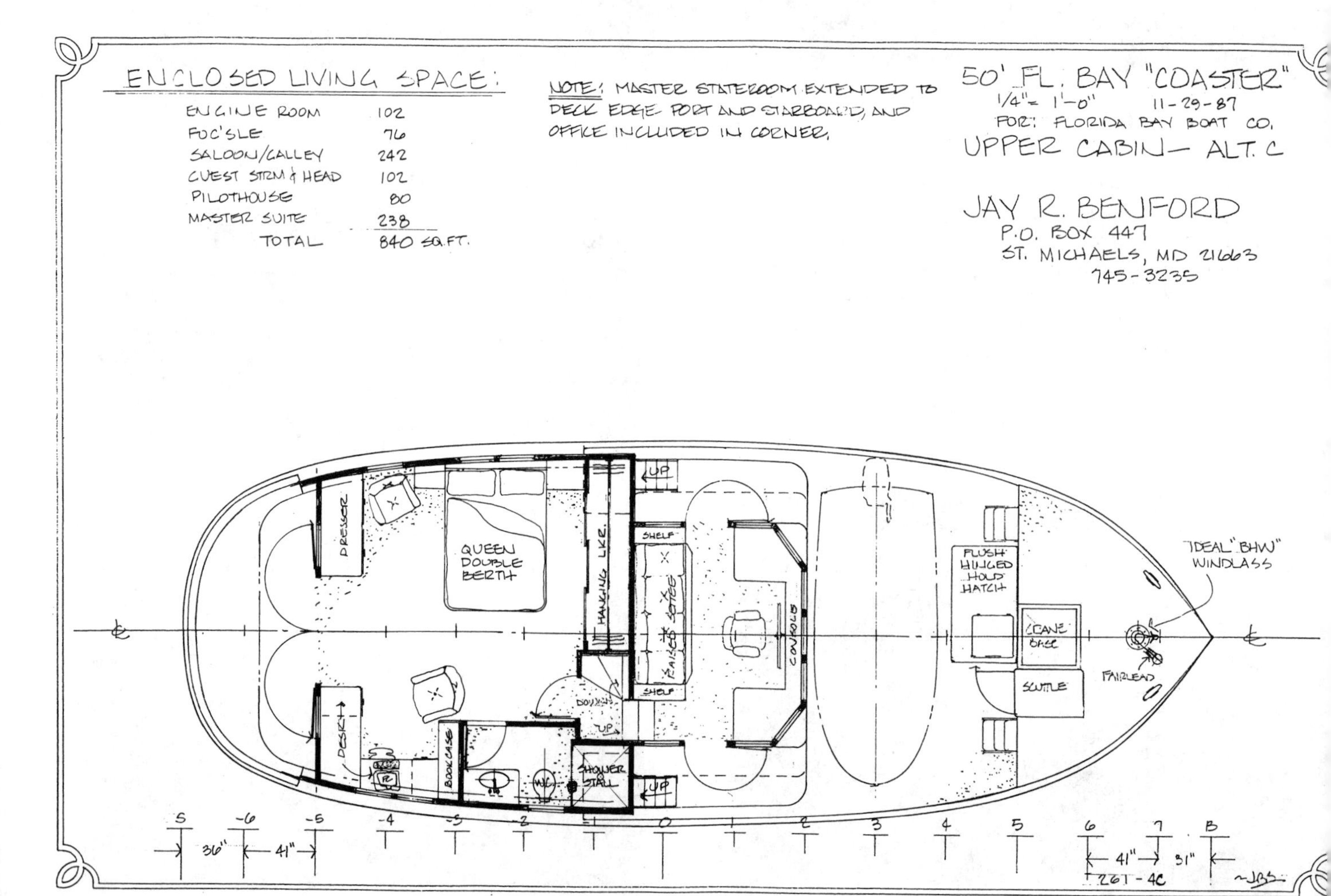

Reuben Trane photo

With the COASTER for your waterfront home, you can enjoy a variety of views in different settings, all without leaving home.

Reuben Trane photos

Six weeks after laying the keel, construction is well underway.

The widely spaced props make for wonderful maneouverability. The props are well protected with struts to the hull and back under the rudders, making for "no problems" groundings while exploring or off-loading the JEEP.

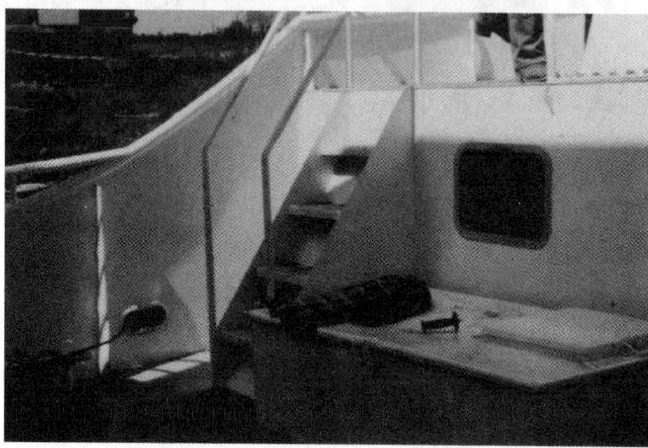

Note entryway to engine room in above photo and hatch to aft hold in photo below.

The COASTER was "walked" from the building shed to the water, just shy of four months from laying the keel, for her launching. She sat comfortably on her keel and skegs. The lifting was done on her mooring cleats and stem, a testament to the ruggedness of her structure.

50' Coaster Variations

(Shown on the following pages)

Long Foc'sle Version

The Long Foc'sle Version of the 50' Coaster is an alternate version for those who don't want to carry a vehicle aboard, but still want the well deck and cargo hold. The foc'sle is longer by three and a half feet and the resulting gain in space makes for a much more spacious layout. A couple alternatives for foc'sle layout would be the one on the Charter Coaster and the one on the preliminary version of ***Key Largo***.

The upper cabin layout shows an optional bridge wing wrapping around the front of the pilothouse and spanning over to the top of the scuttle, with steps down to the foredeck. (This is not shown on the profile view.) It also shows a full width upper stateroom, giving up the walk around side decks and limiting access to the upper aft deck. If it were my own boat, I would not give up the walk around side decks, for they give the boat more workability and maintain the small ship feeling and functionality. There is plenty of storage space in the holds for storage of the out of season clothes and other gear.

The galley has been moved to the mid-level, making the saloon more open, functioning as a combined living room and dining room.

Her Enclosed Living Spaces Are:

Engine Room	102
Long Foc'sle	127
Saloon/Galley	242
Guest Strm.. & Head	102
Pilothouse	80
Master Suite	238
Total	891 Sq. Ft.

Family Version

This version was conceived to provide room for parents and several children to liveaboard. The master stateroom has its own private head and the two kids staterooms share one on the hall. The foc'sle can be a spare stateroom for guests. The foc'sle head is improved from that on the ***Florida Bay*** in having a separate shower stall.

The well deck is raised a bit to provide standing headroom in the cargo hold, opening up the use of that space to a lot of alternative uses.

The entry in the mid-level has a long workbench for doing all sorts of projects, a coat closet and the laundry facilities. The saloon is the full width of the upper deck, making the galley very handy to the pilothouse. The after deck, outside the upper level saloon and galley, is longer than on the ***Florida Bay***, making up for not having one below.

The outside stairs on the port side to the pilothouse gives quick access from the well deck. The starboard side has stairs to the boat deck from the bridge wing.

Caribbean Version

A modification of the Family Version, this one has one less stateroom aft, with the dining area on the lower level. The master stateroom has it's own little aft deck. The galley and laundry are in the mid-level and the upper cabin is solely a saloon.

We increased the freeboard, lowered the profile by eliminating full headroom in the engine room and hold and added fuel tankage. The lower level has big ship style portlights with storm shutters to make the lower level windows less vulnerable at sea. On the 55' ***North Star*** we used my idea of welding on studs to hold Lexan® storm shutters for those times when they're making passages, keeping the larger opening windows that make living aboard so pleasant. This could be done on this version too.

This one could be right at home making passages to the islands of the Caribbean and going further afield. Perhaps a trip down to the Amazon or out to the Pacific Northwest?

Particulars:	***Florida Bay***
Length overall	50'-0"
Length designed waterline	49'-6"
Beam	18'-0"
Draft	4'-0"
Freeboard:	
Forward	7'-4½"
Least	1'-9½"
Aft	2'-2½"
Displacement, cruising trim*	95,000 lbs.
Displacement-length ratio	350
Prismatic coefficient	.595
Water tankage	1,000 Gals.
Fuel tankage	1,000 Gals.
Headroom	6'-7"
GM (from inclining test)	5.2'

***CAUTION:** The displacement quoted here is for the boat in cruising trim. That is, with the fuel and water tanks filled, the crew on board, as well as the crews' gear and stores in the lockers. This should not be confused with the "shipping weight" often quoted as "displacement" by some manufacturers. This should be taken into account when comparing figures and ratios between this and other designs.

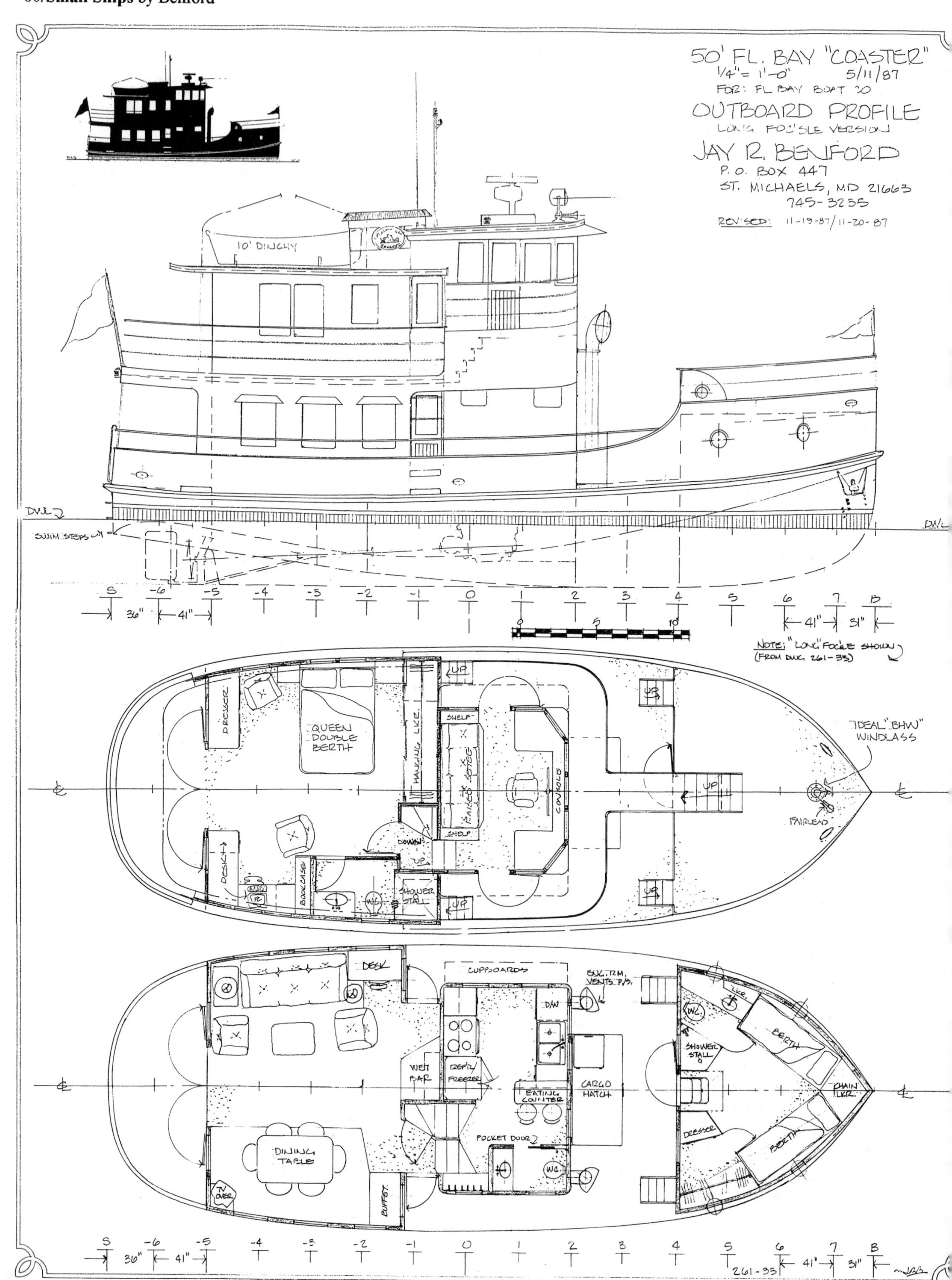

50' FL. BAY "COASTER"
1/4" = 1'-0" 5/11/87
FOR: FL BAY BOAT CO
OUTBOARD PROFILE
LONG FO'C'SLE VERSION
JAY R. BENFORD
P.O. BOX 447
ST. MICHAELS, MD 21663
745-3235
REVISED: 11-19-87/11-20-87
10' DINGHY
DWL
SWIM STEPS
36"
41"
31"
NOTE: "LONG" FOC'LE SHOWN (FROM DWG. 261-33)
DRESSER
QUEEN DOUBLE BERTH
HANGING LKR.
SHELF
RAISED SETTEE
CONSOLE
UP
DOWN
DESK
BOOKCASE
SHOWER STALL
"IDEAL" BHW WINDLASS
FAIRLEAD
DESK
CUPBOARDS
ENG. RM. VENTS P/S.
D/W
WET BAR
REFR/FREEZER
EATING COUNTER
CARGO HATCH
POCKET DOOR
W.C.
DINING TABLE
TV OVER
BUFFET
LKR.
BERTH
SHOWER STALL
CHAIN LKR.
DRESSER
261-33

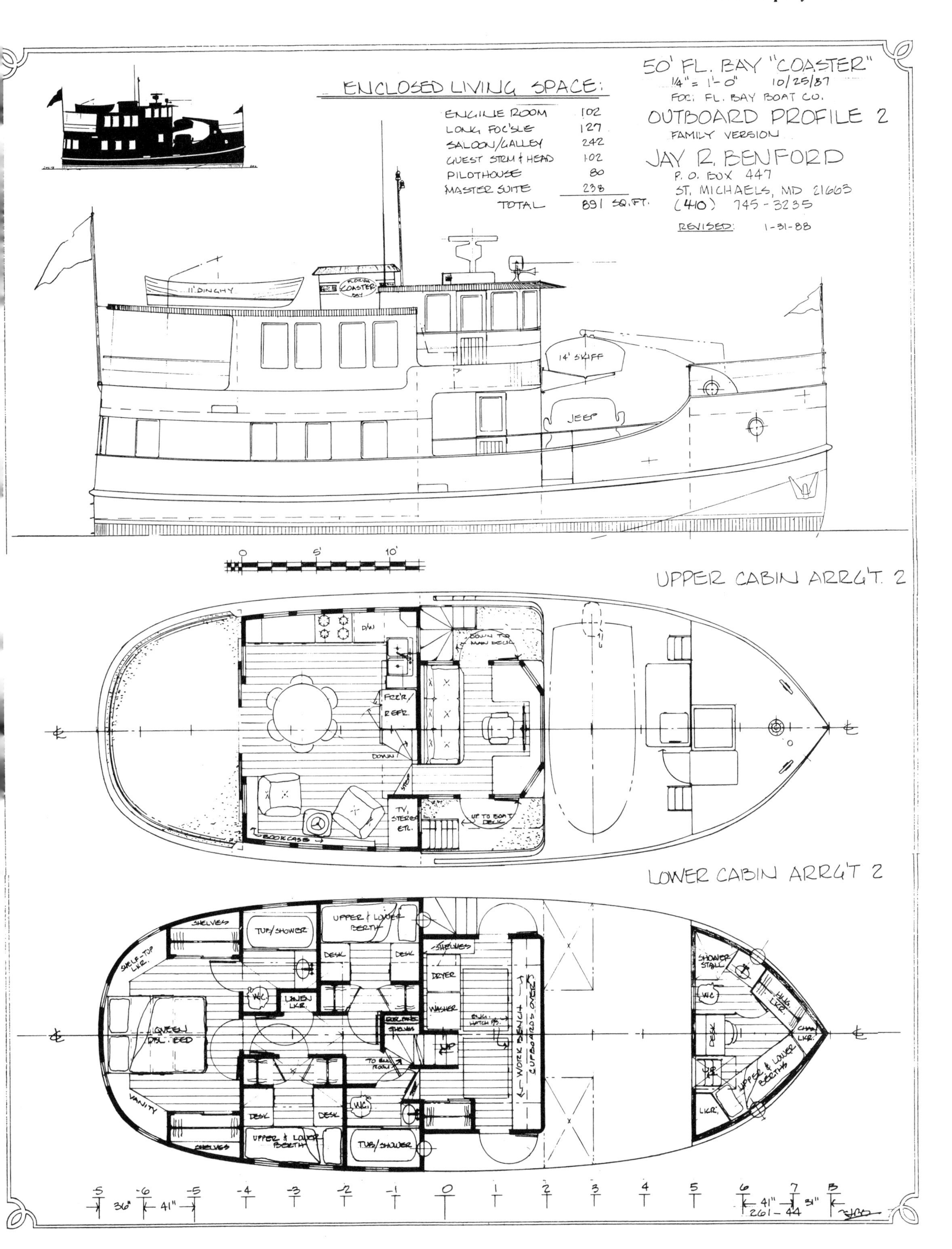
50' FL. BAY "COASTER"
1/4" = 1'-0"   10/25/87
FOR: FL. BAY BOAT CO.
OUTBOARD PROFILE 2
FAMILY VERSION
JAY R. BENFORD
P. O. BOX 447
ST. MICHAELS, MD 21663
(410) 745-3235
REVISED: 1-31-88
ENCLOSED LIVING SPACE:
ENGINE ROOM 102
LONG FOC'SLE 127
SALOON/GALLEY 242
GUEST STRM & HEAD 102
PILOTHOUSE 80
MASTER SUITE 238
TOTAL 891 SQ. FT.
11' DINGHY
14' SKIFF
JEEP
UPPER CABIN ARRG'T. 2
D/W
FRZ'R/ REFR.
DOWN
DOWN TO MAIN DECK
UP TO BOAT DECK
TV, STEREO ETC.
BOOKCASE
LOWER CABIN ARRG'T 2
SHELVES
TUB/SHOWER
UPPER & LOWER BERTH
DESK
SHELF-TOP LKR.
W.C.
LINEN LKR.
DRYER
WASHER
QUEEN DBL. BED
WORK BENCH + OUTBOARDS OVER
TO ENG. ROOM
VANITY
SHOWER STALL
HNG. LKR.
CHAIN LKR.
UPPER & LOWER BERTHS
LKR.
36" 41"
41" 31"
261-44

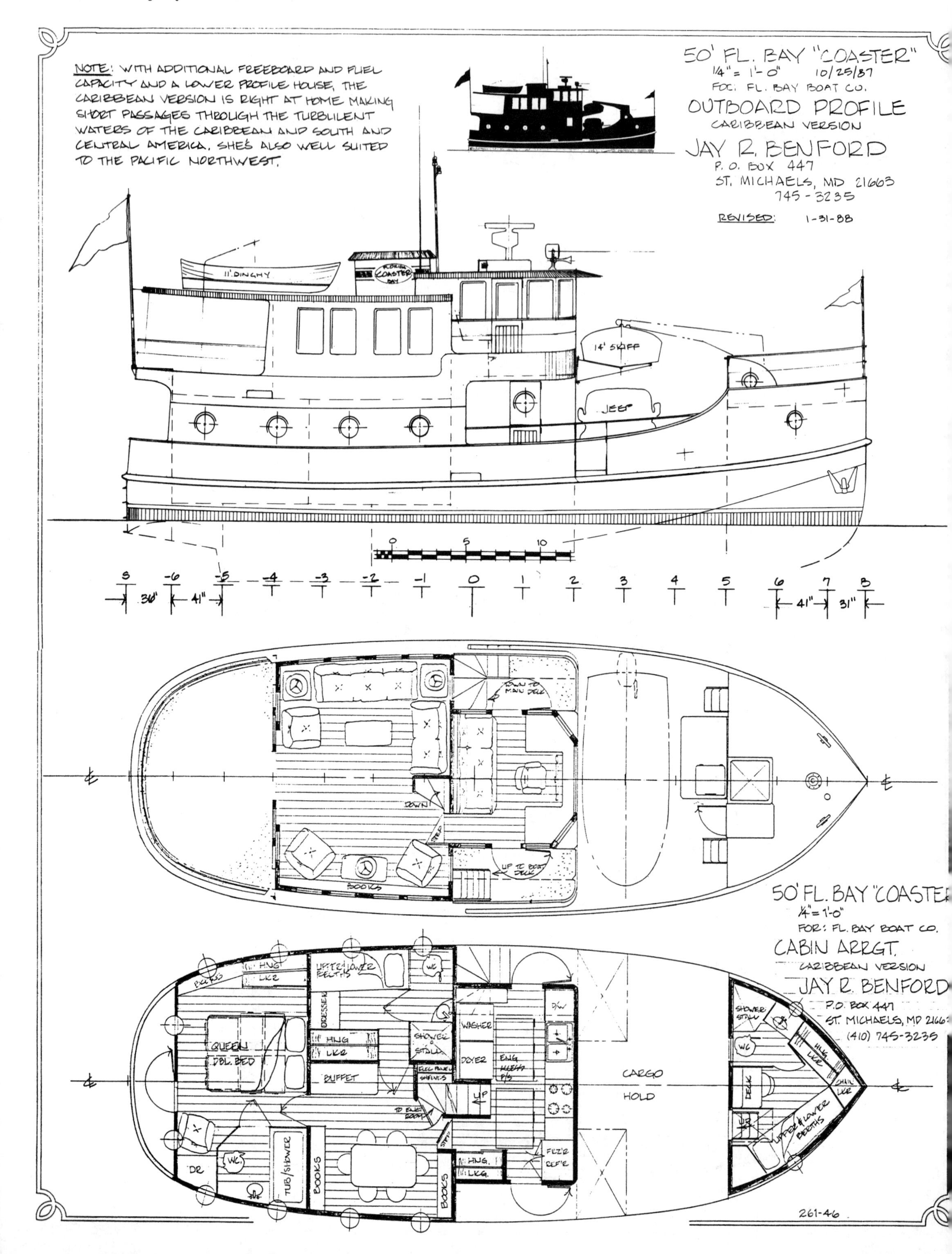
NOTE: WITH ADDITIONAL FREEBOARD AND FUEL CAPACITY AND A LOWER PROFILE HOUSE, THE CARIBBEAN VERSION IS RIGHT AT HOME MAKING SHORT PASSAGES THROUGH THE TURBULENT WATERS OF THE CARIBBEAN AND SOUTH AND CENTRAL AMERICA. SHE'S ALSO WELL SUITED TO THE PACIFIC NORTHWEST.
50' FL. BAY "COASTER"
1/4" = 1'-0" 10/25/87
FOR: FL. BAY BOAT CO.
OUTBOARD PROFILE
CARIBBEAN VERSION
JAY R. BENFORD
P. O. BOX 447
ST. MICHAELS, MD 21663
745-3235
REVISED: 1-31-88
11' DINGHY
14' SKIFF
JEEP
50' FL. BAY "COASTER"
1/4" = 1'-0"
FOR: FL. BAY BOAT CO.
CABIN ARRGT.
CARIBBEAN VERSION
JAY R. BENFORD
P.O. BOX 447
ST. MICHAELS, MD 2166
(410) 745-3235
DOWN TO MAIN DECK
DOWN
UP TO BOAT DECK
BOOKS
HNG LKR
UPPER/LOWER BERTHS
WC
DRESSER
QUEEN DBL. BED
HNG LKR
SHOWER STALL
WASHER
DRYER
ENG ACCESS P/S
D/W
BUFFET
UP
TO ENG ROOM
WC
DR
TUB/SHOWER
BOOKS
BOOKS
HNG. LKG.
FRZ'R REF'R
CARGO
HOLD
SHOWER STALL
WC
HNG LKR
DESK
CHAIN LKR
UPPER & LOWER BERTHS
261-46

50' Liveaboard Coaster & *Seaclusion*

Designs Number 283 & 297
1989 & 1990

This liveaboard version, below and on the next page, was designed to be a deluxe home afloat, with a layout similar to ***Key Largo*** on the smaller 50' x 18' hull. We've left off the cargo hold and vehicle space, joining the forward cabin to the middle level for easy access. The galley, dining, and saloon area are very similar to the 65. The middle level has a study/office that converts to a stateroom. The forward head is shared by the middle and bow staterooms. The "basement" space has the laundry, pantry, freezer and workshop in a full headroom space. The forward end of the engine room also has headroom and direct access from the saloon.

Seaclusion, shown on the second following page, evolved from the 45' ***Sails*** series, with some influence from this liveaboard version. This one has the galley in the mid-level with a forward study at the foc'sle level. This creates an enormous storage locker on deck, under the forward end of the pilothouse. The 55 ***North Star*** shows what can be done in rearranging this space for another stateroom.

Steel Magnolia is a sistership, with a modified interior layout, and is shown on the cover photo at the Florida Bay Coaster yard, alongside ***Key Largo*** and in front of ***North Star*** and another 50-footer under construction in the building shed. She has a delightfully customized interior, with a lot of modern details provided by her owners for their comfort, plus they enjoy entertaining on their yacht club cruises.

Particulars:	**297**
Length overall	50'-0"
Length designed waterline	49'-9"
Beam	18'-0"
Draft	4'-0"
Freeboard:	
Forward	8'-7½"
Waist	4'-1"
Aft	2'-0"
Displacement, cruising trim*	110,000 lbs.
Displacement-length ratio	399
Prismatic coefficient	.64
Pounds per inch immersion	3,800
Water tankage	1,000 Gals.
Fuel tankage	1,000 Gals.
Headroom	6'-7"
GM	3.7'

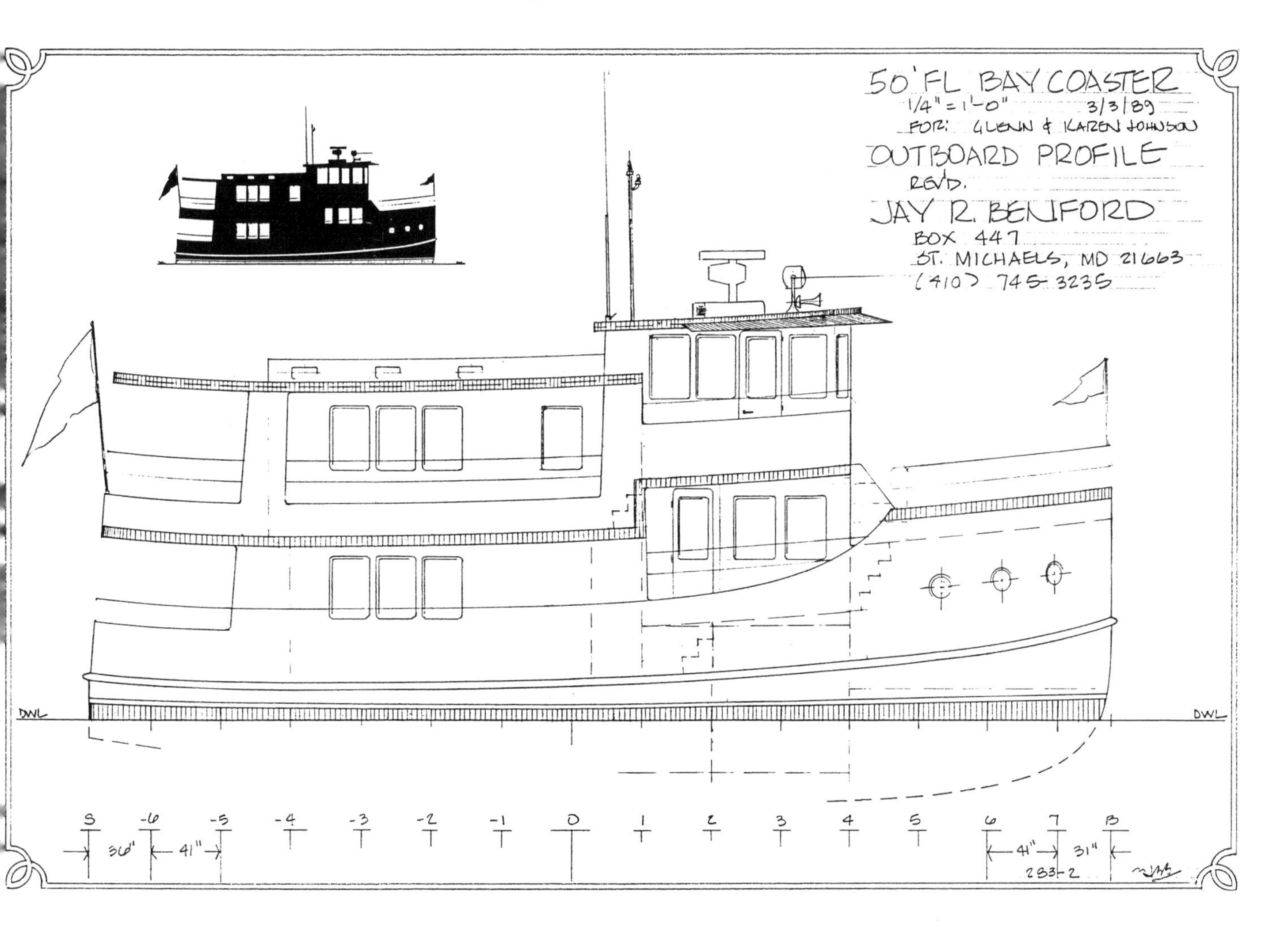

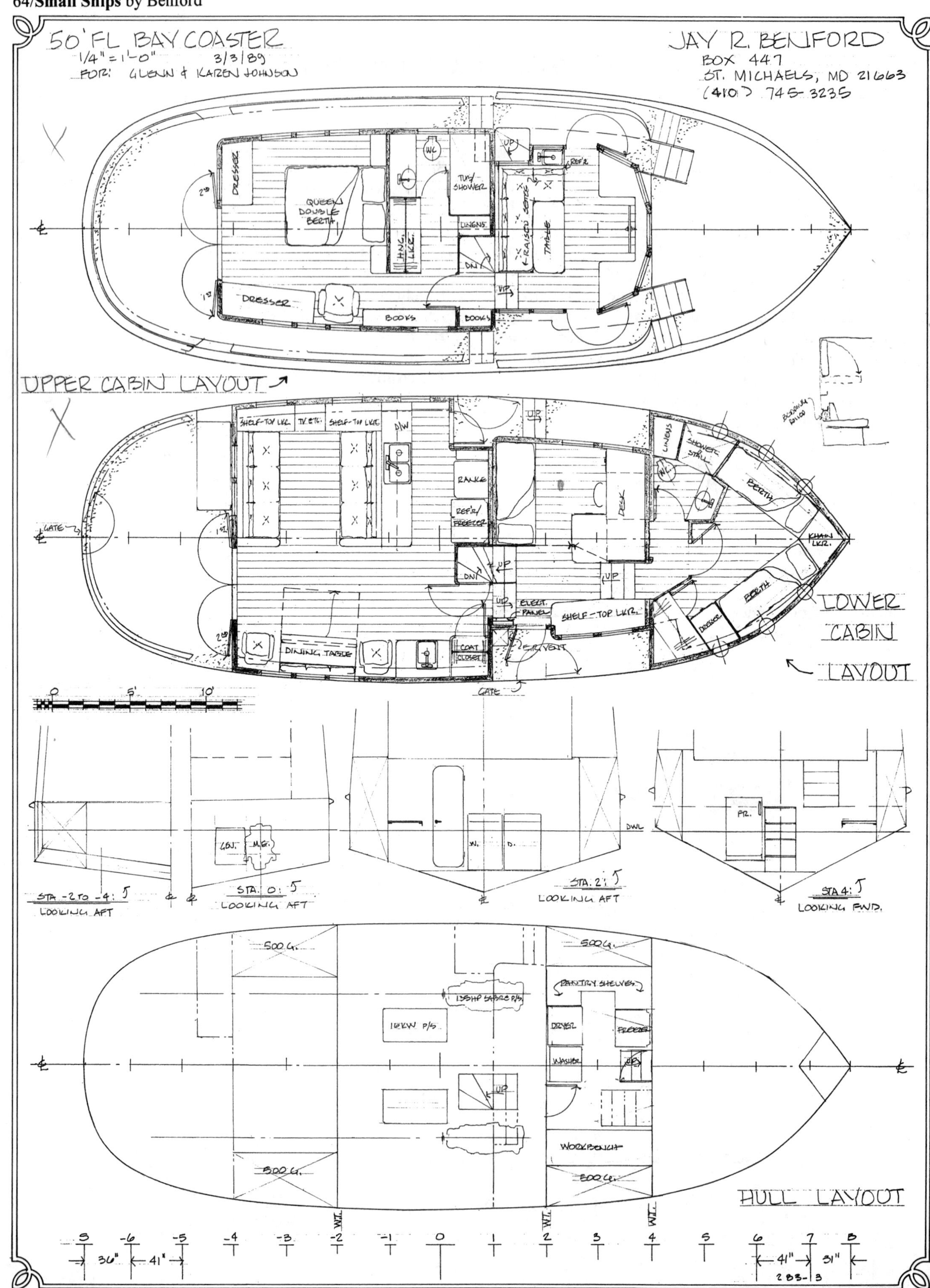

50' FL BAY COASTER
1/4" = 1'-0" 3/3/89
FOR: GLENN & KAREN JOHNSON
JAY R. BENFORD
BOX 447
ST. MICHAELS, MD 21663
(410) 745-3235
DRESSER
QUEEN DOUBLE BERTH
HNG. LKR.
WC
TUB/SHOWER
LINENS
RAISED SETTEE
TABLE
REF'R
DRESSER
BOOKS
BOOKS
UP
DN
UPPER CABIN LAYOUT
SHELF-TOP LKR.
TV ETC.
SHELF-TOP LKR.
D/W
RANGE
REF'R/FREEZER
GATE
DESK
LINENS
SHOWER STALL
WC
BERTH
CHAIN LKR.
BERTH
DRESSER
ELECT. PANEL
SHELF-TOP LKR.
DINING TABLE
COAT CLOSET
SERVENT
GATE
LOWER CABIN LAYOUT
0 5' 10'
GEN.
M.E.
DWL
W.
D.
FR.
STA -2 TO -4: LOOKING AFT
STA. 0: LOOKING AFT
STA. 2: LOOKING AFT
STA 4: LOOKING FWD.
500 G.
500 G.
PANTRY SHELVES
135HP SABRE P/S
16KW P/S
DRYER
FREEZER
WASHER
UP
UP
WORKBENCH
500 G.
500 G.
HULL LAYOUT
W.T.
W.T.
W.T.
-8 -6 -5 -4 -3 -2 -1 0 1 2 3 4 5 6 7 8
36" 41"
41" 31"
283-13

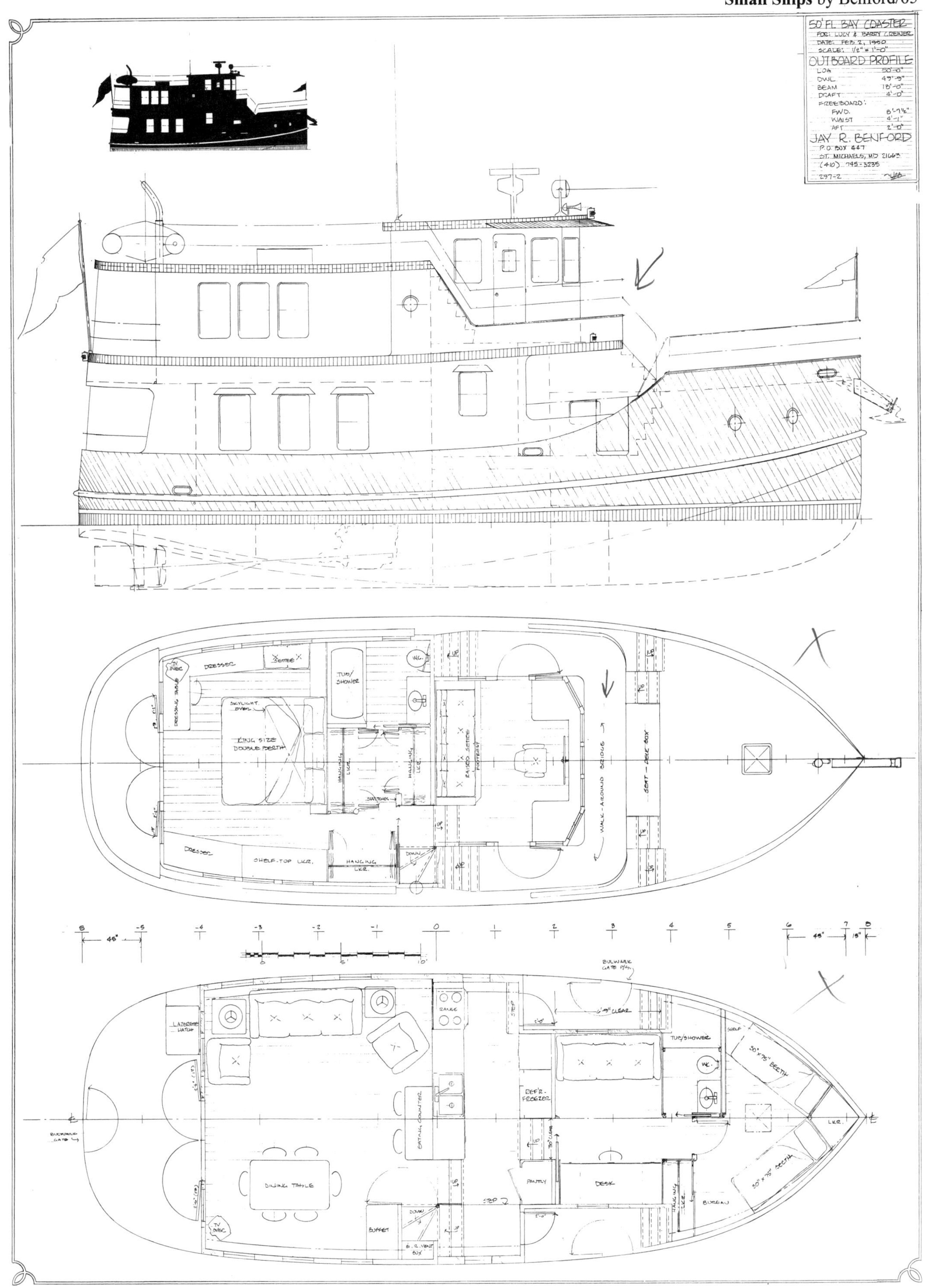
50' FL BAY COASTER
FOR: LUCY & BARRY GREINER
DATE: FEB 2, 1990
SCALE: 1/2" = 1'-0"
OUTBOARD PROFILE
LOA 50'-0"
DWL 49'-9"
BEAM 18'-0"
DRAFT 4'-0"
FREEBOARD:
FWD. 8'-7½"
WAIST 4'-1"
AFT 2'-0"
JAY R. BENFORD
P.O. BOX 447
ST. MICHAELS, MD 21663
(410) 745-3235
297-2
DRESSER
SETTEE
TUB/SHOWER
KING SIZE DOUBLE BERTH
HANGING LKR.
RAISED SETTEE
WALK-AROUND BRIDGE
SEAT - DECK BOX
SHELF-TOP LKR.
RANGE
LAZERETTE HATCH
REF'R. FREEZER
EATING COUNTER
DINING TABLE
BUFFET
PANTRY
DESK
BUREAU
30" x 75" BERTH
LKR.

50' Charter Coaster

Design Number 284
1989

The Coast Guard rules on carrying passengers for hire have for years had a lower limit allowing carrying up to six passengers without having to have the vessel meet all of their Subchapter T rules. This has created a fleet that operate as "six-pack" charter boats.

Thinking that this would be an opportunity to set up a line of mini-cruiseships, we were commissioned to create this concept for a Charter Coaster. For the last several years, the ***Key Largo***, has operated this way. She follows the seasons, going north in the spring, summering in Maine, returning south in the fall, and wintering in Florida. The charter parties love the chance to see various parts of the country. Plus, the last several Coasters have been sold to people who first chartered ***Key Largo*** and loved the experience.

This version has the crew quarters in the bow, with the option of making this into two staterooms which share the head in the extreme bow. Or the door in the centerline bulkhead can be opened and the additional space used for a sitting area with the upper berth hinged down to make a seatback. From the foc'sle there is a second set of stairs that leads down to the laundry, pantry and shop area forward of the engine room. The access to the engine room is from this space.

Amidships is the galley and dining area, with the galley on the crew's end of the space. The dining area is adjacent to the stairs. They lead up to the saloon and down and aft to the three double staterooms, each of which have their own head with a shower stall. The berths are made so that they can be used as twins or moved together to form queen double berths. This option allows a good degree of flexibility in what sorts of charter parties can be booked.

The saloon is over the staterooms, with seating and windows so the whole party can have a good view of the anchorage or the area being cruised.

The pilothouse has a split raised seating layout to get the passengers up where they have a good view forward. The wraparound bridge has extra width forward, which could accommodate some chairs. The after corners of the pilothouse are notched to fit in twin stairs up the boat deck.

The after deck has a staircase leading down to the small after deck. This lower aft deck will be used for swimming, dinghy boarding, and some line handling.

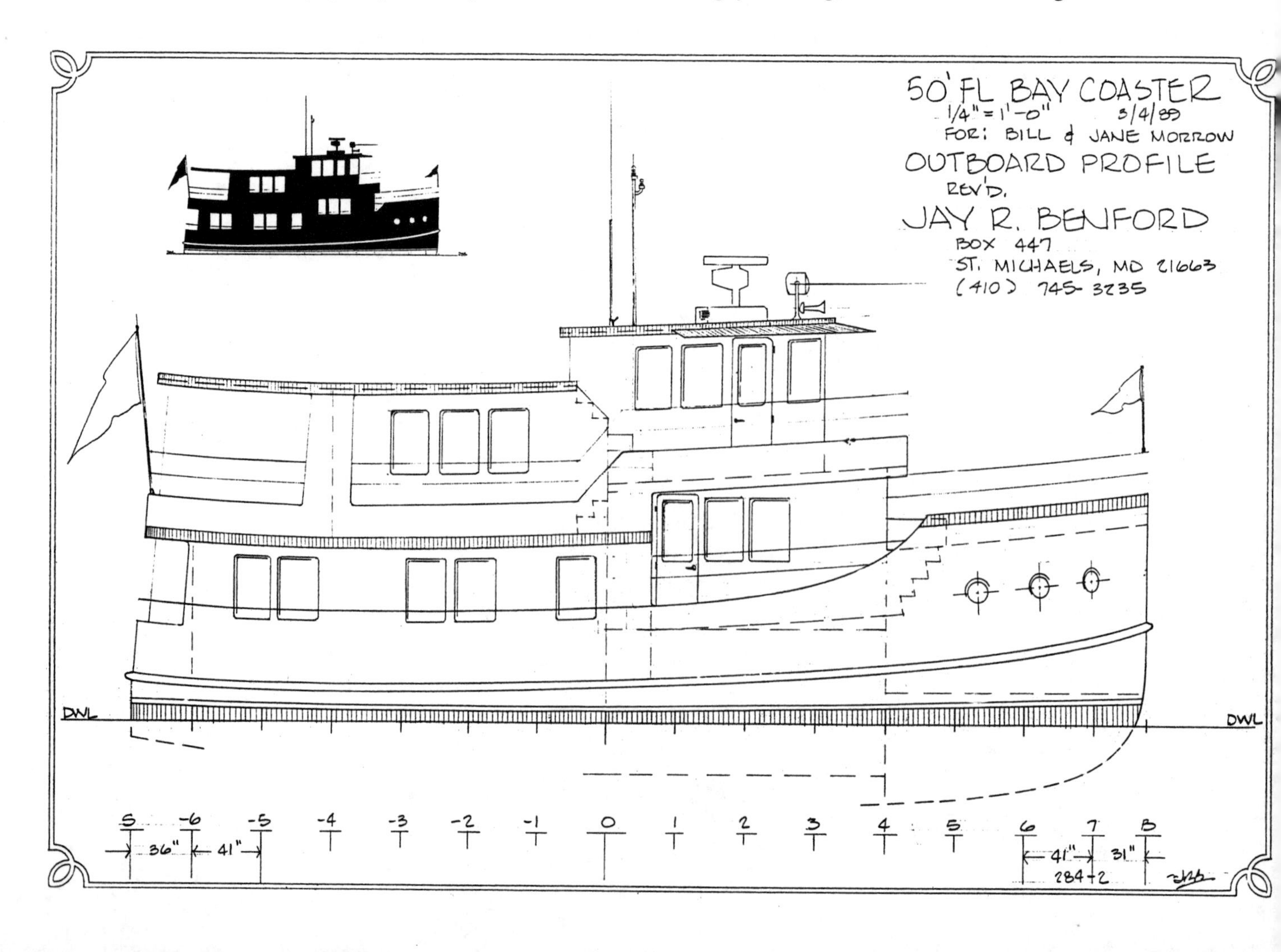

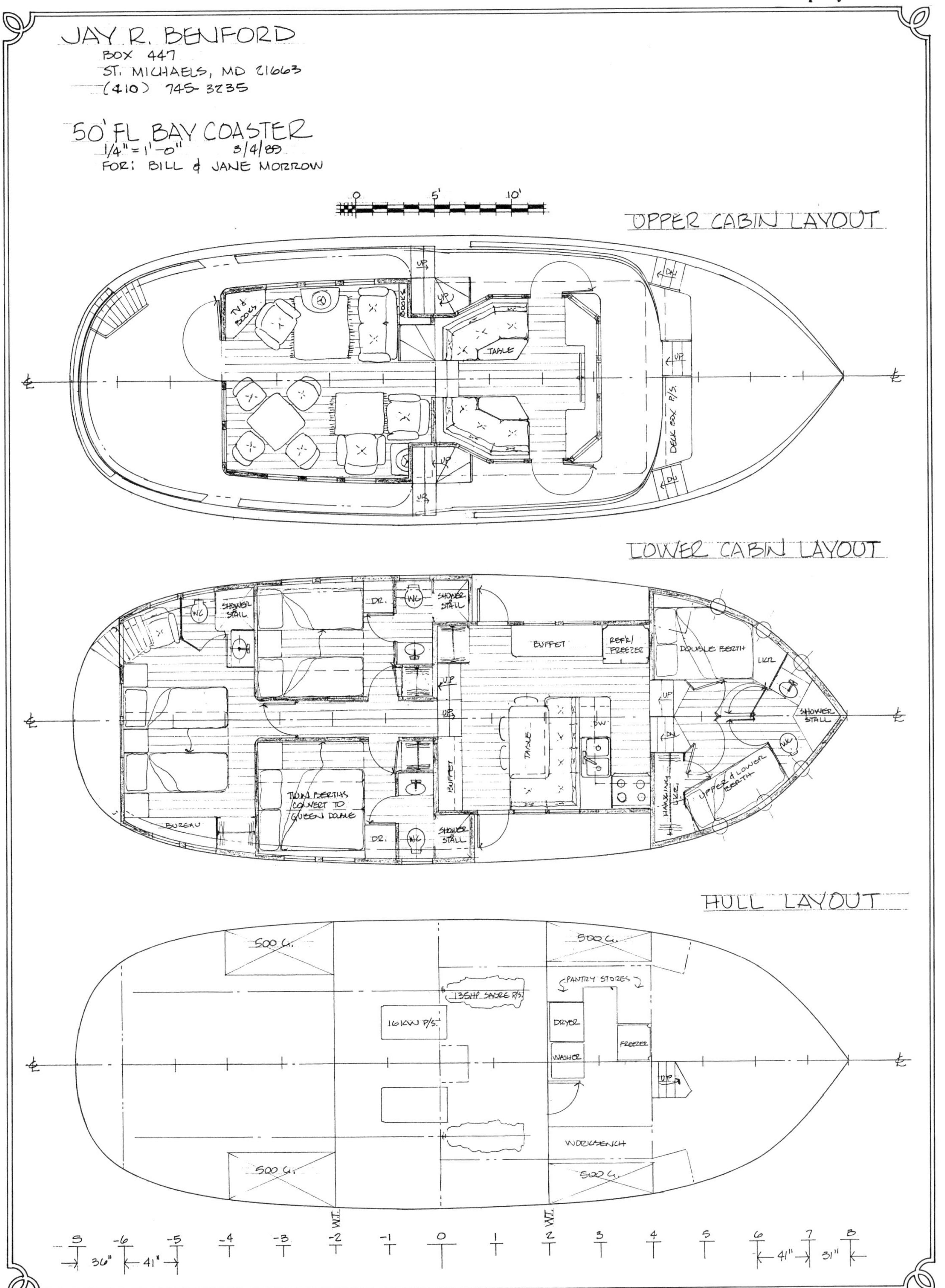

JAY R. BENFORD
BOX 447
ST. MICHAELS, MD 21663
(410) 745-3235
50' FL BAY COASTER
1/4" = 1'-0"
3/4/89
FOR: BILL & JANE MORROW
0
5'
10'
UPPER CABIN LAYOUT
TV & BOOKS
BOOKS
UP
DN
TABLE
DECK BOX P/S.
LOWER CABIN LAYOUT
WC
SHOWER STALL
DR.
BUFFET
REF'R/ FREEZER
DOUBLE BERTH
LKR
TABLE
DW
HANGING LKR.
UPPER & LOWER BERTH
TWIN BERTHS CONVERT TO QUEEN DOUBLE
BUREAU
HULL LAYOUT
500 G.
PANTRY STORES
135HP SABRE P/S.
16KW P/S.
DRYER
WASHER
FREEZER
WORKBENCH
W.T.
36"
41"
31"

50' Tramp Freighter

Design Number 285

1988

The 50' Tramp takes some of the features I liked about the 65' Tramp and recreates them on a smaller boat. The great cabin in the stern is there, but we've eliminated the after open deck space. Being two feet wider, we added a long side deck on the main deck level, joined at each end by the stairs to the end decks.

The traffic from the aft deck flows up and forward alongside the master stateroom and up onto the bridge wings. There are stairs alongside the pilothouse up the boat deck, like we did on ***Key Largo***. There are doors into the after part of the master stateroom and the sides of the pilothouse, both from the side decks.

The cargo hold area has a raised area to hold the car like on ***Key Largo***, which provides an area with standing headroom below it. We've shown additional berthing and storage as options in this hold. Just aft of the hold is the workshop and laundry with access back up to the saloon.

The great cabin aft is the living room, two steps down from the saloon. It could be used as an office space, like on the 65, or to create another stateroom, or as a space for a specialized shop.

Particulars:

Length overall	50'-0"
Length designed waterline	49'-8"
Beam	20'-0"
Draft	4'-0"
Freeboard:	
Forward	8'-9"
Least	2'-0"
Aft	7'-10"
Displacement, cruising trim*	106,000 lbs.
Displacement-length ratio	386
Water tankage	1,300 Gals.
Fuel tankage	1,950 Gals.
Headroom	6'-5" to 6'-8"

***CAUTION:** The displacement quoted here is for the boat in cruising trim. That is, with the fuel and water tanks filled, the crew on board, as well as the crews' gear and stores in the lockers. This should not be confused with the "shipping weight" often quoted as "displacement" by some manufacturers. This should be taken into account when comparing figures and ratios between this and other designs.

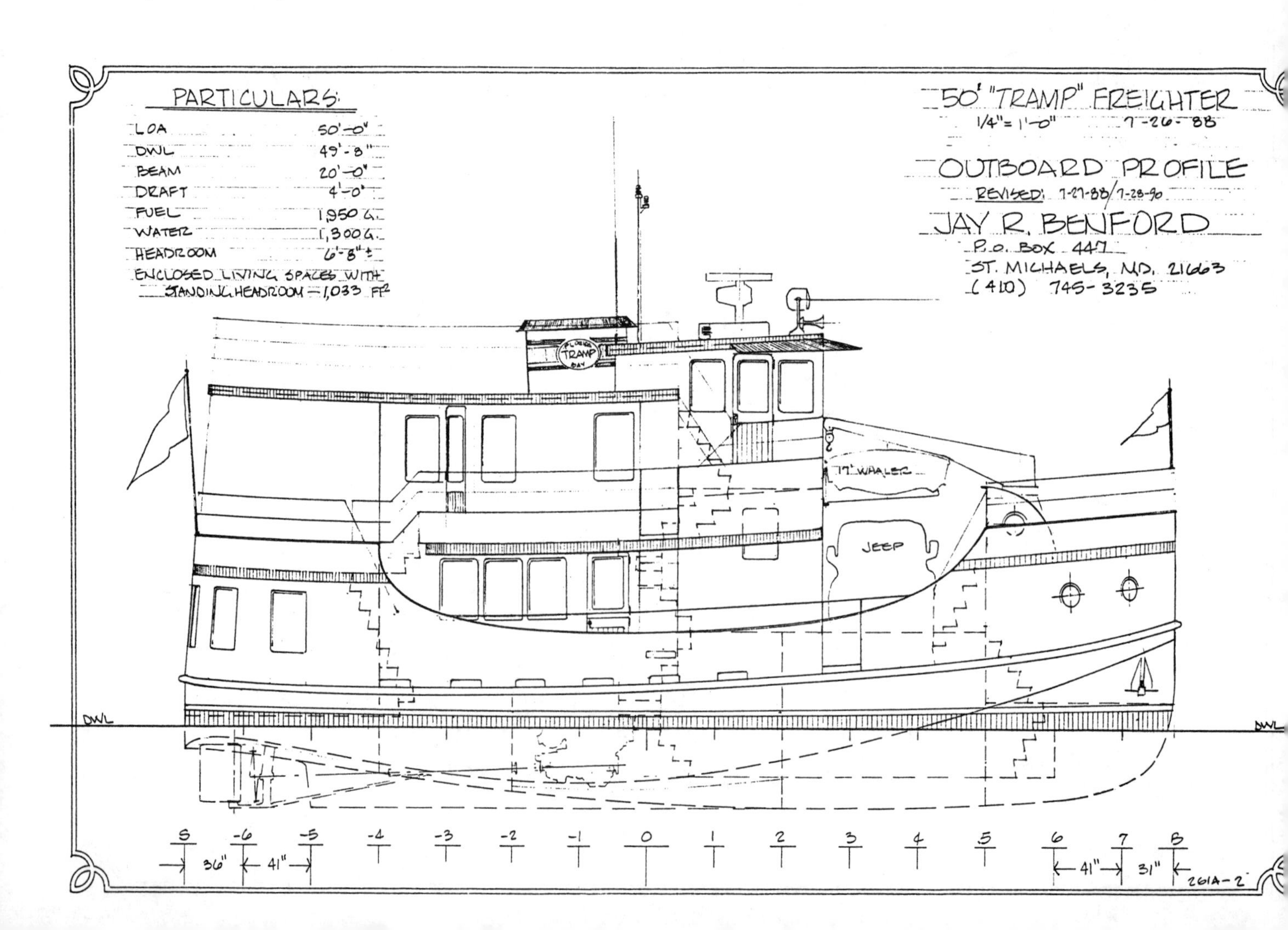

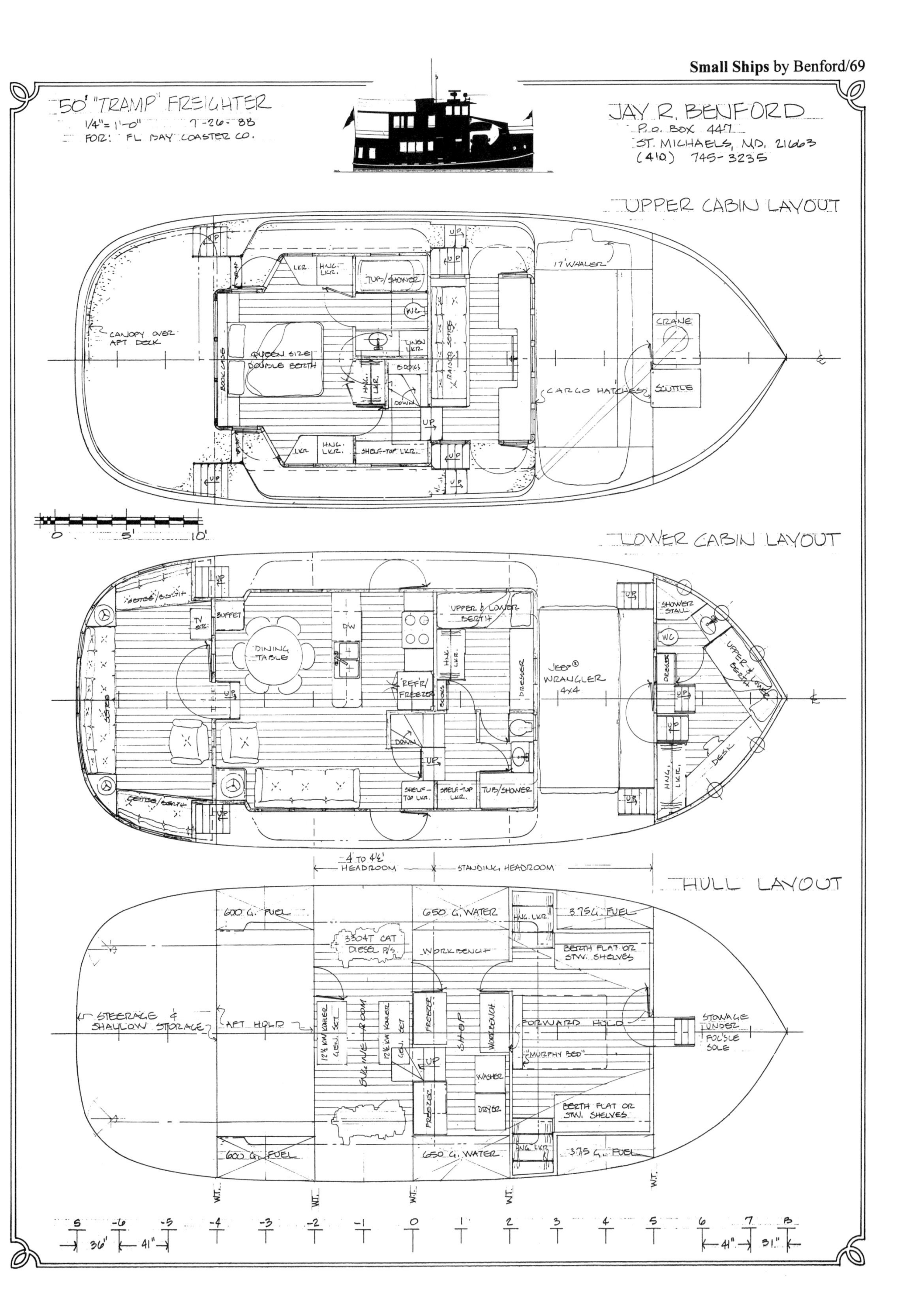
50' "TRAMP" FREIGHTER
1/4" = 1'-0" 7-26-88
FOR: FL BAY COASTER CO.
JAY R. BENFORD
P.O. BOX 447
ST. MICHAELS, MD. 21663
(410) 745-3235
UPPER CABIN LAYOUT
CANOPY OVER AFT DECK
QUEEN SIZE DOUBLE BERTH
BOOK CASE
LKR
HNG. LKR.
TUB/SHOWER
WC
LINEN LKR
BOOKS
DOWN
UP
RAISED SETEE
SHELF-TOP LKR.
17' WHALER
CRANE
CARGO HATCHES
SCUTTLE
0
5'
10'
LOWER CABIN LAYOUT
SETEE/BERTH
TV ETC.
BUFFET
DINING TABLE
DW
REFR/ FREEZER
UPPER & LOWER BERTH
HNG. LKR.
DRESSER
JEEP® WRANGLER 4x4
SHOWER STALL
WC
UPPER & LOWER BERTH
DESK
SETEE
DOWN
SHELF-TOP LKR.
TUB/SHOWER
4 TO 4½' HEADROOM
STANDING HEADROOM
HULL LAYOUT
600 G. FUEL
3304T CAT DIESEL P/S
650 G. WATER
WORKBENCH
HNG. LKR
375 G. FUEL
BERTH FLAT OR STW. SHELVES
STEERAGE & SHALLOW STORAGE
AFT HOLD
12½ KW KOHLER GEN. SET
ENGINE ROOM
FREEZER
SHOP
WORKBENCH
FORWARD HOLD
STOWAGE UNDER FOC'SLE SOLE
"MURPHY BED"
WASHER
DRYER
W.T.
36"
41"
41"
31"

50' x 20' Florida Bay Coaster

Design Number 285

1989

This variation on the Coaster theme is done on a widened hull, like was drawn for the 50' Tramp. The emphasis on this 50-footer is indoor living space, with the outside deck spaces shortened in favor of cabin volume.

The master stateroom is asymmetrical, with a side deck on one side and not the other. Similarly there is a stairway only on the port side of the pilothouse to the boat deck. The saloon is wide open, to bring aboard the living room and dining room furniture that the clients already owned. There is a wet bar built into the forward starboard corner of the saloon.

The berths in the foc'sle are arranged so that they are hinged at their aft ends and swing inboard to give access to the outboard side for ease in making the beds.

The vehicle on deck is an electric golf cart. There's room for a "personal watercraft" instead, or perhaps it could be tucked into the cargo hold if we made the hatch a bit larger.

The dashed lines over the engines are overhead rails with chain hoists. These are for servicing the engines and sliding them forward to lift out through the cargo hatch if they ever needed major servicing or overhaul.

Particulars:

Length overall	50'-0"
Length designed waterline	49'-8"
Beam	20'-0"
Draft	4'-0"
Freeboard:	
Forward	9'-6"
Waist	4'-4"
Aft	2'-3"
Displacement, cruising trim*	106,000 lbs.
Displacement-length ratio	386
Water tankage	1,000 Gals.
Fuel tankage	1,500 Gals.
Headroom	6'-7"

***CAUTION:** The displacement quoted here is for the boat in cruising trim. That is, with the fuel and water tanks filled, the crew on board, as well as the crews' gear and stores in the lockers. This should not be confused with the "shipping weight" often quoted as "displacement" by some manufacturers. This should be taken into account when comparing figures and ratios between this and other designs.

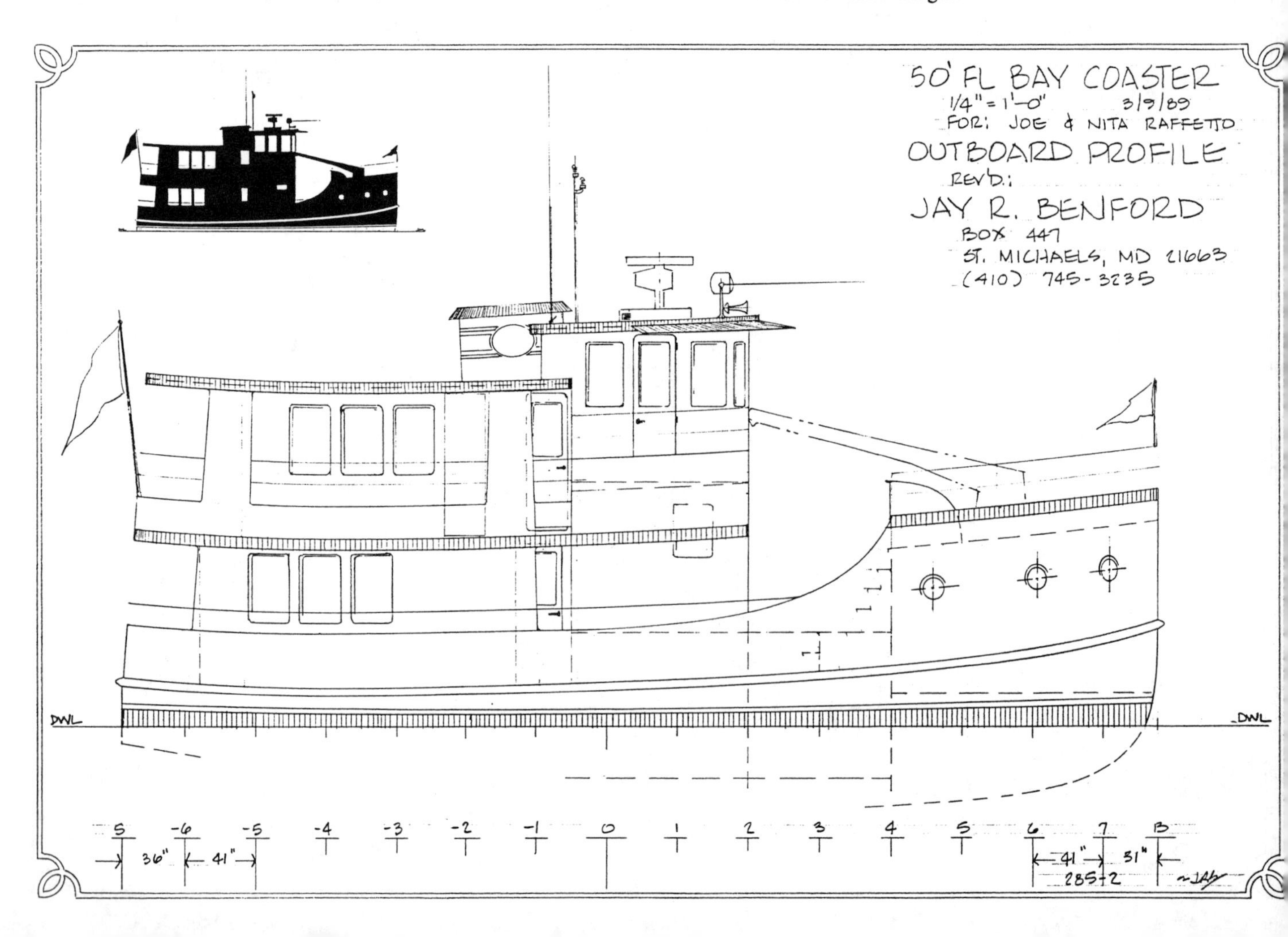

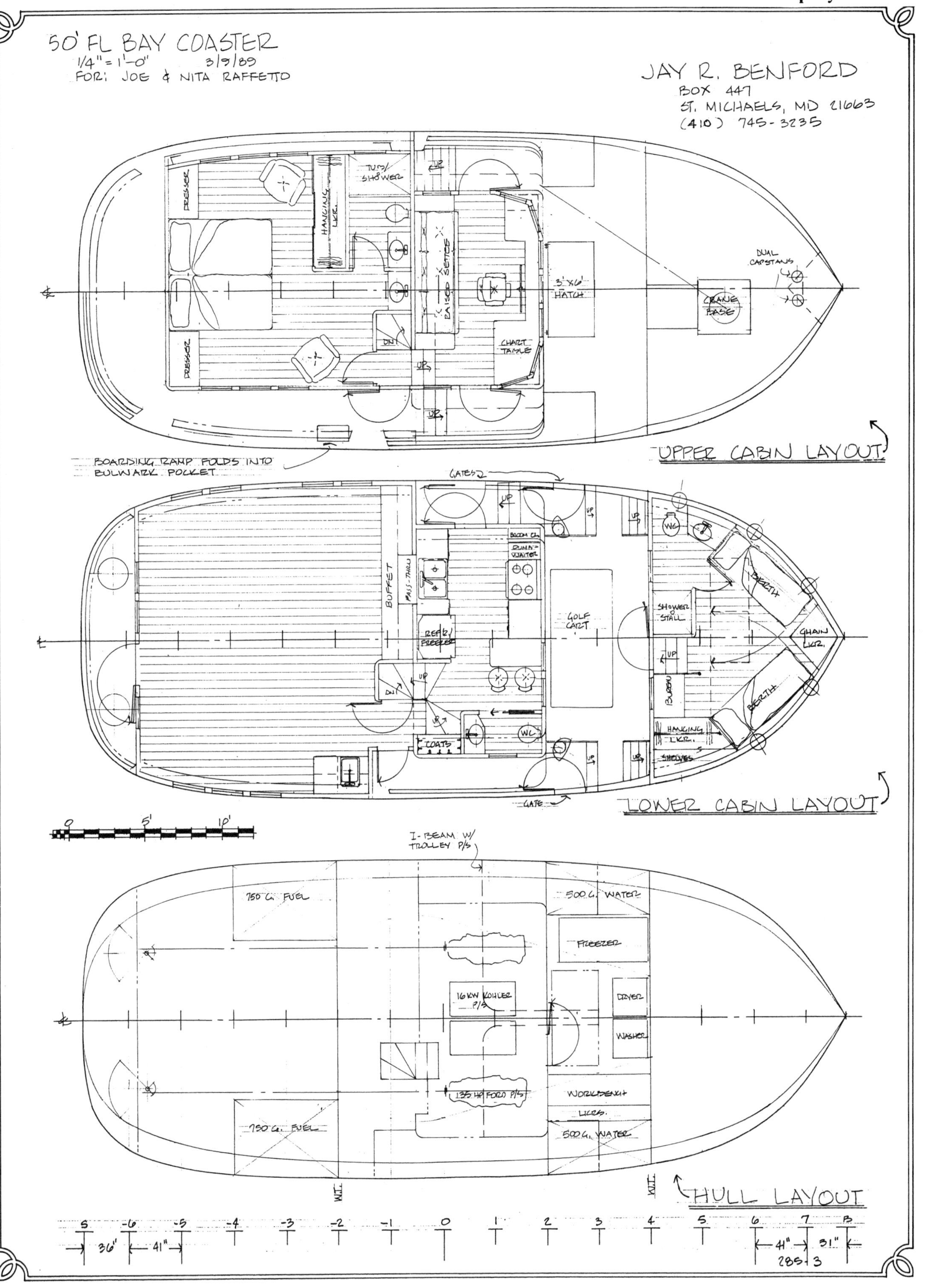

50' FL BAY COASTER
1/4" = 1'-0"   3/9/89
FOR: JOE & NITA RAFFETTO
JAY R. BENFORD
BOX 447
ST. MICHAELS, MD 21663
(410) 745-3235
DRESSER
HANGING LKR.
TUB/SHOWER
UP
RAISED SETTEE
DUAL CAPSTANS
3'x6' HATCH
CRANE BASE
DN
CHART TABLE
BOARDING RAMP FOLDS INTO BULWARK POCKET
UPPER CABIN LAYOUT
GATES
BROOM CL.
DUMB-WAITER
BUFFET
PASS-THRU
REFR/FREEZER
GOLF CART
WC
SHOWER STALL
BERTH
CHAIN LKR.
BUREAU
HANGING LKR.
SHELVES
COATS
GATE
LOWER CABIN LAYOUT
0
5'
10'
I-BEAM W/ TROLLEY P/S
750 G. FUEL
500 G. WATER
FREEZER
16 KW KOHLER P/S
DRYER
WASHER
135 HP FORD P/S
WORKBENCH
LKRS.
HULL LAYOUT
W.L.
36"
41"
31"
285-3

North Star

Design Number 313
1991

Our design for the ***North Star*** started out as a revision to the 50' ***Seaclusion***, which the Florida Bay Coaster Company had just recently built for the Greiners. We added two and a half feet to the stern to make the back porches (aft decks) roomier. We then re-arranged the interior to add a main saloon with the best features from ***Key Largo***, added a middle stateroom and head, aft stairs from the stateroom deck to the lower deck and easier entry to the steerage and lazarette, and turned the basement space into a combination of workshop, laundry, pantry, and captain's stateroom with it's own head. In this process, we revamped the structural design to simplify construction and increased the fuel tankage for some additional range.

During the course of building the boat, Charlie convinced Reuben he'd like to have a longer appearing bow, and thus the stem was moved three feet forward, making the forward stateroom a little bigger and increasing the chain locker space.

On deck, there was the addition of what's sometimes called the Platypus proboscis (and other times called the anchor platform) for the anchor rollers and storage of the anchors. This was combined with some built-in seating at the bow to make a nice and functional layout, facilitating anchoring work. There is seating in front of the bridge and on the bridge, against the front of the pilothouse, both with storage inside.

The boat deck has a stack of three sailing dinghies and a hard bottom inflatable with a crane for launching and retrieving them, chairs for seating and a deck locker in the stack.

North Star was the first Coaster to be fitted with screen doors. These are cleverly arranged (several are sliders) and they add a lot to the comfort of living aboard. The page of color photos near the beginning of this book will give an idea of how nicely ***North Star*** is fitted out.

Jay R. Benford

One subtle intangible that comes with owning a Coaster is the involuntary induction into a society of Coaster aficionados. There are no dues, assessments, papers to fill out, letters of recommendation, ceremonies or membership card. Only a spiritual feeling of loose kinship. Something like a convocation of people with different religious backgrounds, drawn together by a common thread of belief in a higher being but very different in their approach. To some it is a home and recreational dream all in one, to others a means to voyage with the stimulation of the next day's adventure and quiet anchorage. The tie is the character of the Coaster, its appearance, its livability or maybe its distinctiveness. One of our fantasies is a rendezvous of Coasters, at this date nine, rafted, a beautiful sunny day; what a way to end the 20th Century!

Charlie Chapin

The drawing below, done by Reuben Trane, nicely illustrates the interior of ***North Star***.

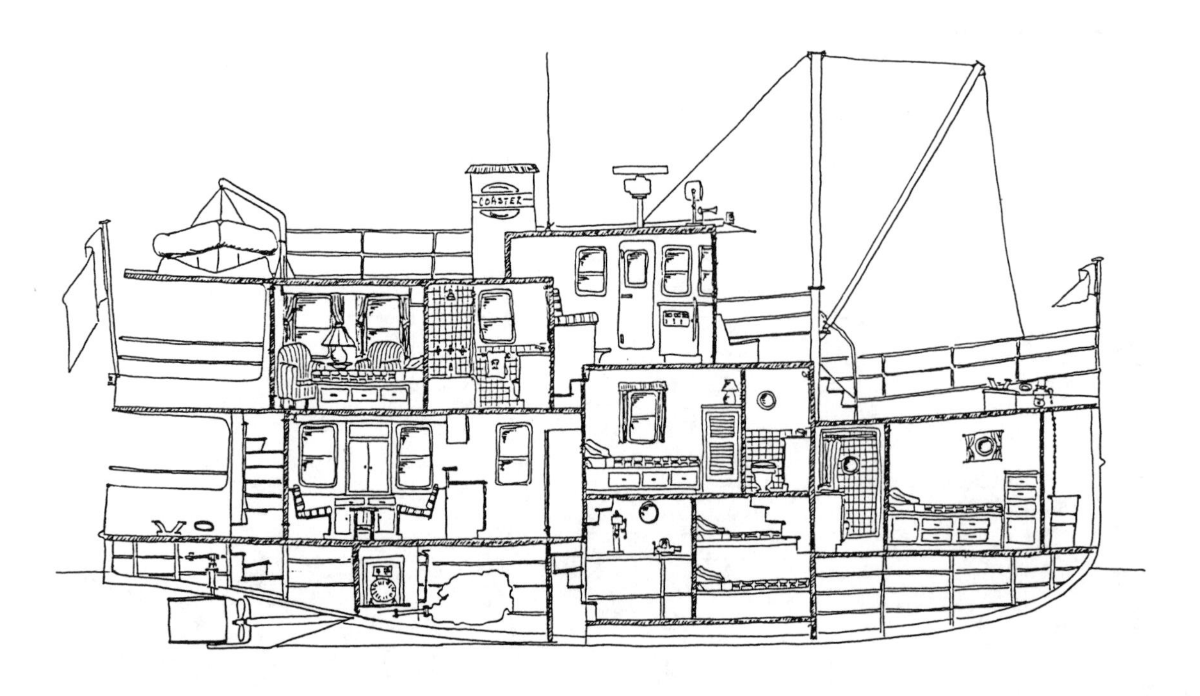

Birthing An Elephant
— or —
Building A Coaster

In biological terms, building a Coaster is like birthing an elephant . . . at least as far as the gestation period is concerned. The other aspects are more pleasurable, particularly if one were the mother elephant. There probably is some other common ground as in being part of creating something special in which you are deeply involved.

The start of the dream occurred at a Newport Wooden Boat Show where I happened upon a rendering of a Coaster in designer Jay Benford's booth. My initial reaction was, "That really looks interesting! I had a Walter Mitty flash of me in command, headed off to distant waters." Reality soon returned and I continued my tour of the show. But, Jay had planted the seed.

Later as plans for retirement were coming close to reality, Kay, my lovely and compatible wife, and I plunged into a serious bout of looking at used boats: Trumpys, Grand Banks, Krogens, Hatteras, etc. while sandwiching in a show visit to ***Key Largo*** and visiting other Coasters; ***Florida Bay***, ***Sails***, ***Seaclusion*** and ***Steelaway,*** along with a couple of yard visits. The more we looked, the more the Coaster came to be the standard. Finally after a short cruise on ***Key Largo*** and a sumo wrestling match with my New England conservatism, we signed — or maybe consummated the deal.

After the pre-signing flurry of activity and lawyers there was a quiet period. A picture of the keel laying from Reuben arrived in time for a raucous boat naming party which did not produce a name, but was an ocean of fun.

Before and during the building process there were consultations with Jay Benford, the designer. He always seemed willing to drop his business at hand to talk on details or concept. Never rushing a conversation and his calm matter of exposition serving as reassurance or aiding a decision. One always had the impression of an analytical and thoughtful response to many less than scientific type questions. Jay always showed great interest in his creation.

After two trips to the yard, we decided to take up residence in Green Cove Springs near the yard to be more involved. In our case, the creation had become a passion. Kay took the lead on the interior planning with me as the peanut gallery (nothing to do with elephants). I focused on the exterior and mechanicals. Kay approached her part as if she were commissioned to do a mansion in Newport while I proceeded to bust the budget with all the gadgets hoping to add character to a boat with character (masts with gold balls at the trucks, smokestack, etc.) and as carefree as possible to operate. As we near a launching date we are both pleased with what our planning has created. Kay's mansion has been reduced somewhat in scale, but not in elegance.

Before moving to Green Cove, we were confined to telephone calls and notes back and forth to Reuben, a somewhat sterile relationship. This all quickly changed with the domicile change. One might even say it was a change of life.

We seemed to be quickly assimilated into the life of the yard, the building crew and Reuben and Gerry Trane. Reuben as the major domo and Grand Pooh Bah of the yard set the tone. Reuben's patience, hand holding, understanding, creativity, guidance, input and good nature has added to our pleasure. In fact, we like to think of him as a helpful friend.

Getting to know the boat yard crew has been an unexpected delight. Damon Runyon, John Steinbeck or Larry McMurtry (**Lonesome Dove**) would have rich material for a major best seller. The character of the people, their joys and sorrows, way of life, pride in their work and personalities form a colorful tapestry of life, a small piece of America. Kay and I have a warm feeling for them all and will leave the yard with a wonderful memory of these fine people, especially Tina Thomas, Reuben's chief of staff, who have been so nice to us.

In perspective, we would not change anything. We are wiser in how to build a boat. Our bank account, like Kay's mansion, is a little shrunken, but our passion is the same. We are looking forward to being afloat but will close the building chapter with a little sadness, happy memories and best of all, new friends.

Does a mother elephant have a sense of pride in her new offspring? We know we do.

Charlie Chapin
North Star

North Star; Our Perfect Boat

Unlimited cruising opportunities
Dignity, character and elegance without affectation
Privilege without ostentation
Grandeur without pretension
North Star our perfect boat.

Kay Chapin

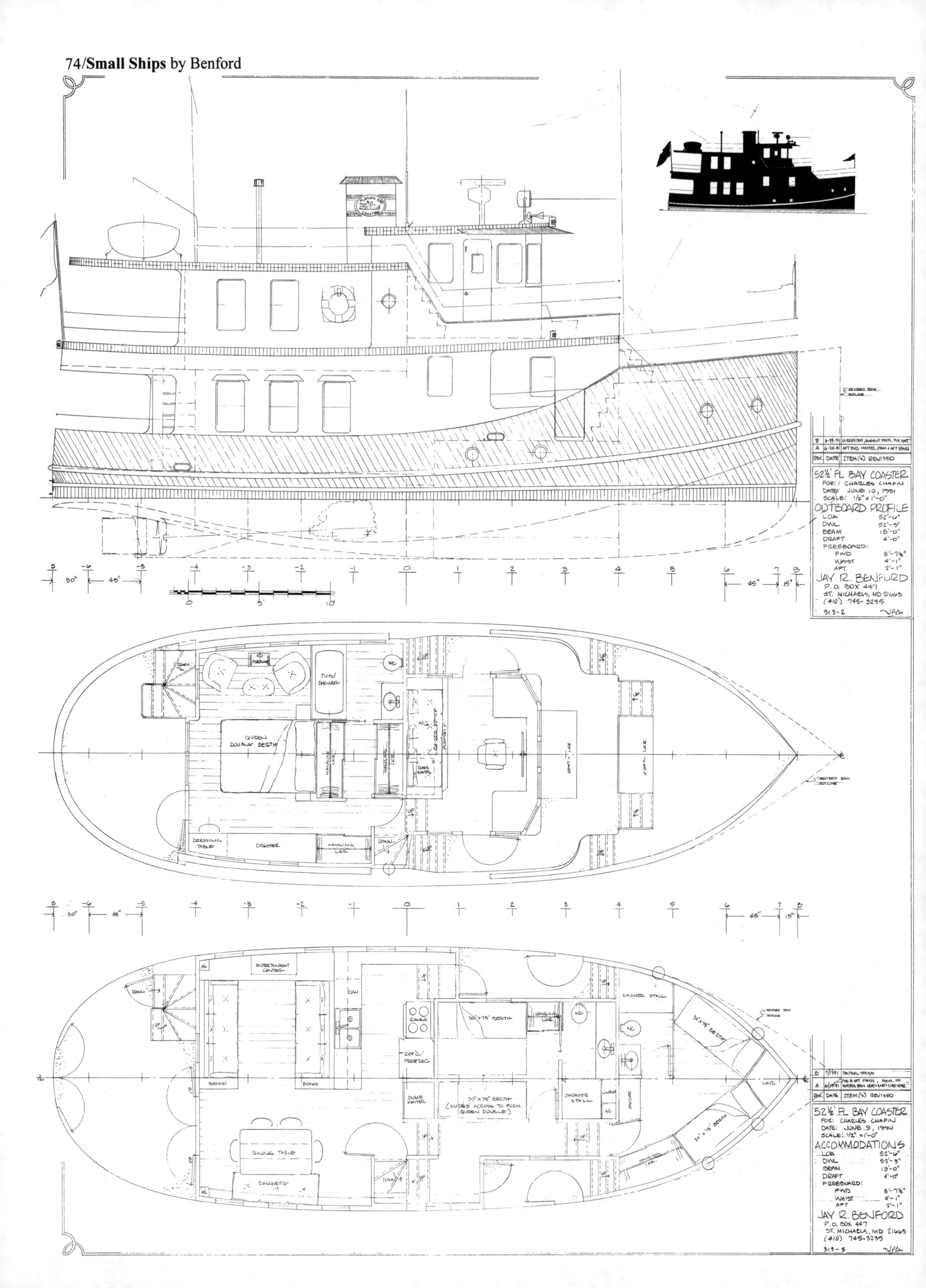
REVISED BOW OUTLINE
B 6-29-91 UNDERBODY, BAGGAGE PORTS, FWD. MAST
A 6-20-91 AFT END MASTER STEM + AFT STAIRS
REV. DATE ITEM(S) REVISED
52½' FL BAY COASTER
FOR: CHARLES CHAPIN
DATE: JUNE 10, 1991
SCALE: 1/2" = 1'-0"
OUTBOARD PROFILE
LOA 52'-6"
DWL 52'-3"
BEAM 18'-0"
DRAFT 4'-0"
FREEBOARD:
FWD 8'-7½"
WAIST 4'-1"
AFT 2'-1"
JAY R. BENFORD
P.O. BOX 447
ST. MICHAELS, MD 21663
(410) 745-3235
313-2
30" 45" 45" 15"
QUEEN DOUBLE BERTH
TUB/SHOWER
HANGING LKR.
DRESSING TABLE
DRESSER
DOWN
UP
DUMB WAITER
SEAT-LKR.
ENTERTAINMENT CENTER
BOOKS
RANGE
REF'G/FREEZER
30"x75" BERTH
30"x75" BERTH (SLIDES ACROSS TO FORM QUEEN DOUBLE)
SHOWER STALL
DINING TABLE
BANQUETTE
LKR.
B 7/1/91 CENTRAL STAIRS
A 6/19/91 FWD & AFT STAIRS, REVS. TO MASTER STRM HEAD + HEADS
REV. DATE ITEM(S) REVISED
52½' FL BAY COASTER
FOR: CHARLES CHAPIN
DATE: JUNE 9, 1990
SCALE: 1/2" = 1'-0"
ACCOMMODATIONS
LOA 52'-6"
DWL 52'-3"
BEAM 18'-0"
DRAFT 4'-0"
FREEBOARD:
FWD. 8'-7½"
WAIST 4'-1"
AFT 2'-1"
JAY R. BENFORD
P.O. BOX 447
ST. MICHAELS, MD 21663
(410) 745-3235
313-3

55' Voyager

Design Number 267
1988

This design was one of the variations we came up with after the first showing of the ***Florida Bay***. At that boatshow we'd had a wide variety of very positive responses to the boat. Of all the people who liked it, each had some variation on the original theme that they wanted to see on "their" Coaster.

This one was for those people who wanted to go long range voyaging under power, in the philosophy pioneered by Bob Beebe. We may have gone a bit overboard in providing tankage for a 10,000 mile range....

The accommodations have a bunk room in the foc'sle for kids, crew, or students along on the passage for an education. The owner's quarters are aft and there is another stateroom aft for two, each with a private head. The dining area is also aft on the lower level, in an area that will have less motion. As a book lover and collector I've made sure there are plenty of book shelves in this area, to function as a library too. The mid-level has the galley and ship's office. The saloon is on the upper aft deck with a good walk-around deck leading up to the pilothouse wing decks and bridge.

The cargo hold, under the vehicle on deck, has room for lots of accumulated treasures, or a marine biology lab.

Particulars:

Length overall	55'-0"
Length designed waterline	54'-3"
Beam	18'-0"
Draft	5'-6"
Freeboard:	
Forward	9'-6"
Least	3'-5"
Aft	3'-6"
Displacement, half-tanked condition*	150,000 lbs.
Displacement-length ratio	419
Water tankage	1,000 Gals.
Fuel tankage	9,000 Gals.
Headroom	6'-7"

***CAUTION:** The displacement quoted here is for the boat in half-tanked condition. That is, with the fuel and water tanks half filled, the crew on board, as well as the crews' gear and stores in the lockers. This should not be confused with the "shipping weight" often quoted as "displacement" by some manufacturers. This should be taken into account when comparing figures and ratios between this and other designs. Add about 36,000 pounds for the fully tanked condition.

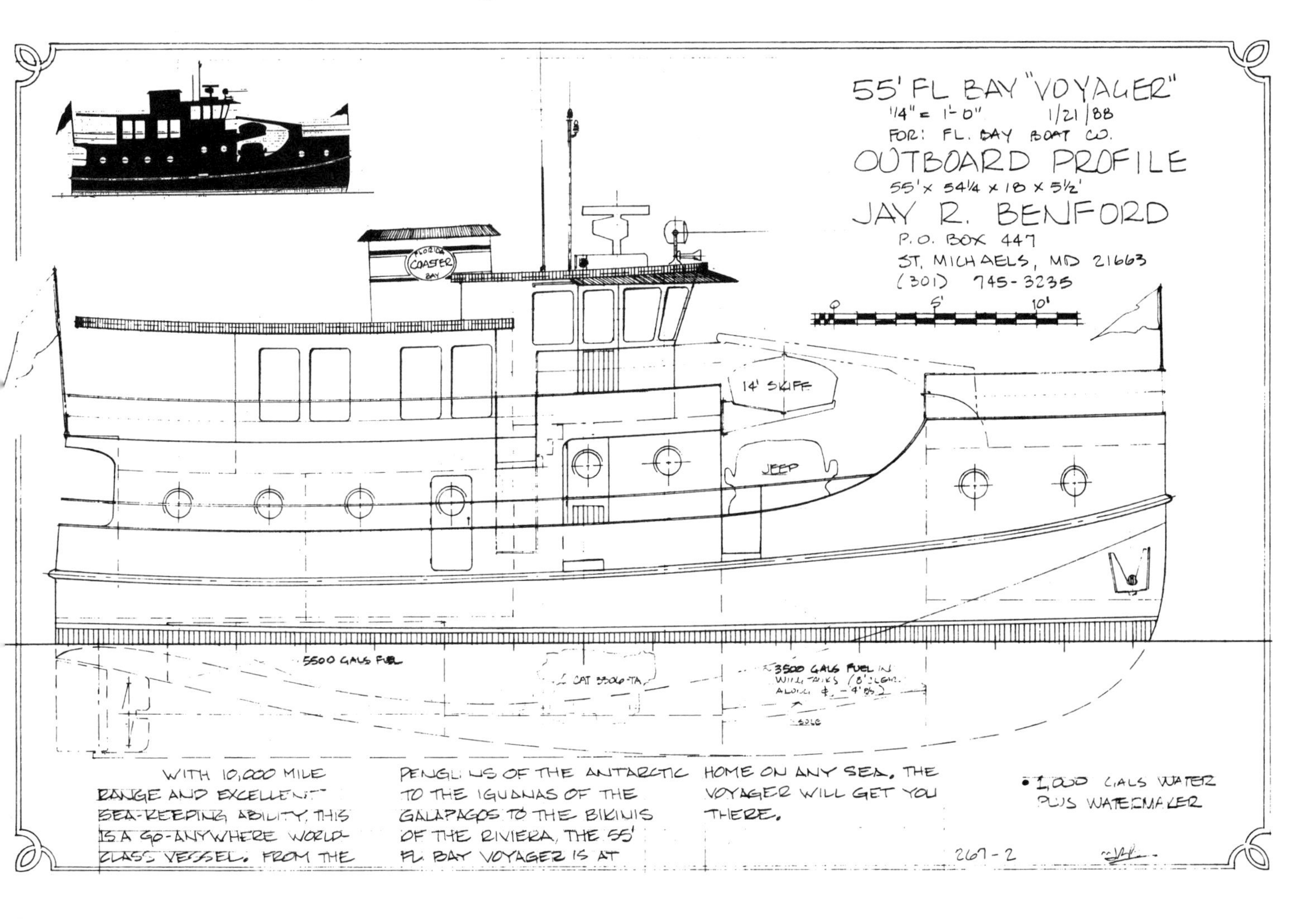

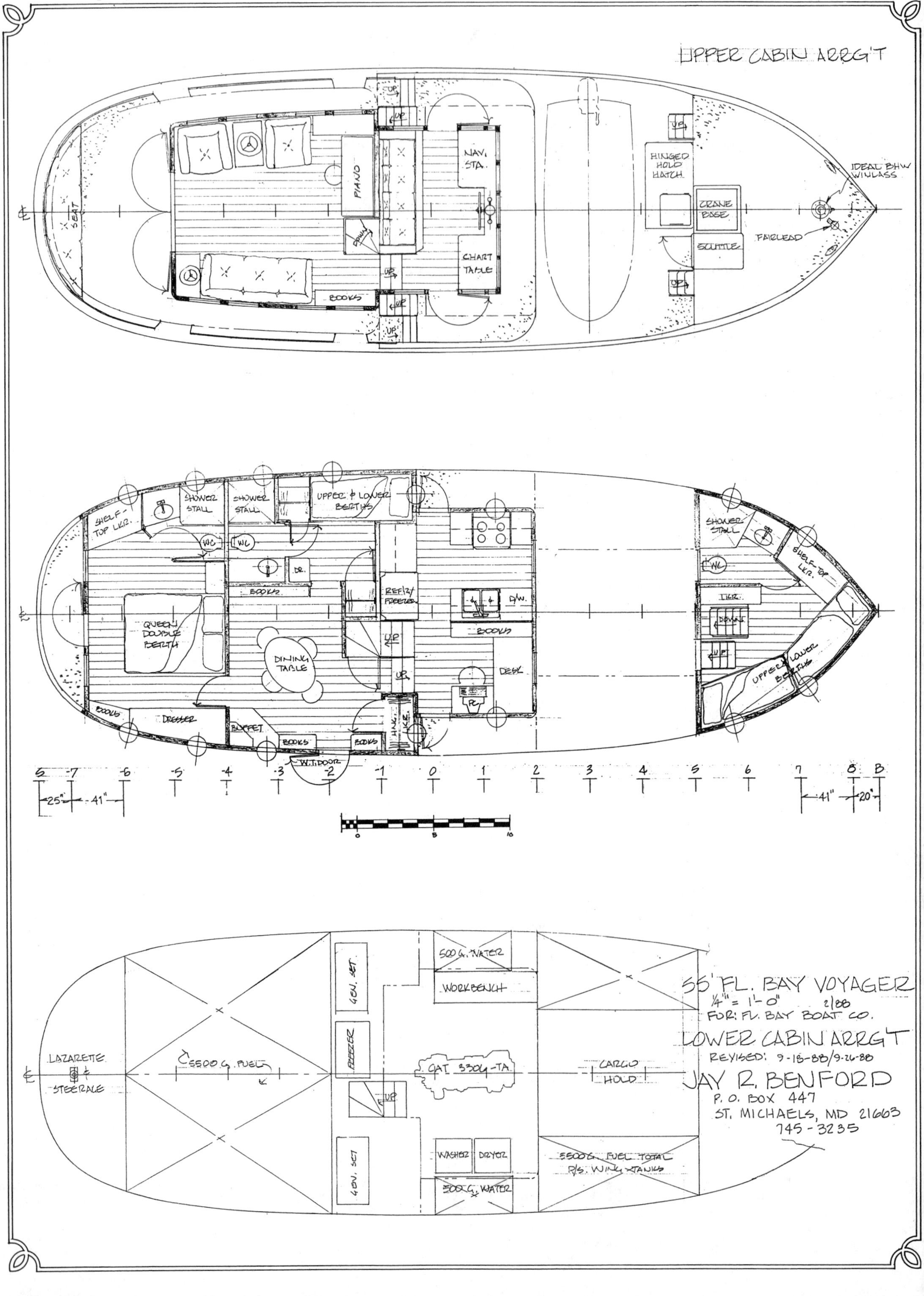

UPPER CABIN ARRG'T

SEAT
PIANO
NAV. STA.
CHART TABLE
BOOKS
DOWN
UP
HINGED HOLD HATCH
CRANE BASE
SCUTTLE
IDEAL BHW WINLASS
FAIRLEAD

SHELF-TOP LKR.
SHOWER STALL
SHOWER STALL
UPPER & LOWER BERTHS
WC
WC
DR.
BOOKS
QUEEN DOUBLE BERTH
DINING TABLE
REFRIG/FREEZER
UP
UP
D/W.
BOOKS
DESK
BOOKS
DRESSER
BUFFET
BOOKS
BOOKS
HNG. LKR.
W.T. DOOR
SHOWER STALL
SHELF-TOP LKR.
WC
LKR.
DOWN
UP
UPPER & LOWER BERTHS

-8 -7 -6 -5 -4 -3 -2 -1 0 1 2 3 4 5 6 7 8 B
25" 41" 41" 20"

0 5 10

500 G. WATER
WORKBENCH
GEN. SET
FREEZER
UP
GEN. SET
LAZARETTE
STEERAGE
5500 G. FUEL
CAT 3304-TA
CARGO HOLD
WASHER
DRYER
500 G. WATER
5500 G. FUEL TOTAL P/S. WING TANKS

55' FL. BAY VOYAGER
1/4" = 1'-0" 2/88
FOR: FL. BAY BOAT CO.
LOWER CABIN ARRG'T
REVISED: 9-18-88/9-26-88
JAY R. BENFORD
P. O. BOX 447
ST. MICHAELS, MD 21663
745-3235

55' West Coaster

Design Number 295
1989

The 55' West Coaster was conceived for a gentleman who lived in California. He wanted a vessel with similar livability to the Florida Bay Coasters plus more ability to deal with open ocean cruising. To achieve this, we considerably modified the hull form, particularly in the bow giving it a finer entry and more flare.

For his cruising, he intended to operate with a couple aboard as paid crew. Their quarters would be in the bow, with ready access to the galley and the stores and utility room and shop below the galley. Thus, the engine room access and service is through the crew's quarters. We've specified one of the very quiet, fuel efficient, low rpm (cruising speed at about 1,000 rpm) Gardner diesels and twin Northern Lights 1,200 rpm generators.

The owner's stateroom is all the way aft, with its own head. Next forward is the guest stateroom again with its own head. Both staterooms have king sized beds.

The galley and dining areas are at the area of amidships entry. Half a level up and aft is the saloon, with a built-in fireplace on the forward bulkhead. On the aft deck, the main area for outdoor socializing, is a wet bar.

Particulars:

Length overall	55'-0"
Length designed waterline	54'-6"
Beam	18'-0"
Draft	5'-3"
Freeboard:	
Forward	9'-6"
Waist	4'-0"
Aft	2'-3"
Displacement, cruising trim*	111,000 lbs.
Displacement-length ratio	306
Water tankage	1,200 Gals.
Fuel tankage	2,000 Gals.
Headroom	7'-0" to 8'-0"

***CAUTION:** The displacement quoted here is for the boat in cruising trim. That is, with the fuel and water tanks filled, the crew on board, as well as the crews' gear and stores in the lockers. This should not be confused with the "shipping weight" often quoted as "displacement" by some manufacturers. This should be taken into account when comparing figures and ratios between this and other designs.

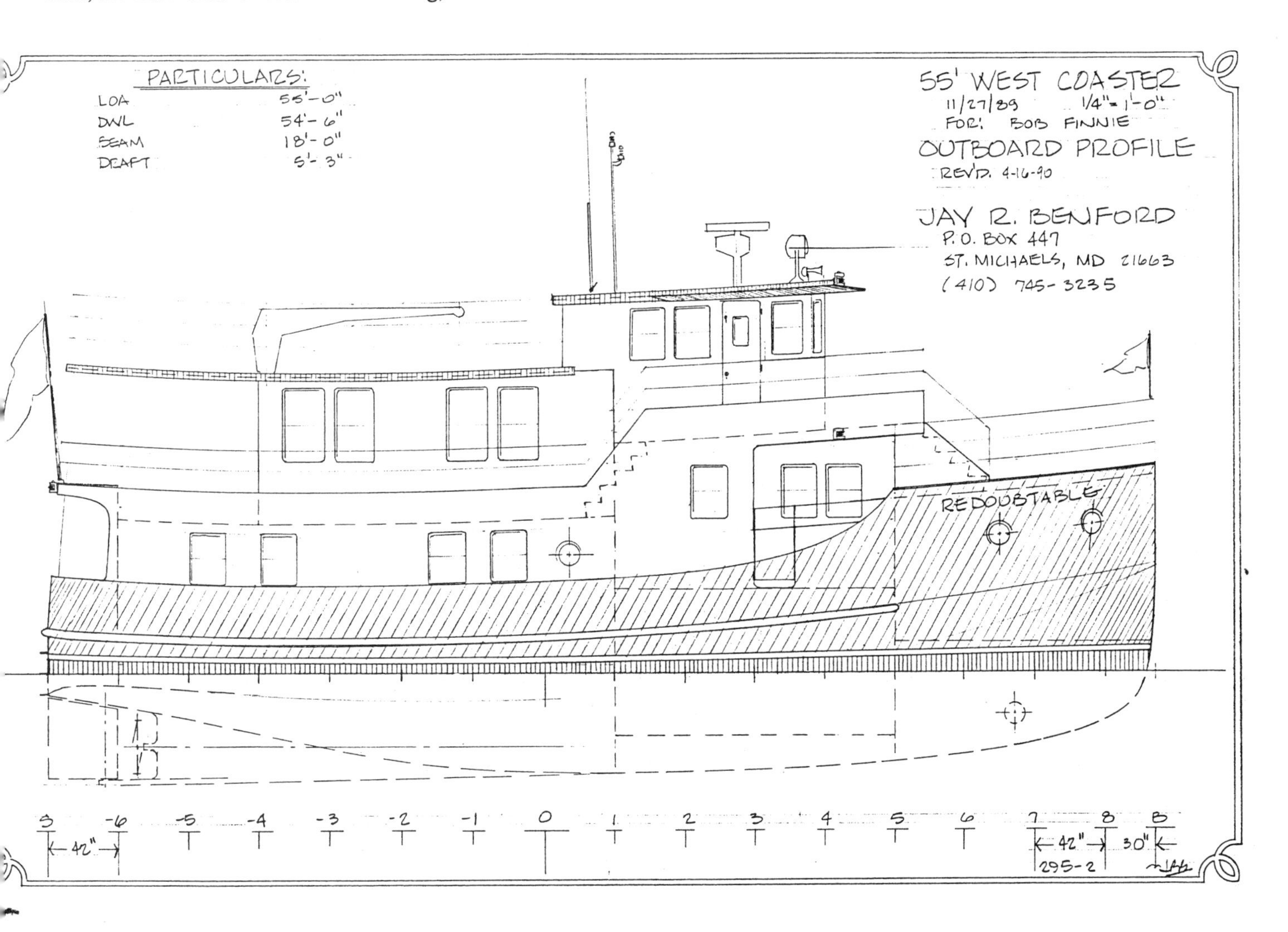

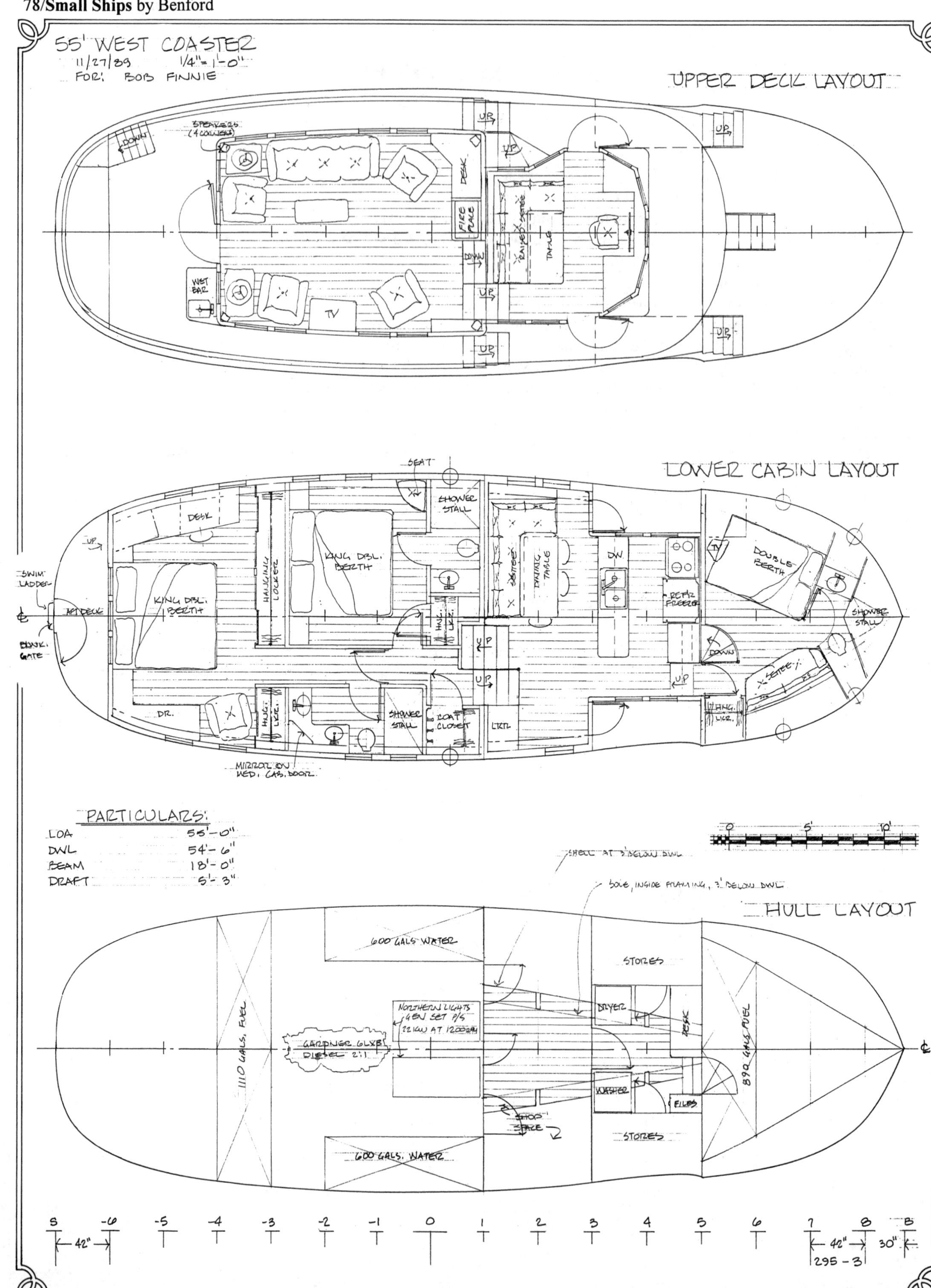
55' WEST COASTER
11/27/89 1/4" = 1'-0"
FOR: BOB FINNIE
UPPER DECK LAYOUT
LOWER CABIN LAYOUT
PARTICULARS:
LOA 55'-0"
DWL 54'-6"
BEAM 18'-0"
DRAFT 5'-3"
HULL LAYOUT
600 GALS WATER
600 GALS. WATER
1110 GALS. FUEL
890 GALS FUEL
GARDNER 6LXB DIESEL 2:1
NORTHERN LIGHTS GEN SET P/S
STORES
DRYER
WASHER
DESK
FILES
SHOP SPACE
KING DBL. BERTH
DOUBLE BERTH
SHOWER STALL
HANGING LOCKER
COAT CLOSET
DINING TABLE
SETTEE
REFR FREEZER
DW.
SWIM LADDER
AFT DECK
MIRROR ON MED. CAB. DOOR
WET BAR
TV
FIRE PLACE
RAISED SETTEE
TABLE
DESK
295-3

57½' Florida Bay Coaster

Design Number 286

1989

The brief for this design was for a boat to have a lot of the good features of the ***Key Largo***, yet without the ten feet of length devoted to carrying the Jeep® on the well deck. I wanted to keep the bow a bit finer to make the boat more easily driven, the results of which can be seen in comparing the plan views of the two boats.

The foc'sle has access to the rest of the boat only from going out onto the side deck and back aft into the saloon. It could have stairs added down to the "basement" area where the bureau is now, which would permit fore and aft traffic without going outside.

The mid-level study has a fireplace, a desk and two lounging chairs. The saloon layout is a variation on the very successful one in ***Key Largo***.

The pilothouse has the stairs to the boat deck notched into its after port corner. The bridge space forward of the pilothouse is quite wide, and has a second helm station fitted for operations in fair weather.

For the ultimate in maneuvering control, in addition to being a twin screw vessel, there are a bow and stern thrusters. The aft one is fitted in a hollow section keel. The engine room has small machine shop facilities in it.

Particulars:

Length overall	57'-6"
Length designed waterline	57'-0"
Beam	20'-0"
Draft	4'-6"
Freeboard:	
Forward	8'-0"
Waist	4'-4"
Aft	2'-0"
Displacement, cruising trim*	153,000 lbs.
Displacement-length ratio	369
Pounds per inch immersion	4,952
Water tankage	1,500 Gals.
Fuel tankage	2,500 Gals.
Headroom	6'-7"

***CAUTION:** The displacement quoted here is for the boat in cruising trim. That is, with the fuel and water tanks filled, the crew on board, as well as the crews' gear and stores in the lockers. This should not be confused with the "shipping weight" often quoted as "displacement" by some manufacturers. This should be taken into account when comparing figures and ratios between this and other designs.

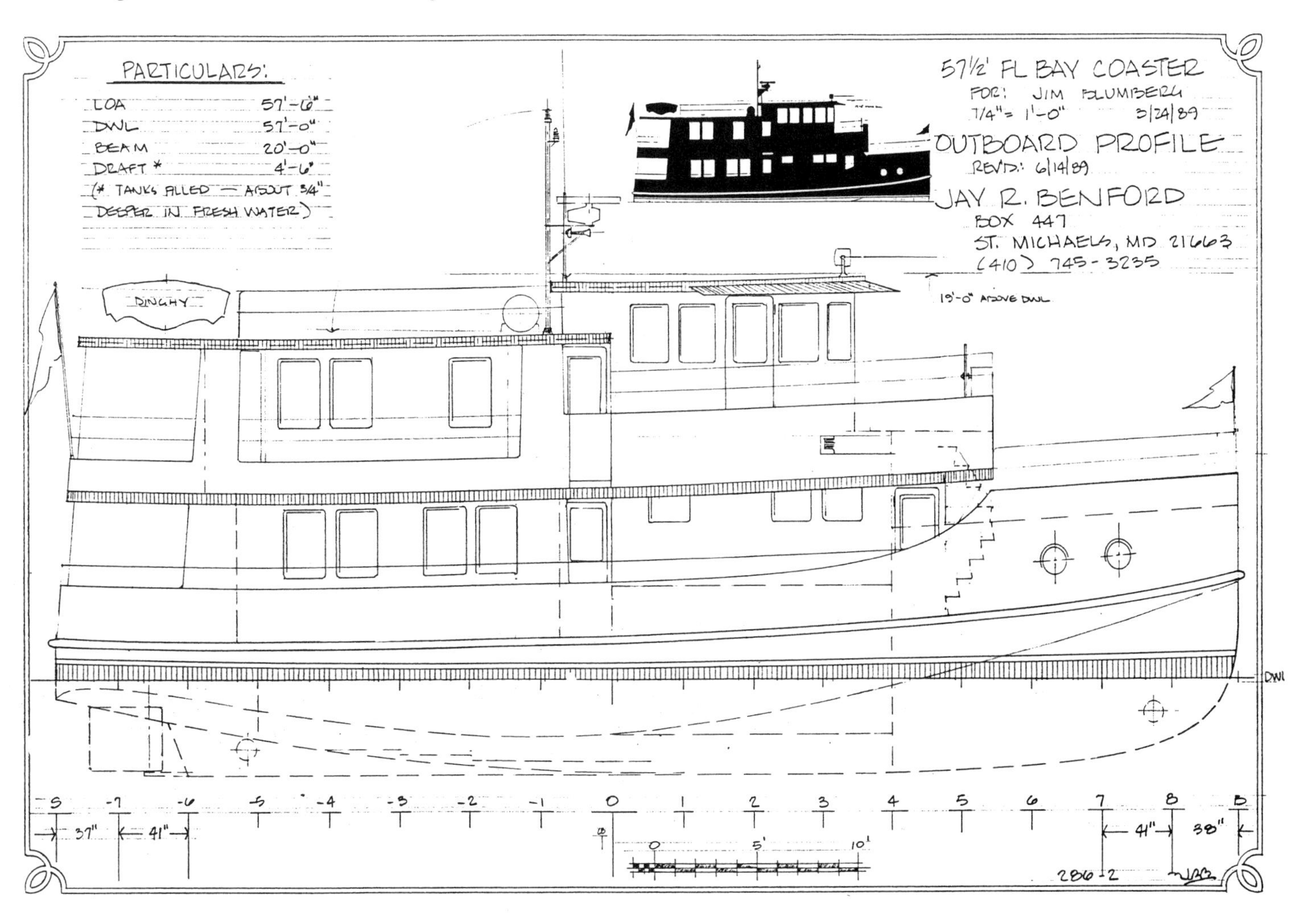

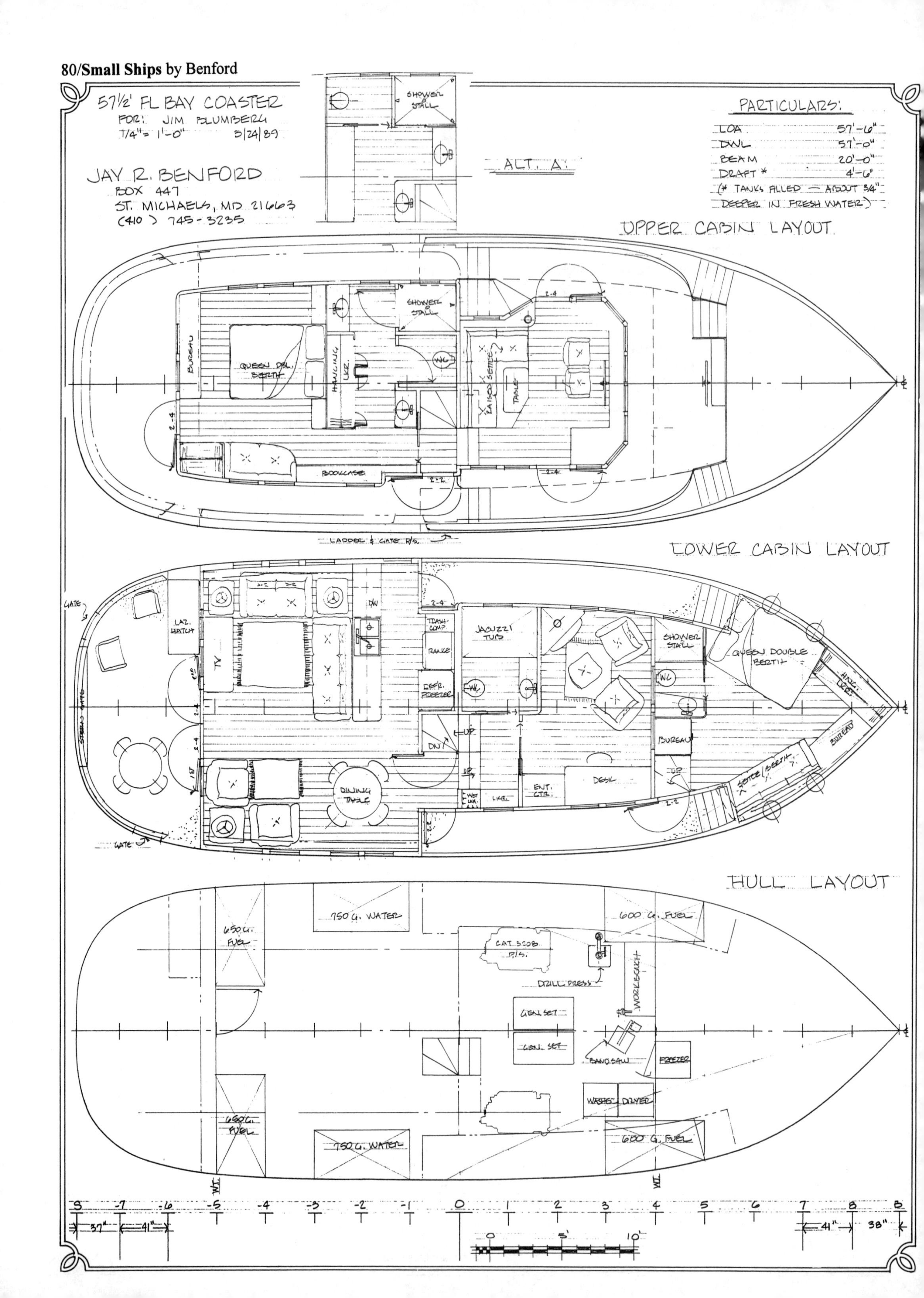

57½' FL BAY COASTER
FOR: JIM BLUMBERG
1/4" = 1'-0" 3/24/89
JAY R. BENFORD
BOX 447
ST. MICHAELS, MD 21663
(410) 745-3235
PARTICULARS:
LOA 57'-6"
DWL 57'-0"
BEAM 20'-0"
DRAFT * 4'-6"
(* TANKS FILLED — ABOUT 3/4" DEEPER IN FRESH WATER)
ALT. A
SHOWER STALL
UPPER CABIN LAYOUT
BUREAU
QUEEN DBL. BERTH
HANGING LKR.
WC
RAISED SETTEE
TABLE
BOOKCASE
LADDER & GATE P/S.
LOWER CABIN LAYOUT
GATE
LAZ. HATCH
TV
DW
TRASH COMP.
RANGE
REFR. FREEZER
JACUZZI TUB
QUEEN DOUBLE BERTH
HNG. LKR.
BUREAU
STERN GATE
DN
UP
DINING TABLE
WET LKR.
LKR.
ENT. CTR.
DESK
SETTEE/BERTH
HULL LAYOUT
750 G. WATER
650 G. FUEL
600 G. FUEL
CAT 3208 P/S.
DRILL PRESS
WORKBENCH
GEN. SET
BANDSAW
FREEZER
WASHER
DRYER
WT
37"
41"
38"
0 5' 10'

The Creation Of The Florida Bay Coasters

A look at the constructive, cooperative and collaborative relationship between designer and builder.

By Reuben Trane, President
Florida Bay Coaster Company

Any large creative project, regardless of its nature, requires a concentrated, well directed, cooperative and collaborative team effort by every member involved in the project. The Florida Bay Coasters, by their very size and complexity, are just such projects.

In the mid-1980's I had the beginnings of an idea for a liveaboard vessel with larger spaces than those normally found on a motor yacht and the capability of carrying both a good sized runabout and a full sized Jeep®. I took this germ of an idea and planted it in the fertile imagination of Jay Benford. I was familiar with much of Jay's work and felt that if anyone could make this concept work, he could.

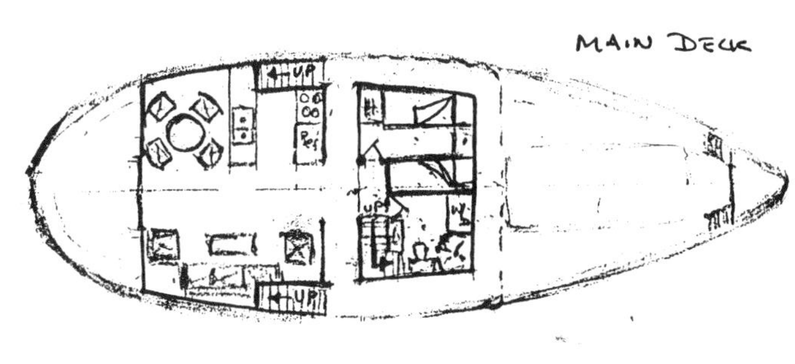

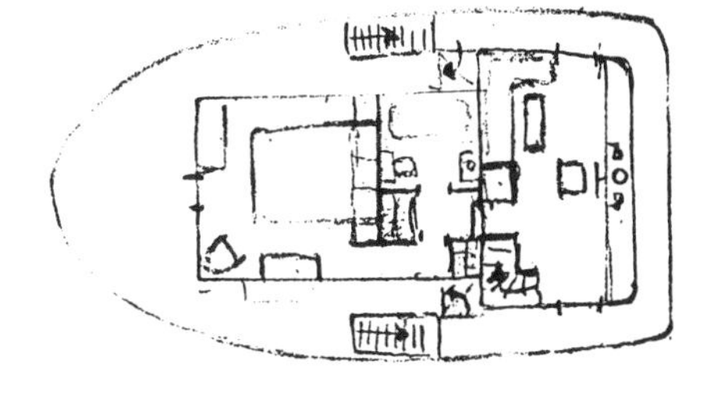

Prior to our first meeting, I sent him the following letter with some preliminary sketches for a river-type vessel, with a low, open forward deck and multi-story house aft. The length was about fifty-five feet, it had two staterooms with the master up behind the pilothouse, saloon with galley, and a stand-up engineroom. The mast and boom at the stem was to handle both the Jeep® and the skiff located in the bow.

September 22, 1986

re: Florida Bay Coaster

Dear Mr. Benford,

I studied with interest your 45' Friday Harbor Ferry. Using her as a starting point, I stretched the bow out 10' and flattened the sheer. Enclosed is a 1/8" sketch of the type of craft I am considering. It turns out not to be too dissimilar from the Chesapeake Buy Boat Conversions.

What I am striving for is as follows:

1. A craft that is ship-like, both in details and construction, rather than yacht or boat-like. (Welded of steel, aluminum or a combination thereof, with a workmanlike exterior finish.)

2. Moderate draft (4' or less).

3. Length and beam as necessary for accommodations. (About 55' x 18')

4. Twin diesel power with substantial generating plant to operate 110v. appliances (i.e. refrigeration, home style central air, stove, tv, washer/dryer, etc.) and be 100% free of marinas.

5. Deck stowage space for a small 4X4 vehicle (i.e. Mitsubishi Montero) and the means to off load it alongside to a wharf or perhaps over the bow to a beach. (i.e. crane or hoist).

6. Deck stowage space for a Rigid Bottom Inflatable and the means to off load same. Probably on the topmost deck abaft the stack.

7. Raised wheel house with seating for 4, with good visibility fore and aft.

8. Main saloon with galley. Household cabinets, appliances and furniture.

9. Master stateroom with queen sized bed, dressers and maximum view and ventilation. Master head with tub.

10. Guest stateroom with twin beds.

11. Second head with stall shower and washer and dryer.

12. Sun deck with room for deck lounges.

13. Covered deck off the main saloon, capable of being enclosed with weather cloths.

14. Exterior and interior ladders (as close to stairs as possible).

15. Engine room with easy access for daily maintenance and standing headroom.

In short, I am looking for a two-bedroom, two-bath home with carport, that I can take with me along the shorelines and rivers of the U.S. Maintenance should be straight forward and kept to a minimum.

There is absolutely nothing like this on the market today (perhaps with good reason), but I feel that properly promoted, and at a reasonable price, a number of units could be sold.

I look forward to talking with you more when I am in Annapolis for the Sailboat Show.

Yours truly,
Reuben Trane

Several of the key parameters were set at this meeting: the boat was to be shallow draft, stable and seaworthy. She was to have a well raised pilothouse (it's been my experience that most owners operate from the fly bridge, regardless of weather), a bright master stateroom on the upper deck with vistas on three sides (this space was modeled on a beach cottage I once rented), a roomy engine room and a great room with full sized kitchen. Most of these features are still found in the Coasters being built today.

Jay started faxing me sketches within the week. I jotted down my ideas and faxed them back. We discussed ideas daily on the phone, the latest sketches in front of us; a process that we still continue on both the creation of new Coasters and the modification of existing ones. The project gradually took shape. The first renderings were for a rather modern looking, mini-freighter with a well deck. Jay insisted that we add a raised foc'sle, both for the lines and the extra freeboard forward. A by-product was the addition of a third cabin up under the raised foredeck. Details now had to be worked out; traffic flow, engine room layout, stair cases, etc. The faxes flew back and forth.

We reached the stage in the design process where we needed more information to continue; the essential creative process was complete... we now needed building plans. Jay and his staff took to the drawing tables and produced a complete set of engineering drawings. As mechanical systems were resolved, other questions surfaced. We sometimes found that certain of our earlier ideas could be improved. A stair needed to be moved a few inches, or a door swing relocated. The faxes buzzed back and forth!

As so often happens in this type of creative process, Jay and I sometimes let our ideas get away from us. When the plans were complete and I priced out the final boat, now a sixty-footer, it was significantly over budget. Back to the drawing table. We both decided that by stacking the skiff over the Jeep®, and turning them athwartships, and knocking off ten feet of length and two feet of beam that we could bring the project back to financial reality. The down-sizing worked fine in both those dimensions, but one thing we couldn't change was height. The resulting fifty-footer was a bit high for her length. It was at this point that Jay suggested we go with more character in her style, and he gave her the look of a turn of the century coaster. And, thus was born the first of the Florida Bay Coasters, the fifty-foot ***Florida Bay***.

The relationship between builder and designer is not over, however, when the plans are delivered and the keel laid. Some things that look great on paper are not just the same in real life. During construction, Jay and I communicated on a regular basis, solving problems, making changes as necessary and incorporating the improvements that we thought up. Whenever possible, Jay would visit the yard to get a first-hand look at how everything was coming together. Many notes were made on how to improve the "next one", when and if we got to do another one.

From her very first exposure at the Fort Lauderdale International Boat Show in October 1987, the ***Florida Bay*** was an immediate popular success. Throngs of curious boaters came through this truly unique vessel. Her final lines carried the classic Jay Benford imprimatur that, added to her stately height and, of course, with the Jeep® on board, made her a stand-out in any crowd. Whereas I had originally intended her to be offered on a limited production basis, the feedback from our many visitors told us otherwise. Many had been waiting for just such a boat to be marketed. As experienced boaters, they all had some of their own ideas as to what would make her the "ideal" liveaboard cruiser.

Shortly after the first showing of the ***Florida Bay***, Jay and I got together and went over the reactions of the boating public to our creation. Some wanted smaller, some larger, some wanted more offshore capability while others wanted more staterooms. We decided to create a study plan booklet reflecting many of these ideas plus some of our own. Jay and I brainstormed, coming up with a number of layouts and profiles. We published a booklet of them that soon sold out. With many more ideas generated between us in the intervening years, the book was soon outdated. Jay has gathered these together and printed them in his book, **Small Ships**.

As it has turned out, we've built a number of Coasters, with each subsequent Coaster having different features; each a custom creation of its own. Jay and I continue to work much the same; sketches fly back and forth on the fax and ideas flow over the phone lines. Added to the mix are the buyers, each with his and her own definitive ideas as to how "their" Coaster is to be designed and built. Today we deal with details from diving compressors to tilt-down stern gates to built-in ovens to dumb waiters to hardwood flooring and trim. Each Coaster is a unique vessel.

And the creative process continues as the Florida Bay Coasters continue to be built, the fleet expanding in both numbers and diversity. The team effort of builder and designer produces for the owners a highly refined, quality vessel, uniquely fitted to their exact needs and desires. This team effort has resulted not only in the fleet of Florida Bay Coasters but a whole new type of multi-story liveaboard vessels, designed to make waterborne living a whole new experience.

60' Freighter Yachts

Design Number 254

1987 & 1988

Reuben's letter on the preceding pages is a good and clear statement of what he wanted. This became the vision for the first Florida Bay Coaster designs. In developing this 60-footer, and on to the 50' ***Florida Bay*** which was the first one built, we had a lengthy and pleasurable exchange of ideas. This has continued with the evolution and development of further Coasters, with each of us having ideas for alternate ways to achieve some specific request made by clients. This diversity has made for a lot of interesting boats.

The sketch below shows some of the evolution of the ideas we looked at in coming up with the final design. It has a different approach to the outside stairs, similar to what we later used on one side of ***Key Largo***.

The hull form we drew for the 60-footer is different than the other Coasters we've built, and it would be interesting to build one and compare how it worked out. We'd initially gone to a different stern on the ***Florida Bay*** in the theory that people would want to drive them nearer the limits of hull speed. As it turns out, they've all proved very comfortable at more modest speeds (speed-length ratios of 1.0 to 1.2) and there is little demand for the boats to run faster. This seems to bear out my theory that once you're aboard one of these boats, you've already arrived where you want to be. Any further travel is for the pleasure of seeing some more of the country. A leisurely (and economical) pace seems to suit this philosophy quite well. The finer stern on this boat might be a good choice for these moderate speeds.

The second 60-footer, the ***Karen Marie II***, is a conceptual study commissioned to look at the idea of making one with a full walk-around lower deckhouse and a full width upper house.

The upper house has the owner's stateroom, head, and home office in it. The after decks are connected by a series of stairs that go from the boat deck down through the accommodation decks to the steering compartment, like we built into the ***Key Largo***.

The bridge around the pilothouse overhangs the car deck, giving a bit of shade and shelter there. The sides of the bridge have stairs to the boat deck and the port side connects back down to the main deck also.

The lower house has notches in the sides for the doors, letting them swing clear inside of the bulwarks, which tumblehome. For my own use, I'd prefer to have the walk-around on the upper deck level and keep the lower house wider to maximize the saloon size. Different boats for different folks....

Particulars:

Length overall	60'-0"
Length designed waterline	58'-8"
Beam	20'-0"
Draft	4'-0"
Freeboard:	
Forward	8'-3"
Least	1'-7"
Aft	2'-0"
Displacement, cruising trim*	120,000 lbs.
Displacement-length ratio	265
Prismatic coefficient	.65
Water tankage	1,000 Gals.
Fuel tankage	1,500 Gals.
Headroom	6'-8"

***CAUTION:** The displacement quoted here is for the boat in cruising trim. That is, with the fuel and water tanks filled, the crew on board, as well as the crews' gear and stores in the lockers. This should not be confused with the "shipping weight" often quoted as "displacement" by some manufacturers. This should be taken into account when comparing figures and ratios between this and other designs.

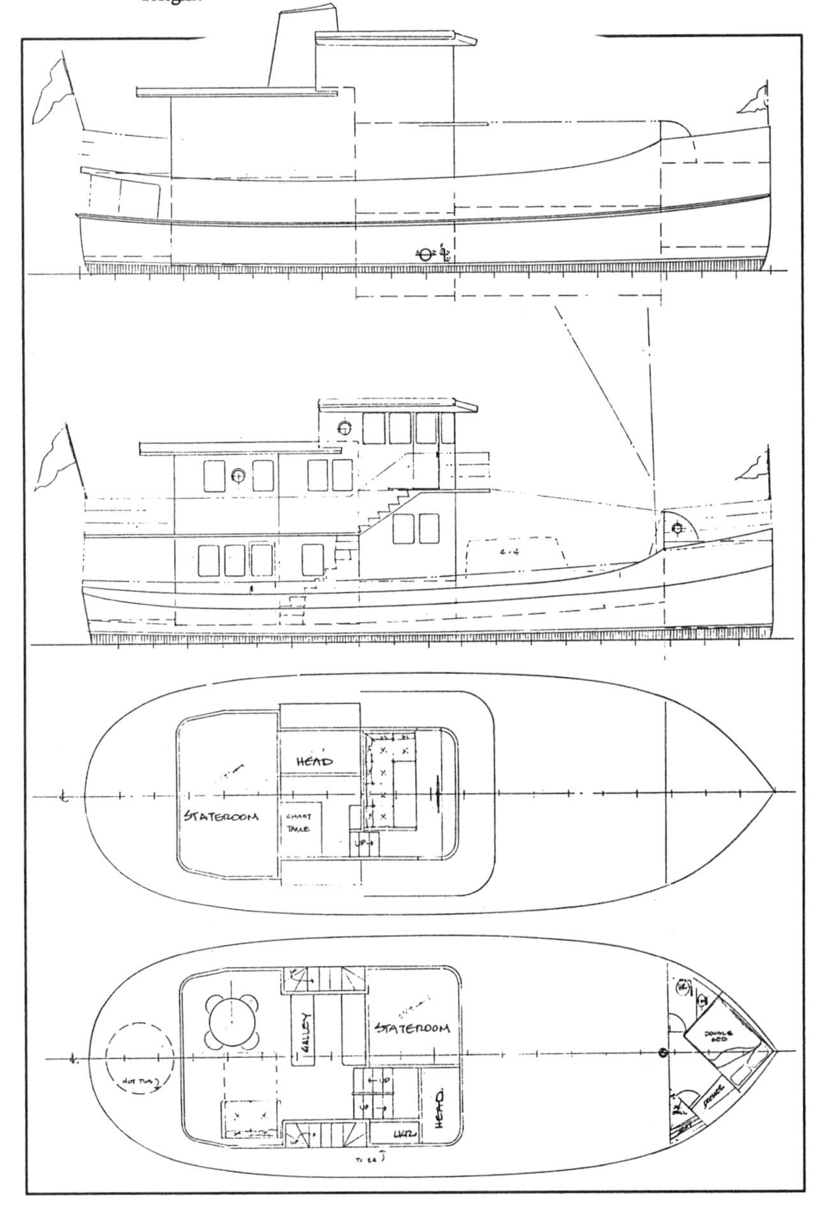

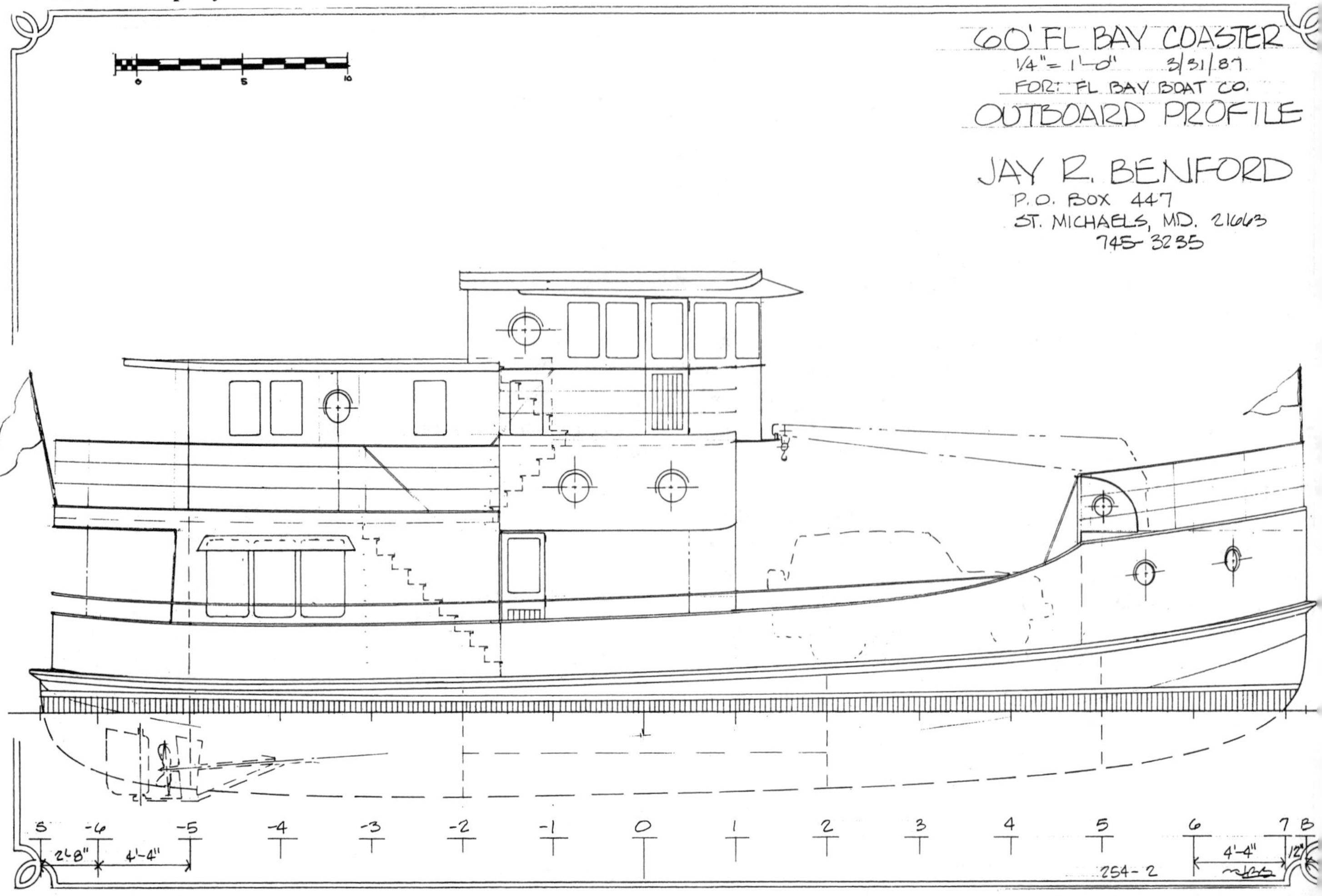

60' FL BAY COASTER
1/4" = 1'-0" 3/31/87
FOR: FL BAY BOAT CO.
OUTBOARD PROFILE

JAY R. BENFORD
P.O. BOX 447
ST. MICHAELS, MD. 21663
745-3235

254-2

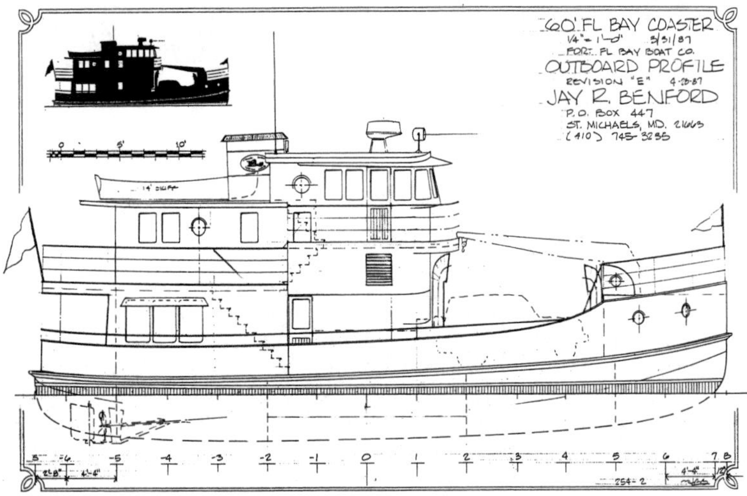

60' FL BAY COASTER
1/4" = 1'-0" 3/31/87
FOR: FL BAY BOAT CO.
OUTBOARD PROFILE
REVISION "E" 4-18-87
JAY R. BENFORD
P.O. BOX 447
ST. MICHAELS, MD. 21663
(410) 745-3235

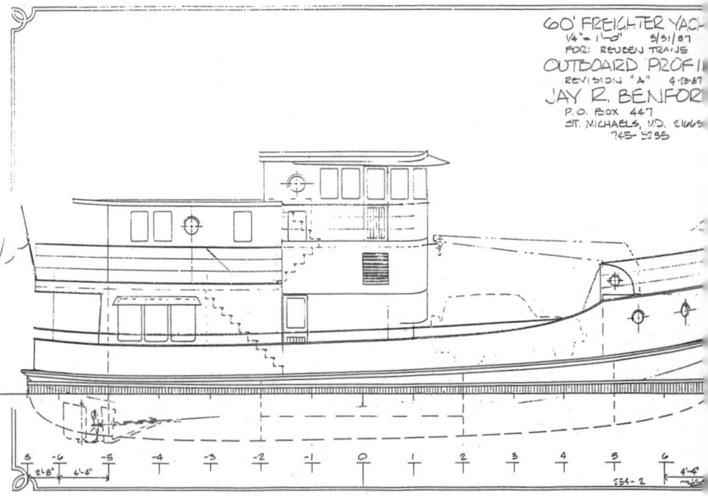

60' FREIGHTER YACHT
1/4" = 1'-0" 3/31/87
FOR: REUBEN TRANE
OUTBOARD PROFILE
REVISION "A" 4-18-87
JAY R. BENFORD
P.O. BOX 447
ST. MICHAELS, MD. 21663
745-3235

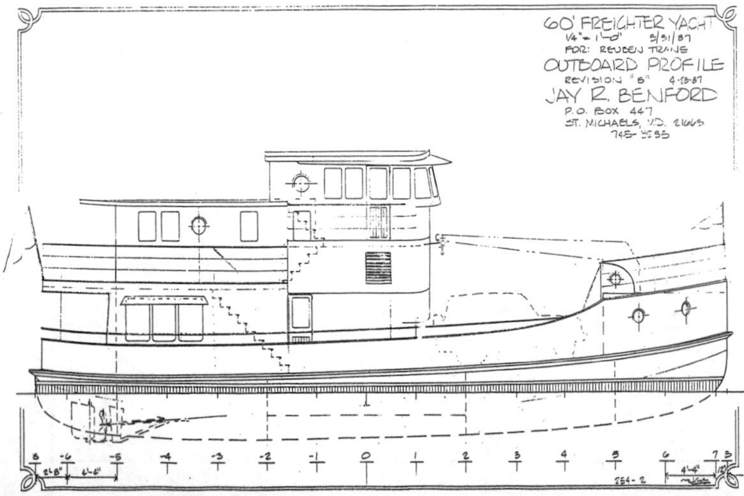

60' FREIGHTER YACHT
1/4" = 1'-0" 3/31/87
FOR: REUBEN TRANE
OUTBOARD PROFILE
REVISION "B" 4-18-87
JAY R. BENFORD
P.O. BOX 447
ST. MICHAELS, MD. 21663
745-3235

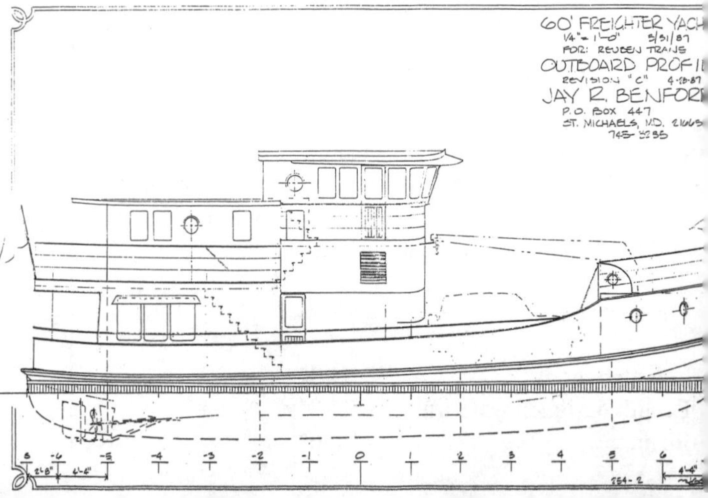

60' FREIGHTER YACHT
1/4" = 1'-0" 3/31/87
FOR: REUBEN TRANE
OUTBOARD PROFILE
REVISION "C" 4-18-87
JAY R. BENFORD
P.O. BOX 447
ST. MICHAELS, MD. 21663
745-3235

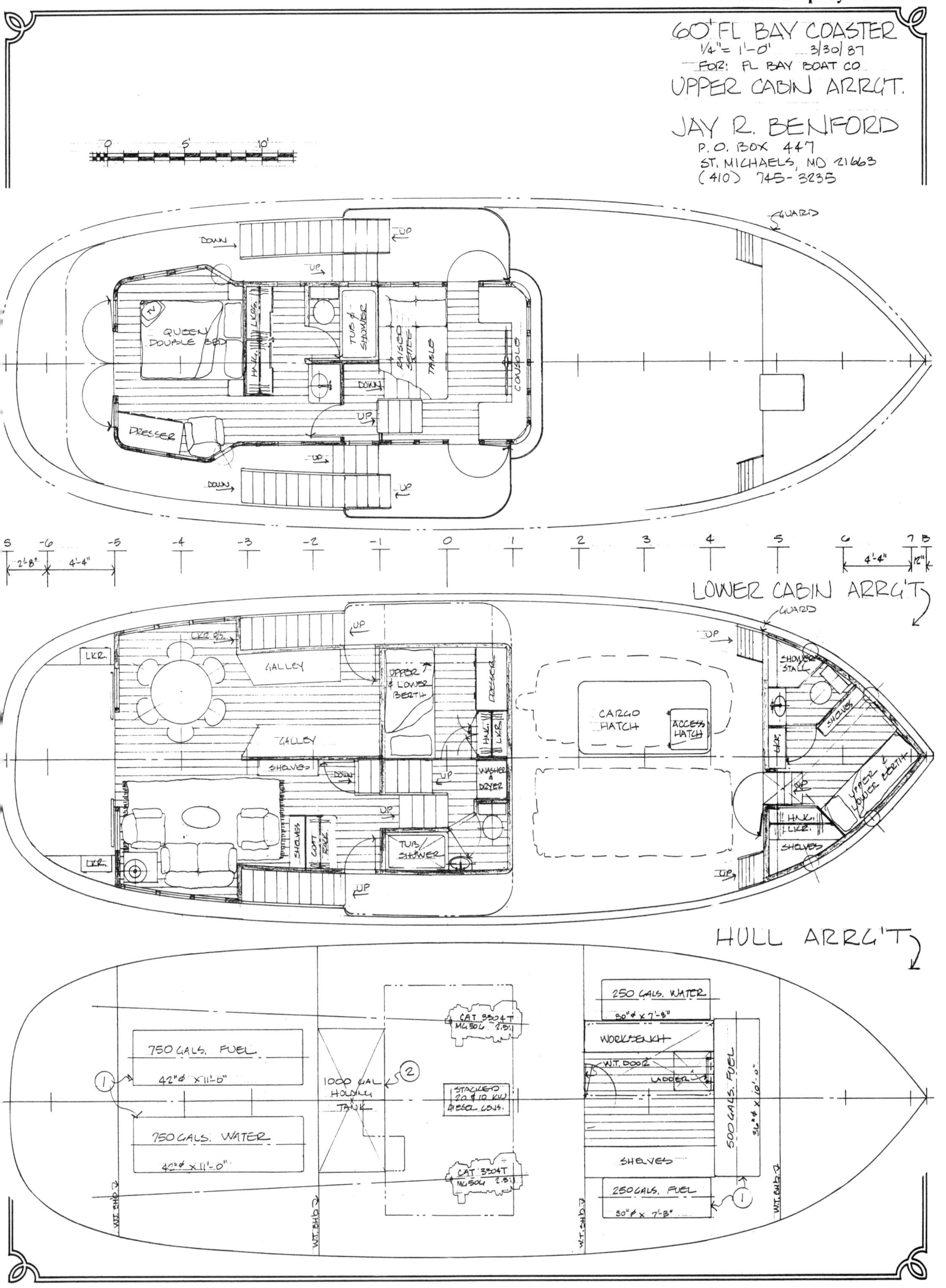
60' FL BAY COASTER
1/4" = 1'-0" 3/30/87
FOR: FL BAY BOAT CO
UPPER CABIN ARRG'T.
JAY R. BENFORD
P. O. BOX 447
ST. MICHAELS, MD 21663
(410) 745-3235
0 5' 10'
GUARD
DOWN
UP
QUEEN DOUBLE BED
HNG. LKRS.
TUB & SHOWER
RAISED SETTEE
TABLE
CONSOLE
DRESSER
LOWER CABIN ARRG'T
GUARD
LKR.
LKR. O/S.
GALLEY
UPPER & LOWER BERTH
DRESSER
CARGO HATCH
ACCESS HATCH
SHOWER STALL
SHELVES
HNG. LKR.
WASHER & DRYER
TUB/SHOWER
COAT RACK
UPPER & LOWER BERTH
HULL ARRG'T
250 GALS. WATER
30"ø x 7'-3"
CAT 3304T MG506 2.5:1
WORKBENCH
750 GALS. FUEL
42"ø x 11'-0"
W.T. DOOR
LADDER
1000 GAL HOLDING TANK
STACKED 20 & 10 KW DIESEL GENS.
500 GALS. FUEL
36"ø x 10'-0"
750 GALS. WATER
42"ø x 11'-0"
SHELVES
CAT 3304T MG506 2.5:1
250 GALS. FUEL
30"ø x 7'-3"
W.T. BHD.

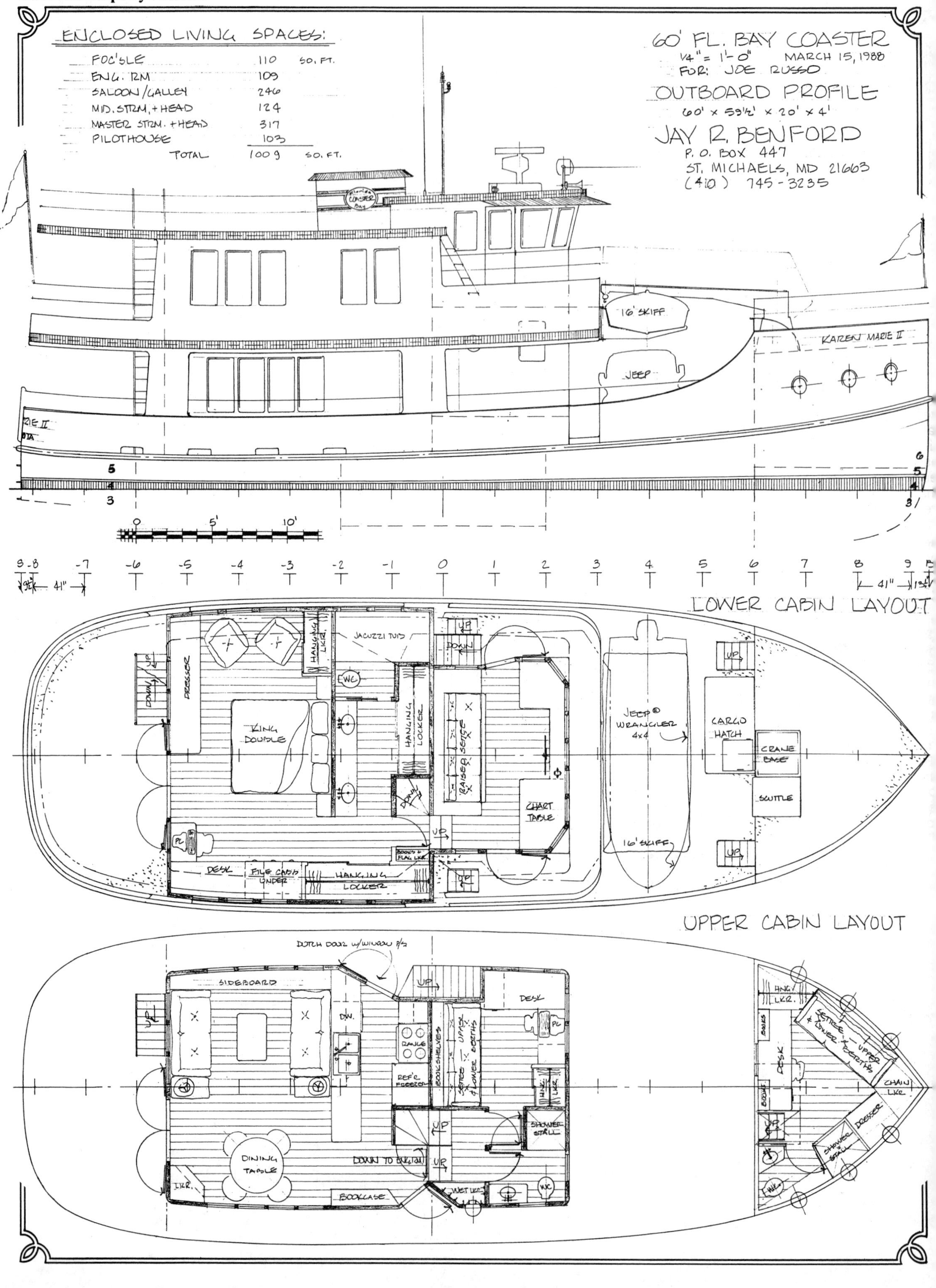

ENCLOSED LIVING SPACES:

FOC'SLE	110	SQ. FT.
ENG. RM	109	
SALOON/GALLEY	246	
MID. STRM. + HEAD	124	
MASTER STRM. + HEAD	317	
PILOTHOUSE	103	
TOTAL	1009	SQ. FT.

60' FL. BAY COASTER
1/4" = 1'-0" MARCH 15, 1988
FOR: JOE RUSSO

OUTBOARD PROFILE

60' x 59½' x 20' x 4'

JAY R. BENFORD
P. O. BOX 447
ST. MICHAELS, MD 21663
(410) 745-3235

16' SKIFF
KAREN MARIE II
JEEP

0 5' 10'

LOWER CABIN LAYOUT

DRESSER
DOWN UP
KING DOUBLE
HANGING LKR.
JACUZZI TUB
WC
HANGING LOCKER
RAISED SETTEE
DOWN
UP
CHART TABLE
JEEP® WRANGLER 4x4
16' SKIFF
CARGO HATCH
CRANE BASE
SCUTTLE
PC
DESK
FILE CABS UNDER
BOOKS + FLAG LKR.
HANGING LOCKER

UPPER CABIN LAYOUT

DUTCH DOOR W/WINDOW P/S
SIDEBOARD
UP
DW.
RANGE
REF'R FREEZER
BOOKSHELVES
SETTEE + LOWER BERTHS
UPPER BERTHS
DESK
PC
HNG. LKR.
SHOWER STALL
DINING TABLE
DOWN TO ENGINE
LKR.
BOOKCASE
WET LKR.
WC
HNG. LKR.
BOOKS
DESK
SETTEE + LOWER - UPPER BERTHS
CHAIN LKR.
DRESSER
SHOWER STALL
WC

Personal Freighters: A New Look At Waterfront Living

by Keith Walters

Combine every possible con- ·nience from your home with ·ery toy you need to enjoy the ·ater, pack them into a steel- ·lled, tramp-steamer looking ·ssel that you can pull up to a ·ach like a landing craft, and ·hat do you have? A "Personal ·eighter."

Jay R. Benford, the designer of ·e such floating home that was ·cked at the Maritime Museum St. Michaels recently, showed · around the 65' x 20' **Key Largo.** naval architect for 27 years, ·nford lived on a 34' topsail ·tch for 10 years. He has de- ·gned everything from working ·gs to power and sail yachts up to ·1 feet. He brokers yachts, ·t only those of his design. He ·o wrote a book, "Sailing ·chts," and several pamphlets his boat designs.

Benford says his designs of ·hort, fat, round live-aboard ·ats like his 'Friday Harbor ·erry' series," prompted a call ·om Reuben Trane, President · the Florida Bay Boat Company. Trane had been building small Mud Hen sailboats, but wanted a retirement boat for himself that included every shoreside convenience in a boat that could be handled by two people. If it suited him, he reasoned, he could build the same boat for others, too.

The result of this designer/builder team effort is the **Key Largo,** an aptly-named vessel that is only slightly smaller than the Florida Key it was named for. The high bow, combined with wrap-around windows in the high pilot house and safety rails all around, make it look like a tramp steamer. Overall, it conjures up visions of coasting freighters.

A hydraulic crane on the bow can off-load the 17' Mako outboard runabout that serves as a dinghy. Parked under the Mako is a Jeep, for shoreside sightseeing or provisioning trips. Where do you find a place to off-load a Jeep? "Boat-launching ramps," says Benford. Simply stick the nose of the **key Largo** into a boat launching ramp and swing the Jeep ashore. The boat only draws 4-½' of water. The jeep is parked on a hatch that covers a king-sized hold reminiscent of ocean-going freighters. But these are all features one can see standing on the dock. For a closer look, Benford toured us through the vessel that stands two stories above the main deck.

We sat on the fantail "deck," much like the deck on a waterfront home, with regular porch furniture, and talked. Benford explained that the transom lowers to the water to become a diving or fishing platform. There is an after deck fresh water shower.

How long did it take to design and build such a craft? "We started in the spring with a blank sheet of paper," he said, "it was completed in October." With all options, the 150,000 lb. displacement steel-hulled boat will deplete your checkbook $750,000

Just forward of the after deck, Benford's wife, Dona, entertained their two daughters, Elizabeth (age 3), and Lorena (18 months), in a bright, airy saloon with two 8' settees. An ingenious table in the saloon pulls out to seat eight diners. There is storage under, over, and behind everything on the boat that permits a cubic foot or two of space.

The galley is open to the saloon, and limed oak galley cabinets repeat the wood finish throughout that makes the boat's interior light and airy. The galley includes every convenience found in shoreside kitchens: Microwave oven, 22 Cu. Ft. refrigerator/freezer, electric range with self-cleaning oven, dishwasher, and a trash compactor.

"Let's go below," Benford offered. In the "basement" of the floating mansion lurks the machinery that makes it move and provides electricity and water. Two Caterpillar 3208 Turbo diesel engines, each turning a 32" prop, sit in sound-muffling compartments on either side of the underdeck 2,5000-gallon fuel tank. Two diesel 20 KW generators sup-

·n a tour of the "Personal Freighter" Key Largo it is easy to imagine ·r away places and hideaway coves with all the comforts.

© photo by Keith Walters

Jay Benford relaxes aboard "Key Largo" tied up at the Chesapeake Bay Maritime Museum dock in St. Michaels.

© photo by Keith Walters

Personal Freighter

ply electricity, though one will provide all the power two people need. Lights on two commercial power panels signal everything is humming smoothly. Two 40-gallon hot water heaters are supplied from the 1,250 gallon fresh water tank replenished by a water-maker at sea, and the 500-gallon holding tank has sanitary treatment. Also below is a clothes washer and dryer, plus a chest freezer.

Each compartment bulkhead below has a watertight ship's door. Forward of the engine room there is a hold big enough to carry almost anything the owners might want on an extended voyage. (The boat has a 1,600-mile cruising range).

Above, the craft has many features designed for comfortable living and entertaining. Forward of the saloon and galley, the study has its own full bathroom with tiled tub and shower, not a cramped "head." A desk and leather sofa-bed complete the study.

On the second level, above the saloon, the owner's suite is as luxurious as any ashore: The king-size bed is below a clear skylight so one can watch the stars or moon on clear nights. The skylight can be covered with cushions to make seating on the open deck above. Mirrors reflect the sofa and dresser, and ample windows make the suite light and airy. A paddle fan above the bed provides air on nights when it is too nice to close up and turn on the air conditioning. The master suite includes a dressing room plus a tiled bath with a whirlpool tub. For those romantic moonlit nights, there is a rear "porch" or deck just aft the suite.

Forward of the master suite is the pilot house. From two stories up, everything on the water below is visible through the ample windows. When sea spray coats the windows there is a spinning "Clear-Vu" screen that whirls the water away, giving at least a small clear view ahead. Every manner of electronic navigation gear is there to help the sailor steer the craft; find his way in day, night, or fog; communicate with others; or determine how much water is under the keel. All in air-conditioned or heated comfort. Behind the captain, a large U-shaped raised settee with a table seats at least six assistant captains. An under-counter refrigerator makes it unnecessary to go below to the galley for refreshments.

Benford explained a feature that makes it possible for only two people to operate the **Key Largo.** There are five stations aboard with engine controls and a toggle for the bow thrusters. In close quarters, or when docking, the rudder is almost useless, so the 36" stainless steel wheel isn't needed to maneuver the boat. Twin shifters control the main engines for fore and aft movement and another lever controls the bow thrusters for steering. One person at the controls, one handling dock lines. Easier than docking a runabout.

At the steering station in the pilot house, one looks over the runabout and the Jeep below it, toward the crane mounted on the bow. Under the crane is the guest cabin, "Comfortable, but not so much so that guests might overstay." Two berths, a dresser and hanging locker, plus a full bath, make this cabin very liveable for guests, or for the owners if they chartered the boat. There are two ladders (stairways) leading to this cabin; one to the outside deck by the Jeep, and one to the cargo bay below it — in case rough weather dictates an inside passage.

I almost forgot the six-person hot tub on the very top deck, with seating next to it on cushions that cover the skylight of the master suite below. There is a gas grill up there, too, should you get hungry from all that hot-tubbing, and safety railings for the kids.

If this all sounds like a luxurious way of gunkholing along coastal waterways while taking everything imaginable with you — well, it is. There is a total of 1,300 square feet of heated, air-conditioned space. Double that space when you add decks. There are wet bars, beds, bath rooms, and refrigerators nearly everywhere.

Benford sees his design as a "waterfront home." Buyers of his Florida Bay Coaster designs, from 45' to 65', will be people who want a second home, or retirees who want to live in the north in summer and south in winter, and "take their home with them."

If one travels north and south along the coast, sooner or later there will be an ocean passage. How does the **Key Largo** handle in rough water? "We were comfortable in eight foot beam seas," Benford says, "and she only rolled 10-degrees." Now, to dye-in-the-wool sailboaters, 10-degrees is only where the fun begins, but power folks like more comfort than rag sailors who enjoy an open cockpit in trashy weather with one rail awash.

Besides, owners of **Key Largo** would object to getting their Jeep sloshed.

The Benford family enjoys showing off the completely equipped bu[t] very comfortable pilot house with its 36" wheel aboard Key Largo.

© photo by Keith Walte[rs]

King size bed, loads of storage, paddle fan and skylight overhea[d] lots of light, water view all around, what more could one wish for in [a] master bedroom, or owner's cabin? © photo by Keith Walte[rs]

Jay Benford hands his wife, Dona a cup of coffee from the galley a[nd] daughters Lorena (18 months), and Elizabeth (age 3) amuse themselve[s] at least as comfortably as they could ashore. © photo by Keith Walte[rs]

65' Florida Bay Coaster *Key Largo*

Design Number 269
1988

The 65' **Key Largo** was the second freighter built by Florida Bay Coaster Company. In her, we took the opportunity to try out many of the ideas for improvements we'd generated in using the 50-footer. All her cabins are quite a bit larger. She has luxury additions like the skylight over the master stateroom and the hot tub on deck in place of the stack. She has full headroom in the hold area, opening up options like a captain's cabin in this space, and greatly increased tankage and a watermaker, making possible an even more extended time away from shore.

The deck space is much greater than on the 50, too, and this allows for the port side stairs amidships from the waist deck to the pilothouse. There are also stairs aft from the lazarette/steering compartment all the way up to the "fun deck" for quick movement when making landings or in moving from one area of the entertaining to another. The "fun deck" has the barbeque, hot tub, picnic tables and chairs and plenty of lounging space.

The ***Key Largo*** is currently operating as a charter vessel, carrying four to six passengers on trips in the Florida area in the winter months and in Maine in the summer, with coastwise trips in transit in between.

The color photos at the beginning of this book show how spacious she is. The Florida Bay Coaster Company has put out some excellent descriptive literature and a VHS video of the boats in use. Drop us a line or give us a call and we'll put you in touch with them.

Particulars:

Length overall	64'-11½"
Length designed waterline	64'-6"
Beam	20'-0"
Draft	4'-6"
Freeboard:	
Forward	9'-1"
Waist	4'-3"
Aft	2'-9"
Displacement, cruising trim*	175,000 lbs.
Displacement-length ratio	291
Prismatic coefficient	.614
Pounds per inch immersion	5,590
Water tankage	1,400 Gals.
Fuel tankage	2,500 Gals.
Headroom	6'-9"
GM (from inclining test)	4.8' (The stability curve below shows how good her stability is; positive all the way to 180°!)

***CAUTION:** The displacement quoted here is for the boat in cruising trim. That is, with the fuel and water tanks filled, the crew on board, as well as the crews' gear and stores in the lockers. This should not be confused with the "shipping weight" often quoted as "displacement" by some manufacturers. This should be taken into account when comparing figures and ratios between this and other designs.

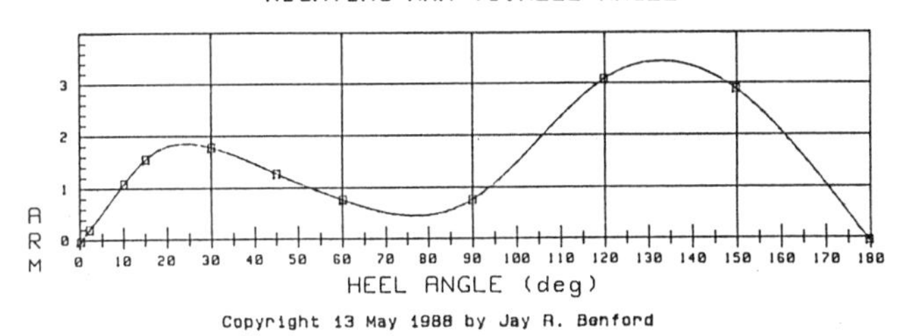

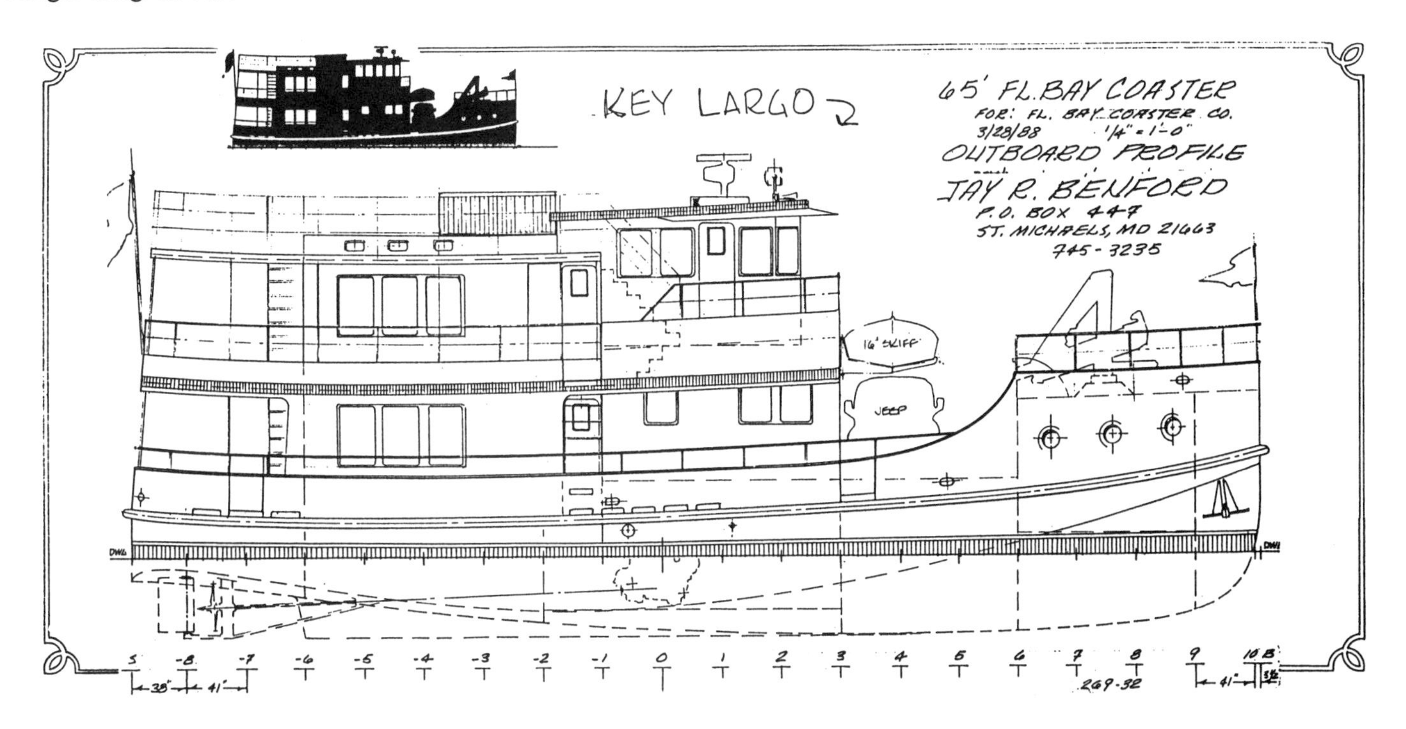

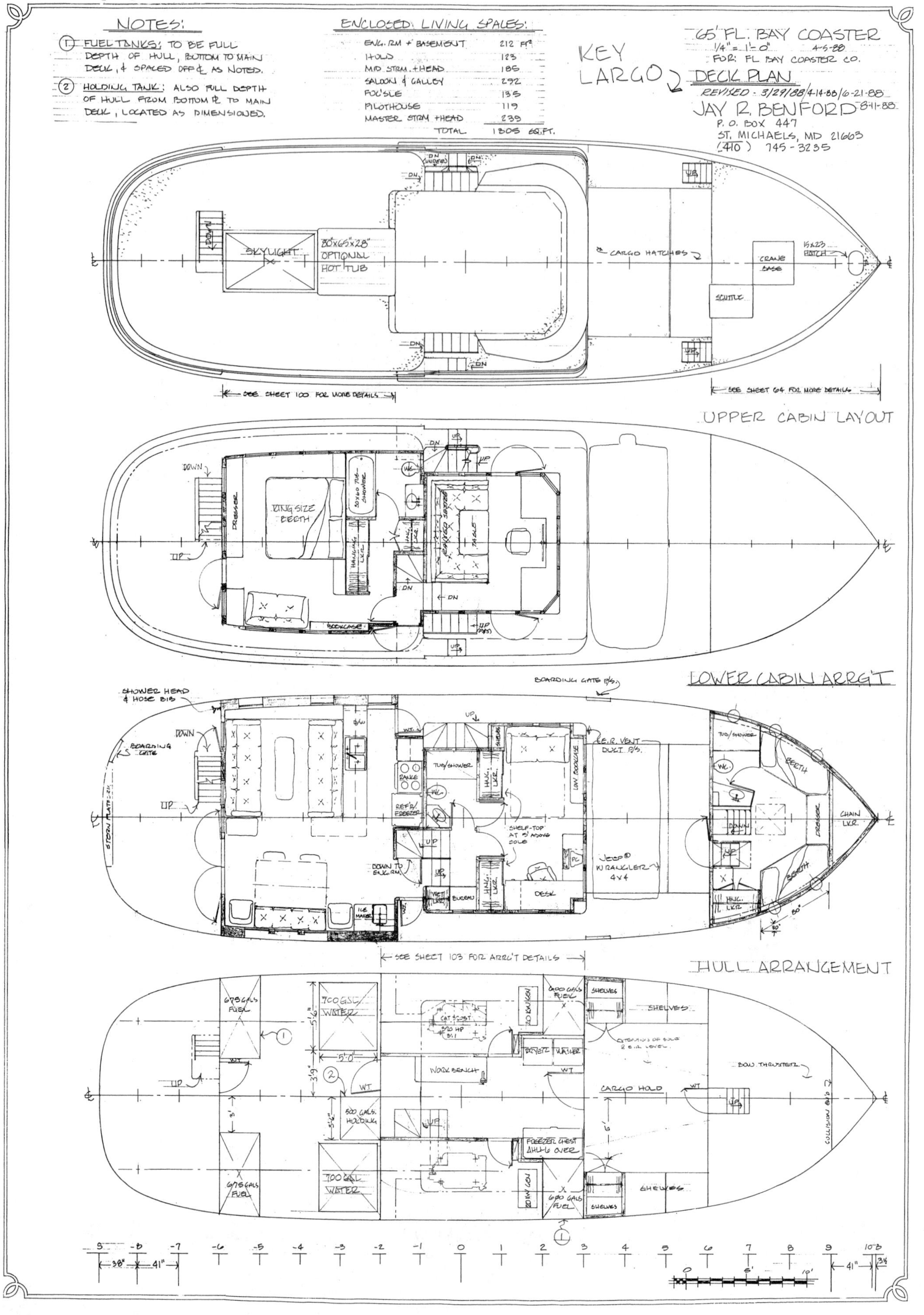

NOTES:

1. FUEL TANKS: TO BE FULL DEPTH OF HULL, BOTTOM TO MAIN DECK, & SPACED OFF ℄ AS NOTED.
2. HOLDING TANK: ALSO FULL DEPTH OF HULL FROM BOTTOM ℄ TO MAIN DECK, LOCATED AS DIMENSIONED.

ENCLOSED LIVING SPACES:

ENG. RM + BASEMENT	212 FT²
HOLD	123
MID STRM. + HEAD	185
SALOON & GALLEY	292
FOC'SLE	135
PILOTHOUSE	119
MASTER STRM + HEAD	239
TOTAL	1305 SQ. FT.

KEY LARGO

65' FL. BAY COASTER
1/4" = 1'-0" 4-5-88
FOR: FL BAY COASTER CO.
DECK PLAN
REVISED - 3/29/88/4-14-88/6-21-88 8-11-88

JAY R. BENFORD
P. O. BOX 447
ST. MICHAELS, MD 21663
(410) 745-3235

SKYLIGHT — 80"x65"x28" OPTIONAL HOT TUB — CARGO HATCHES — CRANE BASE — SCUTTLE — 15x23 HATCH

SEE SHEET 100 FOR MORE DETAILS
SEE SHEET 64 FOR MORE DETAILS

UPPER CABIN LAYOUT

KING SIZE BERTH — DRESSER — 30x60 TUB-SHOWER — HANGING LKR. — BOOKCASE — RAISED SETTEE — TABLE — W.C.

LOWER CABIN ARRG'T

SHOWER HEAD & HOSE BIB — BOARDING GATE — STERN PLATFORM — BOARDING GATE P/S. — RANGE — REFR/FREEZER — TUB/SHOWER — HNG. LKR. — LOW BOOKCASE — E.R. VENT DUCT P/S. — SHELF-TOP AT 5' ABOVE SOLE — DOWN TO ENG. RM. — BUREAU — DESK — JEEP® WRANGLER 4x4 — ICE MAKER — TUB/SHOWER — BERTH — DRESSER — CHAIN LKR. — HNG. LKR.

SEE SHEET 103 FOR ARR'GT DETAILS

HULL ARRANGEMENT

675 GALS FUEL — 700 GAL WATER — 500 GALS HOLDING — CAT 3208T 300 HP 3:1 — WORK BENCH — 20 KW GEN — 600 GALS FUEL — DRYER — WASHER — SHELVES — FREEZER CHEST AHEAD OVER — CARGO HOLD — BOW THRUSTER — COLLISION B/H

-9 -8 -7 -6 -5 -4 -3 -2 -1 0 1 2 3 4 5 6 7 8 9 10

65' Coaster Variations

The 64' Florida Bay Coaster was a study done for a tour operator in Mexico who wanted to carry one small bus load of tourists on the boat. The small guest cabins each had their own head at the entry to the cabin. This would have a shower also. The storage is mainly under the berths. The single cabins could be turned into shared doubles with the opening of the middle panels. The main cabin has seating and galley facilities to feed all the passengers at one sitting. There's plenty of open deck space for outdoor living.

The 65' Florida Bay Coaster is a preliminary version done while we were studying ideas for what became the ***Key Largo***. This one has the 18' beam of the ***Florida Bay***, with 44" added to the after deck space, 24" to the saloon and master stateroom areas, 17" to the middle stateroom and pilothouse, 41" to the hold, and 54" to the foc'sle. The foc'sle has a pair of small staterooms sharing a head in the bow. The pilothouse is just like ***Florida Bay***'s with the addition of the walkway around the front. The other cabins are all larger. Comparing this design to the ***Key Largo*** shows how our thinking evolved.

The 65' Tramp Freighter shows a single screw variation, with a bow thruster for aiding maneuvering. The stern castle (foc'sle is an abbreviation for fore castle) has an office space that can also be used as a guest cabin. The mid-level cabin shows an alternative layout for having an office or study space in there, with the upper and lower berths forming a settee for daytime use. There are forward and aft cargo holds. The small boats are carried on a rack over the aft deck. The masts and booms are used for loading and offloading the cargo, vehicle, and small craft. The aft corner of the pilothouse is notched for stairs to the top deck. Amidships she is fitted with a built-in ladder and upper deck gate, like ***Florida Bay*** and ***Key Largo*** have, for getting on and off at varying level piers. These have proven very practical and an option worth considering.

The 65' Amazon Cruiser is a study for a 24' beam version with the accommodations to make her a small cruiseship. When the new USCG 12 passenger rules go into effect, this boat would make an ideal way to do cruises and tours on the inland rivers and lakes that don't offer this service, or on lesser traveled coastal routes. There are lots of opportunities for pioneering new areas yet.

The 65' Florida Bay Trader is an outline of how a boat of this sort could be used as a traveling store, repair ship, or mini-container ship. This could also be used by someone with a craft business who wanted to move it with the seasons and still have all their working materials with them.

The 65' Florida Bay Clinic shows how one could be set up to function as a traveling hospital ship, dental office, or perhaps a mission boat. Many other professional services could be carried out afloat too. The advent of cellular phones and fax machines add to the opportunities to keep in touch while afloat, making having your office aboard very attractive.

The 65' Florida Bay Packet shows a different approach to making a three cabin charter layout. This one has all the passenger cabins in the stern, each with a private bathroom, and the crew quarters in the bow. Amidships is the cargo hold, which could be made into more cabin spaces if more capacity was needed. The forward pilothouse is elevated for a commanding view. Under it is the galley and dining area, with a half-bath for the passengers. This boat might make for interesting passenger tours on remote coasts or rivers where there is some freight hauling needed too.

The 65' Florida Bay RO-RO, shown below, is an idea for carrying freight trucks between the mainland and a remote island. It could also be used to take a motorhome from one area to another. This motorhome could then be used for exploring inland from the harbors.

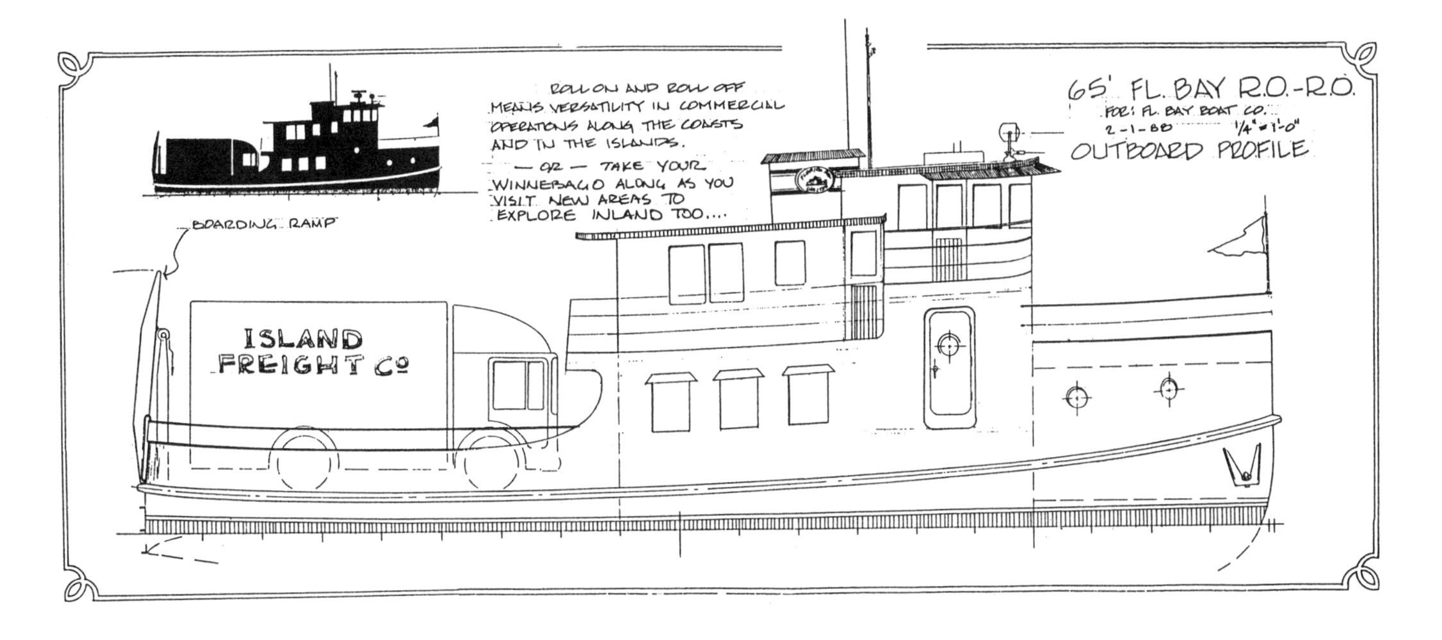

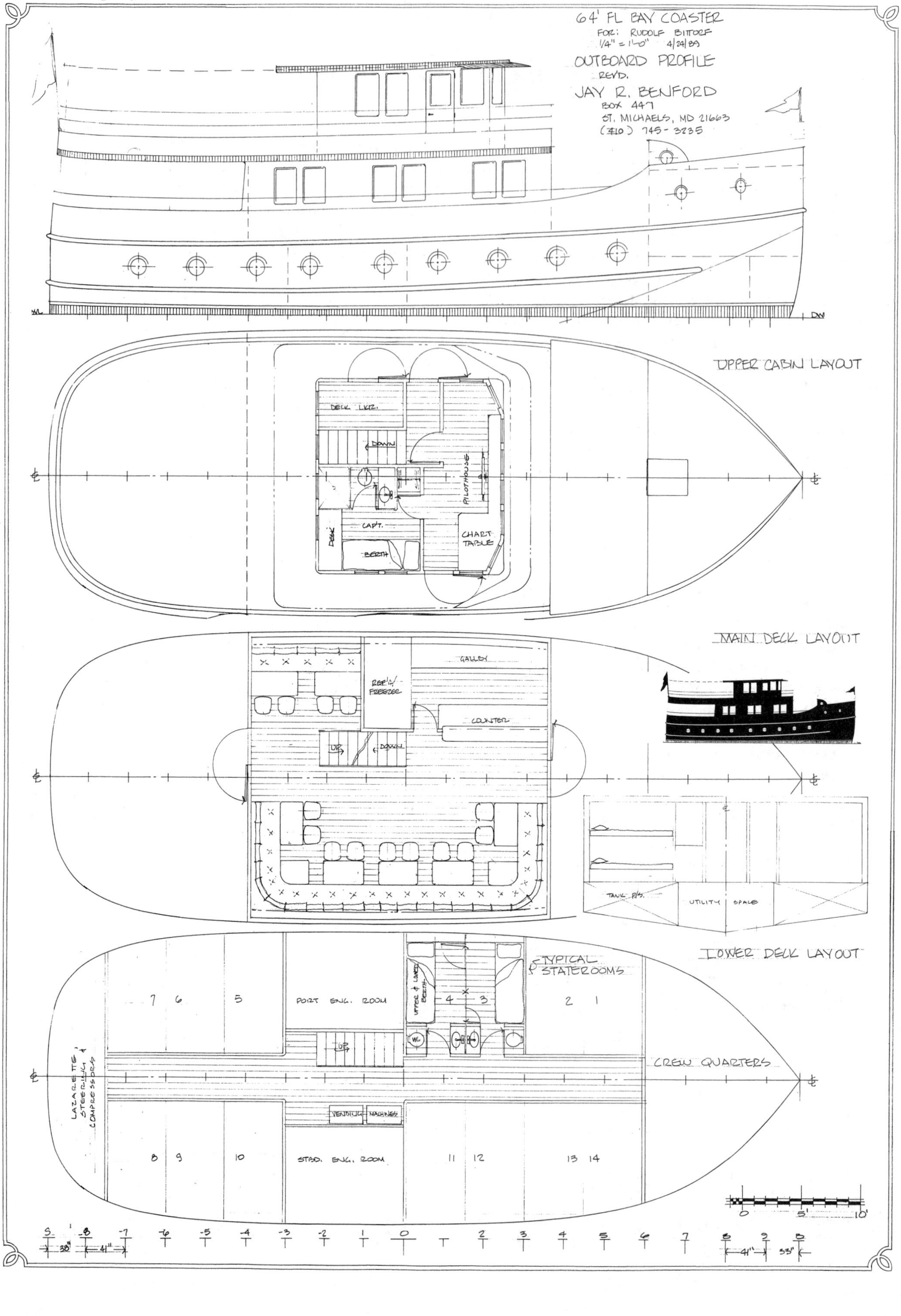

64' FL BAY COASTER
FOR: RUDOLF BITTORF
1/4" = 1'-0" 4/24/89
OUTBOARD PROFILE
REV'D.

JAY R. BENFORD
BOX 447
ST. MICHAELS, MD 21663
(410) 745-3235

UPPER CABIN LAYOUT

DECK LKR.
DOWN
PILOTHOUSE
CAPT.
DESK
BERTH
CHART TABLE

MAIN DECK LAYOUT

GALLEY
REF'G/FREEZER
COUNTER
UP
DOWN
TANK P/S.
UTILITY SPACE

LOWER DECK LAYOUT

TYPICAL STATEROOMS
7 6 5
PORT ENG. ROOM
UPPER & LOWER BERTH
4 3 2 1
WC
UP
CREW QUARTERS
LAZARETTE, STEERING & COMPRESSORS
VENDING MACHINES
8 9 10
STBD. ENG. ROOM
11 12 13 14

0 5' 10'

S -8 -7 -6 -5 -4 -3 -2 -1 0 1 2 3 4 5 6 7 8 9 B
38" 41" 41" 53"

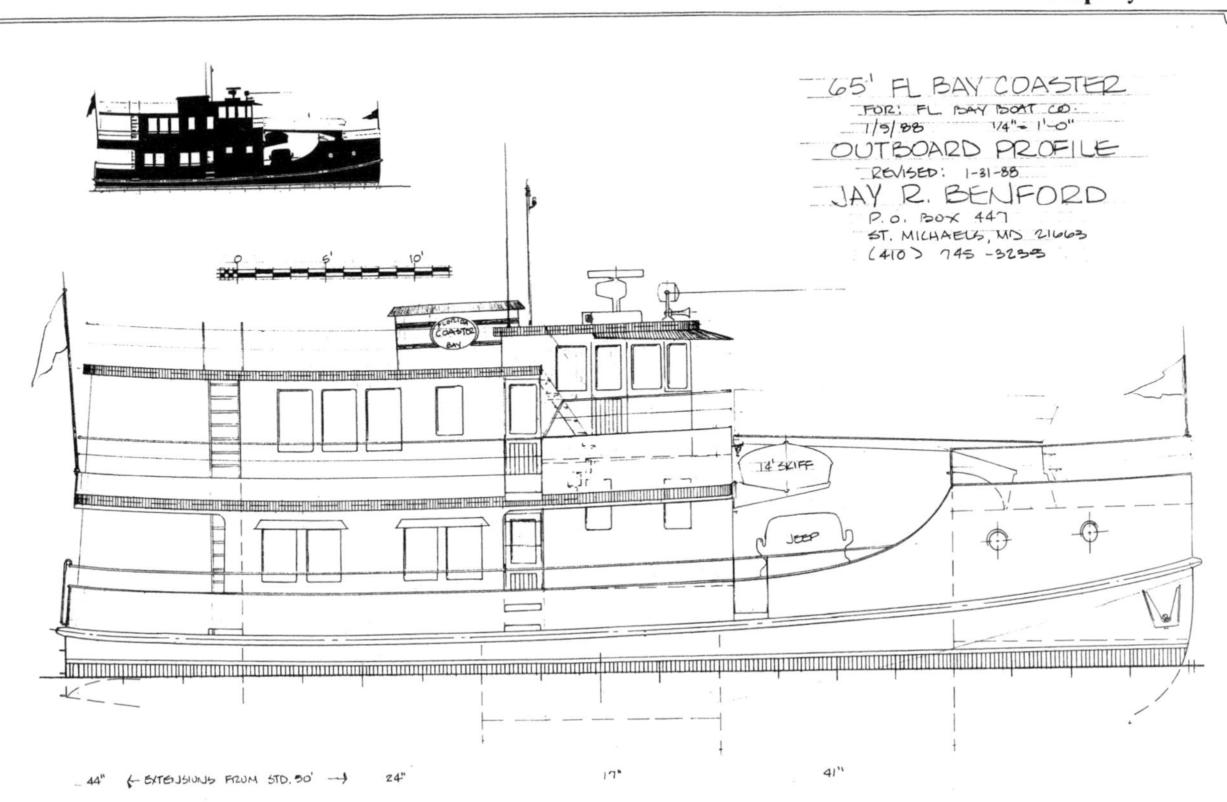

STRETCHED TO 65' THE COASTER COMES INTO HER OWN. ALL THE FEATURES OF THE 50 ARE EXPANDED OUT, CREATING MORE ROOM, MORE STORAGE, MORE SEA-KEEPING ABILITY, & A UNIQUE, HANDSOME APPEARANCE.

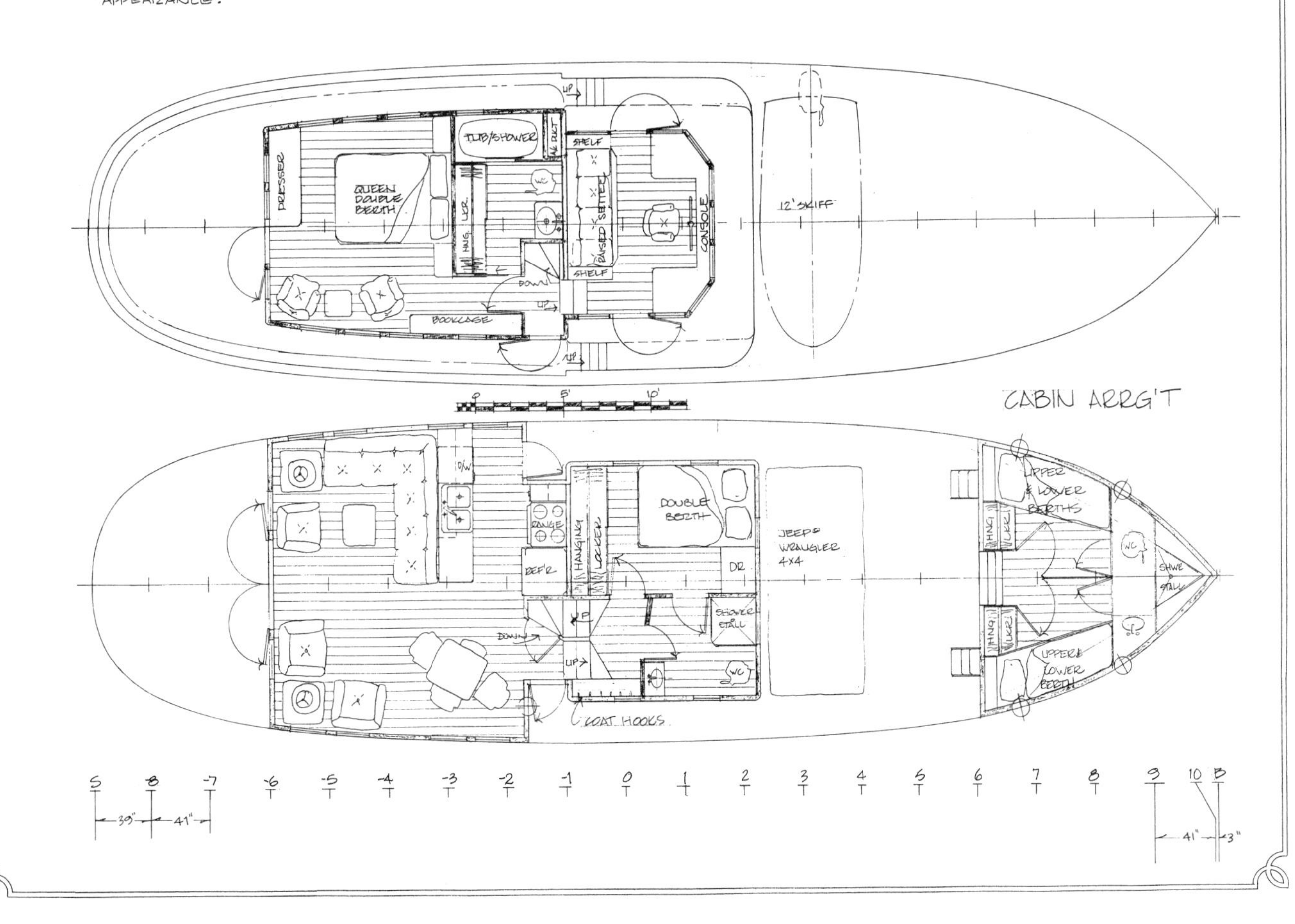

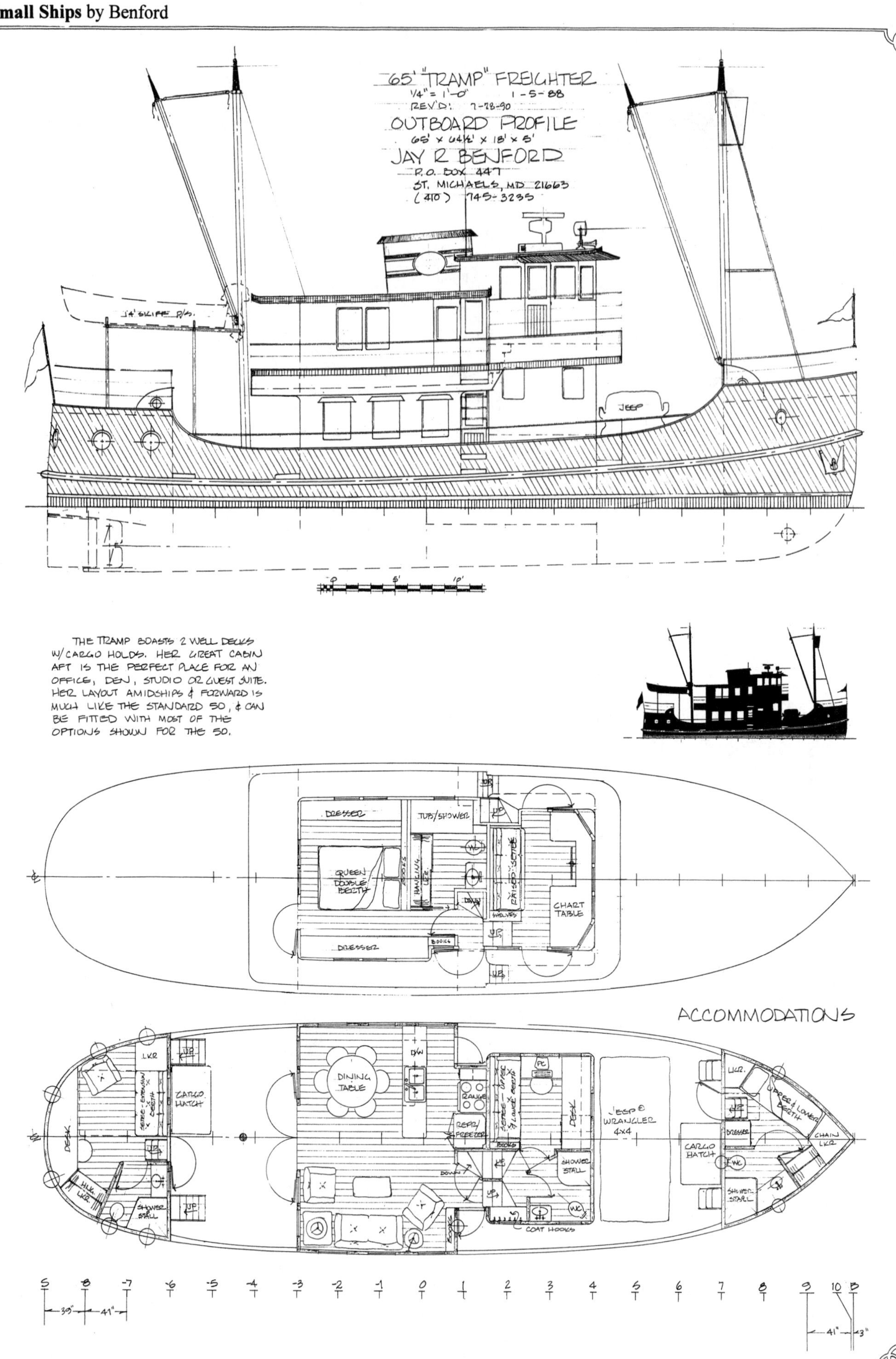

65' "TRAMP" FREIGHTER
1/4" = 1'-0" 1-5-88
REV'D: 7-28-90
OUTBOARD PROFILE
65' x 64 1/2' x 18' x 5'
JAY R BENFORD
P.O. BOX 447
ST. MICHAELS, MD 21663
(410) 745-3235
14' SKIFF P/S.
JEEP
THE TRAMP BOASTS 2 WELL DECKS W/CARGO HOLDS. HER GREAT CABIN AFT IS THE PERFECT PLACE FOR AN OFFICE, DEN, STUDIO OR GUEST SUITE. HER LAYOUT AMIDSHIPS & FORWARD IS MUCH LIKE THE STANDARD 50, & CAN BE FITTED WITH MOST OF THE OPTIONS SHOWN FOR THE 50.
DRESSER
TUB/SHOWER
UP
QUEEN DOUBLE BERTH
BOOKS
HANGING LKR.
RAISED SETTEE
SHELVES
CHART TABLE
DRESSER
BOOKS
ACCOMMODATIONS
LKR
UP
CARGO HATCH
DESK
HNG LKR
SHOWER STALL
DINING TABLE
D/W
RANGE
REFR/FREEZER
DOWN
DESK
JEEP ® WRANGLER 4x4
SHOWER STALL
WC
COAT HOOKS
CARGO HATCH
WC
LKR
UPPER & LOWER BERTH
DRESSER
CHAIN LKR
SHOWER STALL
39"
41"
41"
3"

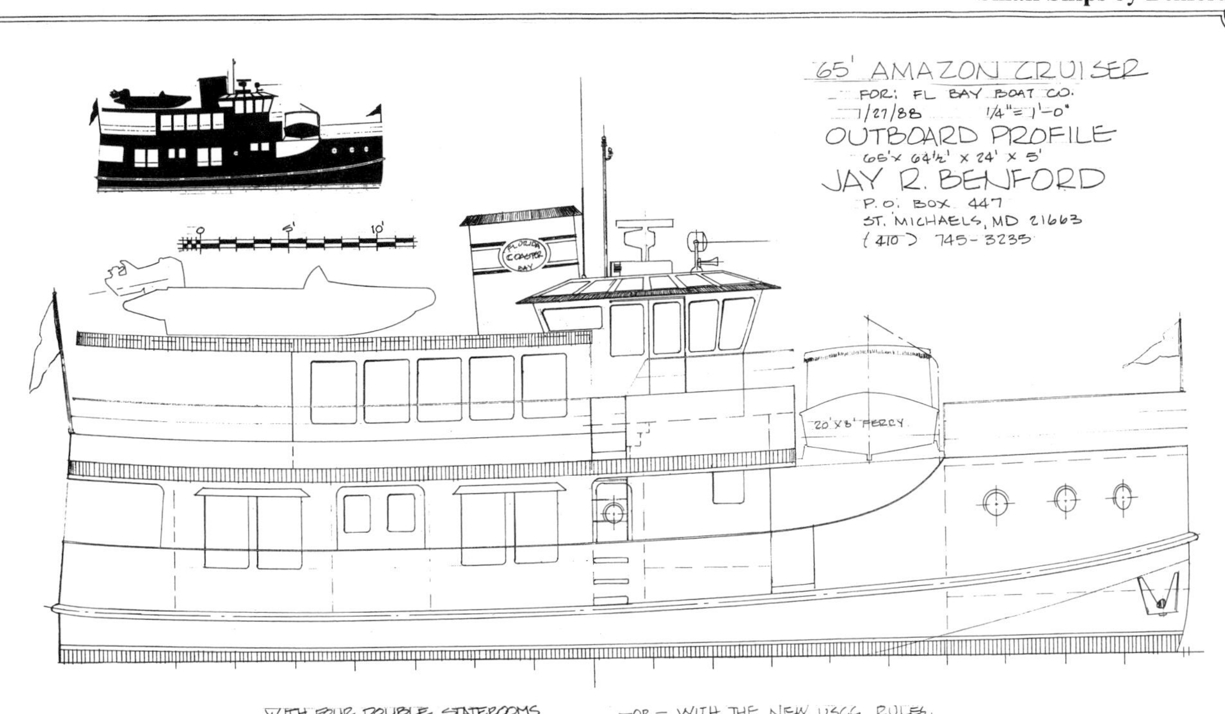

WITH FOUR DOUBLE STATEROOMS (EACH WITH PRIVATE HEAD AND VERANDA) PLUS TWO DOUBLE CREW STATEROOMS, THE FL BAY AMAZON CRUISER IS THE IDEAL RIVER BOAT FOR THE LUXURY CRUISE, COMMERCIAL OPERATOR. GREAT FOR COMFORTABLE SAFARIS UP THE GRAND RIVERS OF THE WESTERN HEMISPHERE

—OR— WITH THE NEW USCG RULES, USE THE BOW STATEROOMS TO RAISE THE CAPACITY TO 12 PASSENGERS. USE THE LARGE VOLUMES IN THE HULL FOR THE CREW QUARTERS & STILL HAVE PLENTY OF HOLD SPACE FOR "TREASURES" FOUND ON THE EXPLORATIONS.

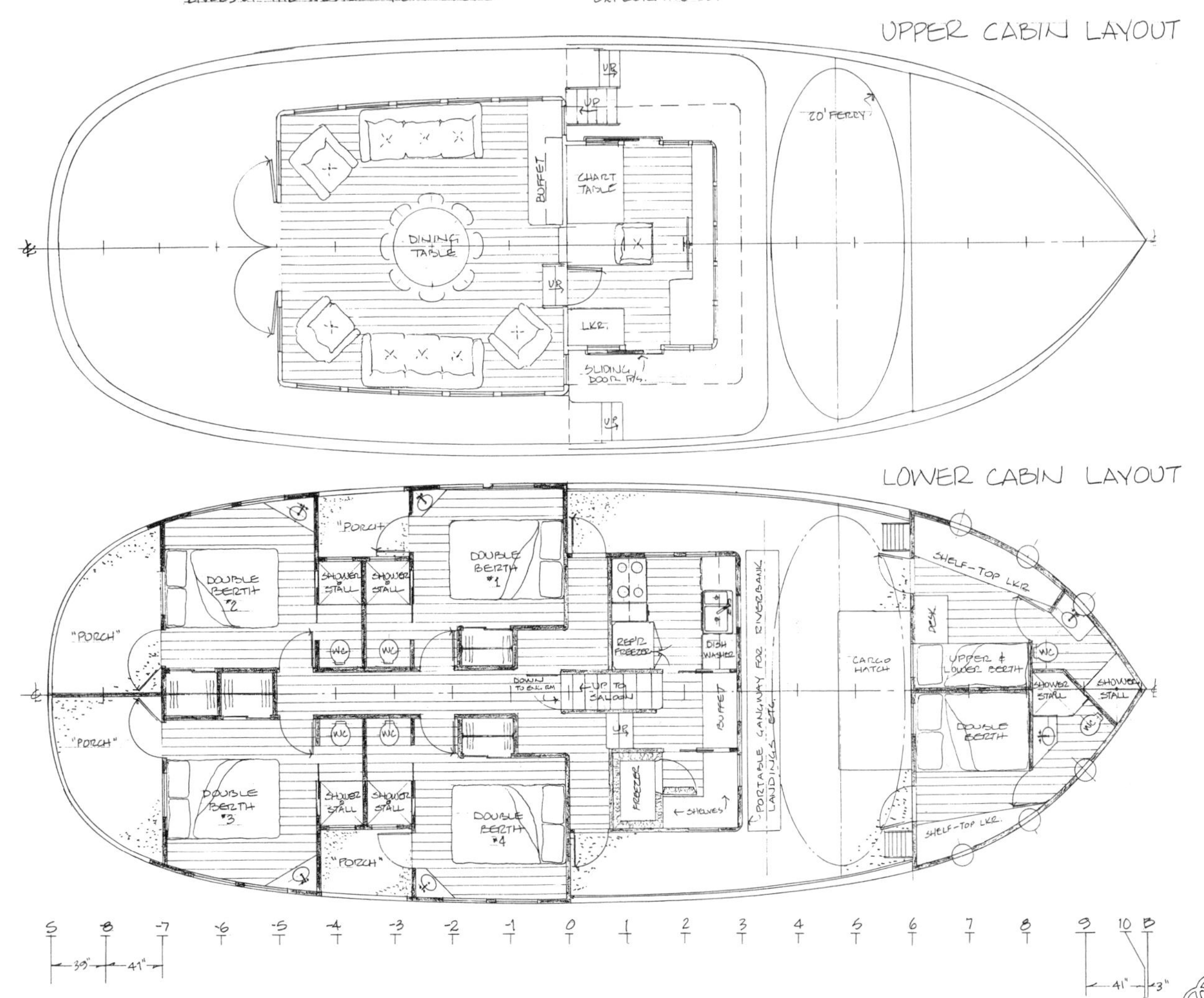

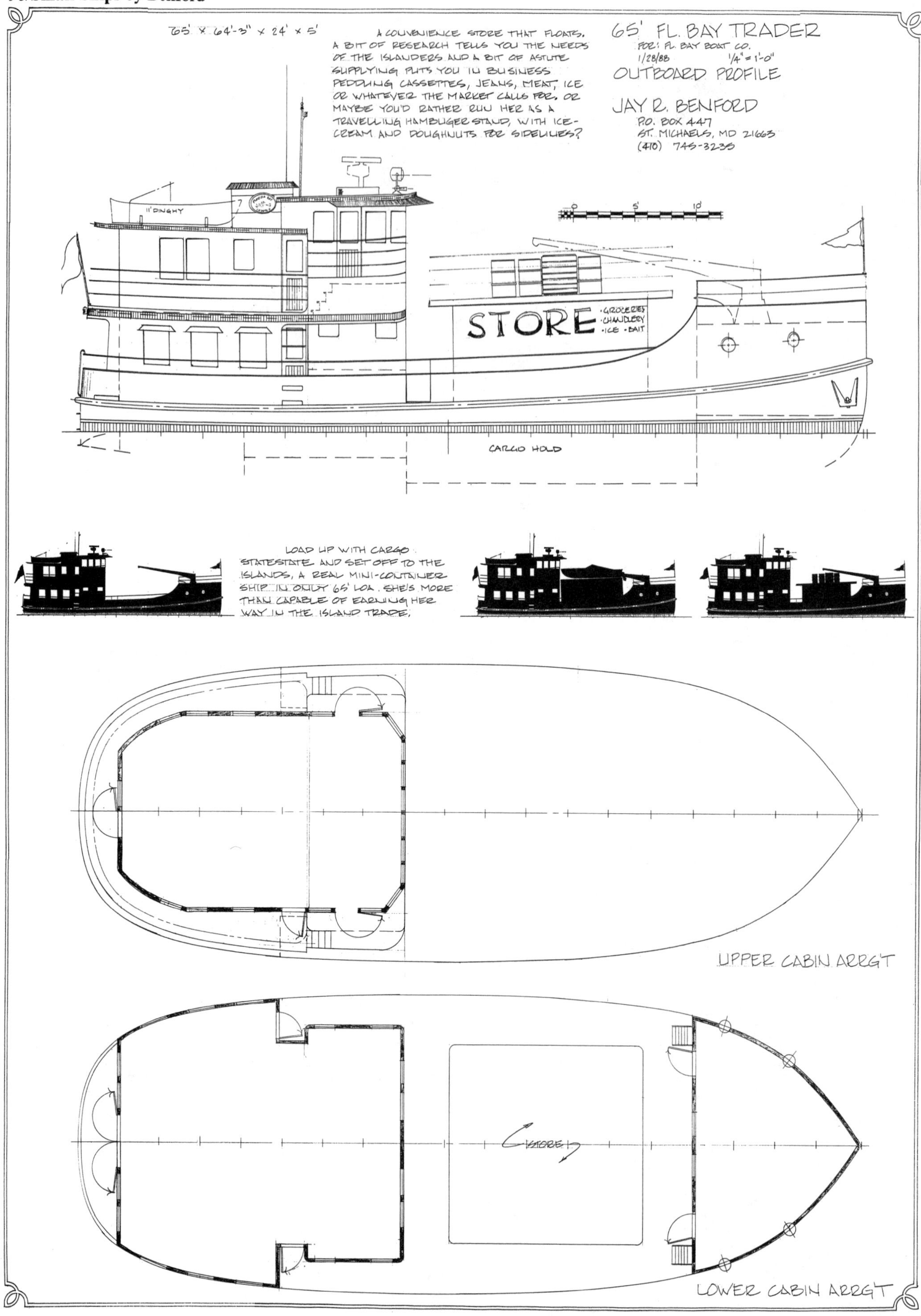
65' x 64'-3" x 24' x 5'
A CONVENIENCE STORE THAT FLOATS. A BIT OF RESEARCH TELLS YOU THE NEEDS OF THE ISLANDERS AND A BIT OF ASTUTE SUPPLYING PUTS YOU IN BUSINESS PEDDLING CASSETTES, JEANS, MEAT, ICE OR WHATEVER THE MARKET CALLS FOR. OR MAYBE YOU'D RATHER RUN HER AS A TRAVELLING HAMBURGER STAND, WITH ICE-CREAM AND DOUGHNUTS FOR SIDELINES?
65' FL. BAY TRADER
FOR: FL. BAY BOAT CO.
1/28/88
1/4" = 1'-0"
OUTBOARD PROFILE
JAY R. BENFORD
P.O. BOX 447
ST. MICHAELS, MD 21663
(410) 745-3235
11' DINGHY
0
5'
10'
STORE
·GROCERIES
·CHANDLERY
·ICE ·BAIT
CARGO HOLD
LOAD UP WITH CARGO STATESTATE AND SET OFF TO THE ISLANDS, A REAL MINI-CONTAINER SHIP IN ONLY 65' LOA. SHE'S MORE THAN CAPABLE OF EARNING HER WAY IN THE ISLAND TRADE.
UPPER CABIN ARRG'T
STORE
LOWER CABIN ARRG'T

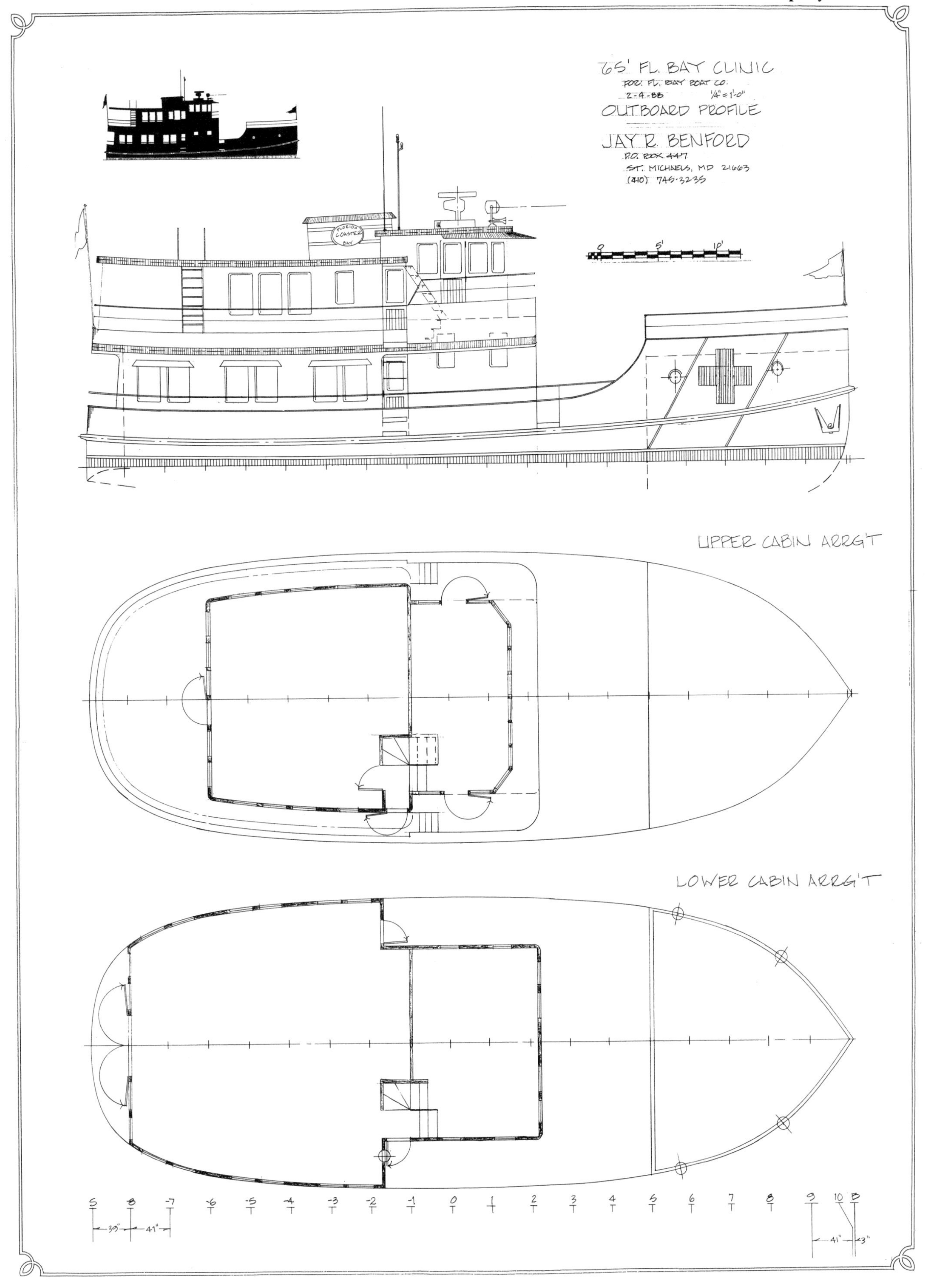
65' FL. BAY CLINIC
FOR: FL. BAY BOAT CO.
2-4-88
1/4" = 1'-0"
OUTBOARD PROFILE
JAY R. BENFORD
P.O. BOX 447
ST. MICHAELS, MD 21663
(410) 745-3235
0
5'
10'
UPPER CABIN ARRG'T
LOWER CABIN ARRG'T
S
-8
-7
-6
-5
-4
-3
-2
-1
0
1
2
3
4
5
6
7
8
9
10
B
39"
41"
41"
3"

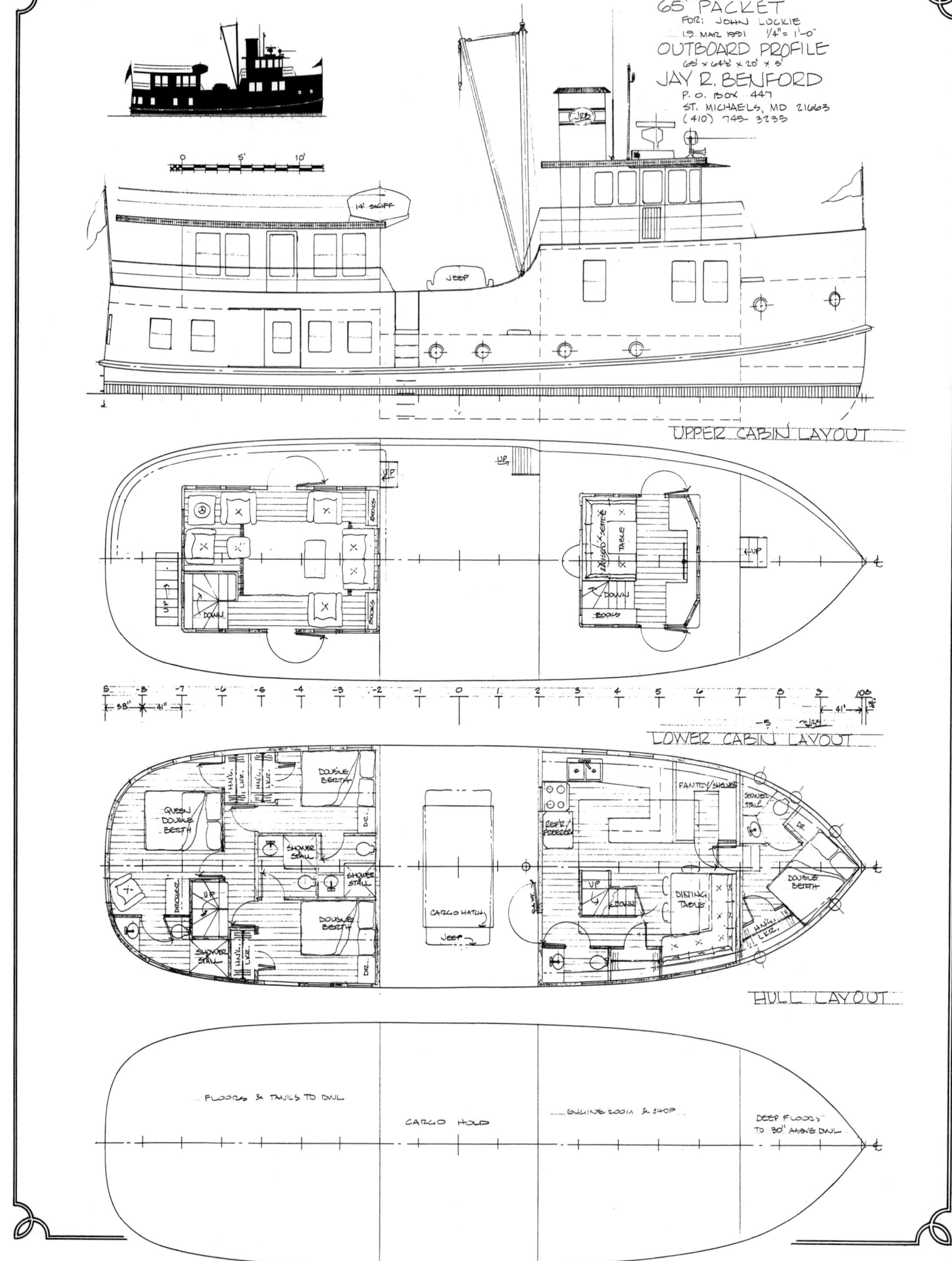
65' PACKET
FOR: JOHN LOCKIE
19 MAR 1991 1/4" = 1'-0"
OUTBOARD PROFILE
65' x 64½' x 20' x 5'
JAY R. BENFORD
P. O. BOX 447
ST. MICHAELS, MD 21663
(410) 745-3235
14' SKIFF
JEEP
UPPER CABIN LAYOUT
BOOKS
DOWN
UP
RAISED SETTEE
TABLE
LOWER CABIN LAYOUT
QUEEN DOUBLE BERTH
DOUBLE BERTH
SHOWER STALL
HNG. LKR.
DRESSER
CARGO HATCH
REEF/FREEZER
PANTRY/SHELVES
DINING TABLE
HULL LAYOUT
FLOORS & TANKS TO DWL
CARGO HOLD
ENGINE ROOM & SHOP
DEEP FLOORS TO 30" ABOVE DWL

Future Freighters

Will a larger freighter better suit your needs? Do you need more room for more staterooms? Would you like even roomier spaces on board? Would you like to have greater cargo capacity?

The design philosophy that has made these boats so successful can readily be extended to other sizes. The sketches shown on this page are some of the ideas we've considered for larger ones.

The U.S. Coast Guard's new regulations, when put into effect, will permit carrying 12 passengers on larger uninspected vessels (yachts), instead of 6 as currently allowed. This will open a world of opportunity to those who would like to run their own mini-cruiseship.

The mini-cruiseship could be run on rivers and canals where a larger vessel would not fit, and where there is lighter demand for tours. It would also make for customization of the cruises, with the charterees participating in setting the itinerary, instead of being taken along with the herd on a large ship.

How about the Seattle to Southeast Alaska route? Or a circumnavigation of southern Florida and the Everglades?

The choice is yours....

Or, would you just like a larger cruising vessel. Let us know what you'd like and we'll be happy to work with you in creating your new small ship.

New Florida Bay Coaster Company!

Since 1993, when the original Florida Bay Coaster Company closed, we've been working with a number of builders to keep our freighter yachts available as we've created more sizes and variations.

Now, Aquiles Faillace has formed a new Florida Bay Coaster Company to build more of these great liveaboard yachts. The first of these is a 58-footer (under construction as we go to press) and it will be followed by a 48-footer. For preliminary drawings of these see pages 274-275 and 278-279. Call or write for more information....

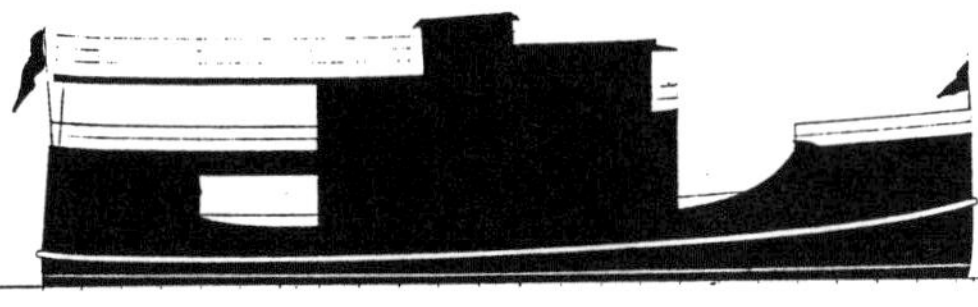

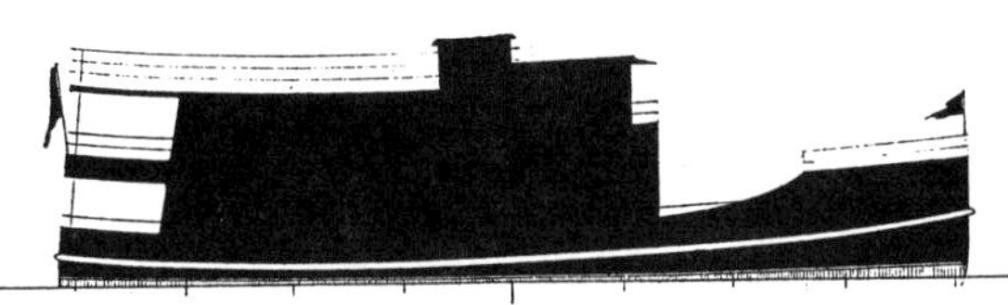

85' FREIGHTER

100' FL. BAY COASTER
1/8" = 1'-0" 1/30/88
FOR: FL. BAY BOAT CO.
OUTBOARD PROFILE

JAY R. BENFORD
P. O. BOX 447
ST. MICHAELS, MD 21663
745-3235

WHY NOT?
YOU'VE GOT THE TOYS AND THE TIME TO ENJOY THEM, THE 100-FOOTER HAS THE ROOM TO CARRY THEM AND YOU ON HIGH ADVENTURE WHETHER DOWN THE COAST OR UP THE RIVERS OR ACROSS THE SEAS.

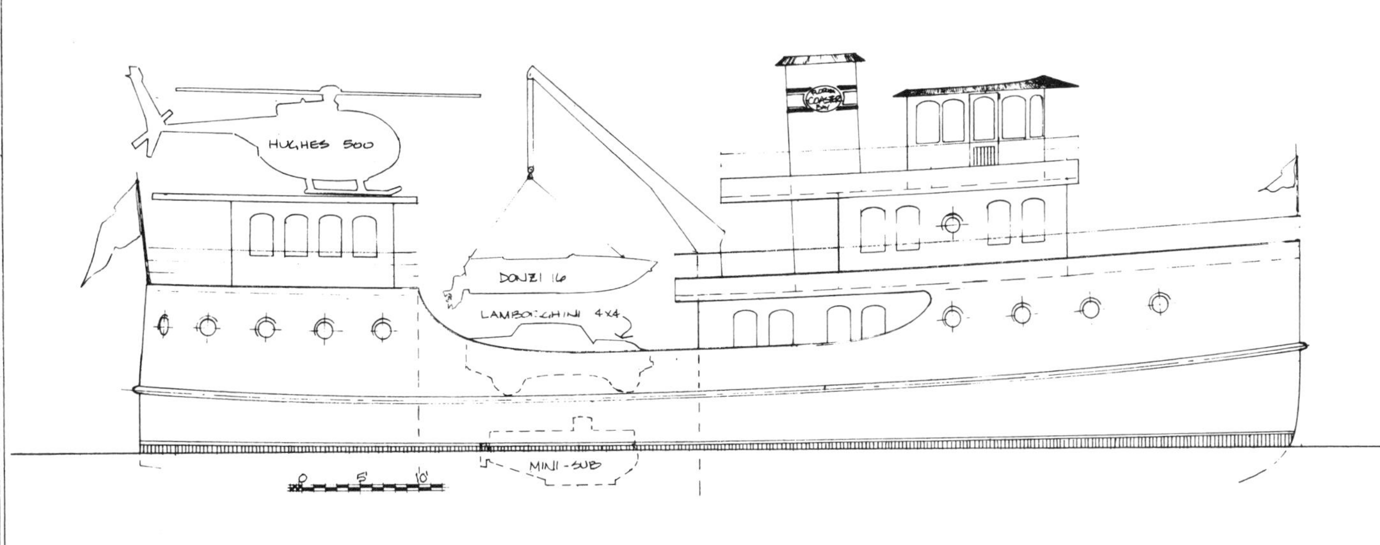

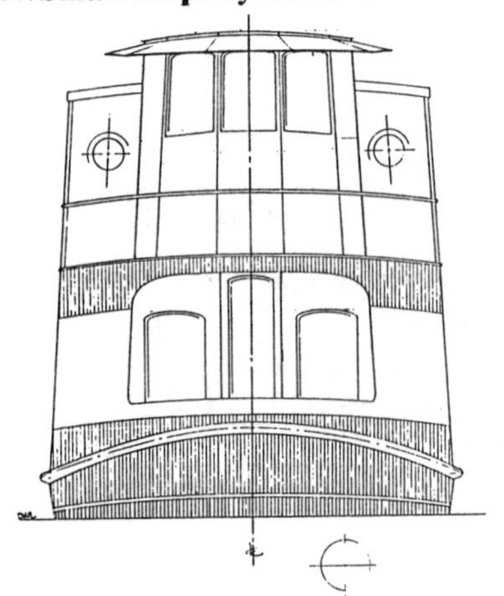

Ferry Yachts

Commentary By Frank Madd

What If?

Twenty years ago, during one of our regular office "what if" sessions, we got into discussing the sad state of what was being offered in the way of houseboats. People who wished a mobile liveaboard boat were offered only one kind of boat. Since several of us live aboard, or have lived aboard, we have a special interest in this lifestyle. The ability to tow waterskiers and consume a gallon or two per mile to go cruising seemed the antithesis of what was really required. What were being called houseboats were really house trailers with a box underneath to keep them afloat. The only quality control seemed to be in getting good four-color printing jobs for the sales brochures.

The Better Mousetrap

Starting with a clean sheet of paper and open minds, we analyzed the requirements of the liveaboard. From this, we laid out the parameters of the ideal houseboat. It would have maximized living accommodations, operate at economical displacement hull speeds, and be built in a manner that would make it a good investment. Also, it must look like it belonged on the water, not like something styled in Detroit for cruising the Interstate highways.

The Stroke Of Genius!

What finally got the project launched was Jay's idea to make it a small ferry boat in style and design. This gave us a boat that looked like a proper little ship, and one that would give the same feeling to those aboard. It also gave great accommodations, by making the cabin practically the full width of the hull. But, most importantly, it was an easily driven hull form that would provide very economical operation. One that could be cruised through the thousands of miles of semi-protected waterways where most people do all their cruising, and one that would have tremendous room for her size and all the features to make her an economical vessel to operate.

The Good Ferry

The result of following these thoughts to their logical conclusion was the Waterbed 30. After Jay designed her in 1972, she was published in quite a number of magazines, and generated more mail than any other design Jay has had published. Thinking we were really onto something, we approached a number of production builders to see about putting her into production. Uniformly, their reaction was negative, ranging from a polite "no thank-you" to loud guffaws. To us, this seemed a sad commentary on the state of the boating business — no one was willing to do something to differentiate himself from his competitors. Here was a whole segment of the market that was being virtually ignored. No wonder so many builders folded during the recent recession; they had no idea how to innovate and offer different products that would set them apart from their competitors and give them a market all to themselves. Since those early negatives from shortsighted builders, we have had a gradually increasing amount of interest in the ferryboat style houseboat idea. A 34' version of the original Waterbed 30 was drawn up for one group. The resulting design has 50 to 100% more living room inside than most 40' houseboats. Her twin skegs protect the props and rudders, and make for graceful, upright groundings. The twin screws make for good maneuvering and the two small diesels provided safe and very economical operation. I think it's the most practical cruising boat I've seen in decades and I hope to live aboard one myself someday.

Jay has also drawn up what he calls the "office version" of the Friday Harbor Ferry. In it, he's got two staterooms, a roomy head with five foot long tub and shower unit and washer and dryer stack set, a good sized office, a large galley and dining and living room combination, and a cozy pilothouse. This 320 square feet of enclosed living space on the lower deck, and 170 in the upper house and 30 in the pilothouse for a total of 520 square feet. (The yacht version has 320 plus 80 on the upper deck for a total of 400 square feet.) If the office version was used for just living aboard, the space the office uses could be added to the living quarters, making it even roomier.

Discussions have ensued with a number of other people interested in a variety of sizes, including a 60-footer, all of which would be most delightful. However, action is what is now needed: bold, fearless and resolute. I am convinced that once this houseboat is built and marketed by a quality-oriented, efficient sales team, we will see this boom into an around-the-world solution for everything from housing shortages to co-op condominiums to charter boat dealers on lake and island resorts to straightforward boating buffs. Our conclusion is simply this: the design came in before its time. The public responded with resounding enthusiasm. The builders, stuck in "the known market", did nothing. After two decades of unwavering and increasing enthusiasm by clientele wanting these boats, the design is ready and waiting. If you are a builder who wants to build a delightful, practical, roomy packet, and you can handle the avalanche of waiting sales, the Good Ferry is here. Don't be hesitant: the public has been waiting for twenty years for this boat and when it's available, response will be high.

Choice Waterfront Homes

Available with panoramic marine views from Puget Sound to Alaska or Florida to Maine. An ideal liveaboard or retirement home or vacation home, the 34' Friday Harbor Ferry offers a majestic variety of unique marine park settings, each available without ever leaving home. These luxury homes offer low maintenance exteriors and warm wood interiors. All utilities are installed including washer and dryer, bathtub and fireplace. They offer two bedrooms, complete kitchen, large sundeck, and two viewing porches. Economical diesel power and large tanks provide for safe and low-cost operation. Neighbors too noisy? Move your house. Tired of mowing the lawn and raking leaves? Try Friday Harbor Ferry boat living for fast relief.

Caution: The Sturgeon Genial warns that Friday Harbor Ferry living may be addictive and habit forming.

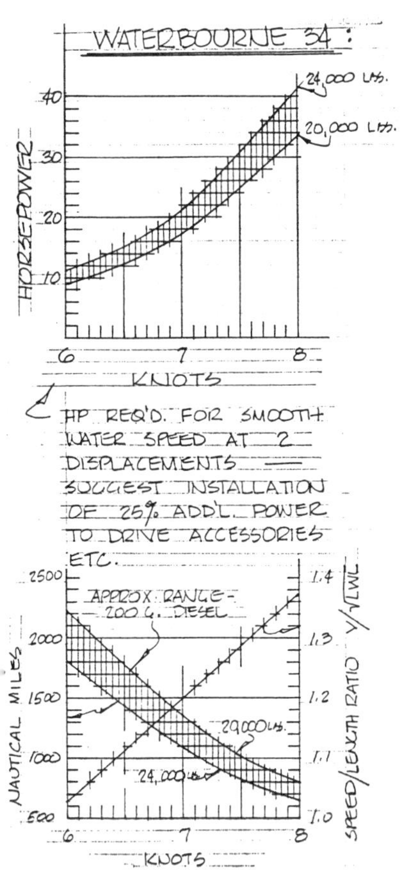

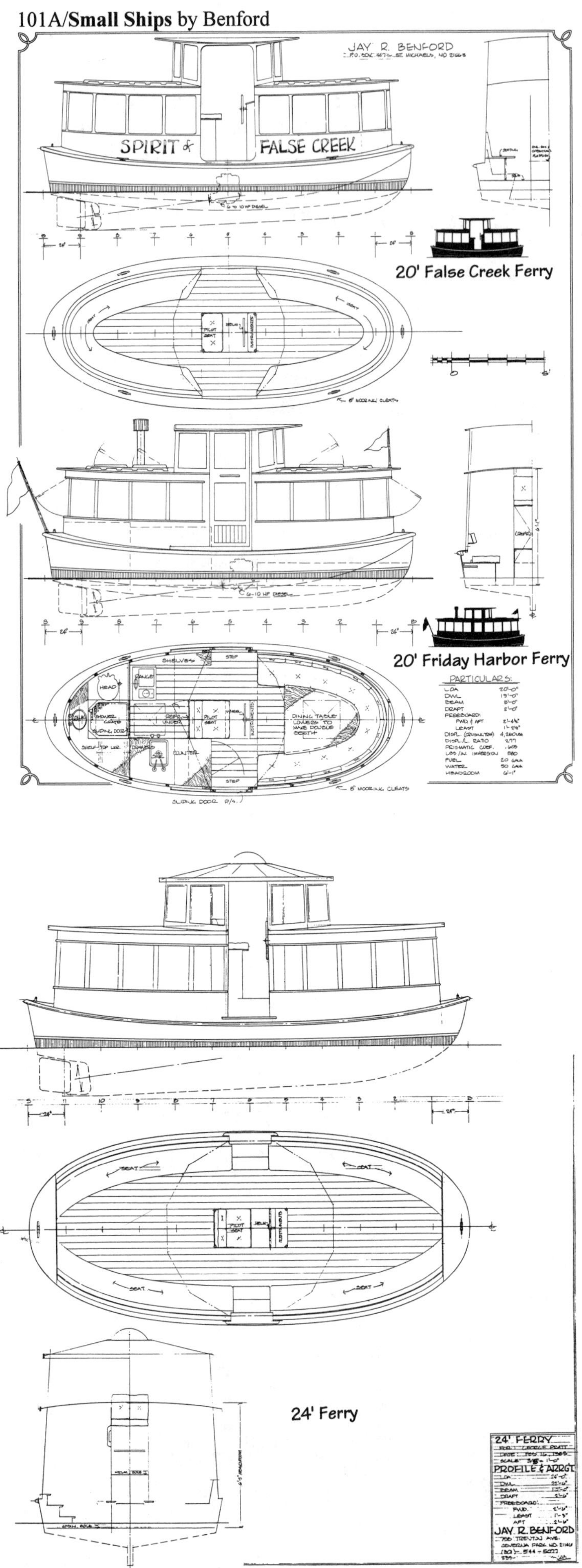

Ferry Yachts

Designs Number 212, 233 & 253
1983, 1985 & 1986

20: Paul Miller of Maple Bay, BC, has now built three fleets of the 20' ferries, which run in Vancouver and Victoria, British Columbia. They're successful workhorses, carrying their load of 12 passengers through all sorts of conditions, and carrying tremendous numbers of them. The two Vancouver fleets did heroic duty during Expo '86 there and were a big hit with the passengers. Having very low operating costs in addition to low construction costs, they are a great success. They could be used for Yacht Club shuttle or service to a private island.

The Friday Harbor version has cruising accommodations for a couple. We've lowered the cabin sole to get standing headroom, put in a galley, and provided for an enclosed head with a shower. She has all the comforts of a small motorhome, without the worry about finding a paved road to the places you want to visit.

24: The 24' Ferry was designed as an enlarged 20-footer, the idea being to double the carrying capacity of the ferry, yet operate with the same, single crew member. The extra size over gives better standing headroom and more room for the seated passengers to stretch their legs. Her power requirement is a bit higher, but the actual installed power will likely be the same since the 20s ended up overpowered in the desire to get smoother 2-cylinder engines in them. Thus, fuel consumption will be a bit higher, but probably a bit less on a per passenger-mile basis.

The 24' Friday Harbor Ferry version has all the comfortable features found on land-based motorhomes. Unlike motorhomes, you can get off the beaten path more easily and find places with privacy to anchor for the night. There are two separate sleeping areas, in the pilothouse and the convertible dinette. The roomy head has a separate tub and shower space. The galley has a refrigerator, sink, stove and storage spaces. The desk makes a nice work space for a computer, a typewriter, or doing chart work. The pilothouse is well lit and provides a commanding view for the helmsman. The raised double bed makes a good seat for daytime operations. The opening to the lower cabin can close off to make it a private stateroom at night. She makes a great liveaboard for a couple or summer home for a small family.

26: Evolved from the 24' Ferry, the 26' North Channel Ferry has a similar layout, with slightly more room due to the extra beam and length. The hull form was modified to have a more conventional bow shape to take the steeper seas expected in cruising the North Channel and going South for the winter.

Particulars:		**212**	**233**	**253**
Length overall		20'-0"	24'-0"	26'-0"
Length designed waterline		19'-0"	22'-6"	25'-3"
Beam		8'-0"	10'-0"	11'-2"
Draft		2'-0"	2'-6"	2'-6"
Freeboard:	Forward	2'-4½"	2'-6"	6'-3"
	Least	1'-2½"	1'-3"	0'-9"
	Aft	2'-4½"	2'-6"	3'-6"
Displacement, cruising trim, lbs.		4,260	9,000	10,000
Displacement-length ratio		277	353	277
Prismatic coefficient		.605	.598	
Pounds per inch immersion		580	859	
Water/Fuel Tankage, Gals.		50/20	132/46	
Headroom		5'-2"/6'-1"	6'-2"	6'-2"

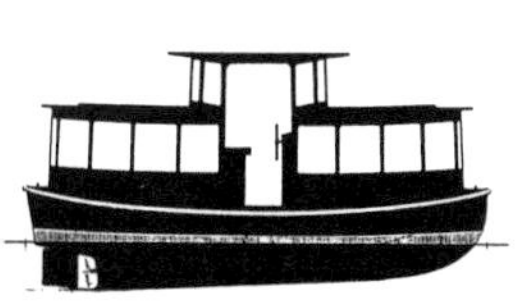
20' False Creek Ferry

24' Ferry

24' Friday Harbor Ferry

26' North Channel Ferry

20' Friday Harbor Ferry

(Below Right) Paul Miller's shop with two of the Vancouver Aquabus fleet under construction, along with two of our Cape Scott 36' double-ended cutters. (Below Left) One of the Victoria Harbor Ferries in service. Photos courtesy of the builder.

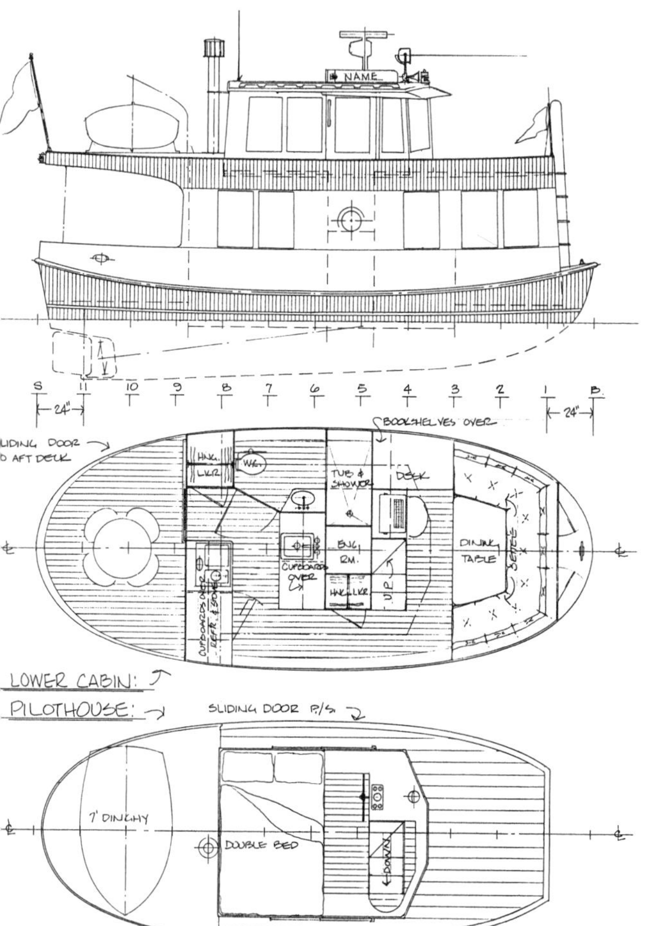

24' Friday Harbor Ferry

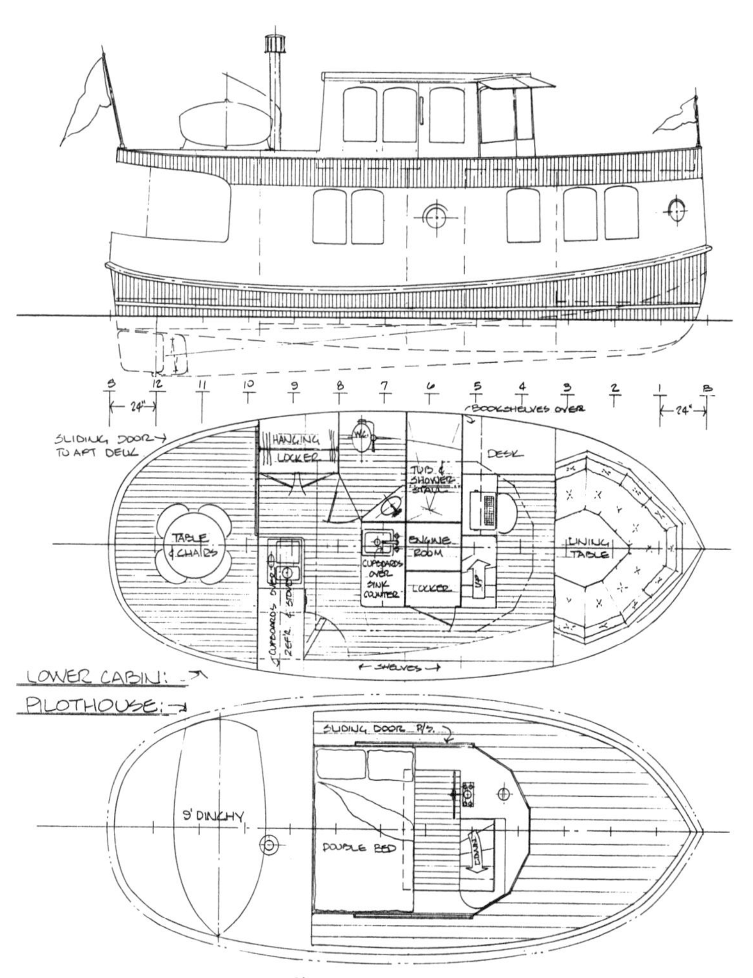

26' North Channel Ferry

The Waterbed 30 & 30' Ferry Yacht

Designs Number 117 & 241
1972 & 1985

Two decades ago I decided that a small boat styled like a ferry boat would have the most useful room in it as a liveaboard boat. It's tremendous volume would permit house-like spaces and get away from the feeling of camping in tight quarters found on many boats.

The styling of the boat like a working ferry would also get away from the stigma usually associated with "house-boats". Too many people think of a houseboat as something like a shoddily built house trailer on a boxy hull. The appearance of a ferry would be more socially acceptable, particularly for someone leaving a sailboat for something with more room in it.

The ability to pull waterskiers at speed, while consuming a gallon or two of fuel per mile does not fit in well with a boat that will have all of one's worldly possessions aboard, for weight is the biggest detriment to speed. Weight does help steady the motion of the boat, and make her more comfortable to be aboard. Thus, we've designed the ferry to be an easily driven displacement hull, keeping her powering requirements modest. She has good capability to carry a load gracefully.

What has started out as something of a lark had in time turned into the concept for a very interesting boat. The original version of this design was called the Waterbed 30, named after the bed in the master stateroom and for the fact that she is a bed upon the water. She was intended to be a liveaboard for a couple, with all the comforts of an efficiency apartment, plus the mobility of being able to unplug and go cruising.

Her design was published in a number of magazines and we had unprecedented response. She struck a chord in people who wanted to live on a boat, but didn't want to have to compromise away the comforts they knew ashore.

The later version, for which construction plans are available for building in plywood and epoxy, is the 30' Ferry Yacht. We've swapped the master stateroom into the pilothouse and created a home office or den in the forward lower cabin. The convertible sofa in the saloon will be a place for over-nighting guests, or the forward cabin could be converted to another stateroom if there are kids in the family.

The section through the pilothouse shows the bank of drawers for chart storage, allowing for literally hundreds of charts to be stored flat and not folded. This should be enough room to keep all the charts you would need to do the full circumnavigation of the Eastern United States, with the Intracoastal Waterway, Great Lakes and inland rivers to be cruised.

Particulars:	**W30**	**30 Ferry**
Length overall	30'-7½"	30'-7½"
Length on guards	30'-0"	30'-0"
Length designed waterline	30'-0"	28'-7"
Beam	12'-0"	12'-0"
Draft	2'-0"	3'-0"
Freeboard:		
Forward	2'-0"	3'-6¾"
Least	1'-3"	2'-0½"
Aft	2'-0"	3'-6¾"
Displacement, cruising trim*	12,600 lbs.	14,000 lbs.
Displacement-length ratio	208	231
Prismatic coefficient	.622	.654
Pounds per inch immersion	975	1316
Water tankage	100 Gals.	175 Gals.
Fuel tankage	100 Gals.	120 Gals.
Headroom	6'-7"	6'-5" to 6'-7"

***CAUTION:** The displacement quoted here is for the boat in cruising trim. That is, with the fuel and water tanks filled, the crew on board, as well as the crews' gear and stores in the lockers. This should not be confused with the "shipping weight" often quoted as "displacement" by some manufacturers. This should be taken into account when comparing figures and ratios between this and other designs.

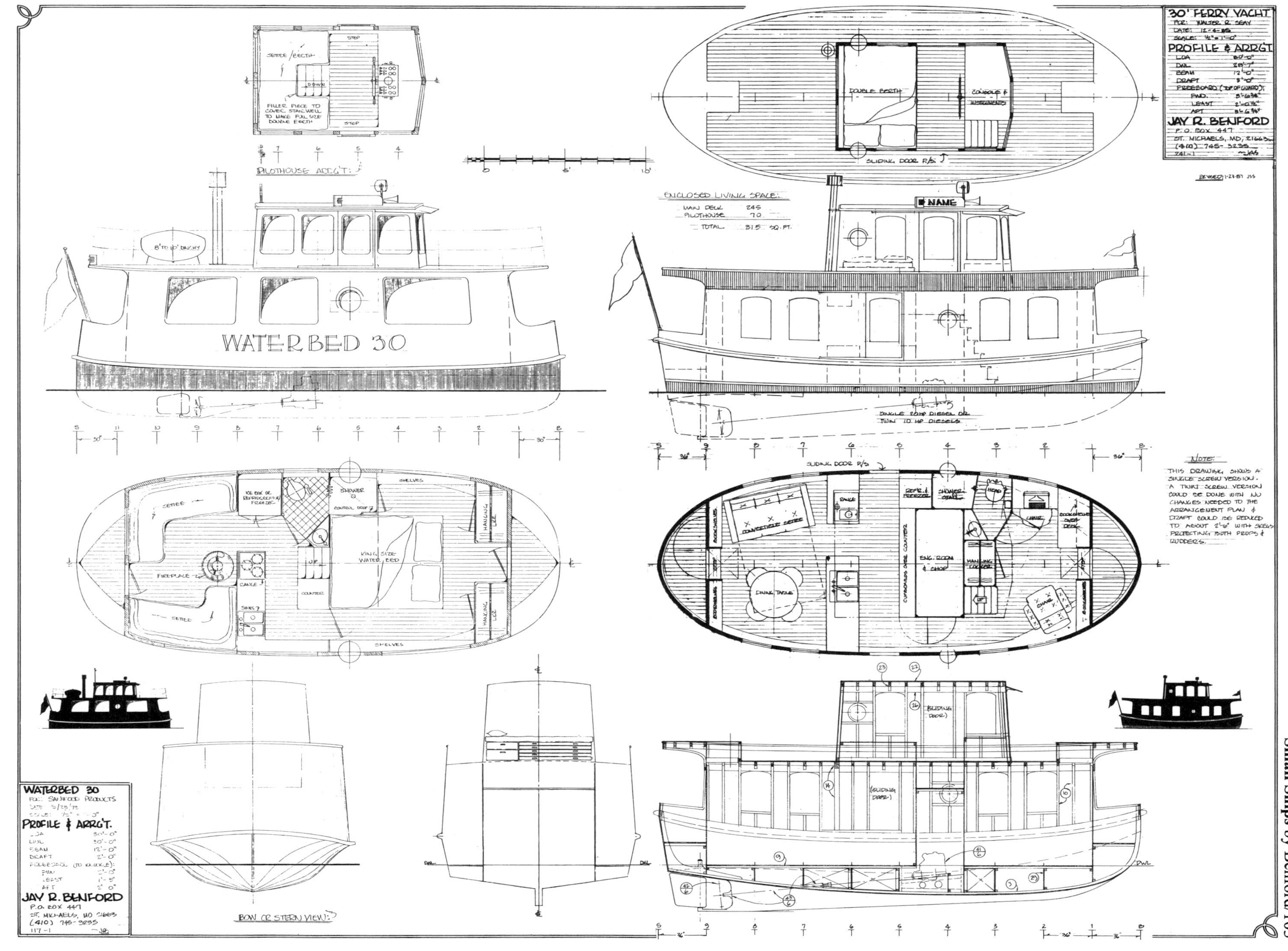
PILOTHOUSE ARR'G'T.
SETTEE / BERTH
DOWN
FILLER PIECE TO COVER STAIRWELL TO MAKE FULL SIZE DOUBLE BERTH
STEP
WATERBED 30
8' TO 10' DINGHY
SHOWER
SHELVES
HANGING
KING SIZE WATER BED
COUNTER
FIREPLACE
SETTEE
BOW OR STERN VIEW
WATERBED 30
PROFILE & ARR'G'T.
LOA 30'-0"
LWL 30'-0"
BEAM 12'-0"
DRAFT 2'-0"
JAY R. BENFORD
P.O. BOX 447
ST. MICHAELS, MD 21663
(410) 745-3235
117-1
30' FERRY YACHT
PROFILE & ARRGT.
JAY R. BENFORD
P.O. BOX 447
ST. MICHAELS, MD, 21663
(410) 745-3235
241-1
DOUBLE BERTH
CONSOLE & INSTRUMENTS
SLIDING DOOR P/S
ENCLOSED LIVING SPACE:
MAIN DECK 245
PILOTHOUSE 70
TOTAL 315 SQ. FT.
NAME
SINGLE 20HP DIESEL OR TWIN 10 HP DIESELS
NOTE:
THIS DRAWING SHOWS A SINGLE SCREW VERSION. A TWIN SCREW VERSION COULD BE DONE WITH NO CHANGES NEEDED TO THE ARRANGEMENT PLAN & DRAFT COULD BE REDUCED TO ABOUT 2'6" WITH SKEGS PROTECTING BOTH PROPS & RUDDERS.
DINING TABLE
CONVERTIBLE SETTEE
ENG. ROOM & SHOP
HANGING LOCKER
SHOWER STALL
HEAD
SLIDING DOOR

34' Friday Harbor Ferry

Designs Number 183 & 224
1979 & 1985

The 34' Friday Harbor Ferry design resulted from a request for a roomier version of the 30-footer, for more comfortable living aboard. The increase in size from 30 to 34' in length and from 12 to 14' of beam makes a great difference in spaciousness.

The head has room for a bathtub with shower and a washer and dryer stackset too. A house-size side-by-side refrigerator and freezer can be run off shore power, a small generator, or an invertor.

The version with the beach umbrella on the aft deck shows another usage for these versatile boats. This is a concept for floating hotel rooms, with two independent rooms. They share a common entry and galley on the lower level. Each unit has its own private head with shower stall and its own private aft deck.

We could have made a version that was to be lightly outfitted and driven at high speed. However, we felt that the operating philosophy should be that once you were aboard you were already where you wanted to be. Moving the boat might be more for a change of scenery, although cruising in one would be fun too.

Look over the following drawings and imagine how it would be to live on one yourself....

Particulars:	**183**	**224**
Length — over guards	34'-7½"	34'-7½"
molded hull	34'-0"	34'-0"
waterline	34'-0"	32'-9"
Beam — over guards	14'-7½"	14'-7½"
molded hull	14'-0"	14'-0"
waterline	13'-0"	13'-0"
Draft, cruising trim*	2'-3"	3'-0"
Freeboard:		
Forward	2'-6"	3'-0"
Least	1'-6"	1'-6"
Aft	2'-6"	3'-0"
Displacement, cruising trim*	23,000 lbs.	26,375lbs.
Displacement-length ratio	261	335
Prismatic coefficient	0.54	
Pounds per inch immersion	1,635	1,808
Water tankage	300 gals.	300 Gals.
Fuel tankage	200 gals.	240 Gals.
Power	twin 18-27 hp diesels	
Headroom	6'-5"	6'-7"
Enclosed living spaces — (office version)		
lower house	320 sq. ft.	
upper house	170 sq. ft.	
pilothouse	30 sq. ft.	
total	520 sq. ft.	
GM (est. — very stiff)**	5'	5'

***CAUTION:** The displacement quoted here is for the boat in cruising trim. That is, with the fuel and water tanks filled, the crew on board, as well as the crews' gear and stores in the lockers. This should not be confused with the "shipping weight" often quoted as "displacement" by some manufacturers. This should be taken into account when comparing figures and ratios between this and other designs.

**Stability on this boat is excellent. It would take about 40 knots of side wind to heel her to her guard rail. Two adults stepping aboard at her side door should heel her only about one degree.

The first of the 34' Friday Harbor Ferries was built by professional builder Garry Parenteau and is home for his family. These photos show the roomy saloon, galley and forward staterooms.

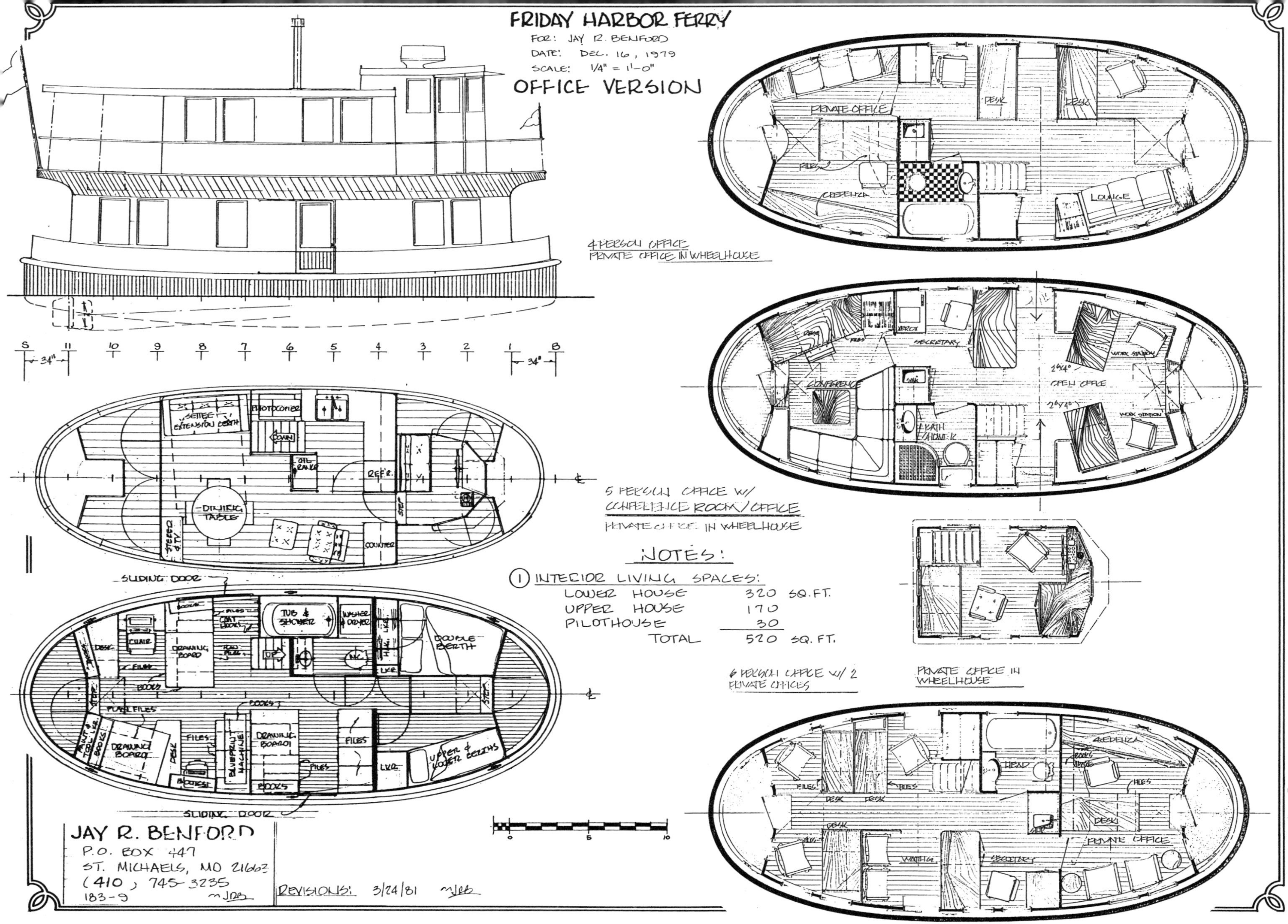
FRIDAY HARBOR FERRY
FOR: JAY R. BENFORD
DATE: DEC. 16, 1979
SCALE: 1/4" = 1'-0"
OFFICE VERSION
4 PERSON OFFICE
PRIVATE OFFICE IN WHEELHOUSE
5 PERSON OFFICE W/
CONFERENCE ROOM/OFFICE
PRIVATE OFFICE IN WHEELHOUSE
NOTES:
1 INTERIOR LIVING SPACES:
LOWER HOUSE 320 SQ.FT.
UPPER HOUSE 170
PILOTHOUSE 30
TOTAL 520 SQ.FT.
6 PERSON OFFICE W/ 2
PRIVATE OFFICES
PRIVATE OFFICE IN
WHEELHOUSE
PRIVATE OFFICE
DESK
LOUNGE
CREDENZA
FILES
SECRETARY
XEROX
WORK STATION
OPEN OFFICE
CONFERENCE
BATH
SHOWER
HEAD
WAITING
SETTEE/
EXTENSION BERTH
PHOTOCOPIER
DOWN
OIL RANGE
REF'R.
STEP
COUNTER
DINING TABLE
STEREO & TV
SLIDING DOOR
TUB & SHOWER
WASHER & DRYER
DOUBLE BERTH
CHAIR
DRAWING BOARD
BOOKS
UP
LKR
PLAN FILES
BLUEPRINT MACHINE
UPPER & LOWER BERTHS
JAY R. BENFORD
P.O. BOX 447
ST. MICHAELS, MD 21663
(410) 745-3235
183-9
REVISIONS: 3/24/81

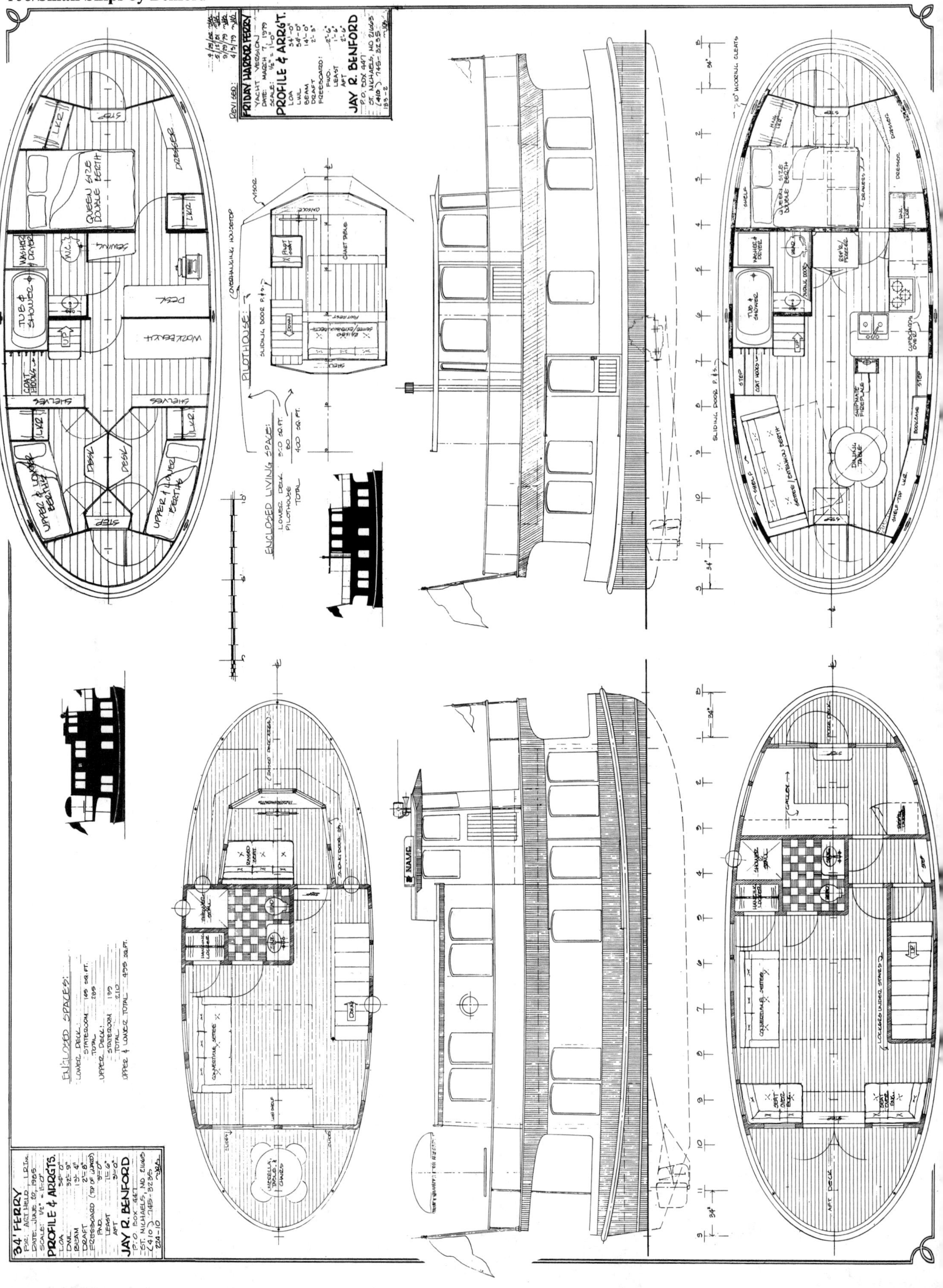

34' FERRY
FOR: ART HELD, LR Inc.
DATE: JUNE 20, 1985
SCALE: 1/2" = 1'-0"
PROFILE & ARRGTS.

LOA	34'-0"
DWL	32'-9"
BEAM	13'-4"
DRAFT	2'-8"
FREEBOARD (TOP OF GUARD)	
FWD.	3'-0"
LEAST	1'-6"
AFT	3'-0"

JAY R. BENFORD
P.O. BOX 447
ST. MICHAELS, MD 21663
(410) 745-3235
224-10 ~JRB

ENCLOSED SPACES:

LOWER DECK:	
STATEROOM	145 SQ. FT.
TOTAL	285
UPPER DECK:	
STATEROOM	135
TOTAL	210
UPPER & LOWER TOTAL	495 SQ. FT.

REVISED: 4/29/82 ~JRB, 5/12/81 ~JRB, 9/29/79 ~JRB, 4/3/79 ~JRB

FRIDAY HARBOR FERRY
YACHT VERSION
DATE: MARCH 7, 1979
SCALE: 1/2" = 1'-0"
PROFILE & ARRG'T.

LOA	34'-0"
LWL	34'-0"
BEAM	14'-0"
DRAFT	2'-3"
FREEBOARD:	
FWD.	2'-6"
LEAST	1'-6"
AFT	2'-6"

JAY R. BENFORD
P.O. BOX 447
ST. MICHAELS, MD 21663
(410) 745-3235
183-2 ~JRB

ENCLOSED LIVING SPACE:

LOWER DECK	320 SQ. FT.
PILOTHOUSE	80
TOTAL	400 SQ. FT.

PILOTHOUSE:

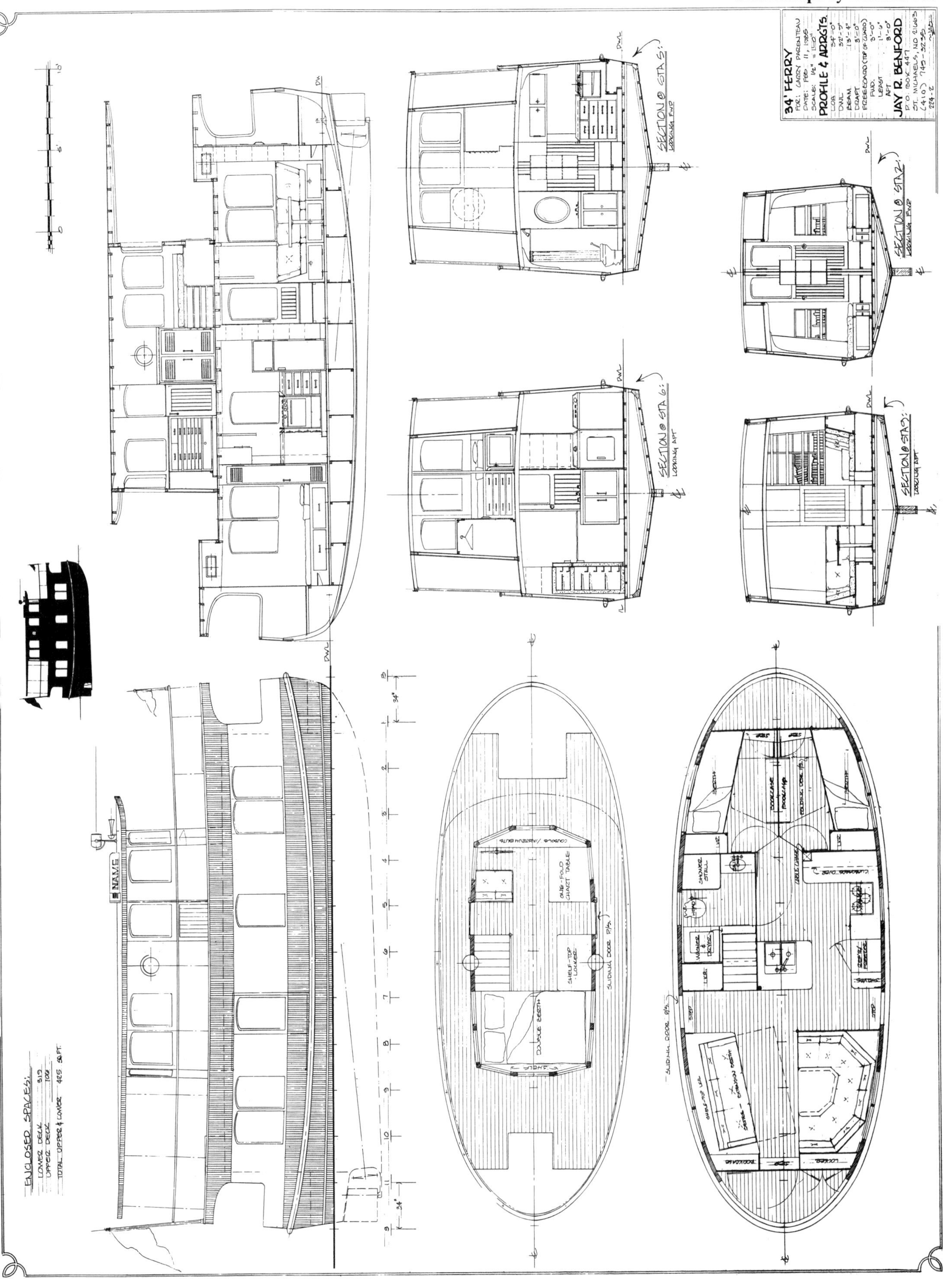
34' FERRY
FOR: GARRY PARENTEAU
DATE: FEB 11, 1985
SCALE: 1/2" = 1'-0"
PROFILE & ARRGTS.
LOA 34'-0"
DWL 32'-5"
BEAM 13'-4"
DRAFT 3'-0"
FREEBOARD (TOP OF GUARD)
FWD. 3'-0"
LEAST 1'-6"
AFT 3'-0"
JAY R. BENFORD
P.O. BOX 447
ST. MICHAELS, MD 21663
(410) 745-3235
224-2
SECTION @ STA 5: LOOKING FWD
SECTION @ STA 2: LOOKING FWD
SECTION @ STA 6: LOOKING AFT
SECTION @ STA 9: LOOKING AFT
ENCLOSED SPACES:
LOWER DECK 319
UPPER DECK 106
TOTAL UPPER & LOWER 425 SQ. FT.
NAME
ONE-FOLD CHART TABLE
SHELF-TOP LOCKER
SLIDING DOOR P/S
DOUBLE BERTH
SHOWER STALL
WASHER & DRYER
SLIDING DOOR P/S
SETTEE - EXTENSION BERTH

45' Friday Harbor Ferry, 48' Packet & 60' Schoolship

Designs Number 210, 235, 264 & 108
1983, 1985, 1987 & 1973

As spacious as a shoreside two or three bedroom home, the 45' Friday Harbor Ferry is a luxury home afloat. The 16' beam version (#210) was done originally with the idea of cold-molding or fiberglass construction. The 18' beam version (#235) was done later on, and plans have been completed for building in plywood with epoxy gluing and sealing.

Her intended power is a pair of 50 to 60 horsepower diesels, with over 3:1 reduction, so they will swing good sized props and have sufficient thrust to move her well in a breeze. This will give reasonable fuel consumption, even at cruising speed. She has substantial tankage, as befits a liveaboard, with much more generous water tankage than is common on yachts of this size.

The 48' Coastal Packet has room for lots of luxury touches, like the hot tub let in flush under the aft deck. Her layout has three private staterooms, each with a private head and shower stall. This is a layout that would be suited to both living aboard and use in charter work.

The 60' Schoolship is one of the oldest of our ferry designs. She is an idea we worked up in conjunction with a company that operated a charter boat out of Seattle. A large part of their work was taking out marine biology groups, and they thought they might be able to expand their work if they had a bigger boat that was set up to be a better working platform. They got caught up in the school districts' budget cuts, so it was never funded and, thus, we never got to try this one out.

Particulars:	**45' (#210)**	**45' (#235)**	**48'**	**60'**
Length overall	45'-0"	45'-0"	48'-0"	60'-0"
Length designed waterline	45'-0"	44'-0"	47'-6"	60'-0"
Beam	16'-0"	18'-0"	18'-0"	24'-0"
Draft	3'-0"	3'-6"	4'-0"	4'-0"
Freeboard:				
Forward	3'-9"	5'-0"	4'-6"	4'-9"
Least	1'-6"	3'-0"	1'-4"	3'-3"
Aft	3'-9"	5'-0"	1'-10"	4'-9"
Displacement, cruising trim*	34,125	45,000	75,000	97,550
Displacement-length ratio	167	236	312	202
Prismatic coefficient	.586	.672	.736	.569
Water tankage	500 Gals.	640 Gals.	800 Gals.	1,500 Gals.
Fuel tankage	500 Gals.	620 Gals.	600 Gals.	1,000 Gals.
Headroom	6'-6"	6'-6"	6'-7"	6'-9"

***CAUTION:** The displacement quoted here is for the boat in cruising trim. That is, with the fuel and water tanks filled, the crew on board, as well as the crews' gear and stores in the lockers. This should not be confused with the "shipping weight" often quoted as "displacement" by some manufacturers. This should be taken into account when comparing figures and ratios between this and other designs.

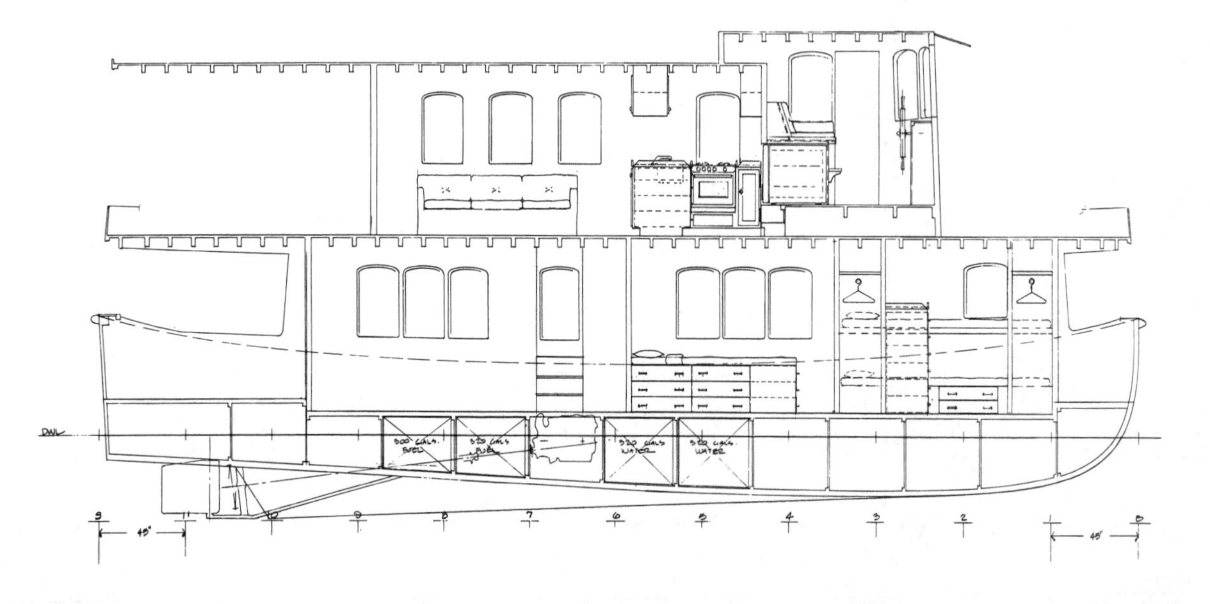

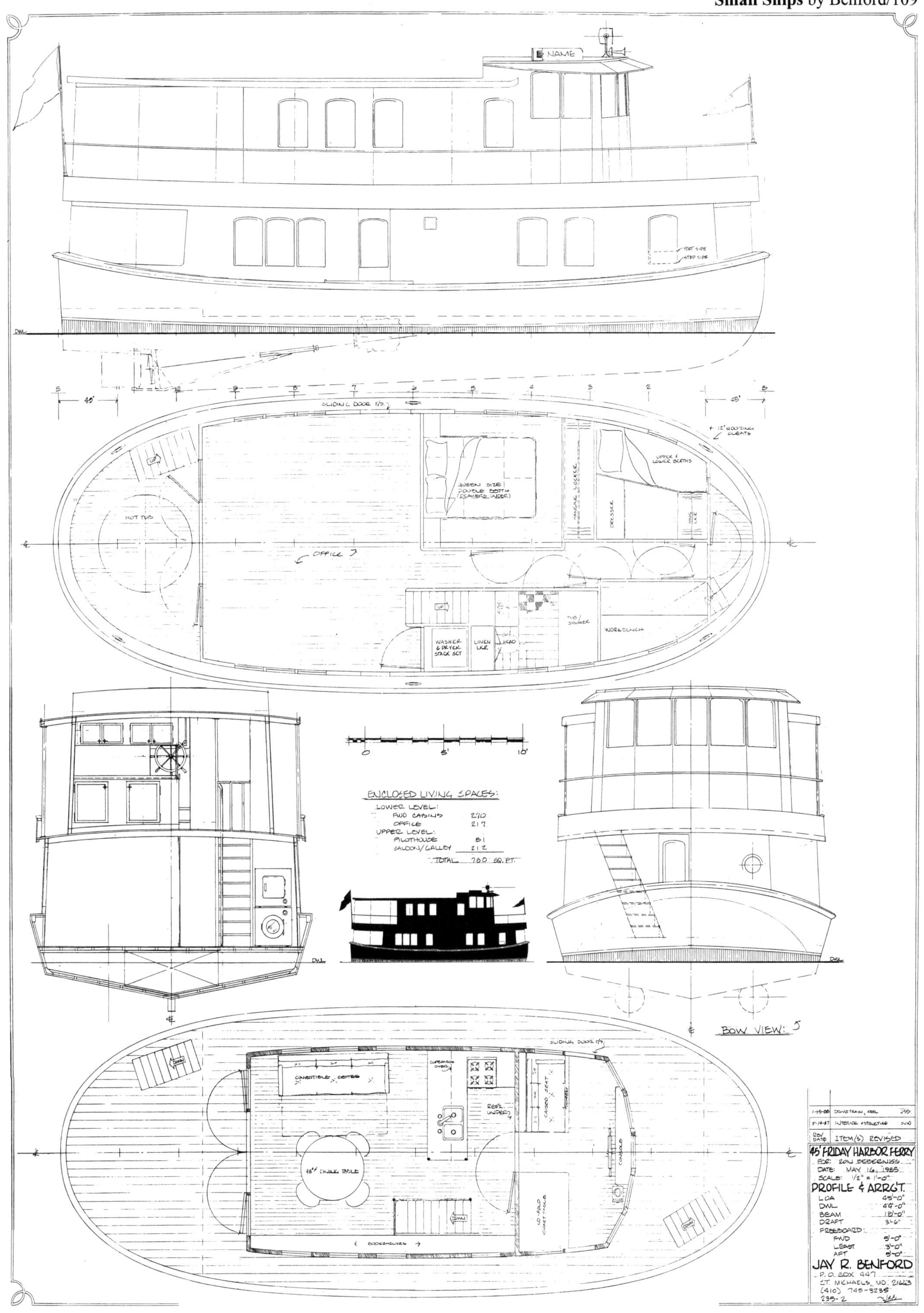

NAME
DWL
SLIDING DOOR P/S
12" MOORING CLEATS
UPPER & LOWER BERTHS
QUEEN SIZE DOUBLE BERTH (DRAWERS UNDER)
HANGING LOCKER
DRESSER
HOT TUB
OFFICE
UP
WASHER & DRYER STACK SET
LINEN LKR.
TUB/SHOWER
WORKBENCH
0
5'
10'
ENCLOSED LIVING SPACES:
LOWER LEVEL:
FWD CABINS 270
OFFICE 217
UPPER LEVEL:
PILOTHOUSE 81
SALOON/GALLEY 212
TOTAL 780 SQ.FT.
BOW VIEW
DOWN
CONVERTIBLE SETTEE
48" DINING TABLE
REFR. UNDER
CONSOLE
BOOKSHELVES
1-15-88 DRIVETRAIN, KEEL
5-14-87 INTERIOR & STRUCTURE
REV DATE ITEM(S) REVISED
45' FRIDAY HARBOR FERRY
FOR: RON BEBERNISS
DATE: MAY 16, 1985
SCALE: 1/2" = 1'-0"
PROFILE & ARRG'T.
LOA 45'-0"
DWL 44'-0"
BEAM 18'-0"
DRAFT 3'-6"
FREEBOARD:
FWD 5'-0"
LEAST 3'-0"
AFT 5'-0"
JAY R. BENFORD
P.O. BOX 447
ST. MICHAELS, MD. 21663
(410) 745-3235
235-2

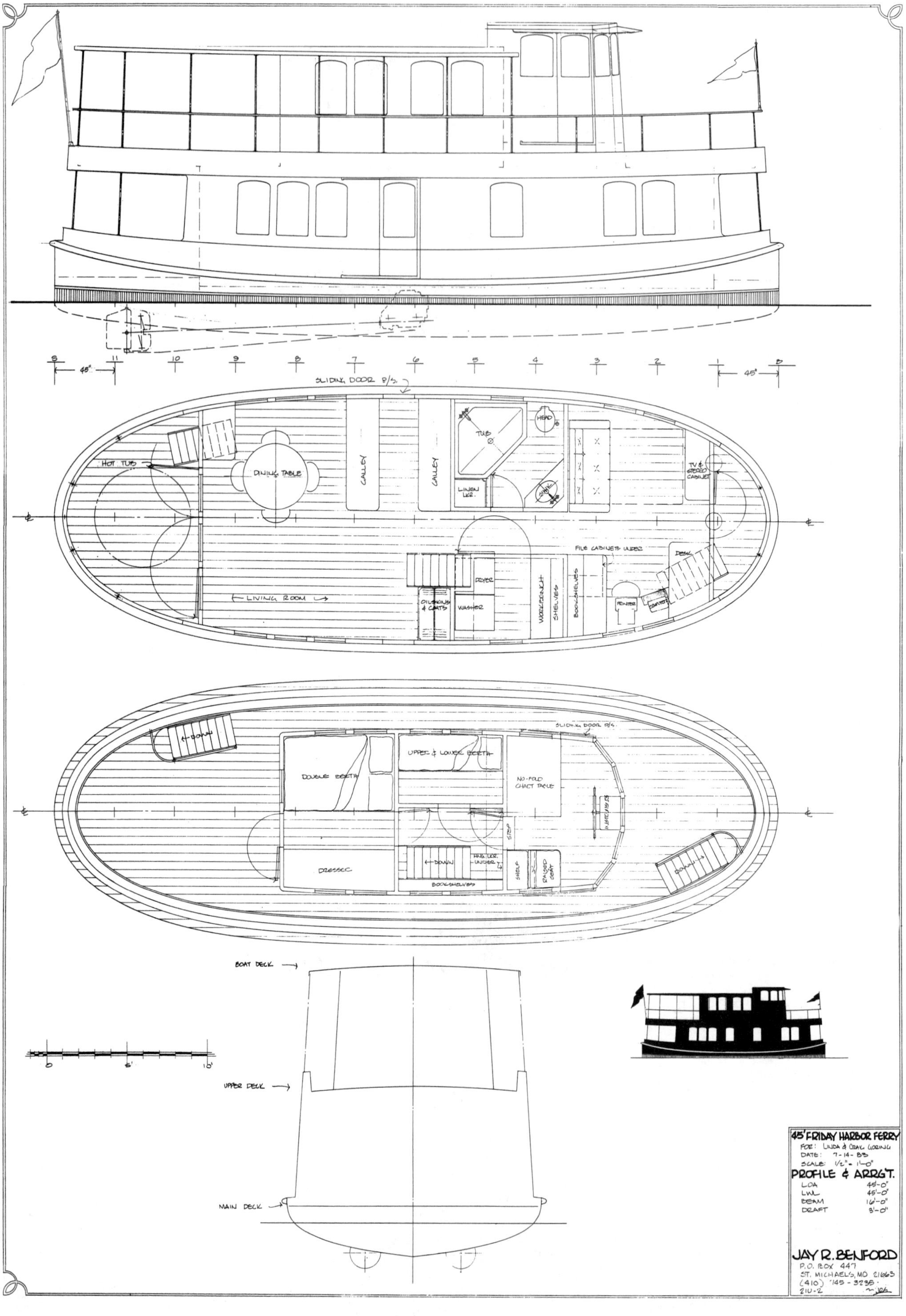

SLIDING DOOR P/S.

HOT TUB

DINING TABLE

GALLEY

GALLEY

TUB

HEAD

LINEN LKR.

SINK

TV & STEREO CABINET

FILE CABINETS UNDER

DESK

LIVING ROOM

DRYER

OILSKINS & COATS

WASHER

WORKBENCH

SHELVES

BOOKSHELVES

PRINTER

COMPUTER

DOWN

SLIDING DOOR P/S

UPPER & LOWER BERTH

DOUBLE BERTH

NO-FOLD CHART TABLE

STEP

DOWN

HNG. LKR. UNDER

SHELF

RAISED SEAT

DRESSER

BOOKSHELVES

DOWN

BOAT DECK

UPPER DECK

MAIN DECK

0 5' 10'

45' FRIDAY HARBOR FERRY

FOR: LINDA & CRAIG GORING

DATE: 7-14-83

SCALE: 1/2" = 1'-0"

PROFILE & ARRG'T.

LOA	45'-0"
LWL	45'-0"
BEAM	16'-0"
DRAFT	3'-0"

JAY R. BENFORD

P.O. BOX 447

ST. MICHAELS, MD 21663

(410) 745-3235

210-2

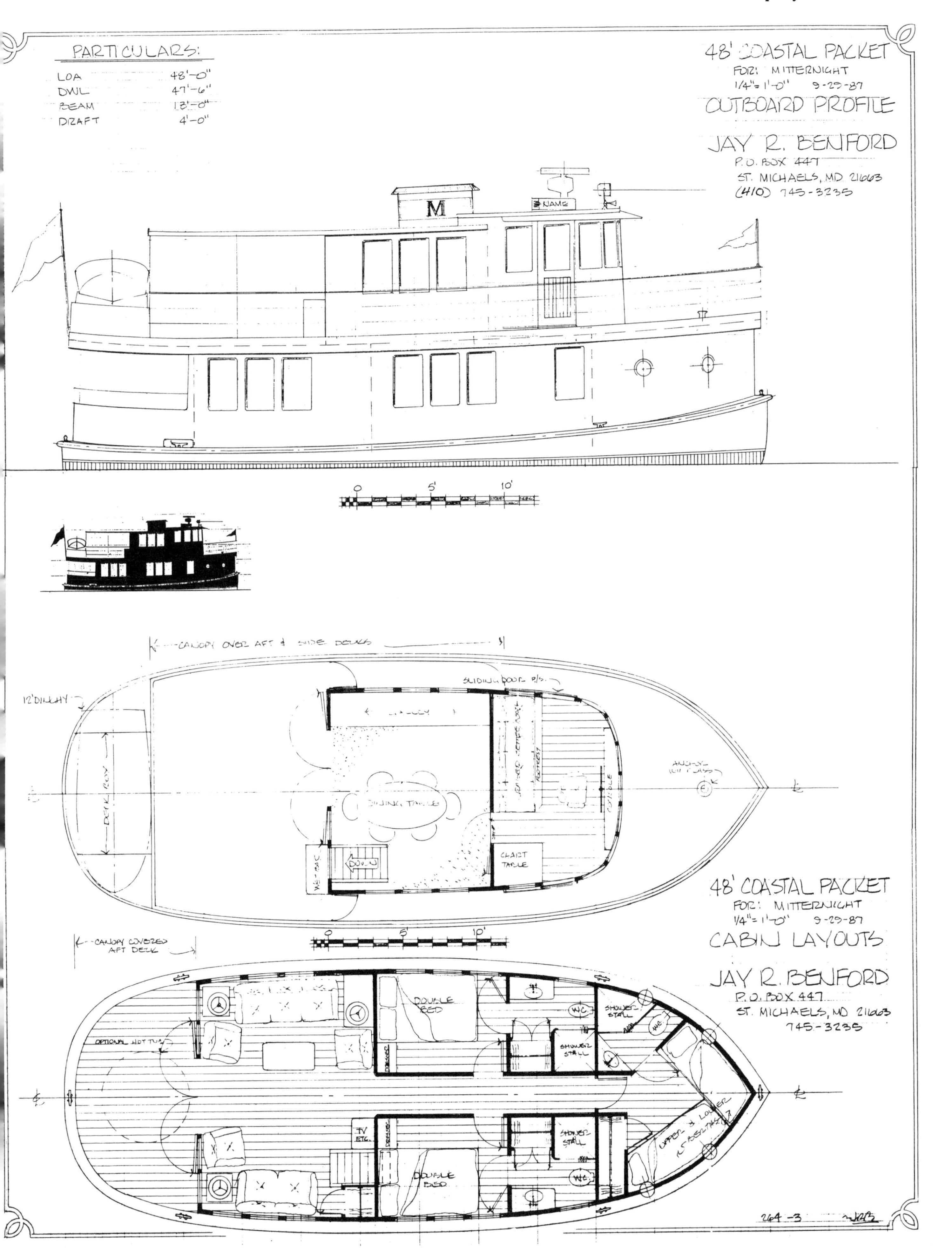
PARTICULARS:
LOA 48'-0"
DWL 47'-6"
BEAM 13'-0"
DRAFT 4'-0"
48' COASTAL PACKET
FOR: MITTERNIGHT
1/4"=1'-0" 9-29-87
OUTBOARD PROFILE
JAY R. BENFORD
P.O. BOX 447
ST. MICHAELS, MD 21663
(410) 745-3235
NAME
M
0
5'
10'
CANOPY OVER AFT & SIDE DECKS
SLIDING DOOR P/S
12' DINGHY
DINING TABLE
DOWN
CHART TABLE
ANCHOR WINDLASS
48' COASTAL PACKET
FOR: MITTERNIGHT
1/4"=1'-0" 9-29-87
CABIN LAYOUTS
JAY R. BENFORD
P.O. BOX 447
ST. MICHAELS, MD 21663
745-3235
CANOPY COVERED AFT DECK
OPTIONAL HOT TUB
DOUBLE BED
W.C.
SHOWER STALL
TV ETC.
UPPER & LOWER
264-3

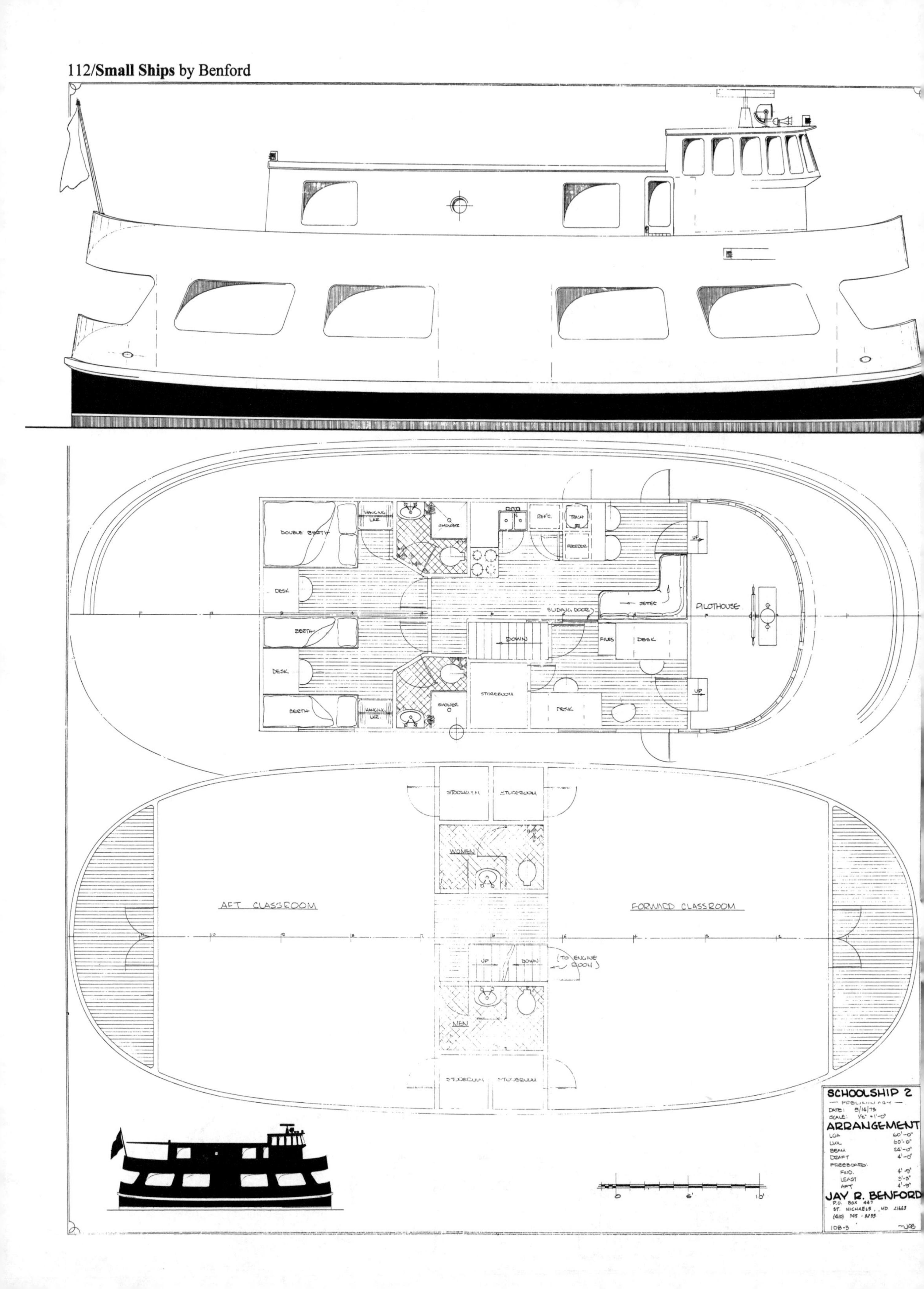

DOUBLE BERTH
HANGING LKR.
SHOWER
REF'G.
TRASH
FREEZER
DESK
SETTEE
PILOTHOUSE
SLIDING DOOR
UP
BERTH
DOWN
FILES
DESK
STOREROOM
SHOWER
HANGING LKR.
BERTH
STOREROOM
STOREROOM
WOMEN
AFT CLASSROOM
FORWARD CLASSROOM
UP
DOWN
(TO ENGINE ROOM)
MEN
STOREROOM
STOREROOM
SCHOOLSHIP 2
— PRELIMINARY —
DATE: 5/14/73
SCALE: 1/2" = 1'-0"
ARRANGEMENT
LOA 60'-0"
LWL 60'-0"
BEAM 24'-0"
DRAFT 4'-0"
FREEBOARD:
FWD. 4'-9"
LEAST 3'-3"
AFT 4'-9"
JAY R. BENFORD
P.O. BOX 447
ST. MICHAELS, MD 21663
(410) 745 - 3235
10B-3
JRB
0
5'
10'

65' Excursion Boats

Designs Number 294 & 317
1989 & 1991

The ***Patriot*** of St. Michaels is a 65' excursion boat, now in service in St. Michaels, Maryland. Her primary services are scenic and historic river tours for up to 210 people and dinner cruises for about 90. The ***Patriot*** shows another facet of our design work specializing in small ships and cruising yachts. She's built of steel, with the pilothouse and attached house built of aluminum. Her lower house has the heads, stairs, and service area all aft, with the rest of the house a wide-open space of about 940 square feet. With this versatility in adapting to different service needs she can be used for meals, dances, weddings, live music groups, or small theater productions — all in a climate controlled space.

The upper deck features a classic round front pilothouse, with an elevated bridge around it, for excellent visibility.

We spent extra effort in creating a vessel that would almost double the prior boat's capacity (210 vs. 118), with eight feet more beam, and still be driven a little faster; all with lowered fuel consumption and a reduced wake. She admeasured 50 gross tons, has fuel tankage for 2,400gals. and water tankage for 1,300gals.

The following drawing shows one of the things we do as a part of designing a developable hull surface, whether for steel, aluminum, plywood, or fiberglass panels. It's part of the sophisticated computer design software that we use and allows us to "unwrap" the surfaces to a flat panel.

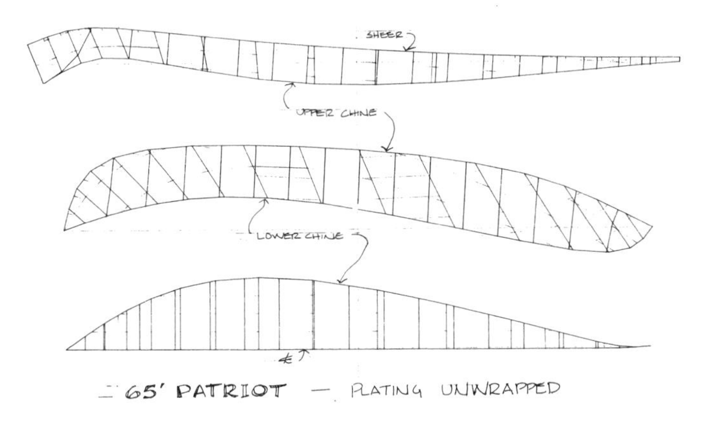

During the design process, we use it as a check against proposed materials sizes, whether it be eight foot wide sheets of steel or plywood or some other size chosen by us or the builder. It's a real pain, let alone expense, for the builder to have to splice on a few extra inches in width to make up a panel. If we can design around the standard material sizes while it's easy to make changes either on paper or in the computer, it is a considerable savings for the builder, which can translate to profits to the builder and a reasonable price to the owner.

The process also is a help in the design work, allowing us to have accurate areas for doing weight calculations. Computers have certainly changed the way we do design work, and let us do more work in less time. This translates into better and more accurate design work and results in boats that perform better and can be built in less time.

Designs like the ***Patriot*** can be done in almost any size. Do you need a 50, 100, or 500 passenger excursion boat? This type of simple, yet elegant, classic working vessel can be a good basis for a boat that will produce good earnings. If you look in the section about our 35' Packet you will see some ideas for smaller versions.

Particulars:	**294**	**317**
Length overall	64'-10"	64'-11"
Length designed waterline	62'-9"	63'-0"
Beam	26'-0"	22'-0"
Draft	5'-0"	4'-6"
Freeboard:		
Forward	5'-0"	6'-6"
Least	3'-4"	3'-0"
Aft	4'-3"	4'-0"
Displacement, cruising trim*	198,000 lbs.	170,000
Displacement-length ratio	358	304
Prismatic coefficient	.59	
Pounds per inch immersion	6,180	
Water tankage	1,300 Gals.	1,800 Gals.
Fuel tankage	2,400 Gals.	2,000 Gals.
Headroom	6'-9"	6'-7"
GM	9.4' (with all passengers on upper deck)	

***CAUTION:** The displacement quoted here is for the boat in working trim. That is, with the fuel and water tanks filled, the crew and passengers on board, as well as the crews' gear and stores in the lockers. This should not be confused with the "shipping weight" often quoted as "displacement" by some manufacturers. This should be taken into account when comparing figures and ratios between this and other designs.

Also included are some preliminary studies for a couple of other 65-foot variations. These were drawn up to explore the possibility of creating a boat that looked like the Chesapeake Bay Buy Boats. The capacity would likely be more in the 100 to 150 range in the layouts as shown. The captain and crew would have staterooms and heads in the area under the pilothouse and galley.

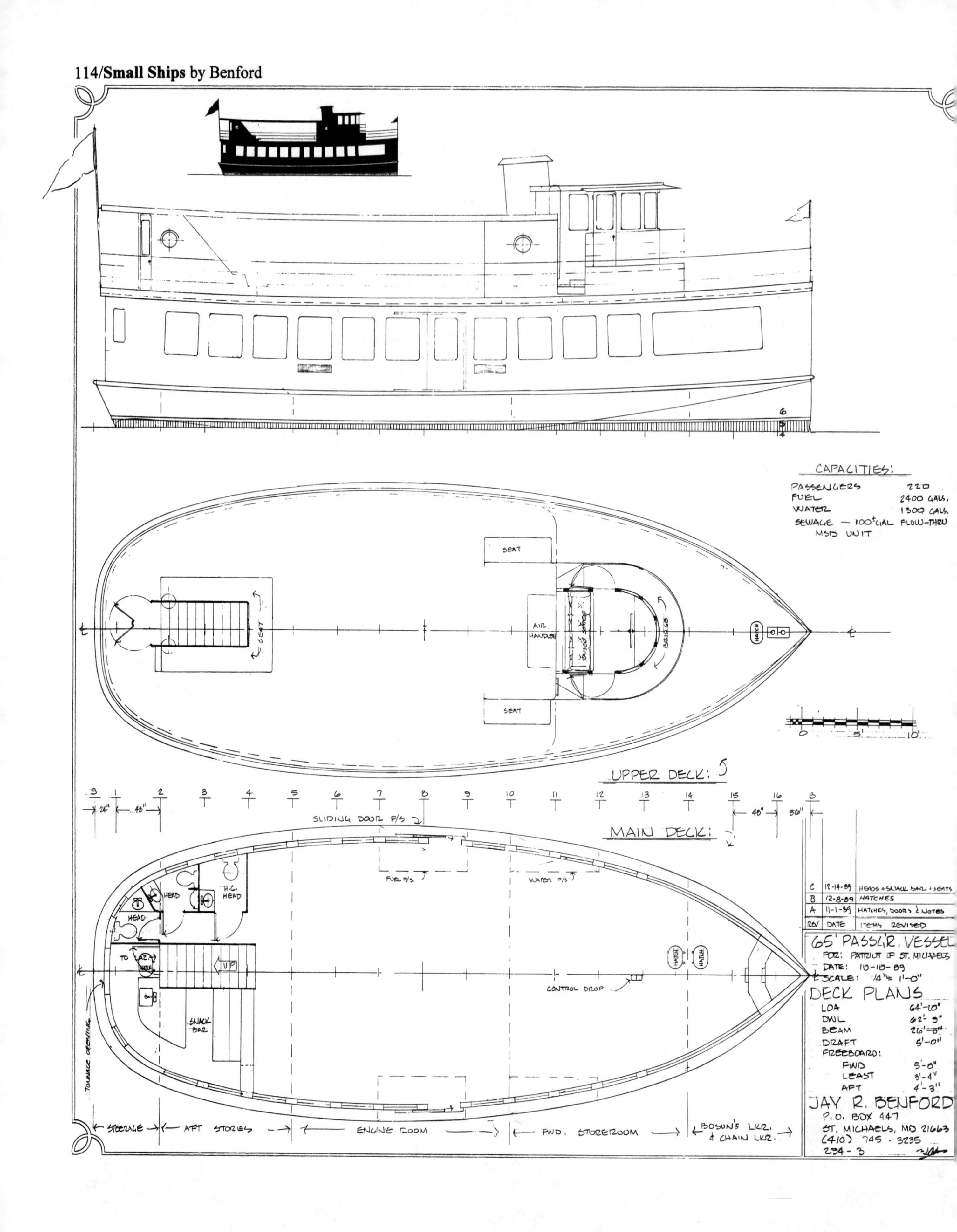
CAPACITIES:
PASSENGERS 220
FUEL 2400 GALS.
WATER 1300 GALS.
SEWAGE — 100± GAL FLOW-THRU MSD UNIT
SEAT
AIR HANDLER
BRIDGE
HATCH
SEAT
0 5' 10'
UPPER DECK:
MAIN DECK:
SLIDING DOOR P/S
FUEL P/S
WATER P/S
HEAD
H.C. HEAD
HEAD
UP
SNACK BAR
HATCH
HATCH
CONTROL DROP
STEERAGE
AFT STORES
ENGINE ROOM
FWD. STOREROOM
BOSUN'S LKR. & CHAIN LKR.
C 12-14-89 HEADS + SNACK BAR + SEATS
B 12-8-89 HATCHES
A 11-1-89 HATCHES, DOORS & NOTES
REV DATE ITEMS REVISED
65' PASS'G'R. VESSEL
FOR: PATRIOT OF ST. MICHAELS
DATE: 10-10-89
SCALE: 1/4" = 1'-0"
DECK PLANS
LOA 64'-10"
DWL 62'-9"
BEAM 26'-8"
DRAFT 5'-0"
FREEBOARD:
FWD 5'-6"
LEAST 3'-4"
AFT 4'-3"
JAY R. BENFORD
P.O. BOX 447
ST. MICHAELS, MD 21663
(410) 745-3235
234-3

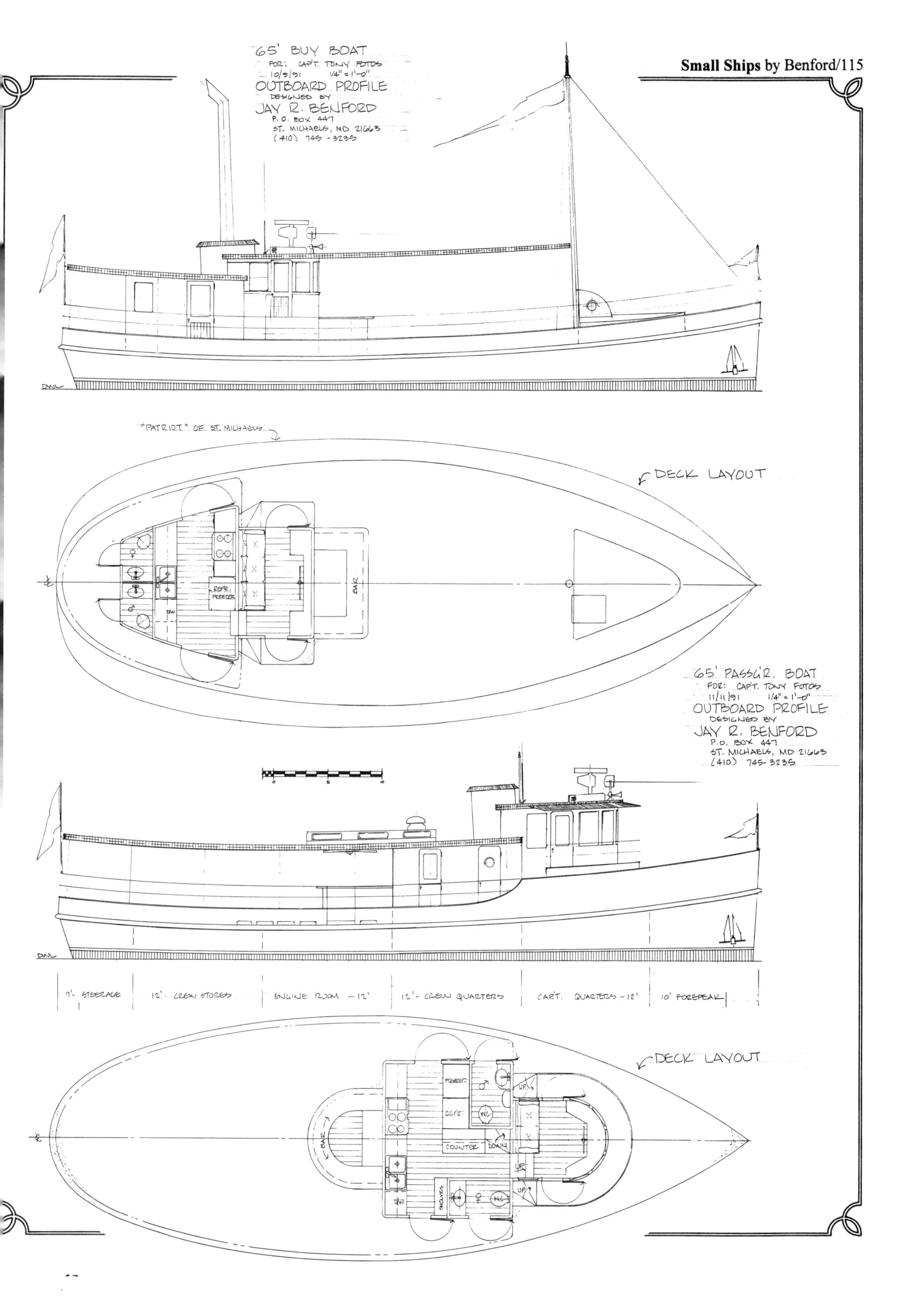
65' BUY BOAT
FOR: CAP'T. TONY FOTOS
10/3/91 1/4" = 1'-0"
OUTBOARD PROFILE
DESIGNED BY
JAY R. BENFORD
P.O. BOX 447
ST. MICHAELS, MD 21663
(410) 745-3235
DWL
"PATRIOT" OF ST. MICHAELS
DECK LAYOUT
REFR. FREEZER
D/W
BAR
65' PASSG'R. BOAT
FOR: CAP'T. TONY FOTOS
11/11/91 1/4" = 1'-0"
OUTBOARD PROFILE
DESIGNED BY
JAY R. BENFORD
P.O. BOX 447
ST. MICHAELS, MD 21663
(410) 745-3235
0 5 10
DWL
7' - STEERAGE
12' - CREW STORES
ENGINE ROOM - 12'
12' - CREW QUARTERS
CAP'T. QUARTERS - 12'
10' FOREPEAK
DECK LAYOUT
FREEZER
REFER
W.C.
COUNTER
DOWN
UP
BAR
D/W
SHELVES

72' CE2

Design Number 234
1985

The brief for this design was for a boat to operate as a conference and entertainment (thus CE) facility for a waterfront hotel. With ever increasing demands on waterfront hotels for room capacity, this would let the usual space allocated for conference rooms be used for more revenue producing bedrooms. Additionally, the boat would provide a feature that would attract groups to meet on it as a change of pace from the typical and mundane.

The people using the facility could thus have the choice of whether they wanted an uninterrupted day afloat, away from the demands of ringing phones, while enjoying a continually changing vista with the boat underway. Or, they could stay tied to shore, to let the participants come and go during the meeting or party.

Particulars:

Length overall	72'-0"
Length designed waterline	69'-0"
Beam	32'-6"
Draft	6'-0"
Freeboard:	
Forward	6'-0"
Least	4'-0"
Aft	6'-0"
Displacement, cruising trim*	196,000 lbs.
Displacement-length ratio	266
Prismatic coefficient	.64
Water tankage	1,600 Gals.
Fuel tankage	2,000 Gals.
Headroom	7'-0" to 8'-0"

***CAUTION:** The displacement quoted here is for the boat in cruising trim. That is, with the fuel and water tanks filled, the crew on board, as well as the crews' gear and stores in the lockers. This should not be confused with the "shipping weight" often quoted as "displacement" by some manufacturers. This should be taken into account when comparing figures and ratios between this and other designs.

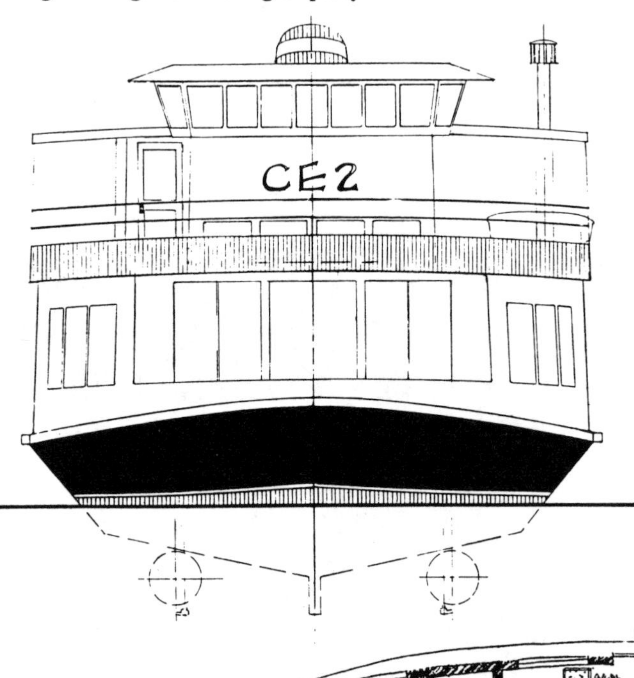

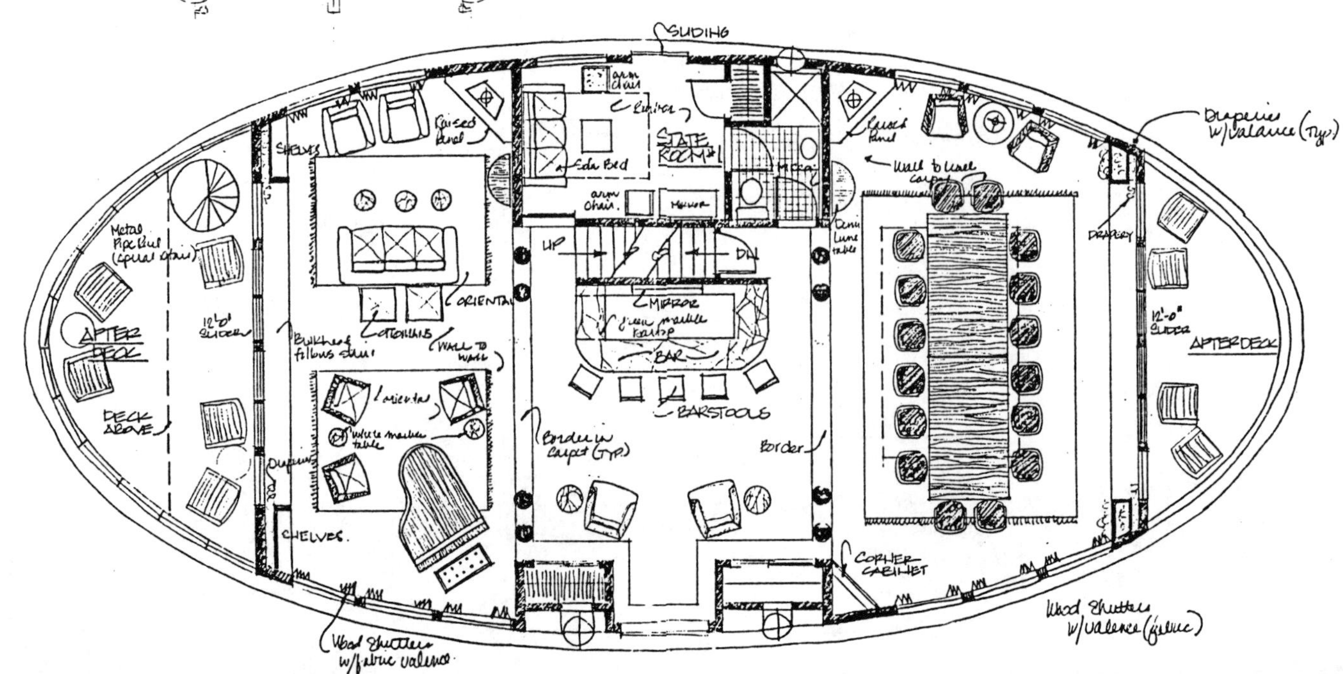

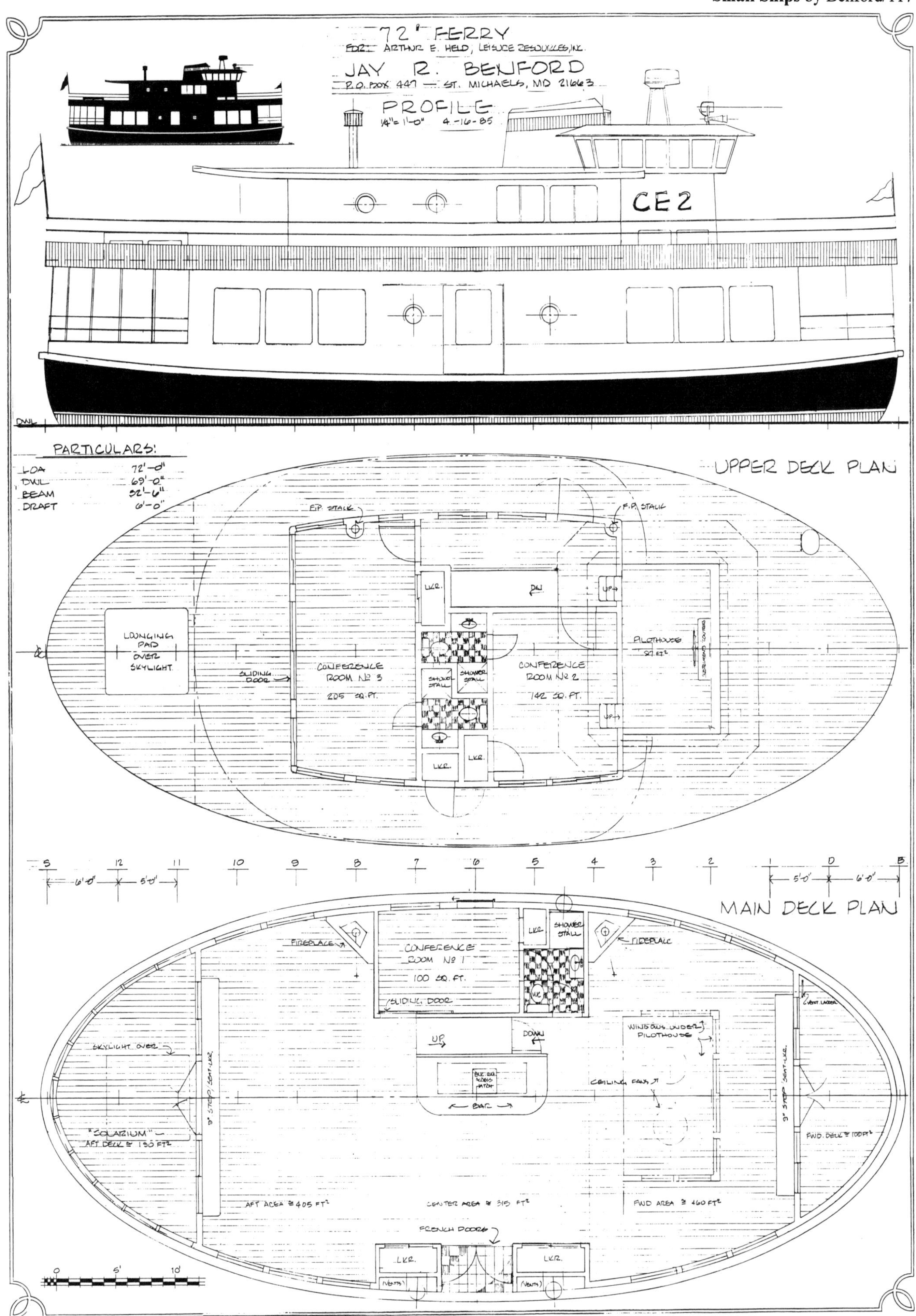

72' FERRY
FOR: ARTHUR E. HELD, LEISURE RESOURCES, INC.
JAY R. BENFORD
P.O. BOX 447 — ST. MICHAELS, MD 21663
PROFILE
1/4" = 1'-0" 4-16-85
CE2
DWL
PARTICULARS:
LOA 72'-0"
DWL 69'-0"
BEAM 32'-6"
DRAFT 6'-0"
UPPER DECK PLAN
F.P. STALK
LKR.
DN
UP
LOUNGING PAD OVER SKYLIGHT
SLIDING DOOR
CONFERENCE ROOM No 3
205 SQ. FT.
SHOWER STALL
CONFERENCE ROOM No 2
142 SQ. FT.
PILOTHOUSE
MAIN DECK PLAN
FIREPLACE
CONFERENCE ROOM No 1
100 SQ. FT.
SLIDING DOOR
WINDOWS UNDER PILOTHOUSE
DOWN
SKYLIGHT OVER
CEILING FANS
BAR
3' STEP SEAT LKR.
"SOLARIUM" AFT DECK ≅ 130 FT²
FWD. DECK ≅ 100 FT²
AFT AREA ≅ 405 FT²
CENTER AREA ≅ 315 FT²
FWD AREA ≅ 460 FT²
FRENCH DOORS
(VENTS)
0 5' 10'

26' Day Boat

Design Number 6

1965

Three decades ago, when I was apprenticing with John Atkin, we talked about the great number of boats that seemed to be used only as cocktail lounges, sitting in the marinas all year and never doing anything more than being a place for entertaining guests.

With this thought in mind, I later drew up this 26' Day Boat. She has all the requisites that John and I discussed; there's a wet bar, refrigerator, plenty of room for sitting around, and an enclosed head with a shower stall.

Looking back at the design, with this separation of time, I'd revamp the profile, raising the housetop a bit for some additional headroom and extend it all the way to the stern. I'd omit the powerplant entirely, using the dinghy to tow her to her haulouts, and think about turning the forward end of the boat into a sleeping cabin. If power was really wanted, I'd go for an outboard, which could also be used on a larger dinghy for harbor tours.

Particulars:

Length overall	26'-0"
Length designed waterline	20'-0"
Beam	12'-0"
Draft	1'-2"
Freeboard:	
Forward	4'-7"
Least	2'-7"
Aft	3'-2"
Displacement, cruising trim	5,860 lbs.
Water tankage	100 Gals.
Fuel tankage	96 Gals.
Headroom	6'-0" to 6'-5"

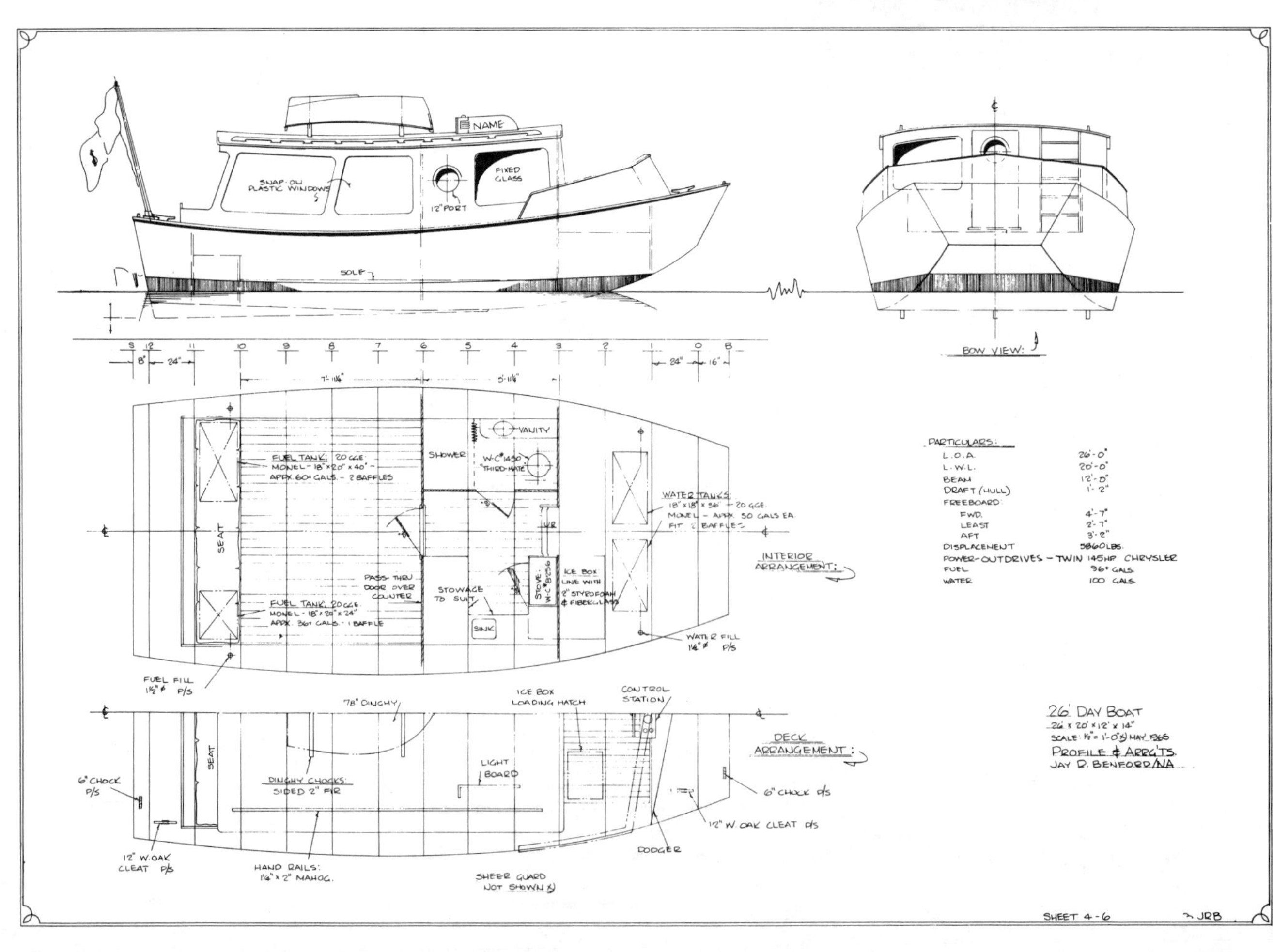

36' House Barge

Design Number 1

1964

This is a conceptual study for simple liveaboard, done while apprenticing with John Atkin, about three decades ago. My concept/fantasy was to work on the idea of what might be done in the way of a boat to live aboard in reasonable comfort and have a little office space. The lower deck level had all the living spaces on it. I'd add a washer and dryer to the layout to make living aboard more convenient.

The upper cabin was to be a small study/office, with drawing board, desk, files and a couple chairs in which to sit and talk boats. She's intended to be built from commonly available lumberyard materials. Construction would be similar to stick-built house building, with plywood used instead of sheet rock and more attention paid to fits and gussets on the framing connections. I'd build the whole boat with plywood and sheath the hull with epoxy and cloth to take the abrasion and local impact loads.

With no power plant of her own, she could be towed by one of our small tugs, like the 14 or 20-footers in our **Pocket Cruisers & Tabloid Yachts** book or one of the smaller tugs in this book. This tug could also serve as a shore boat, if living at anchor or on a mooring. The tug could also serve for doing some cruising and exploring.

Looking at the design today, I'd revise the upper house top to parallel the lower deck housetop and extend it to become more of a canopy and shelter for the upper deck, to make it a great place for having social gatherings, picnics, and just sitting out and enjoying the weather.

Particulars:

Length overall	36'-0"
Length designed waterline	32'-0"
Beam	14'-0"
Draft	1'-0"
Displacement, cruising trim*	27,800 lbs.
Displacement-length ratio	378
Prismatic coefficient	.938
Water tankage	500 Gals.
Headroom	6'-7"

***CAUTION:** The displacement quoted here is for the boat in cruising trim. That is, with the fuel and water tanks filled, the crew on board, as well as the crews' gear and stores in the lockers. This should not be confused with the "shipping weight" often quoted as "displacement" by some manufacturers. This should be taken into account when comparing figures and ratios between this and other designs.

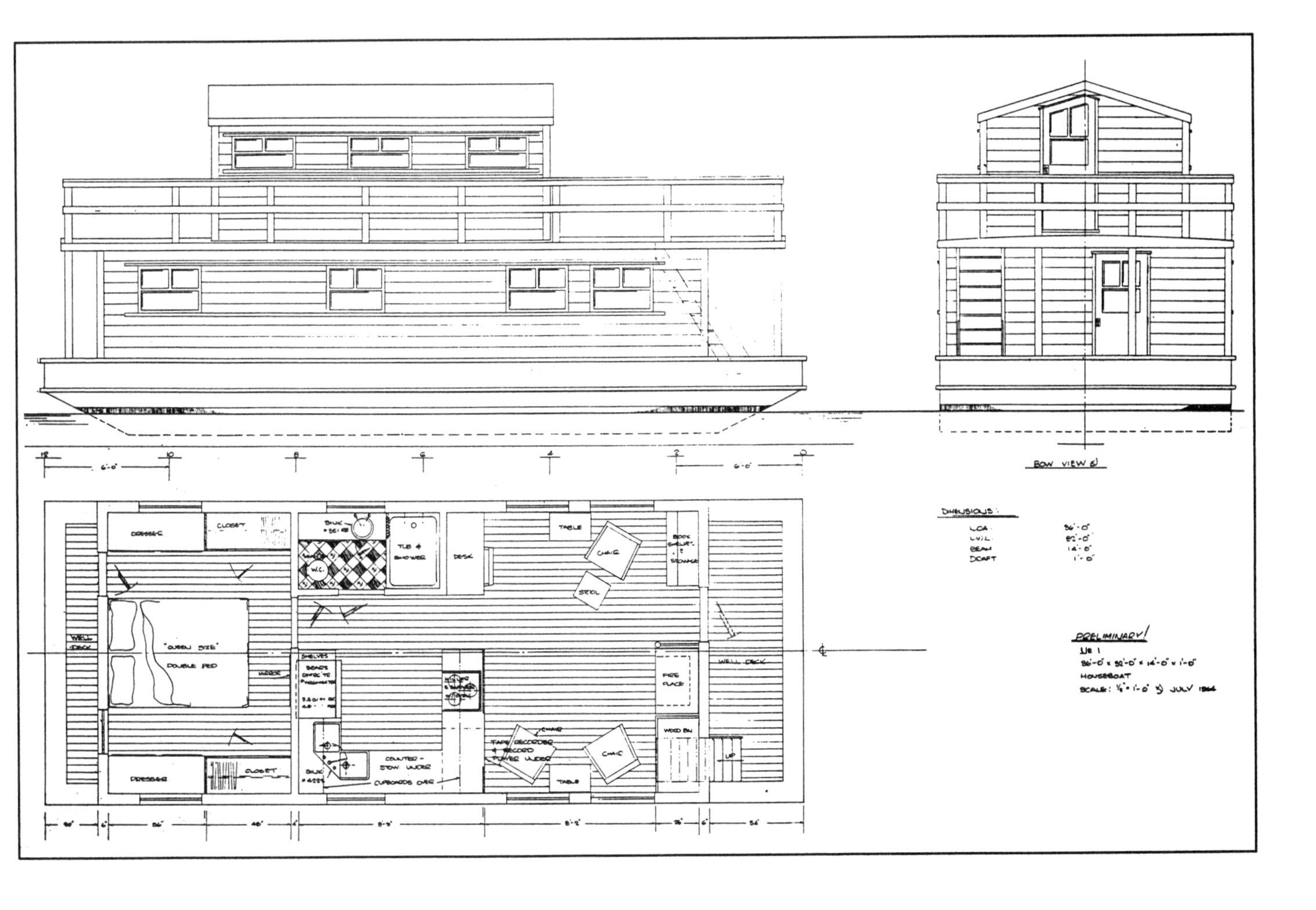

Square One

Design Number 288, 1989

This design might readily be thought of as putting a sailing pram on steroids. Her living accommodations are maximized within her hull.

The project began with Reuben Trane, president of the Florida Bay Coaster Company, being contacted by his old friend, author Sloan Wilson. He had written up a description of the boat he dreamed of having to liveaboard and Reuben and I collaborated to turn it into something that might be professionally built with economy.

We looked at the historic sailing scows and took them as a point of departure in creating this idea to be built of steel and contemporary building materials techniques.

We came up with two different ideas on a simple rig for the boat. The junk or lug rig is a very low-tech solution and is hinged at the pilothouse top for bridge clearance. The other rig is a large jib on furling gear, with a mast also in a tabernacle, this time folding forward. In both cases, we're assuming that the boat will be motored to windward and use the rig off the wind. If most cruising sailors were honest enough, they'd admit that this is how they really do cruise....

We've put in lots of windows for light and ventilation. The aft sofa and the pilothouse seating provide guest berthing and the second head off the passageway can serve them.

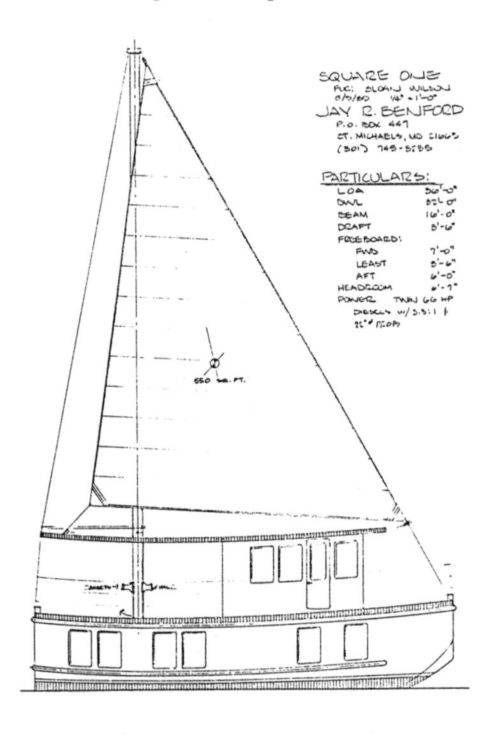

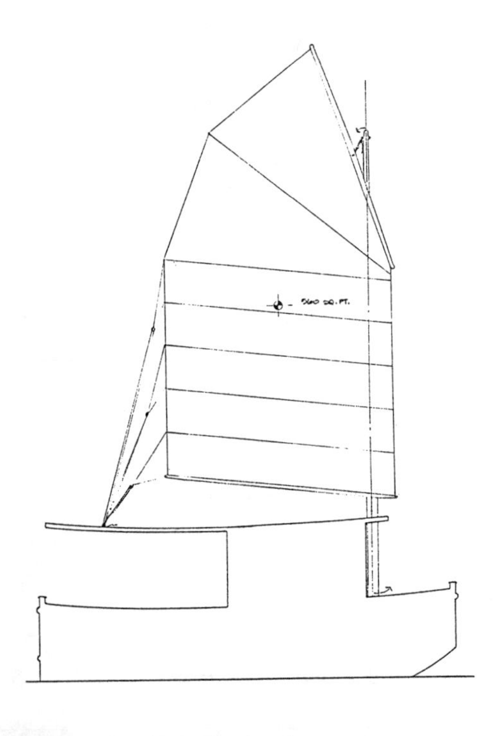

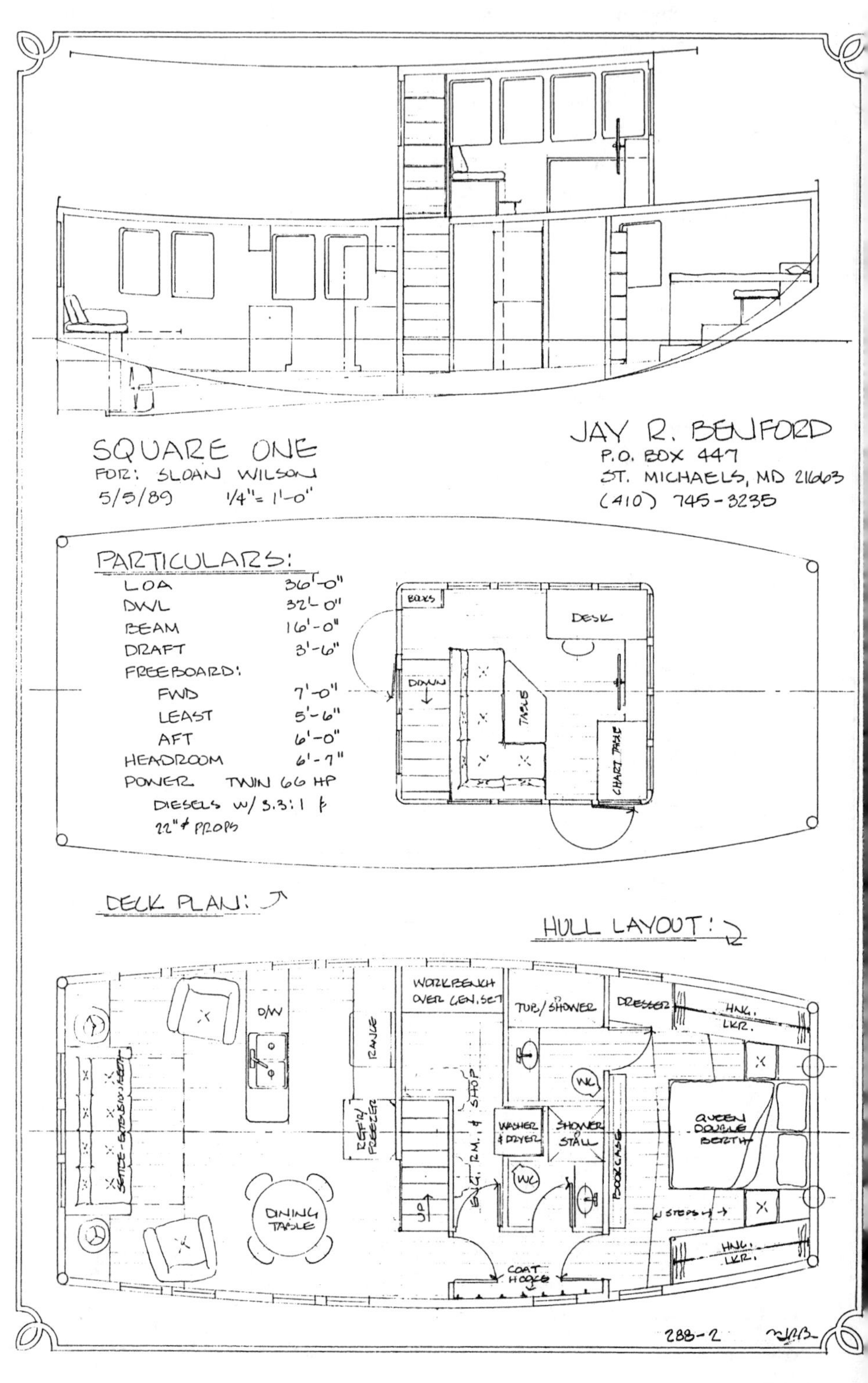

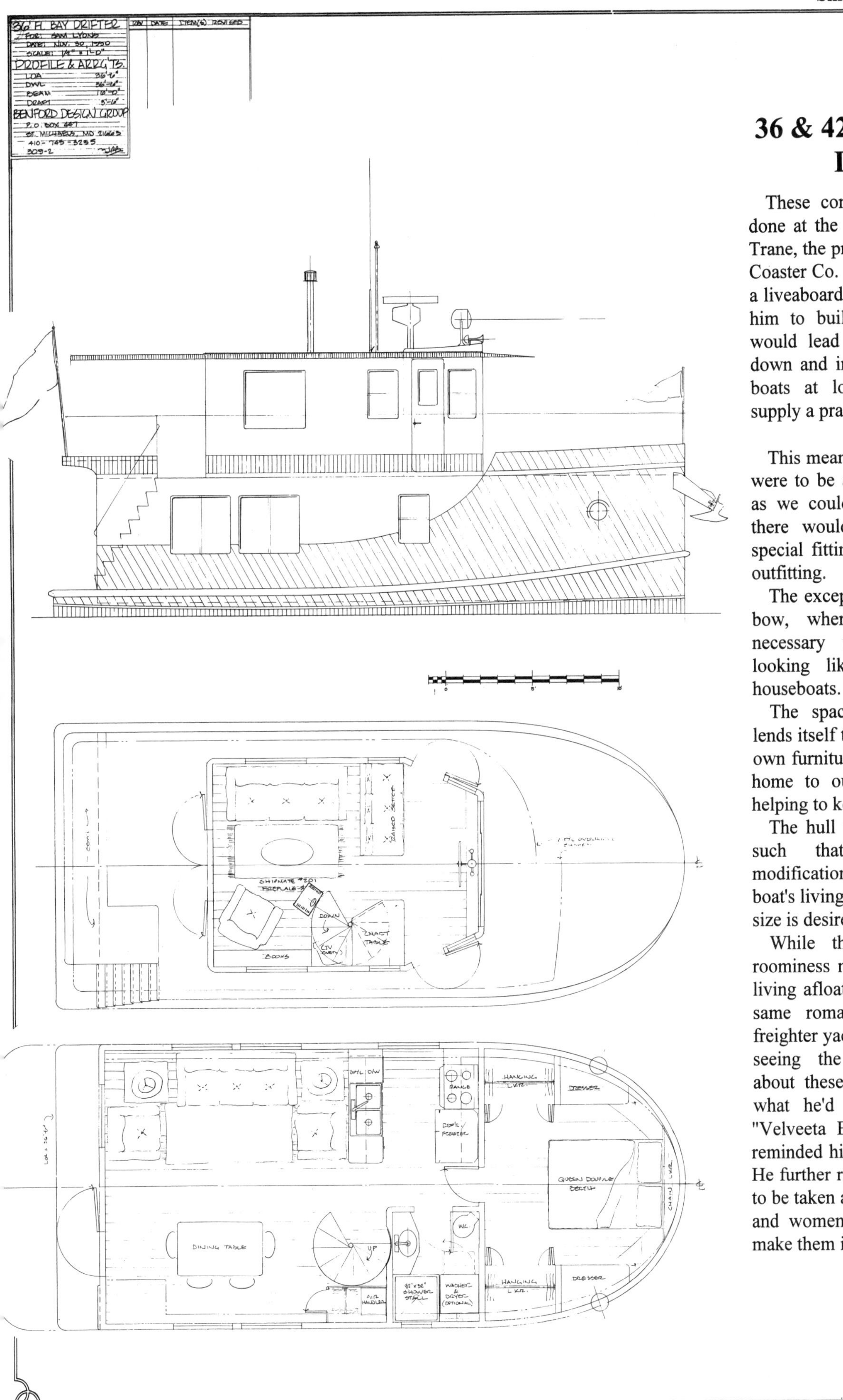

36 & 42' Florida Bay Drifters

These conceptual designs were done at the suggestion of Reuben Trane, the president of Florida Bay Coaster Co. His idea was to create a liveaboard that was as simple for him to build as possible. This would lead to keeping his costs down and in turn let him sell the boats at lower prices and still supply a practical liveaboard.

This meant that the living spaces were to be as close to rectangular as we could make them, so that there would be a minimum of special fitting work needed in the outfitting.

The exception to this was in the bow, where some shape was necessary to keep them from looking like the square ended houseboats.

The spaciousness of the boat lends itself to people bringing their own furniture and appliances from home to outfit the boat, further helping to keep costs in hand.

The hull form, as conceived, is such that we can make modifications to it and stretch the boat's living spaces a bit, if a larger size is desired.

While these boats offer the roominess needed for comfortable living afloat, they do not have the same romantic appeal that our freighter yachts offer. A friend, on seeing the drawings, remarked about these boats that they were what he'd taken to calling the "Velveeta Boats" for their shape reminded him of a box of cheese. He further remarked, not intending to be taken as a sexist, that in boats and women, "It's the curves that make them interesting"....

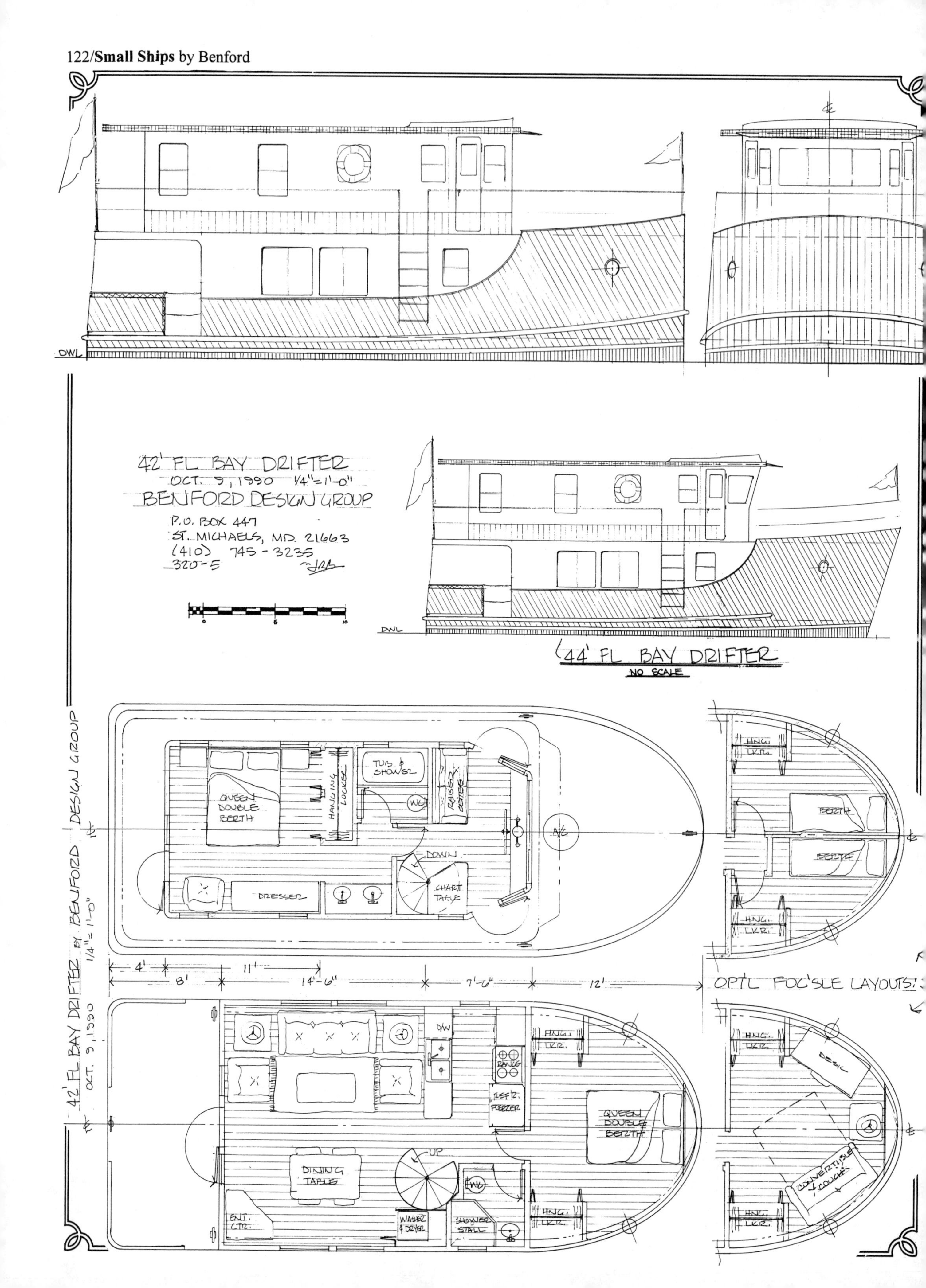

DWL
42' FL BAY DRIFTER
OCT. 9, 1990 1/4"=1'-0"
BENFORD DESIGN GROUP
P.O. BOX 447
ST. MICHAELS, MD. 21663
(410) 745-3235
320-5
0
5
10
DWL
44' FL BAY DRIFTER
NO SCALE
42' FL BAY DRIFTER BY BENFORD DESIGN GROUP
1/4" = 1'-0"
OCT. 9, 1990
QUEEN DOUBLE BERTH
HANGING LOCKER
TUB & SHOWER
RAISED BOTTLES
W.C.
A/C
DOWN
CHART TABLE
DRESSER
HNG. LKR.
BERTH
4'
11'
8'
14'-6"
7'-6"
12'
OPT'L FOC'SLE LAYOUTS
D/W
RANGE
REFR. FREEZER
QUEEN DOUBLE BERTH
DINING TABLE
UP
W.C.
ENT. CTR.
WASHER & DRYER
SHOWER STALL
DESK
CONVERTIBLE COUCHES

60' Floating Lodge

Design Number 293

1989

Our clients for this project are the successful operators of a charter boat operation at Sioux Falls on Lake Of The Woods in western Ontario. They had a number of smaller boats in service and wanted to add a larger version to their operation, by building their own boats instead of buying stock houseboats.

The operation consists mostly of people coming to the area to go fishing and enjoy the scenery. Thus, a larger boat like this will often leave with several smaller fishing skiffs in tow. The mothership/lodge would move to a location where they wanted to try out the fishing, tie off to the shore and send the crew out in the small boats to enjoy fishing. Thus, the aptness of the Floating Lodges name for the company.

Shown below and on the bottom of the following page is the first preliminary study we did. The drawing on the upper part of the following page is the final version, with a number of refinements and changes that are evident in comparing the two versions.

The private sleeping cabins can be used for a couple, with one of the double beds being used, or by a pair of fishermen with both beds in use.

The layout has three separate bathrooms, one of which has a whirlpool tub, plus a separate shower room. There is a sitting room in the pilothouse area and at the forward end of the lower cabin. The kitchen has plenty of room to whip up meals for a gang of hungry fishermen and the outdoor barbecue gives a way to cook the fish while they're really fresh. The gas can locker on the outside of the cabin keeps the cans safely out of the way. The large locker opening onto the aft deck is a place to keep all the fishing rods and tackle boxes.

The foundation for these lodges are catamaran floats, of a simplified shape to be built in steel. They are powered by outdrives in each hull.

Particulars:

LOA	60'-0"
LWL	60'-0"
Beam	18'-0"
Draft (average load)	2'-0"
Freeboard, main deck	1'-8"

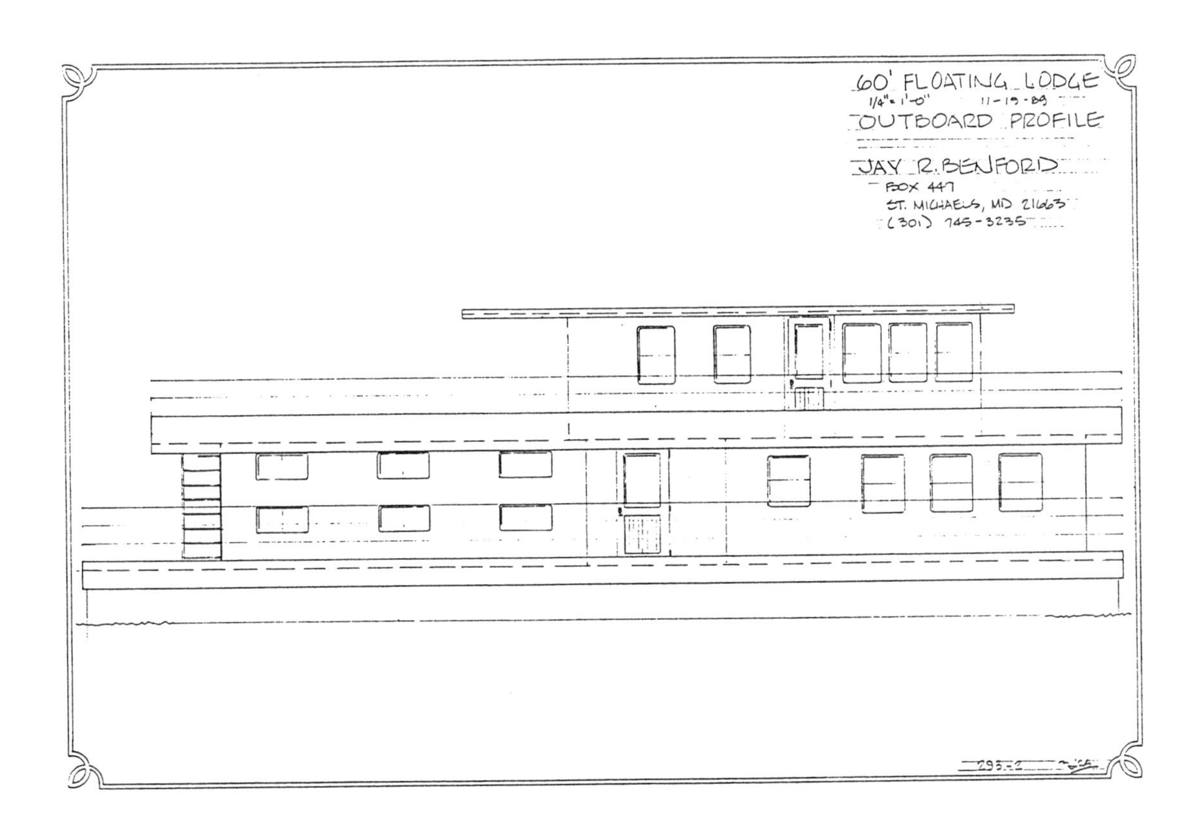

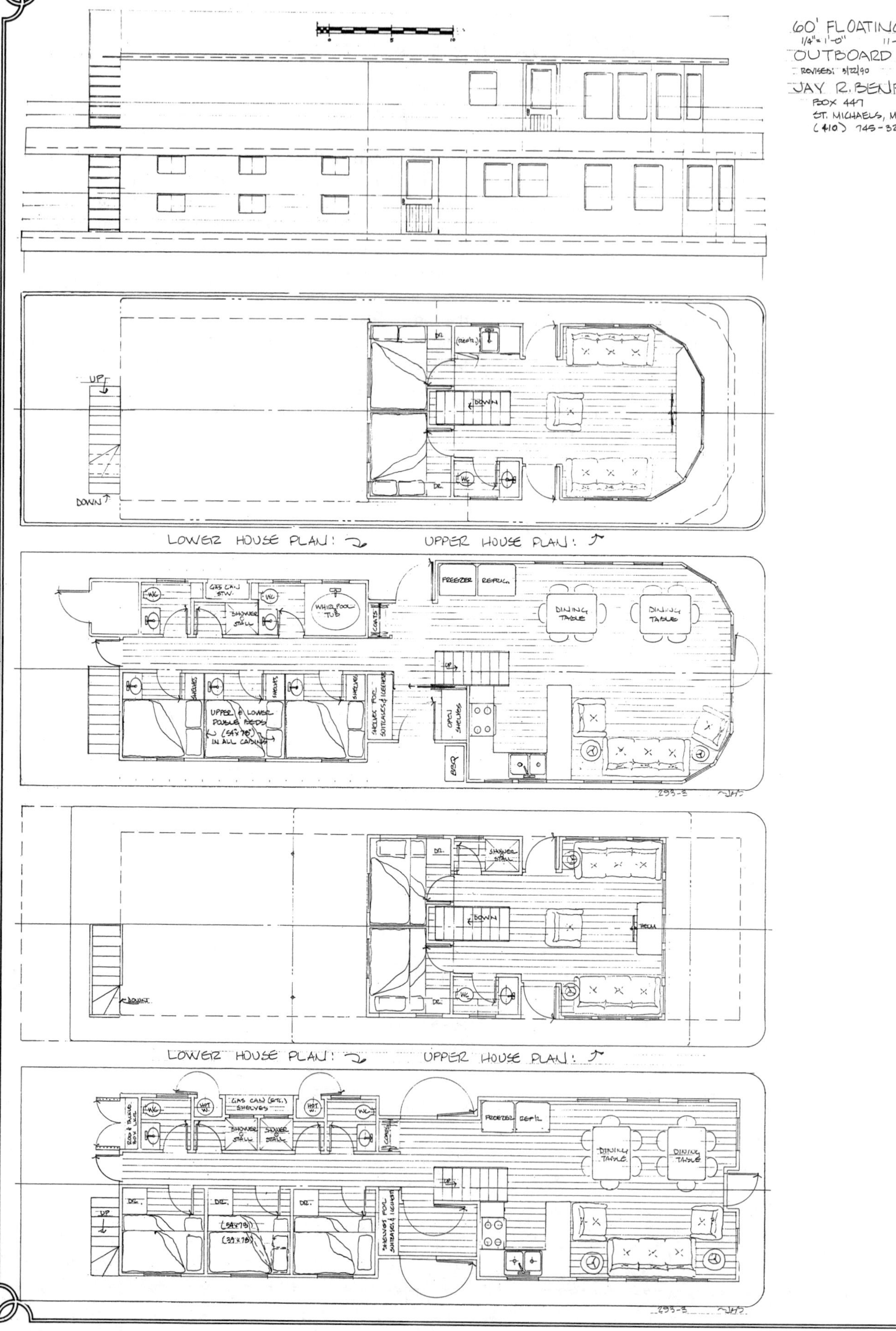

60' FLOATING LODGE
1/4" = 1'-0" 11-19-89
OUTBOARD PROFILE
REVISED: 3/22/90
JAY R. BENFORD
BOX 447
ST. MICHAELS, MD 21663
(410) 745-3235
LOWER HOUSE PLAN:
UPPER HOUSE PLAN:
FREEZER
REFRIG.
DINING TABLE
WHIRLPOOL TUB
SHOWER STALL
UPPER & LOWER DOUBLE BEDS (54"x78") IN ALL CABINS
SHELVES FOR SUITCASES & LICKERS
OPEN SHELVES
DOWN
UP
HELM

Tugs & Tug Yachts

Early in my career in yacht design I spent a couple years (1966 to 1968) as the staff naval architect for Foss Launch & Tug in Seattle. As I recall, at the time Foss operated about a hundred tugs and a couple hundred barges doing all sorts of services. I was fortunate to be part of the design and creation of several tugs while with them, from small harbor tugs to larger, ocean going tugs. The row of silhouettes below are similar to the tugs designed during my time with Foss.

Since that time I've had the opportunity to put this experience with real working tugs to good use creating a number of other tugs, both for work and pleasure. As will be seen in the following section of this book, I've continued to enjoy looking at and creating new tugs. Many of the more recent tug designs we've done have been yachts with modest power for economical cruising and more spacious accommodations than would be found aboard a working vessel.

Most of the tug yachts can be powered to do useful work, whether as a harbor tug, yarding boats around for service work, or towing equipment for diving, salvage, or repair work.

The ***Martha Foss***, shown in the above photo is 80' long and has a 27' beam. She is single screw, powered with a D399 Caterpillar diesel driving an 88" diameter propeller. This prop operated in a steering Kort nozzle with very close tip clearance. This provides the thrust equal to a much larger power plant with a free running prop. She's proven an excellent worker, doing log towing, ship-handling and general towing work.

*Left Above: The completed steel hull and deck for Harland Domke's 32' Tug is hauled off to the outfitting site. Left Below: Bill Burtis's **Maggie** does some charter work. Below: Bob Bichan's **Kristin B** has lots of room for carrying a group of friends. Photos courtesy of the owners.*

25' Tug Yacht

Design number 325

1994

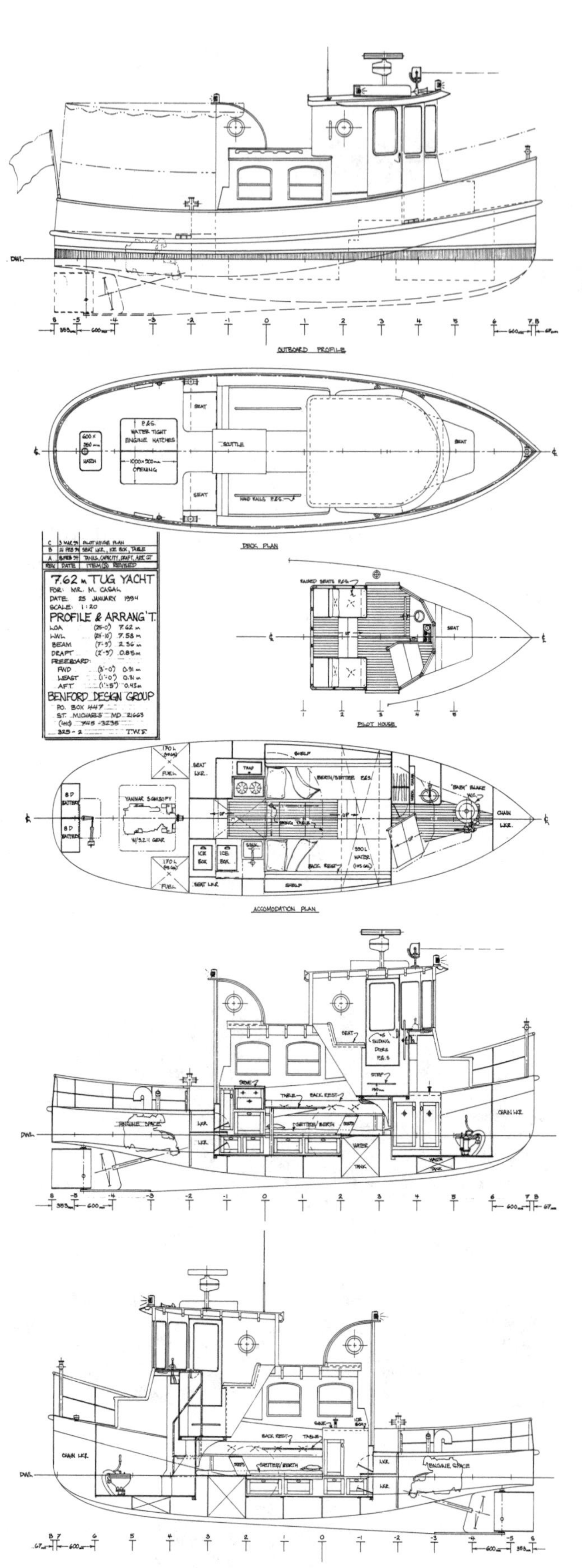

The design brief for this boat was to create a vessel that would be useful for overnighting and day cruises on a lake in Switzerland. The pilothouse has two raised seats for the Skipper and Mate to see where they're going. Down a couple steps forward are the head and sink, in an area that can be separated for privacy from the rest of the boat. The main cabin has two settee berths for sitting inside out of the weather. The center section forward of them has a piece that pulls aft and makes them into a wide double berth. The galley is across the after end of this saloon. The icebox has two hatches for access from in the saloon and out in the cockpit. The optional canopy over the aft deck will provide both sun and rain shelter.

The design has a single chine hull form with developable surfaces. Construction is robust with conservative steel scantlings. She has integral tankage for both fuel and water. Power is a Yanmar 27hp diesel — more than enough to drive her to hull speed and just right for very economical operation.

The 25' Tug plans are in Metric scales (mostly 1:20) and well detailed, including a set of computer faired and drawn patterns for the frames, shown small scale below. Plating panel "unwraps" are also part of the plan package.

Particulars:

Length overall	25'-0"
Length designed waterline	24'-10"
Beam	7'-9"
Draft	2'-9"
Freeboard:	
Forward	3'-0"
Least	1'-0"
Aft	1'-5"
Displacement, cruising trim	10,000 lbs
Displacement-length ratio	290
Prismatic coefficient	.564
Water tankage	103 Gal.
Fuel tankage	90 Gal.
Headroom	6'-8"

***CAUTION:** The displacement quoted here is for the boat in cruising trim. That is, with the fuel and water tanks filled, the crew on board, as well as the crews' gear and store in the locker. This should not be confused with the "shipping weight" often quoted as "displacement" by some manufacturers. This should be taken into account when comparing figures and ratios between this and other designs.

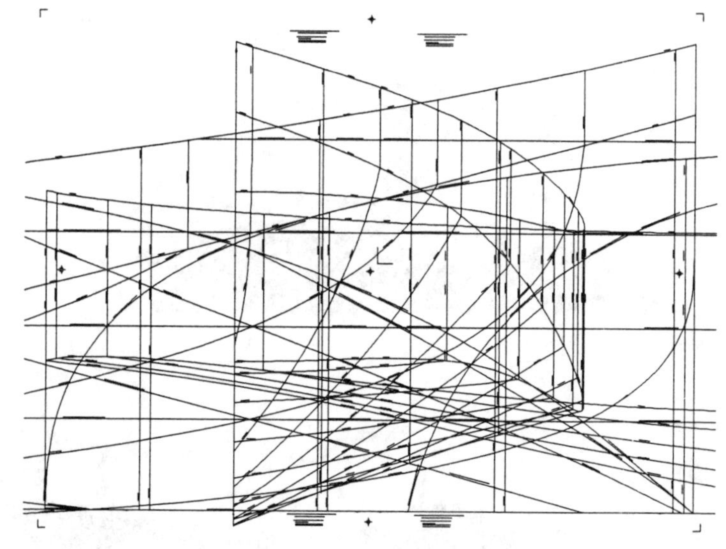

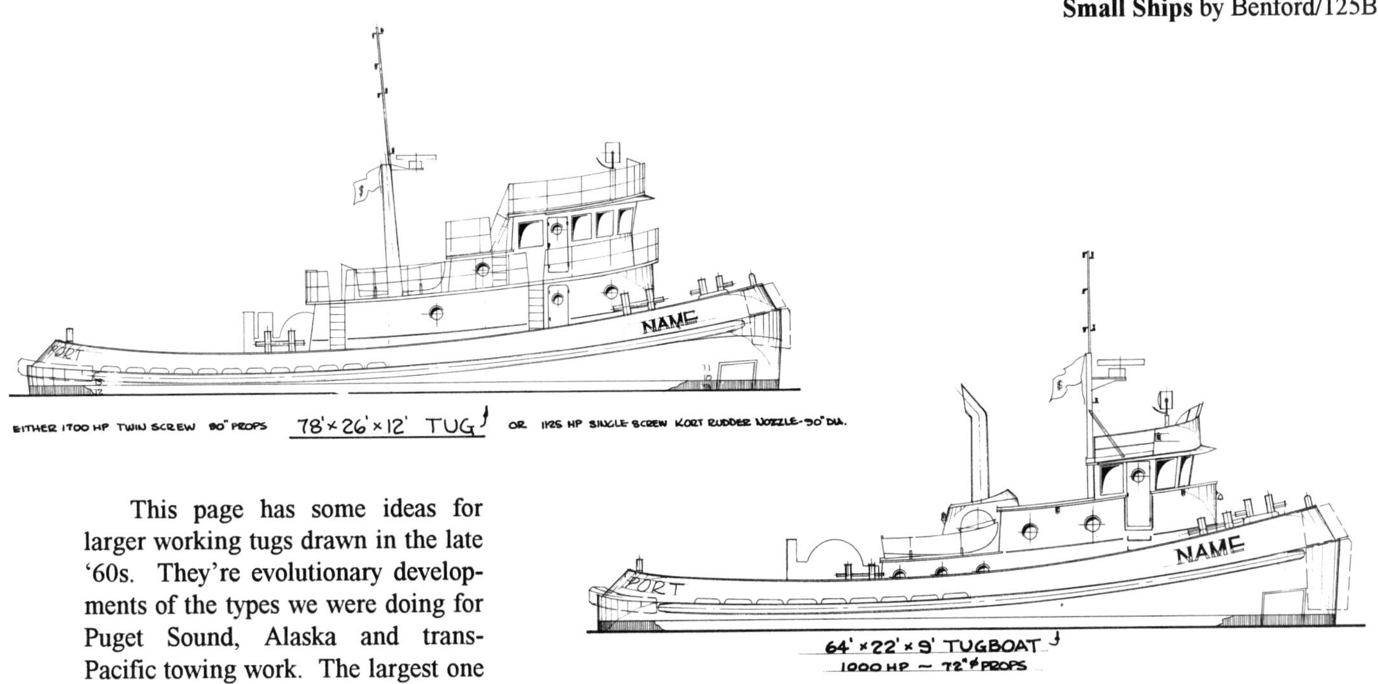

This page has some ideas for larger working tugs drawn in the late '60s. They're evolutionary developments of the types we were doing for Puget Sound, Alaska and trans-Pacific towing work. The largest one might work offshore and the others coastwise and ship-handling.

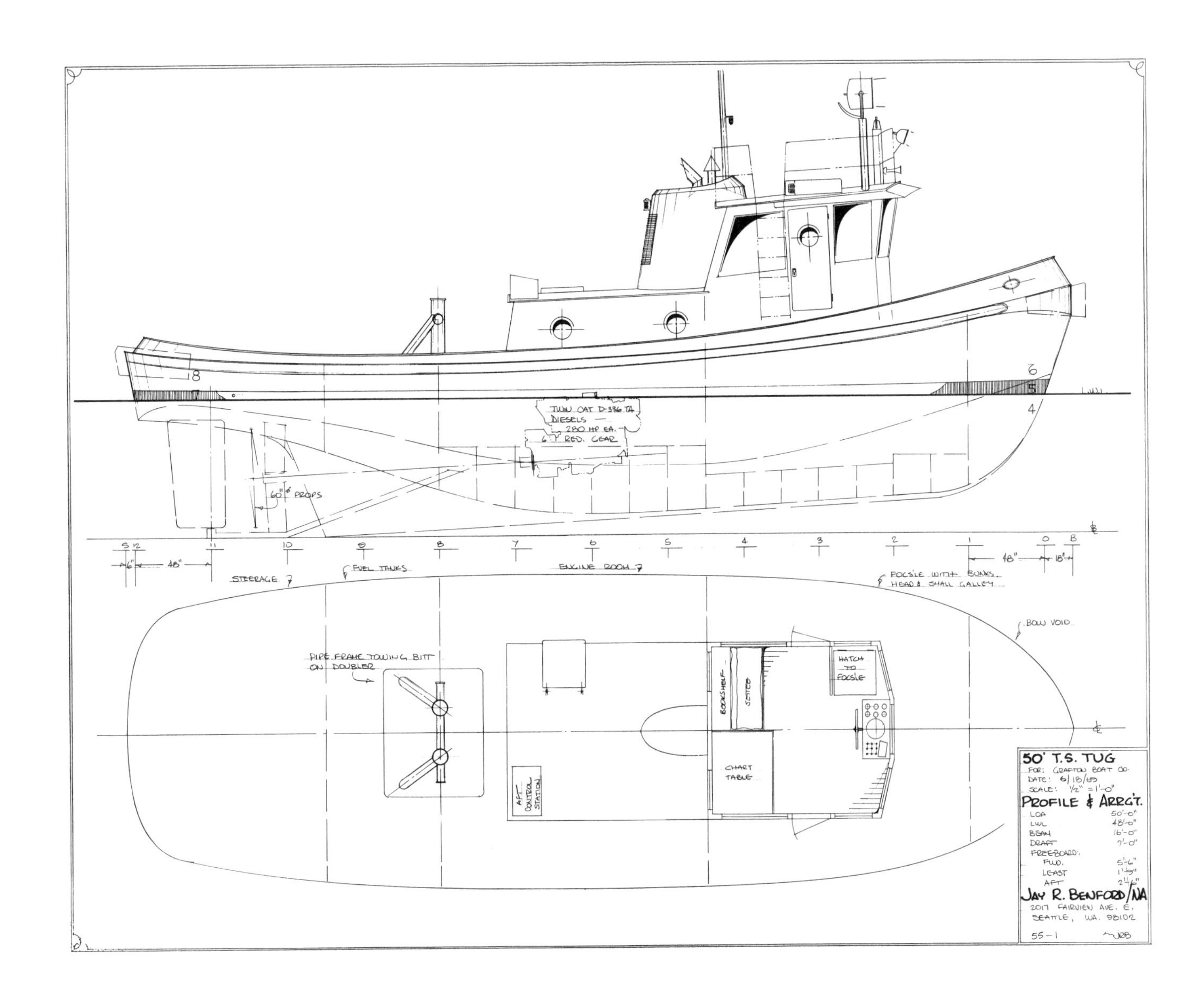

31' Tug Yacht

Design number 223

1984

After the 38' Tugs were being built, it became obvious to me that there were a lot of people who would love to own one, but found the price too high. Not wanting to miss out on the possibility of making more sales, the builder commissioned us to do a couple rounds of conceptual work on both a smaller and larger version.

The 31-footer shown here was the resulting smaller version. One of the concerns expressed to me after seeing the accommodation plan for the 31 was to the effect that why should anyone buy the 38 when they had almost the same interior on the 31....

Of course there is a little bit more space both inside and on deck on the 38-footer, but the essence of the 38's layout is in the 31' Tug. For someone wanting a practical cruising tug yacht for a crew of four, this design has a lot to recommend it. It has been used in building a variation on the 32' Tug's hull.

The accommodations show another variation on the theme we've developed with the other 32, 35 and 37' Tug Yachts. All have elements of their layouts that might be interchanged to create further variations to suit individual needs.

The head is located in between both staterooms, with a separate shower stall.

The galley is laid out on the passageway out to the aft deck, with a refrigerator recessed in over the engine.

The profile shows a shortened pilothouse from that shown on the accommodation plan for that area. They are indicative of the two variations we looked at in doing this conceptual study. The raised settee has room for four to sit and have a meal, and is high enough for the crew to watch where they're going. The sliding doors, port and starboard, give access to the sides of the boat and let the helmsman readily look out and see what's happening in making a landing.

The bow stateroom has good sized twin berths. It's right under a built-in seat/trunk cabin, which makes for good headroom in the cabin and a wonderful sitting place on deck.

Particulars:

Length overall	31'-0"
Length designed waterline	29'-6"
Beam	13'-0"
Draft	3'-0"
Freeboard:	
Forward	4'-6"
Least	1'-3"
Aft	2'-0"
Displacement, cruising trim	13,850lbs.
Displacement-length ratio	241
Ballast	2,000 lbs.
Prismatic coefficient	.626
Water tankage	105 Gal.
Fuel tankage	105 Gal.
Headroom	6'-2" to 6'-10"

***CAUTION:** The displacement quoted here is for the boat in cruising trim. That is, with the fuel and water tanks filled, the crew on board, as well as the crews' gear and store in the locker. This should not be confused with the "shipping weight" often quoted as "displacement" by some manufacturers. This should be taken into account when comparing figures and ratios between this and other designs.

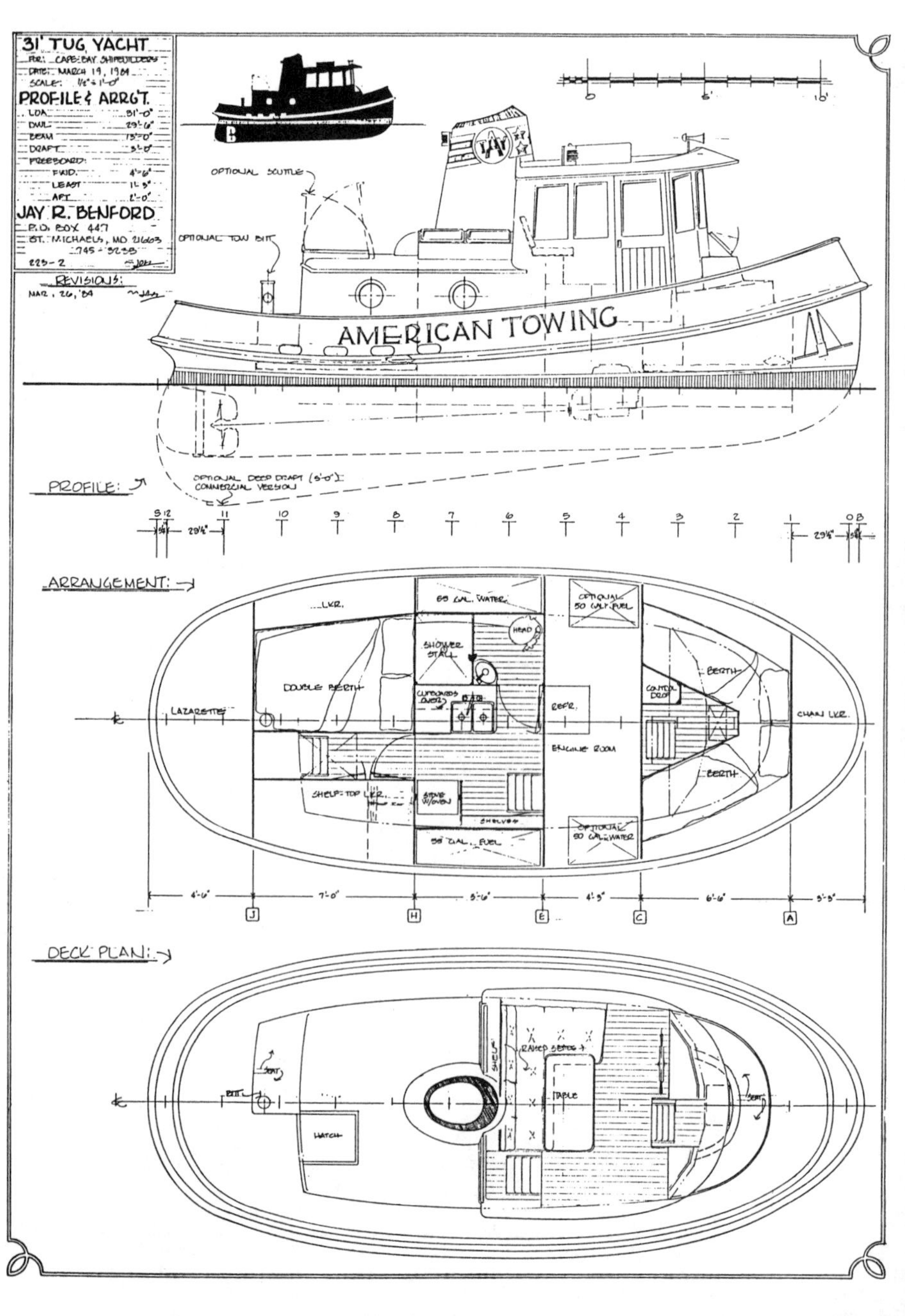

32' Tug Yacht

Design number 227

1984

Shortly after the 31' Tug had been drawn, we had a request for a steel tug of about the same size. The resulting design is shown on this page and several have been built. Building in steel has allowed for some variations in custom tailoring the boats to the individual buyers which is not possible with a set of fiberglass molds. The same would be true for doing variations by building in wood, whether in plywood and epoxy, cold-molding, or carvel planking.

The standard 32 is laid out with the engine well aft, under a portion of the galley counter. The double berth aft has comfortable sitting headroom over it, being under the raised pilothouse seating.

The two variations shown in smaller scale are a couple of ideas on how this basic design could be modified to have a different sort of yacht layout.

Particulars:

Length overall	32'-0"
Length designed waterline	30'-0"
Beam	13'-0"
Draft	3'-0"
Freeboard:	
Forward	4'-6"
Least	1'-3½"
Aft	2'-0½"
Displacement, cruising trim*	18,150lbs.
Displacement-length ratio	300
Prismatic coefficient	.61
Water tankage	130 Gal.
Fuel tankage	190 Gal.
Headroom	6'-5" to 6'-9"

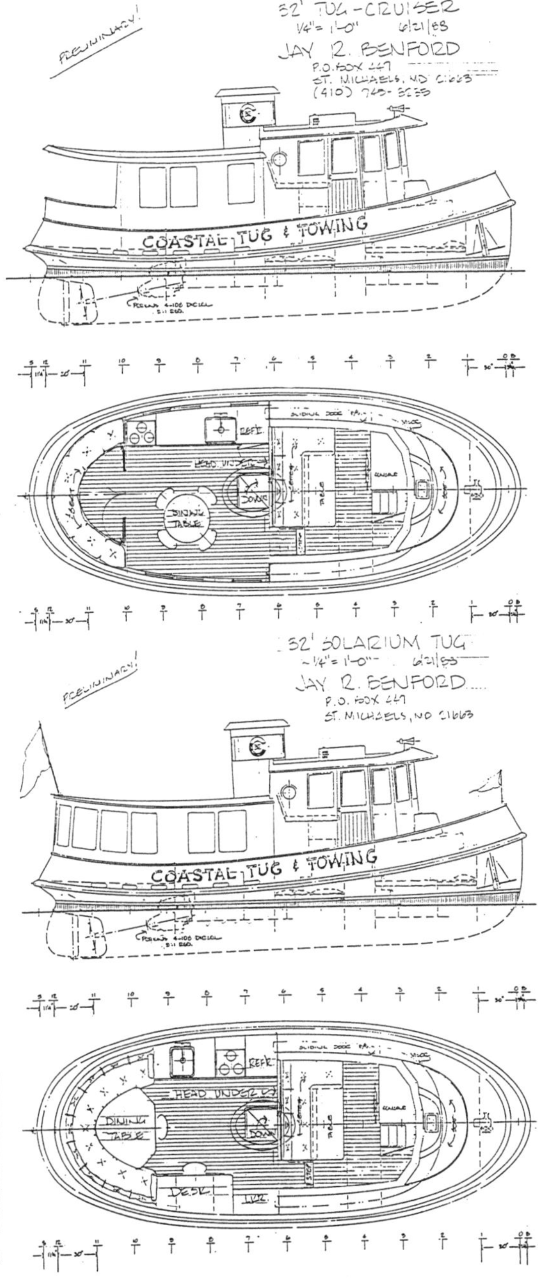

35' Tug Yacht

Design number 231

1984

This design was commissioned by a friend who had been building wooden boats for decades, using traditional plank-on-frame techniques. He wanted to try out cold-molding while he created a small working and salvage tug.

Later, we were asked to modify the design for building in foam cored 'glass and thus another version came to be.

The accommodations show another variation on the theme we've developed with the 31, 32 and 37' Tug Yachts. All have elements of their layouts that might be interchanged to create further variations to suit individual needs.

The head is generously sized, with a separate shower stall with a built-in seat. The galley is along the passageway to the aft deck, and has counter space over the engine box and a refrigerator elevated over the engine too. The diesel stove will be good for northern climates, providing heat to the cabin, an alternative source for heating the shower water, and a way to bake bread and roast turkeys.

The dinette is under the raised settee in the pilothouse, with good sitting headroom over both seating areas, above and below. Either could be built in a manner to convert to another double berth, if additional crew berthing is needed.

The bow stateroom has a good sized double berth. It's right under a lovely traditional skylight, which will make a wonderful addition to this sleeping cabin. You'll be able to look up and see what the weather is before you get out of bed and watch the stars at night. The skylight also adds to the feeling of space by adding natural light and ventilation to the stateroom.

The pilothouse has plenty of windows to provide good visibility and ventilation. The port and starboard doors give quick access to both sides of the boat for line handling and checking on how close in you are in maneuvering in tight quarters.

All in all, the 35' Tug Yacht is a straight-forward and practical cruising or working tug.

Particulars:

Length overall	35'-0"
Length designed waterline	33'-0"
Beam	13'-7½"
Draft	4'-0"
Freeboard:	
Forward	5'-0"
Least	1'-9"
Aft	2'-9"
Displacement, cruising trim	19,200lbs.
Displacement-length ratio	239
Ballast	3,000 lbs.
Ballast ratio	16%
Prismatic coefficient	.603
Water tankage	150 Imp.Gal.
Fuel tankage	180 Imp.Gal.
Headroom	6'-8"

***CAUTION:** The displacement quoted here is for the boat in cruising trim. That is, with the fuel and water tanks filled, the crew on board, as well as the crews' gear and store in the locker. This should not be confused with the "shipping weight" often quoted as "displacement" by some manufacturers. This should be taken into account when comparing figures and ratios between this and other designs.

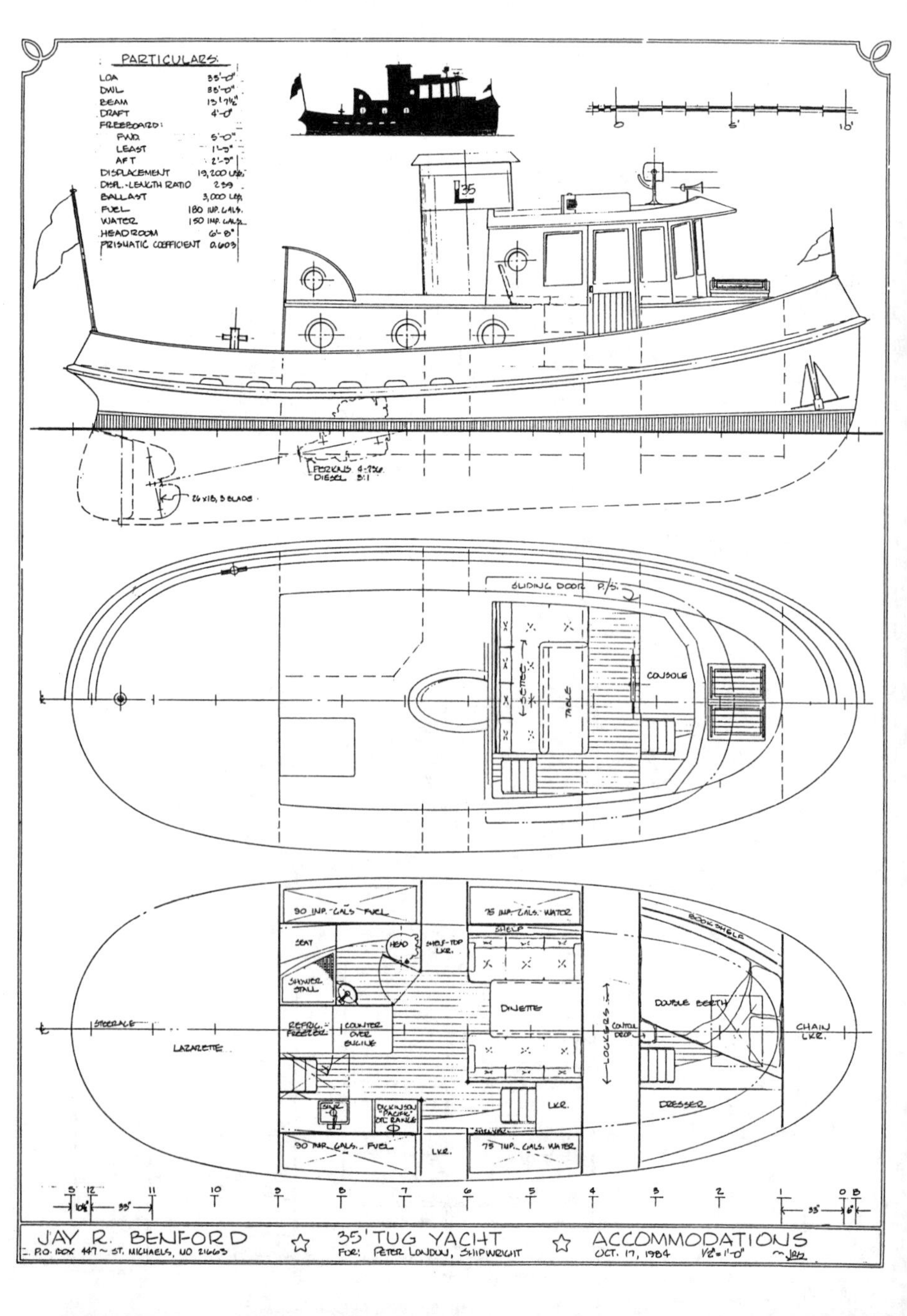

37' Tug Yachts

Design number 251 & 282

1986 & 1989

The first 37' Tug Yacht design, as shown on this page, was a commission from a builder doing some of the 32's in steel, who had a client who wanted a boat for some salvage and diving work.

We took the concept of the 32' Tug, added five feet of deck aft, flattened the run and added twin screws to get her into the semi-displacement speed range. The deckhouse is the same shape and the interior variations are interchangeable between the 32 and 37. The panel of skylight hatches on top of the after house gives wonderful light to this space. We first did this on the ATY-38's and it was a great success.

The engines are under the after deck, accessed through hatches on deck, plus from behind the companionway stairs and a removable panel in the shower stall.

The second 37' Tug Yacht, shown on the following page, is 20" wider, with a full width saloon in the after house. It is also a single screw and the installation details for the twin or single engine can be used in either hull.

The wider 37 was intended for taking small groups of people on day trips or excursions. The wide open saloon has the galley and guest head right adjacent to the saloon. The owner's stateroom and private head are in the bow. The elevated pilothouse has seating and a table for some of the passengers to join in and watch the operation of the boat. The upper aft deck would have tall rails, like the foredeck. The forward trunk cabin and skylight is sculpted to make a comfortable seat for sitting forward.

With a bit of rearrangement, this boat could be made into a two stateroom liveaboard, swapping the second head for more berthing and putting the galley in the after house with the living room and dining room furniture, creating a saloon not unlike our 35' Pacquette.

These designs, originally done for steel, would be ideal candidates for building in plywood and epoxy. Or for building in aluminum, if this is your preference. We could convert any of our steel designs for this type of construction. Let us know if this is of appeal.

Particulars:	**251/282**
Length overall	37'-0"
Length designed waterline	34'-11"/35'-0"
Beam, over hull	12'-4"/14'-0"
Beam, over guards	13'-0"/14'-8"
Draft	3'-4"/3'-6"
Freeboard:	
Forward	4'-7¾"/6'-9"
Least	1'-8"/1'-9"
Aft	2'-3¼"/2'-2½"
Displacement, cruising trim*	25,300lbs.
Wide version	28,600lbs.

Displacement-length ratios	265/300
Prismatic coefficient	.70
Water tankage	130/230 Gal.
Fuel tankage	190/580 Gal.
Headroom	6'5"
GM (original/wide version)	4'/6'

***CAUTION:** The displacement quoted here is for the boat in cruising trim. That is, with the fuel and water tanks filled, the crew on board, as well as the crews' gear and store in the locker. This should not be confused with the "shipping weight" often quoted as "displacement" by some manufacturers. This should be taken into account when comparing figures and ratios between this and other designs.

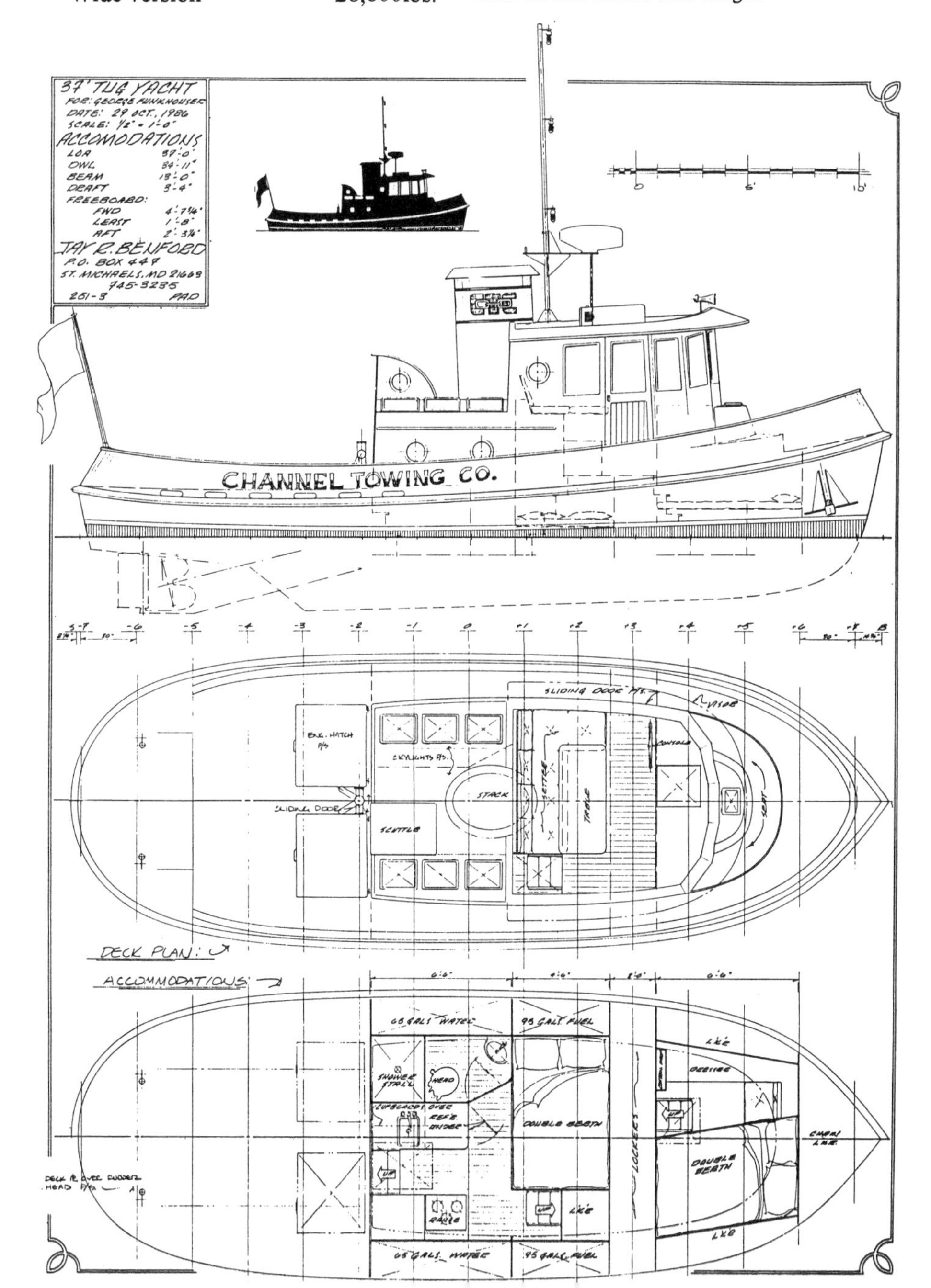

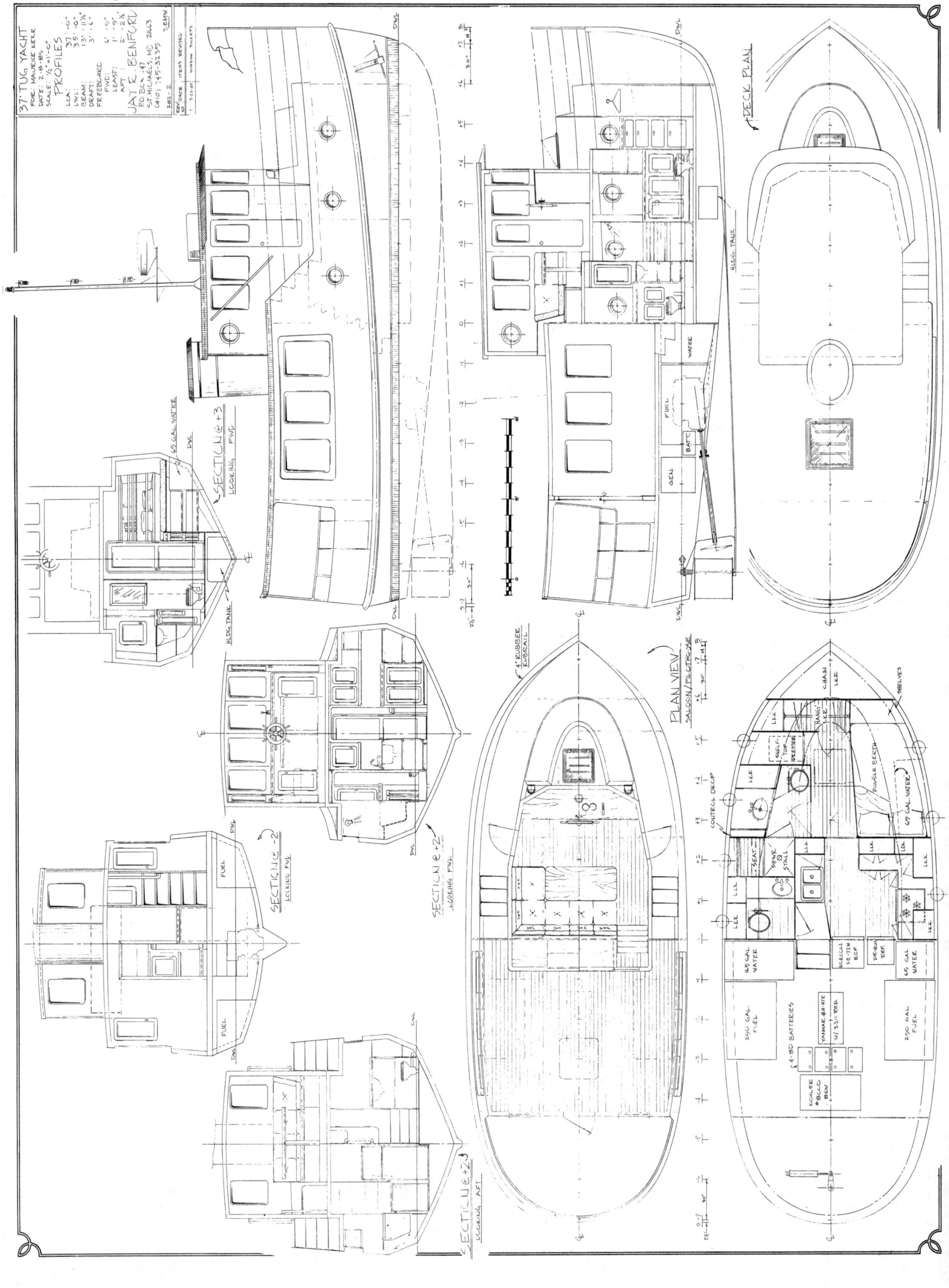
37' TUG YACHT
PROFILES
JAY R. BENFORD
DECK PLAN
PLAN VIEW
SALOON/PILOTHOUSE
SECTION @ +3
LOOKING FWD
SECTION @ +2
LOOKING FWD
SECTION @ -2
LOOKING FWD
SECTION @ -2
LOOKING AFT
FUEL
WATER
GEN.
BATT.
HLDG. TANK
4" RUBBER RUBRAIL
CHAIN LKR
SHELVES
DOUBLE BERTH
65 GAL. WATER
250 GAL. FUEL

38' Tug Yacht

Design Number 234

1985

My late friend Ron Brown and I evolved this design over a period of time. The more we talked about it, the more ideas we came up with for evolutionary refinements.

Ron liked the way this boat exuded the "true tug" look of the working tugs we'd seen in the major seaports. He felt that too many of the boats being offered as tug yachts were simply trawler yachts with a stack tacked on for looks. I agreed, and drew on the tugs I'd worked on for Foss during the '60's to get the right hull form, sheer, and house proportions.

By paying particular attention to the spacial relations of the interior elements we wanted, I was able to come up with several workable layouts, depending on how the client wanted to use the boat. We were able to keep all these within the deckhouse that had the right "look". The twin skylights either side of the stack not only give great light and ventilation to the galley, but provide extra headroom so we could keep the shafts of the twin screw version under the cabin sole.

The raised settee in the pilothouse puts the passengers at a height where they can see all around. It also creates a space under it where we can have a full headroom shower stall and head.

Along with the twin screw version we also got a commission for the 40' aluminum barge. This barge has integral tankage for 4,000 gallons of diesel fuel, and can readily carry a couple dozen drums on deck too, or a myriad of other supplies. The deck structure was designed to be able to carry a D5 Caterpillar for special projects. The 12' landing ramp makes for easy off-loading.

The barge and tug are connected with a special push linkage that we designed. It is articulated so that the tug and barge can pitch and roll independently and still let the tug keep good steering control. The tug and barge work very well together as an island support team.

In the process of working on the tug and barge, we had a chance to do an inclining test. The results, when run through our computer hydrostatics program, indicated she has positive stability all the way to 180° — a very stiff and safe boat. This highly desirable condition makes her a good basis for venturing well out of sight of land.

The company Ron Brown founded built four of these tugs before his death. I've been very pleased with how the design worked out in real life.

Particulars:

Length overall	38'-0"
Length designed waterline	36'-0"
Beam	14'-0"
Draft	4'-0"
Freeboard:	
Forward	4'-9½"
Least	1'-3"
Aft	2'-0¼"
Displacement, cruising trim*	23,400 lbs.
Displacement-length ratio	224
Ballast	2,000 lbs.
Prismatic coefficient	.64
Pounds per inch immersion	1,650
Water tankage	1,600 Gals.
Fuel tankage	2,000 Gals.
Headroom	7'-0" to 8'-0"
GM (from inclining test)	4.8'

***CAUTION:** The displacement quoted here is for the boat in cruising trim. That is, with the fuel and water tanks filled, the crew on board, as well as the crews' gear and stores in the lockers. This should not be confused with the "shipping weight" often quoted as "displacement" by some manufacturers. This should be taken into account when comparing figures and ratios between this and other designs.

Loafer (above and right in Monica Brown's photos) is one of the cruising versions of the 38' Tug Yacht. ***Tug*** and ***Barge*** (below and prior page) are the twin screw version of the tug and the 40' barge. The tug's wonderfully springy sheer is well evident and she looks every bit the descendant of the classic working tugs. Ron and I were delighted with the results of our work in making the "look" of these boats just right. The tug and barge work as an integrated island support team, hauling fuel oil and a myriad of other supplies.

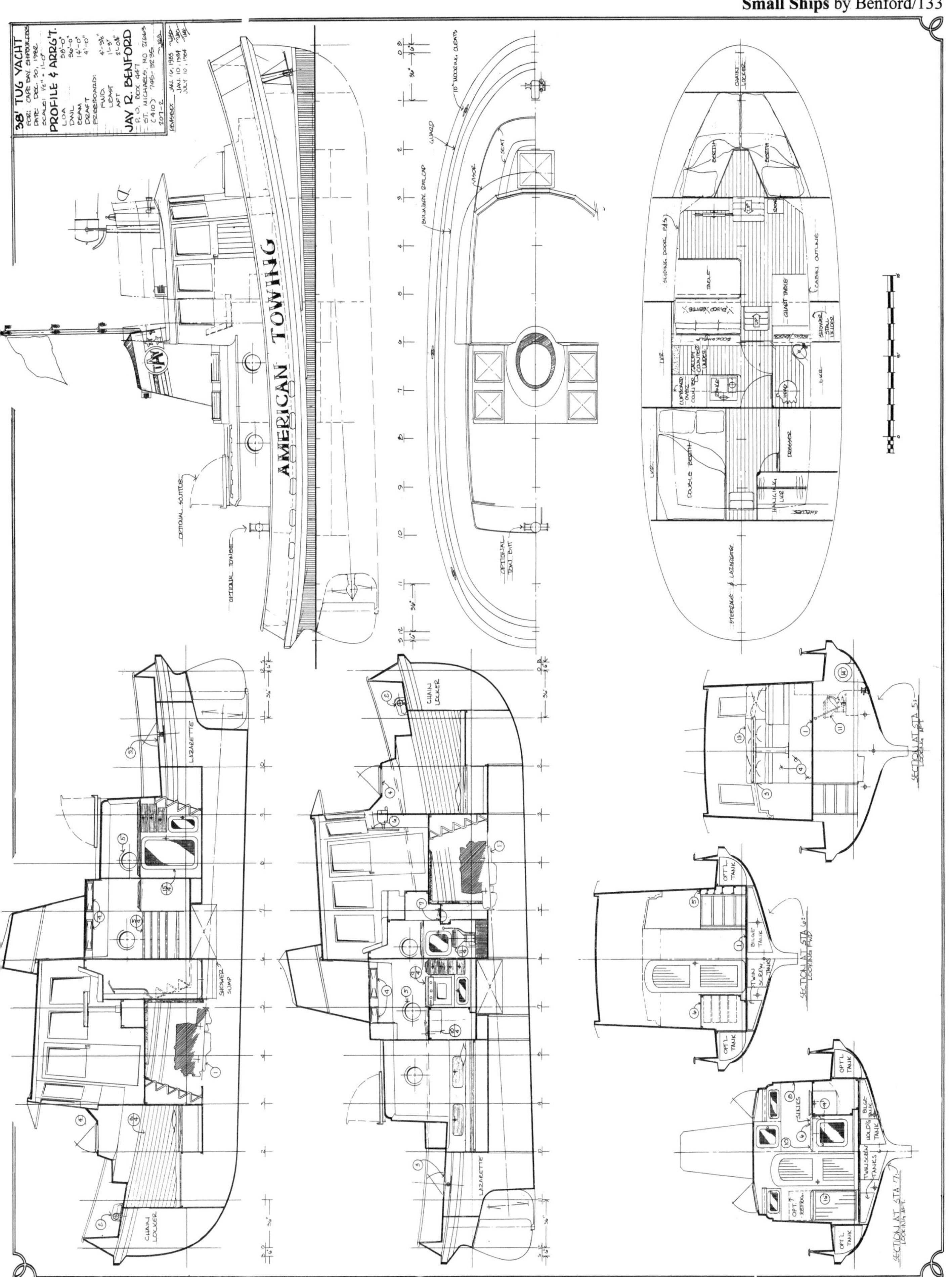
38' TUG YACHT
FOR: CAPE BAY SHIPBUILDERS
DATE: DEC. 30, 1982
SCALE: 1/2" = 1'-0"
PROFILE & ARR'G'T.
LOA 38'-0"
DWL 36'-0"
BEAM 14'-0"
DRAFT 4'-0"
FREEBOARD:
FWD. 4'-3 3/8"
LEAST 1'-5"
AFT 2'-0 1/4"
JAY R. BENFORD
P. O. BOX 447
ST. MICHAELS, MD 21663
(410) 745-3235
207-2
REVISED: JAN. 16, 1983
JAN. 10, 1984
JULY 10, 1984
AMERICAN TOWING
OPTIONAL SCUTTLE
OPTIONAL TOWBEE
CHAIN LOCKER
LAZARETTE
SHOWER SUMP
BULWARK RAILCAP
GUARD
10" MOORING CLEATS
VISOR
SEAT
OPTIONAL TOW BITT
LKR.
DOUBLE BERTH
CUPBOARD OVER COUNTER
GALLEY COUNTER UNDER
RANGE
TABLE
SLIDING DOOR P&S
BERTH
CHAIN LOCKER
STEERING & LAZARETTE
HANGING LKR
DRESSER
HEAD
CHART TABLE
SHOWER STALL UNDER
CABIN OUTLINE
OPT'L TANK
TWIN SCREW TANKS
BILGE TANK
SECTION AT STA 6: LOOKING FWD
OPT. REFRIG.
SINKS
HOLD'G TANK
SECTION AT STA 7: LOOKING AFT
SECTION AT STA 5: LOOKING AFT

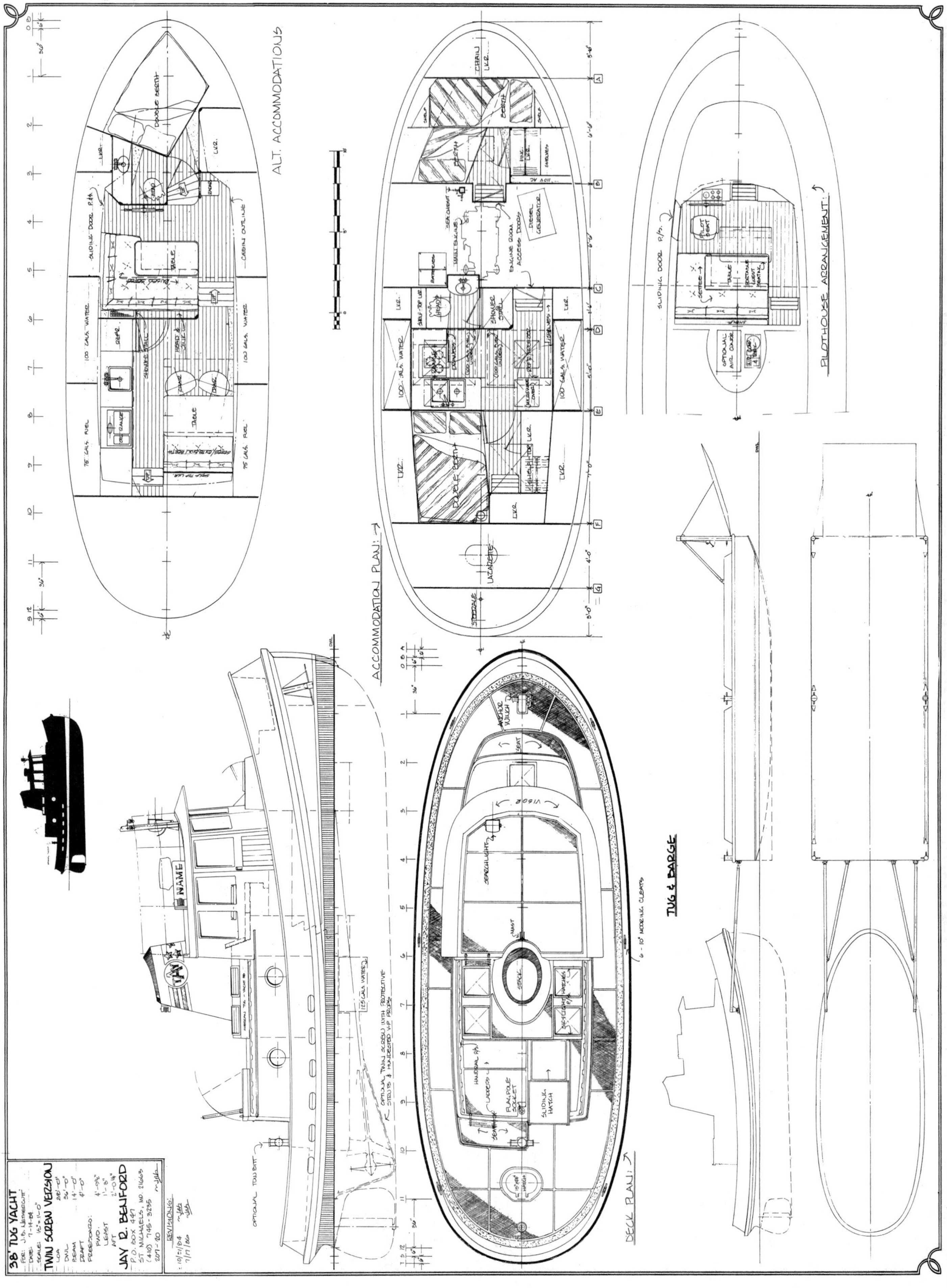

ALT. ACCOMMODATIONS

ACCOMMODATION PLAN

PILOTHOUSE ARRANGEMENT

DECK PLAN

TUG & BARGE

38' TUG YACHT
FOR: J.P. WEINBRECHT
DATE: 7-14-84
SCALE: 1/2" = 1'-0"
TWIN SCREW VERSION
LOA 38'-0"
DWL 36'-0"
BEAM 14'-0"
DRAFT 4'-0"
FREEBOARD:
FWD. 4'-9 1/2"
LEAST 1'-8"
AFT 2'-0 1/4"
JAY R. BENFORD
P.O. BOX 447
ST. MICHAELS, MD 21663
(410) 745-3235
207-20
REVISIONS:

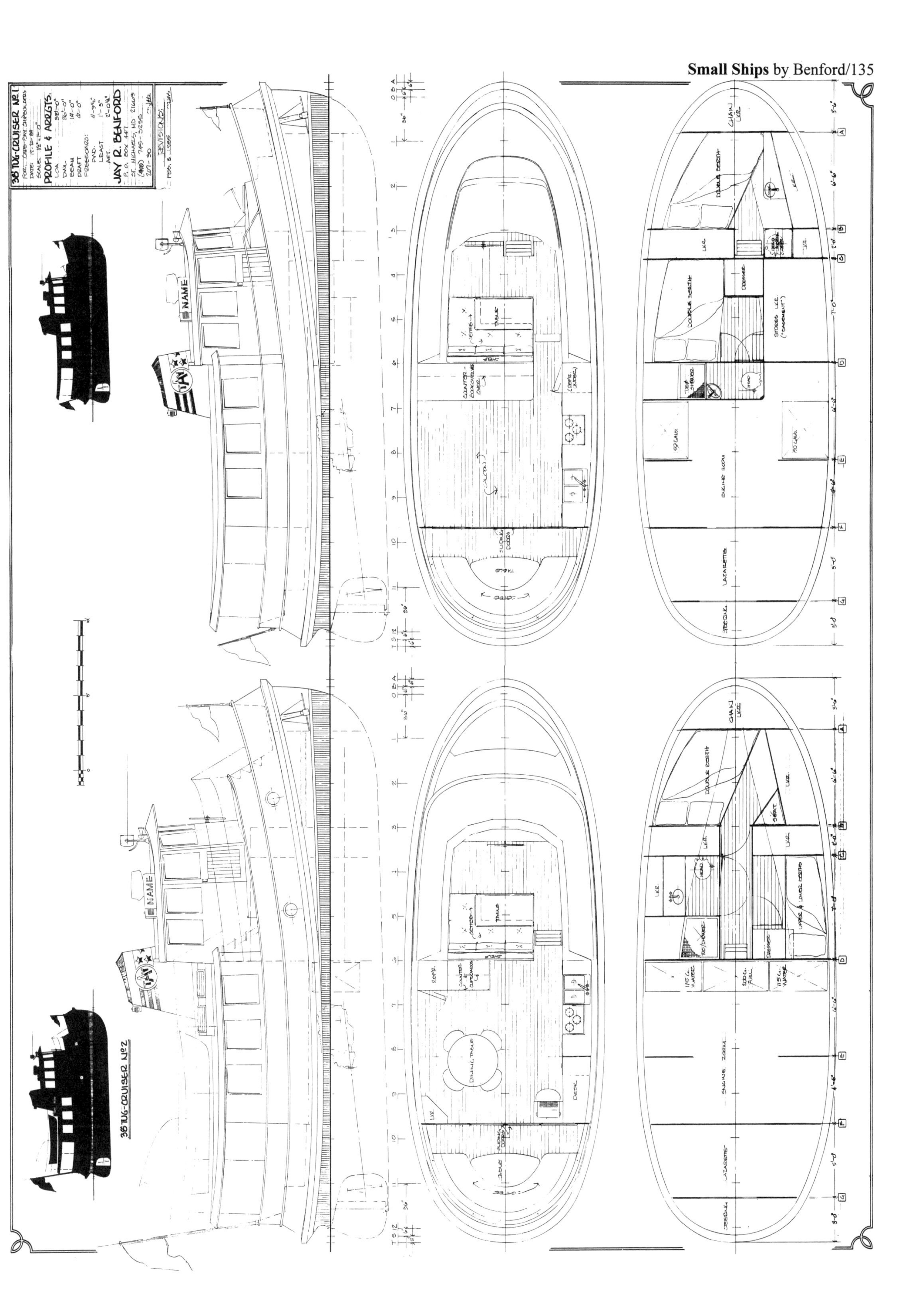
35' TUG-CRUISER No 1
PROFILE & ARRGTS.
JAY R. BENFORD
NAME
35' TUG-CRUISER No 2
NAME
DOUBLE BERTH
ENGINE ROOM
LAZARETTE
STEERING
TABLE
DINING TABLE
DESK
CHAIN LKR.
LKR.
DRESSER
SLIDING DOORS

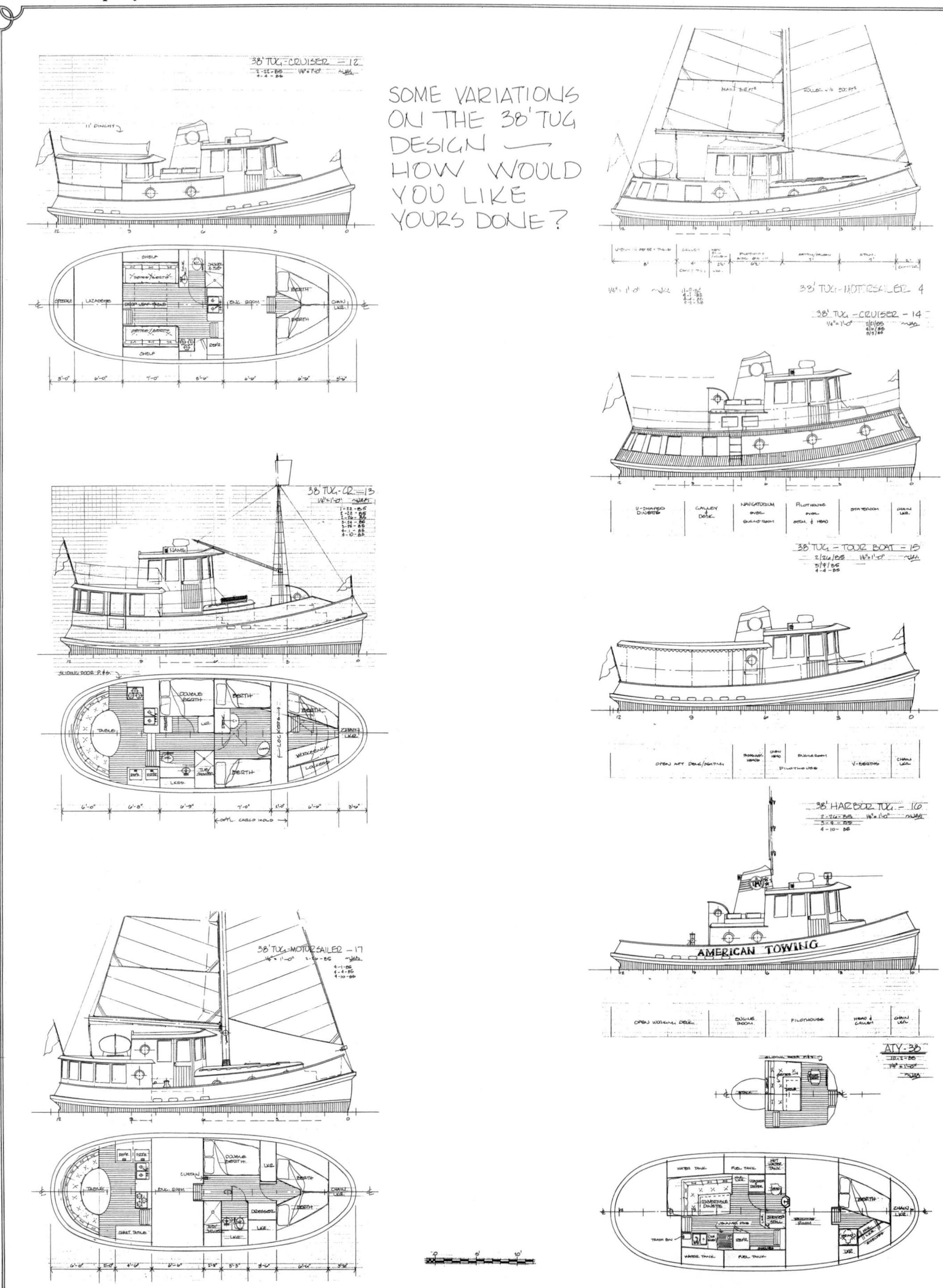
SOME VARIATIONS ON THE 38' TUG DESIGN — HOW WOULD YOU LIKE YOURS DONE?
38' TUG-CRUISER - 12
38' TUG-CR. - 13
38' TUG-MOTORSAILER - 17
38' TUG-MOTORSAILER 4
38' TUG-CRUISER - 14
38' TUG - TOUR BOAT - 15
38' HARBOR TUG - 16
AMERICAN TOWING
ATY-38

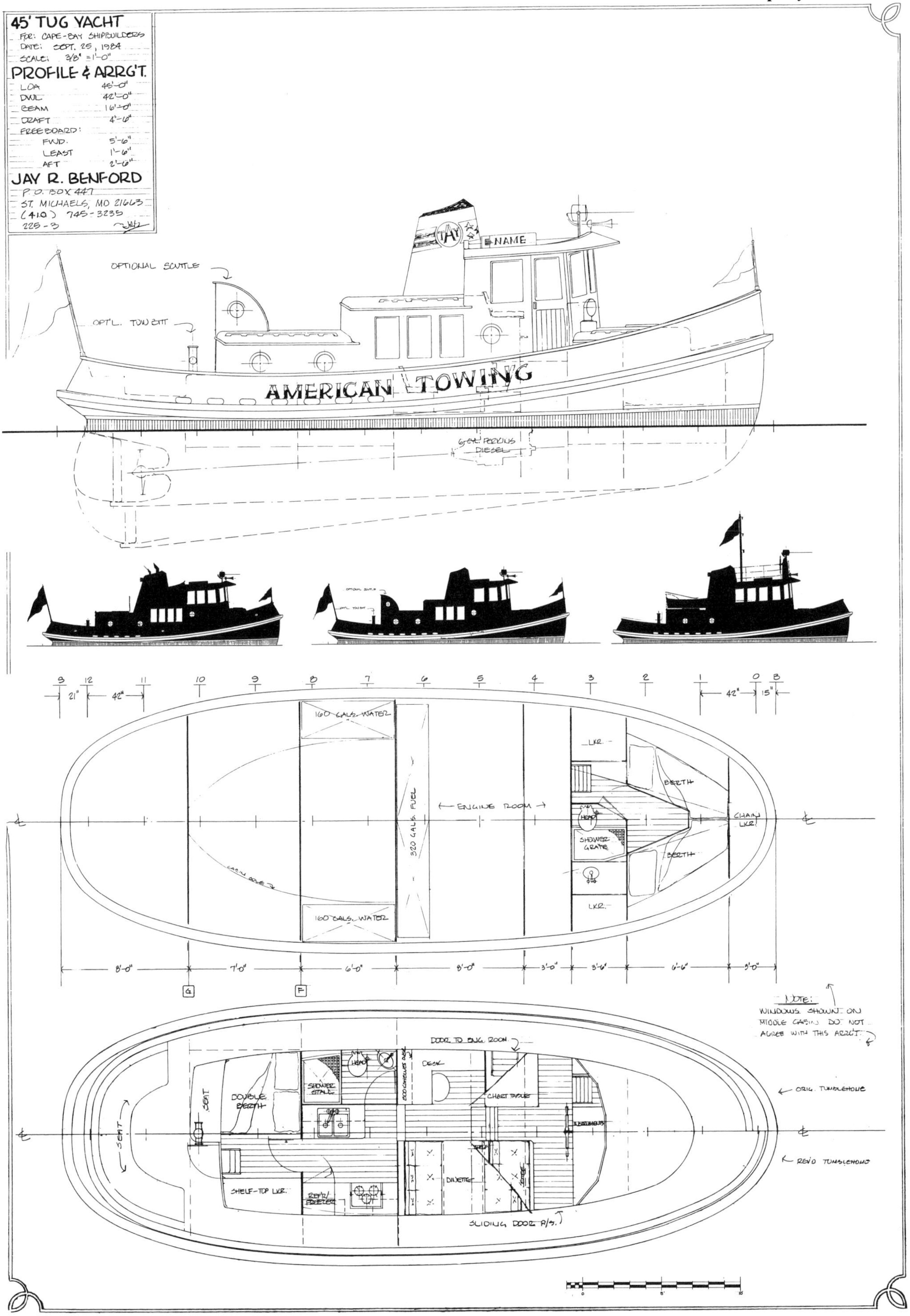

45' TUG YACHT
FOR: CAPE-BAY SHIPBUILDERS
DATE: SEPT. 25, 1984
SCALE: 3/8" = 1'-0"

PROFILE & ARRG'T.

LOA	45'-0"
DWL	42'-0"
BEAM	16'-0"
DRAFT	4'-6"
FREEBOARD:	
FWD.	5'-6"
LEAST	1'-6"
AFT	2'-6"

JAY R. BENFORD
P.O. BOX 447
ST. MICHAELS, MD 21663
(410) 745-3235
225-3

OPTIONAL SCUTTLE
OPT'L. TOW BITT
NAME
AMERICAN TOWING
6 CYL. PERKINS DIESEL

OPTIONAL SCUTTLE
OPT'L. TOW BITT

160 GALS. WATER
320 GALS. FUEL
← ENGINE ROOM →
160 GALS. WATER
LKR.
BERTH
HEAD
SHOWER GRATE
BERTH
CHAIN LKR.
LKR.

8'-0" 7'-0" 6'-0" 8'-0" 3'-0" 3'-6" 6'-6" 5'-0"
21" 42" 42" 15"

NOTE:
WINDOWS SHOWN ON MIDDLE CABIN DO NOT AGREE WITH THIS ARRG'T.

DOOR TO ENG. ROOM
HEAD
SHOWER STALL
DESK
DOUBLE BERTH
SEAT
CHART TABLE
SEAT
SHELF-TOP LKR.
REFR/FREEZER
DINETTE
SLIDING DOOR P/S.
← ORIG. TUMBLEHOME
← REV'D TUMBLEHOME

48' Tug Yacht *Elijah Curtis*

Design Number 292

1989

Designed as a liveaboard on which a couple can follow the seasons up and down the East Coast, this 48-footer is a husky small ship. She has an open walkaround at main deck level, making line handling and boarding easy. The housetop of the saloon overhangs the side and after decks, for extra shelter from sun and rain. The pilothouse is raised a few steps, and has a big ship style bridge around it.

The saloon is roomy and well lit, with windows on three sides. It contains the lounging area, the dining area, and the galley. Just a few steps down leads to the two staterooms and two heads. The forward stateroom is the master, with a private entry head. The tub and shower stall is shared between the two heads, making for one quite roomy space instead of two small stalls.

A few steps up from the saloon is the office or navigation area and the pilothouse. The oilskin locker also will serve as a coat closet in day-to-day living aboard.

The water tanks, which will be drawn down and refilled most often, are located amidships where they will have practically no effect on fore and aft trim. We've given her fuel tanks forward, in the double bottom under the cabin sole, and aft in the corners of the engine room. By judiciously watching how these are drawn down, good fore and aft transverse trim can be maintained, even with varying loads as the boat is cruising and other things are consumed or added to her cargo.

The scantling section is typical of how our freighters are done also, and will give an idea of the heft of a steel boat of this size.

Particulars:

Length overall	48'-0"
Length designed waterline	42'-9"
Beam	17'-0"
Draft	3'-6"
Freeboard:	
Forward	6'-6"
Least	2'-6"
Aft	3'-4"
Displacement, cruising trim*	67,400 lbs.
Displacement-length ratio	385
Water tankage	1,000 Gals.
Fuel tankage	1,000 Gals.
Headroom	6'-5"

***CAUTION:** The displacement quoted here is for the boat in cruising trim. That is, with the fuel and water tanks filled, the crew on board, as well as the crews' gear and stores in the lockers. This should not be confused with the "shipping weight" often quoted as "displacement" by some manufacturers. This should be taken into account when comparing figures and ratios between this and other designs.

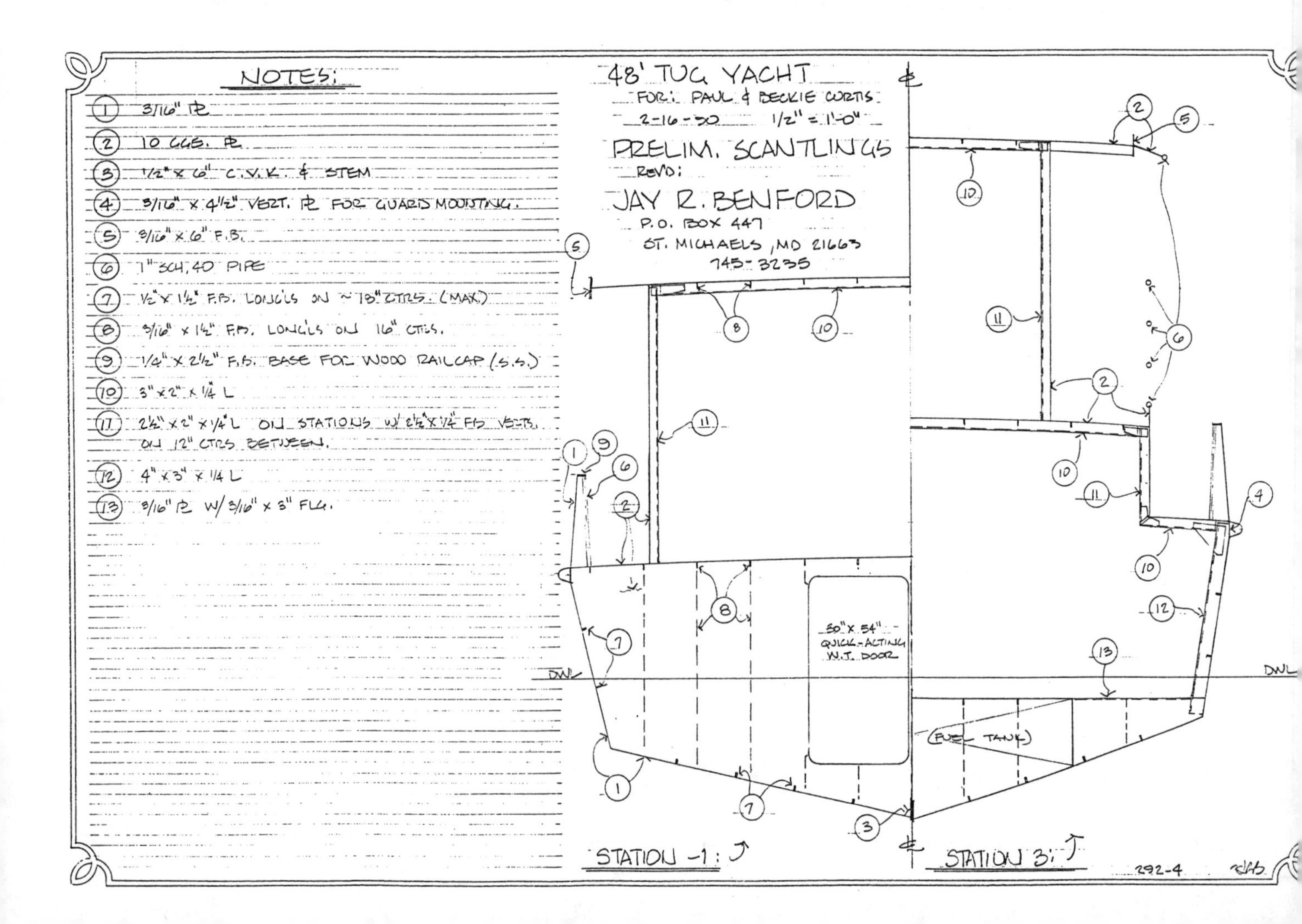

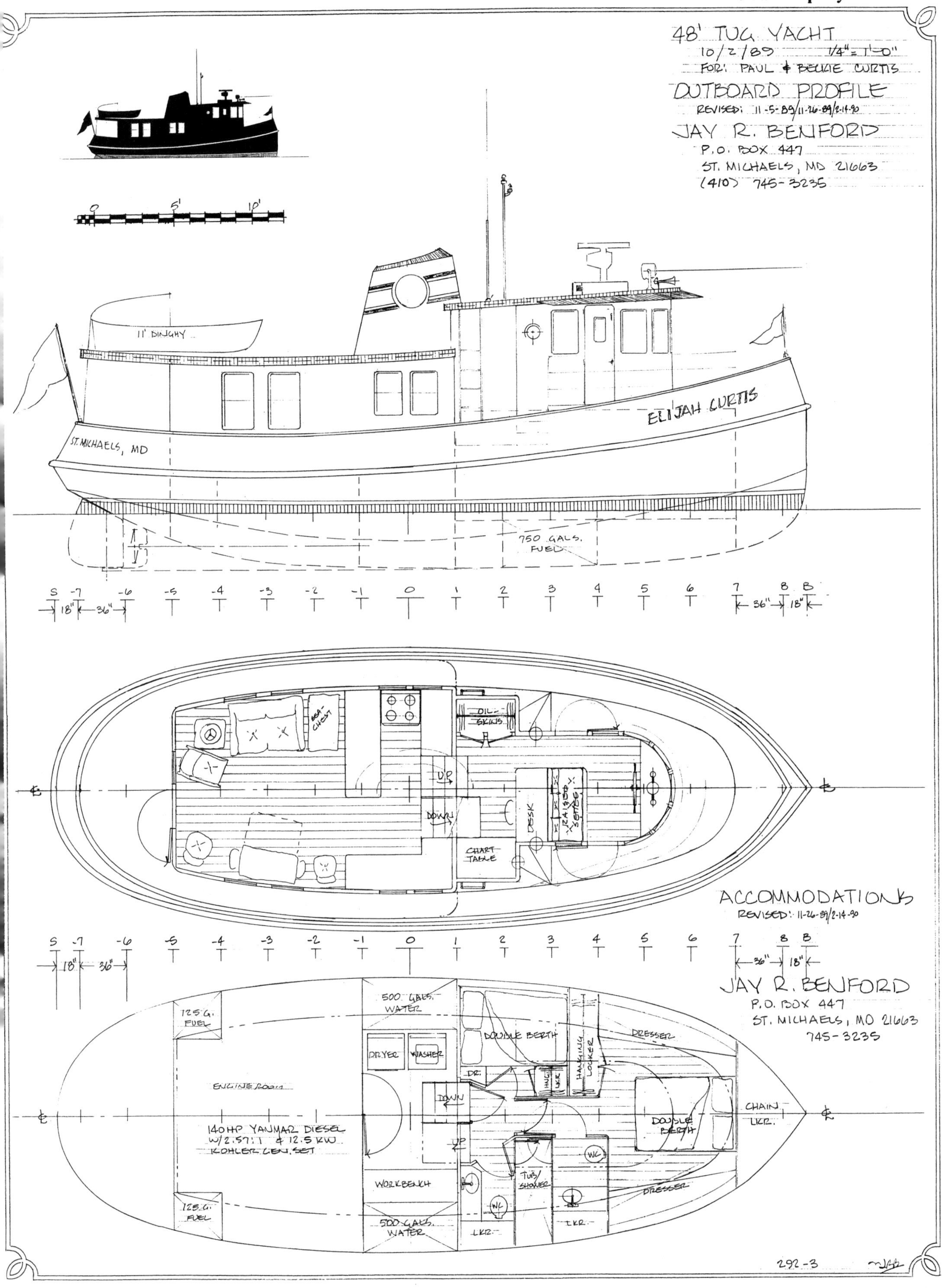
48' TUG YACHT
10/2/89 1/4" = 1'-0"
FOR: PAUL & BECKIE CURTIS
OUTBOARD PROFILE
REVISED: 11-5-89/11-26-89/2-14-90
JAY R. BENFORD
P.O. BOX 447
ST. MICHAELS, MD 21663
(410) 745-3235
11' DINGHY
ELIJAH CURTIS
ST. MICHAELS, MD
750 GALS. FUEL
SEA CHEST
OIL SKINS
UP
DOWN
DESK
RAISED SETTEE
CHART TABLE
ACCOMMODATIONS
REVISED: 11-26-89/2-14-90
JAY R. BENFORD
P.O. BOX 447
ST. MICHAELS, MD 21663
745-3235
125 G. FUEL
500 GALS. WATER
DOUBLE BERTH
DRESSER
DRYER
WASHER
HANGING LOCKER
ENGINE ROOM
DOWN
CHAIN LKR.
140HP YANMAR DIESEL
W/2.57:1 & 12.5 KW
KOHLER GEN. SET
UP
WC
WORKBENCH
TUB/SHOWER
LKR.
292-3

32' St. Pierre & 36' Power Dory

Designs Number 113 & 301

1974 & 1990

32' St. Pierre Dory. Designed for construction in plywood, this traditional St. Pierre Power Dory has proven a practical fishing vessel. She has comfortable living accommodations for her size, a good fish hold, and a working trolling cockpit. Her ability to get out to the fishing grounds quickly yet with reasonable comfort, and to keep to sea in a variety of conditions makes her an attractive investment. Her plans use commonly available lumber yard materials, making her a practical choice for a working boat on a budget.

36' Power Dory. This design evolved from our 36' Sailing Dory design. We left off the ballast keel and revised the plans for building in steel. She could also be built in plywood or aluminum like the original sailing version, with a significant reduction in weight and some ballast added to make up some of the difference.

The extra four feet of length has been put to good use, with a more spacious pilothouse and better cruising accommodations. Two alternative pilothouse styles are shown on the drawings. The windows raked forward at their top will have less reflected internal light at night, making this a practical choice.

Particulars:	**32'**	**36'**
Length overall	32'-0"	36'-0"
Length designed waterline	25'-6"	31'-0"
Beam	10'-10½"	11'-0"
Draft	2'-8"	3'-1"
Freeboard:		
Forward	4-8½"	5'-1"
Least	2'-0¼"	3'-1"
Aft	4'-2½"	4'-4"
Displacement, cruising trim*	8,660 lbs.	12,000 lbs.
Displacement-length ratio	233	180
Pounds per inch immersion	742	983
Water tankage	100 Gals.	50 Gals.
Fuel tankage	100 Gals.	260 Gals.
Headroom	5'-7" to 6'-6"	6'-0" to 6'-7"

***CAUTION:** The displacement quoted here is for the boat in cruising trim. That is, with the fuel and water tanks filled, the crew on board, as well as the crews' gear and stores in the lockers. This should not be confused with the "shipping weight" often quoted as "displacement" by some manufacturers. This should be taken into account when comparing figures and ratios between this and other designs.

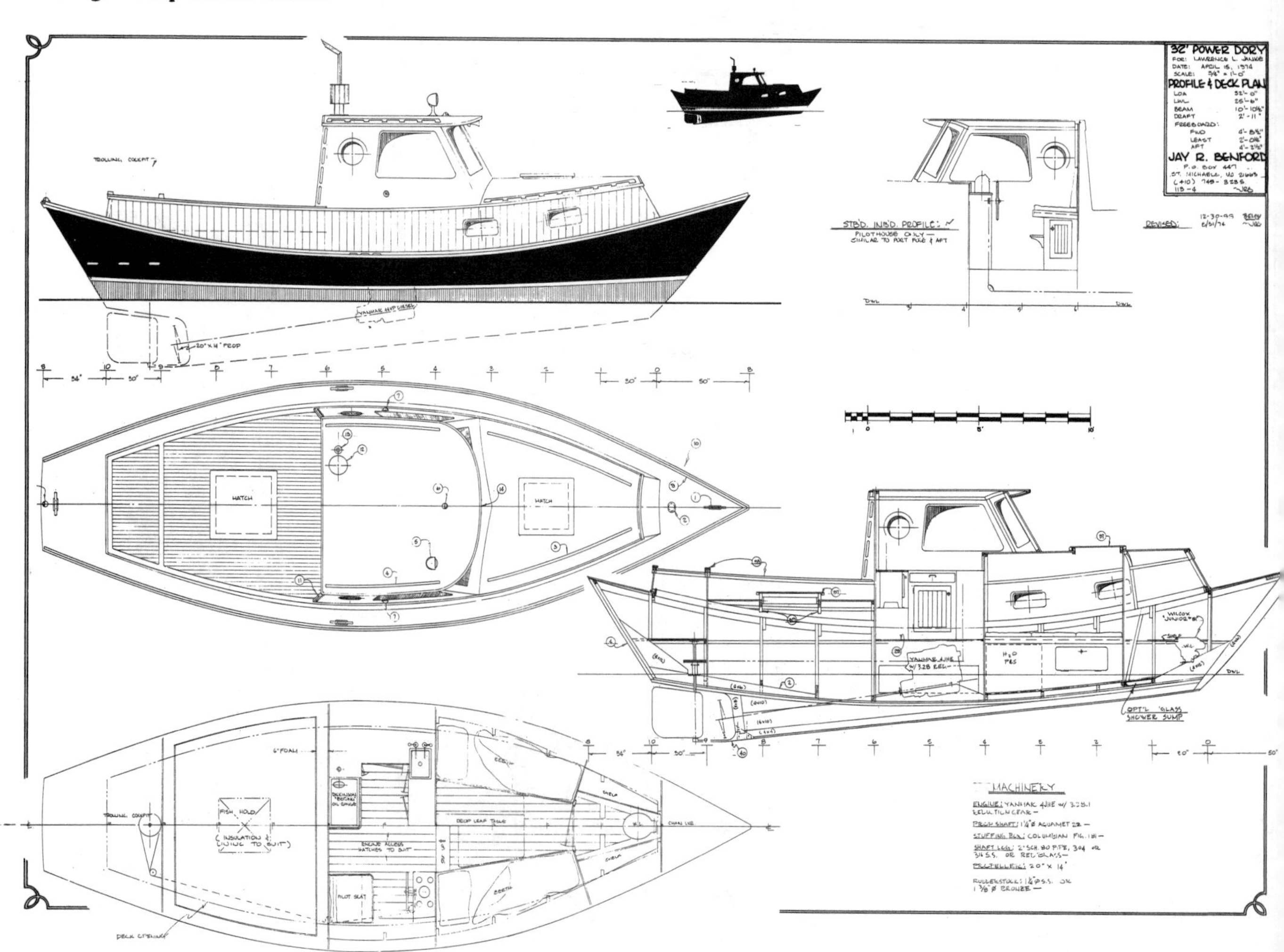

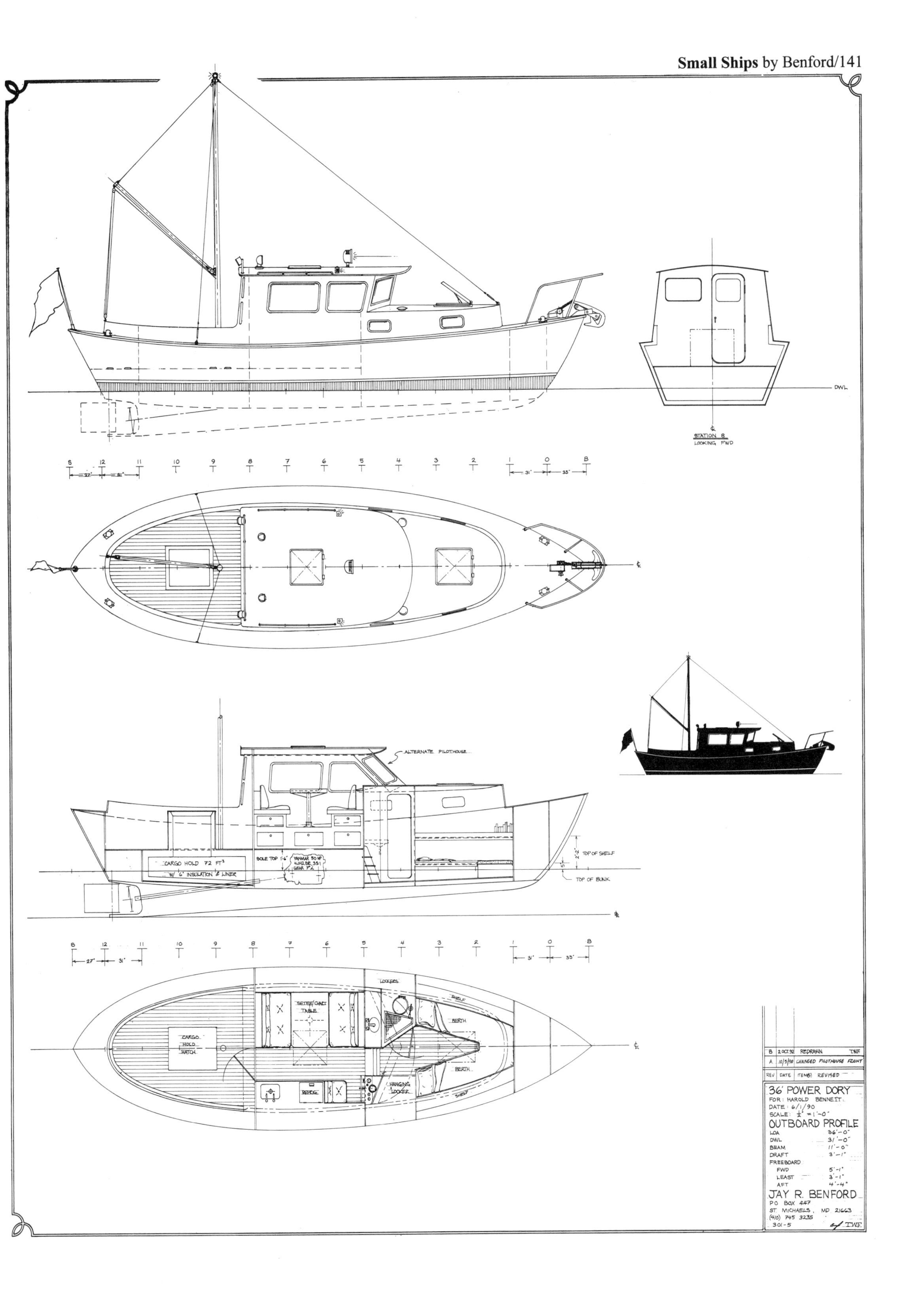
DWL
STATION 8
LOOKING FWD
ALTERNATE PILOTHOUSE
CARGO HOLD 72 FT³
W/ 6" INSULATION & LINER
TOP OF SHELF
TOP OF BUNK
LOCKERS
SHELF
BERTH
CARGO HOLD HATCH
REFRIG.
HANGING LOCKER
B | 2 OCT. 92 | REDRAWN | T.W.F.
A | 10/5/90 | CHANGED PILOTHOUSE FRONT
REV | DATE | ITEM(S) REVISED
36' POWER DORY
FOR: HAROLD BENNETT
DATE: 6/1/90
SCALE: ½" = 1'-0"
OUTBOARD PROFILE
LOA 36'-0"
DWL 31'-0"
BEAM 11'-0"
DRAFT 3'-1"
FREEBOARD
FWD 5'-1"
LEAST 3'-1"
AFT 4'-4"
JAY R. BENFORD
PO BOX 447
ST. MICHAELS, MD 21663
(410) 745 3235
301-5
T.W.F.

42', 46', 52', & 56' Fishing Vessels

Designs Number 81, 120, 84, 193
1972, 1975, 1971, & 1981

The 42, 46, and 52-foot designs were originally drawn and built in wood construction in the 1940's by George Calkins. By special arrangement with George Calkins, we have converted these designs with these alternate layouts.

The 56-footer was designed for an entrepreneur who liked the look of the 52, with its schooner style house, and wanted to build the larger one in steel. His concept was to build the boat with his machine shop in the hold space. He could then follow the fishing fleet around and provide on-the-spot repairs and work on the boats without too much lost time during the short fishing seasons. Using these designs with other sorts of businesses on board opens a wide range of opportunities. Something to think about....

***CAUTION:** The displacement quoted here is for the boat in half-load condition. That is, with the fuel and water tanks filled, the crew on board, as well as the crews' gear and stores in the lockers, plus enough fish and/or ice in the hold to load her down to this level. This should not be confused with the "shipping weight" often quoted as "displacement" by some manufacturers. This should be taken into account when comparing figures and ratios between this and other designs.

Particulars:	**42'**	**46'**	**52'**	**56'**
Length overall	42'-0"	45'-9¾"	52'-9"	56'-0"
Length designed waterline	38'-6"	41'-3½"	48'-4"	50'-0"
Beam	12'-0"	13'-4"	14'-6"	14'-6"
Draft	5'-6"	5'-3"	6'-0"	6'-4"
Freeboard:				
Forward	7'-5½"	8'-2"	9'-7"	9'-7½"
Least	3'-7½"	4'-7¾"	4'-8"	4'-8"
Aft	4'-2½"	5'-1½"	5'-7"	5'-7"
Displacement, half-load*	51,300	57,000	87,800	92,425
Displacement-length ratio	401	361	347	330
Water tankage, gals.	200	250	540	565
Fuel tankage, gals.	1,000	1,250	1,740	2,000
Headroom	6'-3"	6'-3"	6'-3"	6'-4"

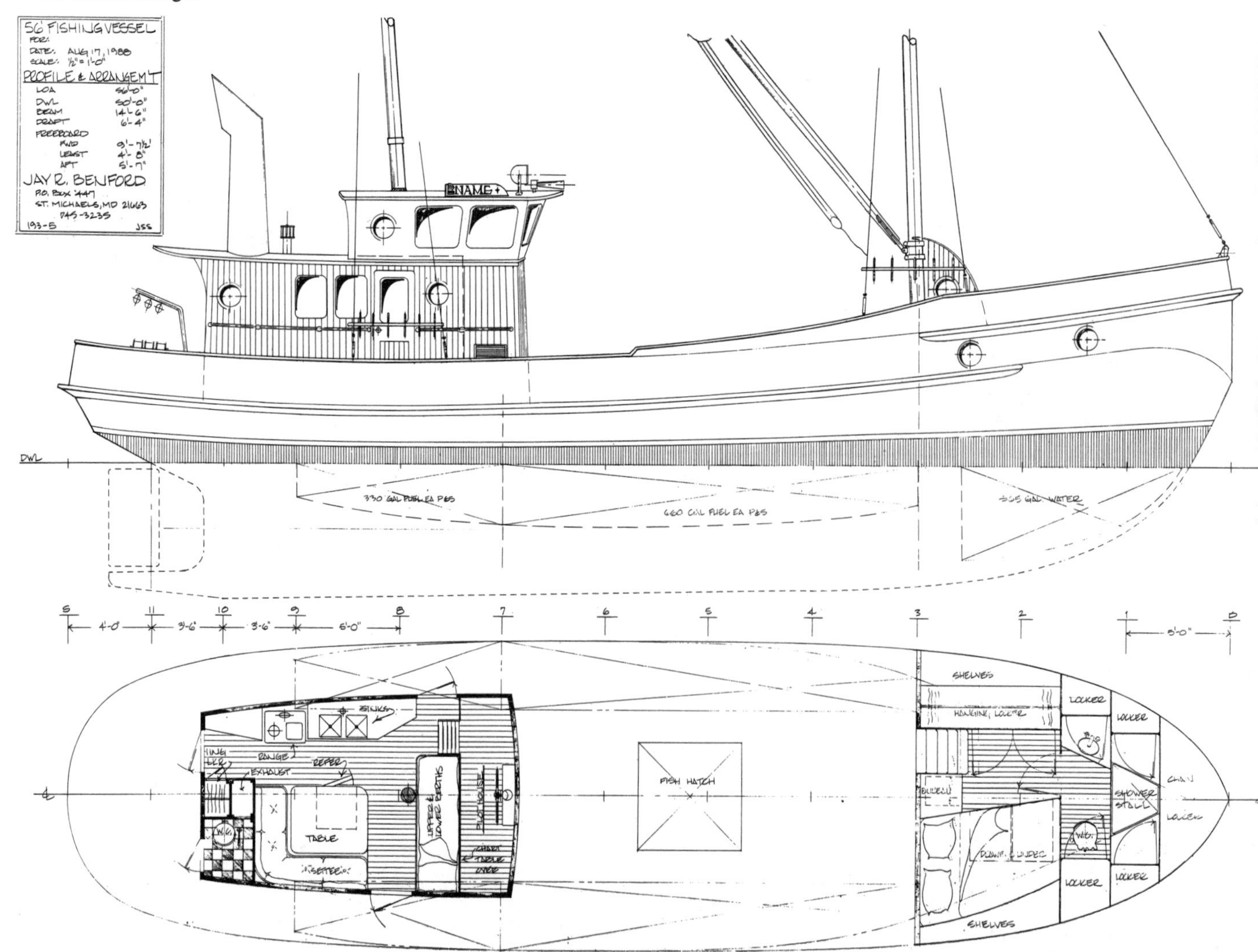

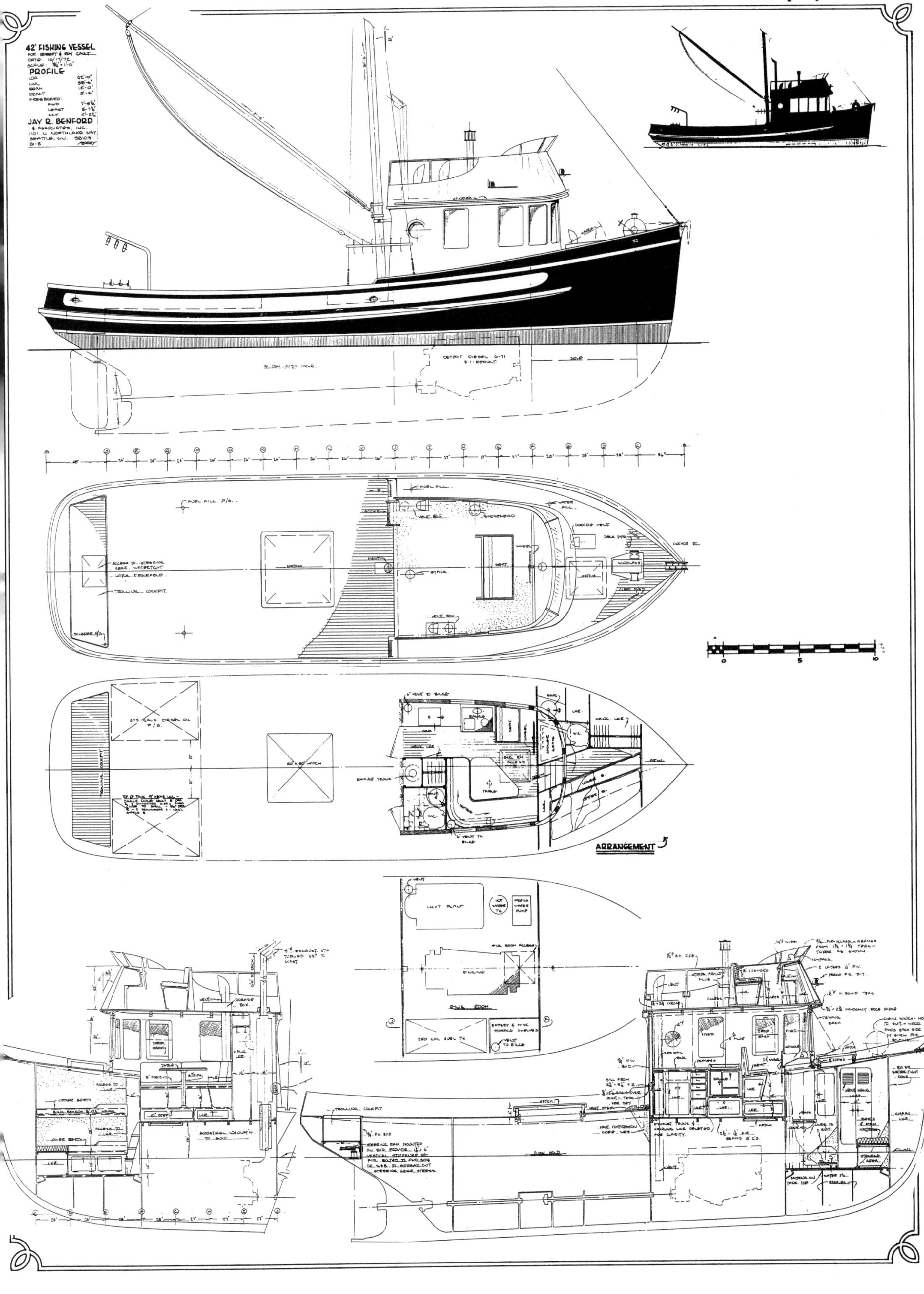

42' FISHING VESSEL
DATE 10/17/72
PROFILE

LOA	42'-0"
LWL	38'-6"
BEAM	12'-0"
DRAFT	5'-6"
FREEBOARD:	
FWD.	7'-8½"
LEAST	5-7½"
AFT	4'-2½"

JAY R. BENFORD
& ASSOCIATES, INC.
1101 N NORTHLAKE WAY
SEATTLE, WA. 98103
81-3

ARRANGEMENT

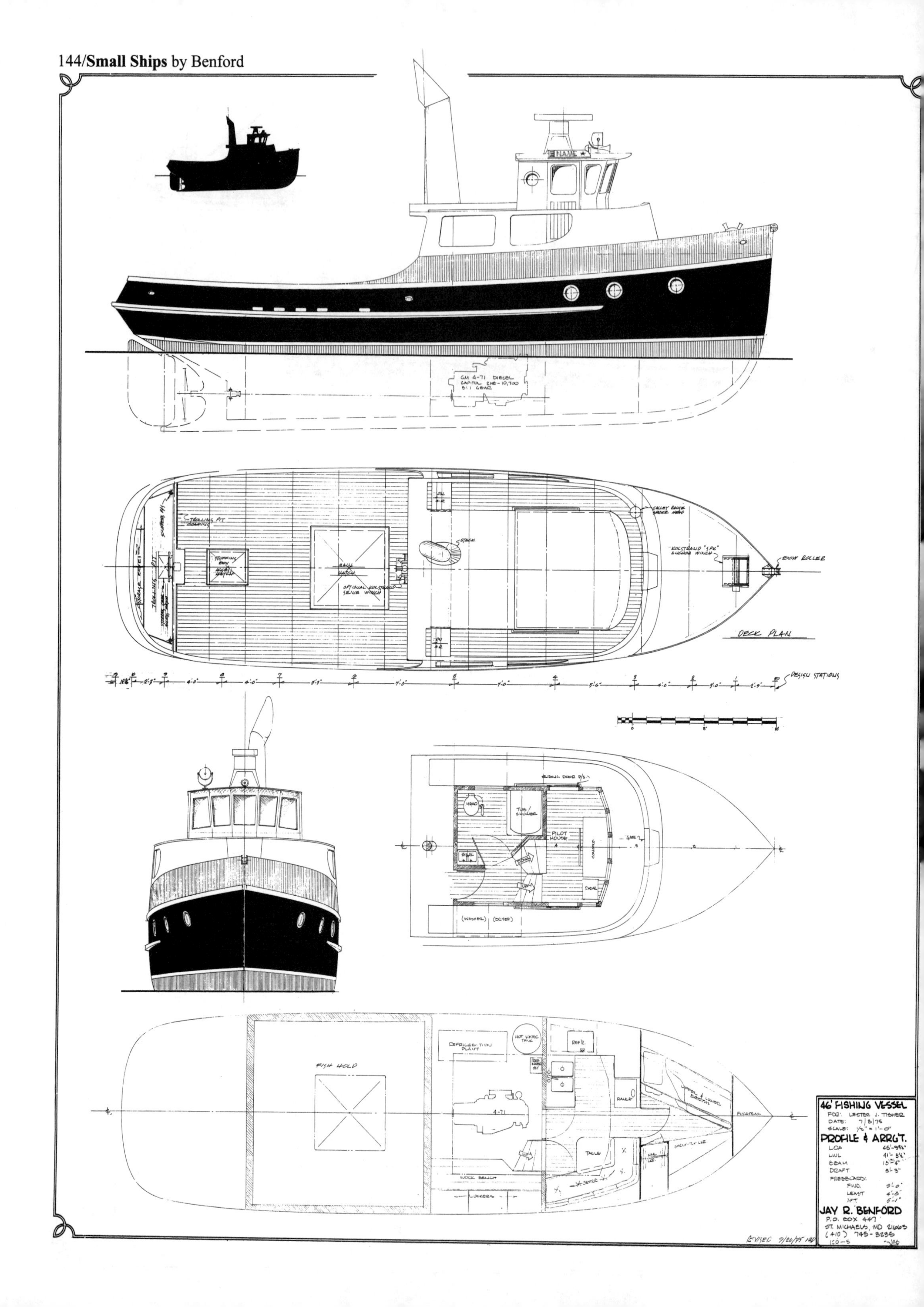

GM 4-71 DIESEL
CAPITOL 2HS-10,700
DECK PLAN
BOW ROLLER
DESIGN STATIONS
PILOT HOUSE
(WASHER) (DRYER)
FISH HOLD
4-71
WORK BENCH
LOCKERS
46' FISHING VESSEL
FOR: LESTER J. TISHER
DATE: 7/8/75
SCALE: 1/2" = 1'-0"
PROFILE & ARRG'T.
LOA 45'-9¾"
LWL 41'-3¼"
BEAM 13'-4"
DRAFT 5'-3"
FREEBOARD:
FWD. 9'-0"
LEAST 4'-8"
AFT 5'-1"
JAY R. BENFORD
P.O. BOX 447
ST. MICHAELS, MD 21663
(410) 745-3235

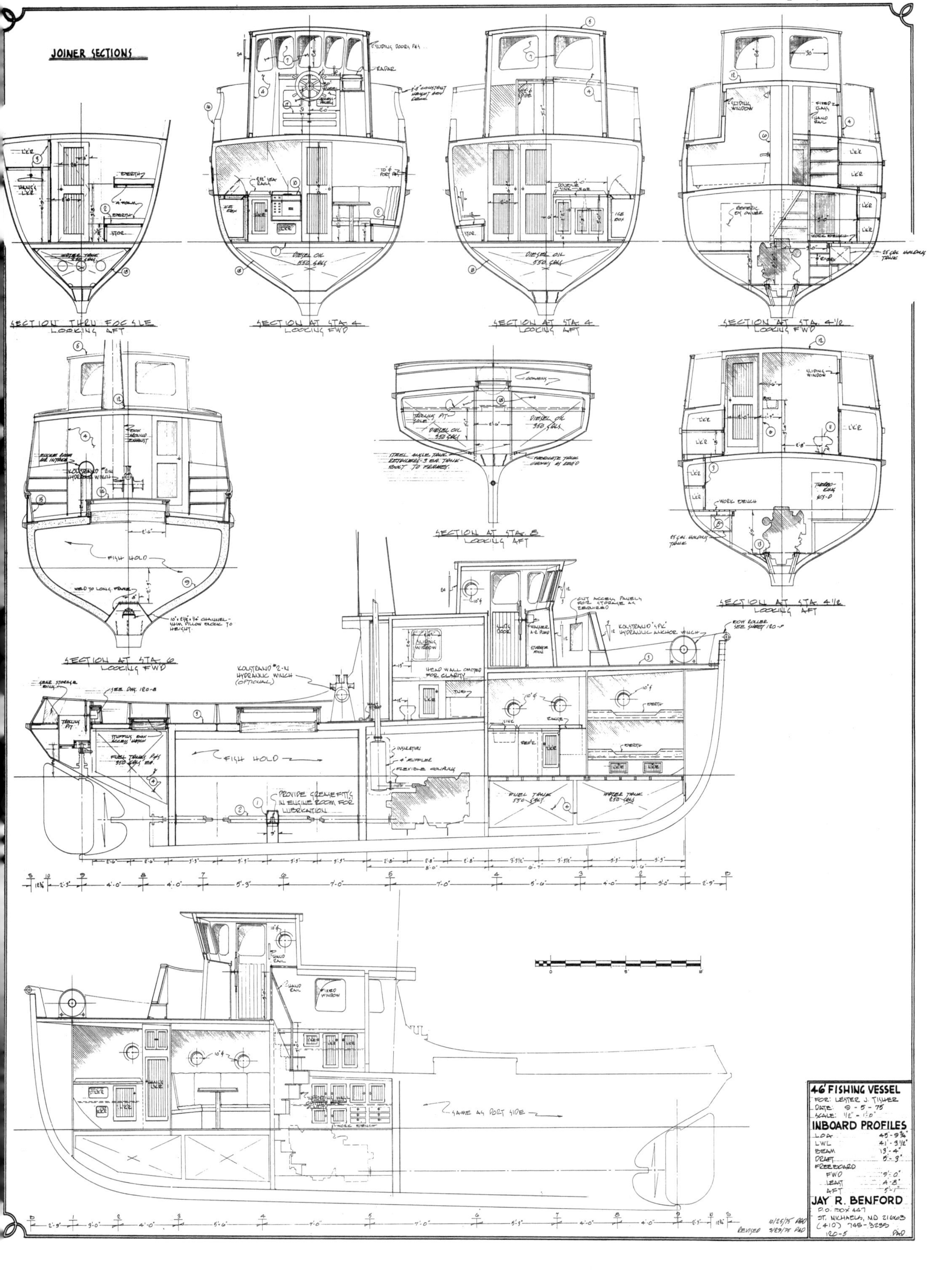

JOINER SECTIONS

SECTION THRU FOC'SLE
LOOKING AFT

SECTION AT STA. 4
LOOKING FWD

SECTION AT STA. 4
LOOKING AFT

SECTION AT STA. 4 1/2
LOOKING FWD

SECTION AT STA. 6
LOOKING FWD

SECTION AT STA. 8
LOOKING AFT

SECTION AT STA. 4 1/2
LOOKING AFT

DIESEL OIL 550 GALS

FISH HOLD

KOLSTRAND #2-N HYDRAULIC WINCH (OPTIONAL)

KOLSTRAND 'SPK' HYDRAULIC ANCHOR WINCH

BOW ROLLER SEE SHEET 120-P

HEAD WALL OMITTED FOR CLARITY

PROVIDE GREASE FITTINGS IN ENGINE ROOM FOR LUBRICATION

FUEL TANK 550 GALS

WATER TANK 250 GALS

SAME AS PORT SIDE

46' FISHING VESSEL
FOR: LESTER J. TISHER
DATE: 9 - 5 - 75
SCALE: 1/2" = 1'-0"

INBOARD PROFILES

LOA	45'-9 3/4"
LWL	41'-3 1/2"
BEAM	13'-4"
DRAFT	5'-3"
FREEBOARD	
FWD	9'-0"
LEAST	4'-8"
AFT	5'-1"

JAY R. BENFORD
P.O. BOX 447
ST. MICHAELS, MD 21663
(410) 745-3235
120-5 PAD

REVISED 10/25/75 PAD
3/29/75 PAD

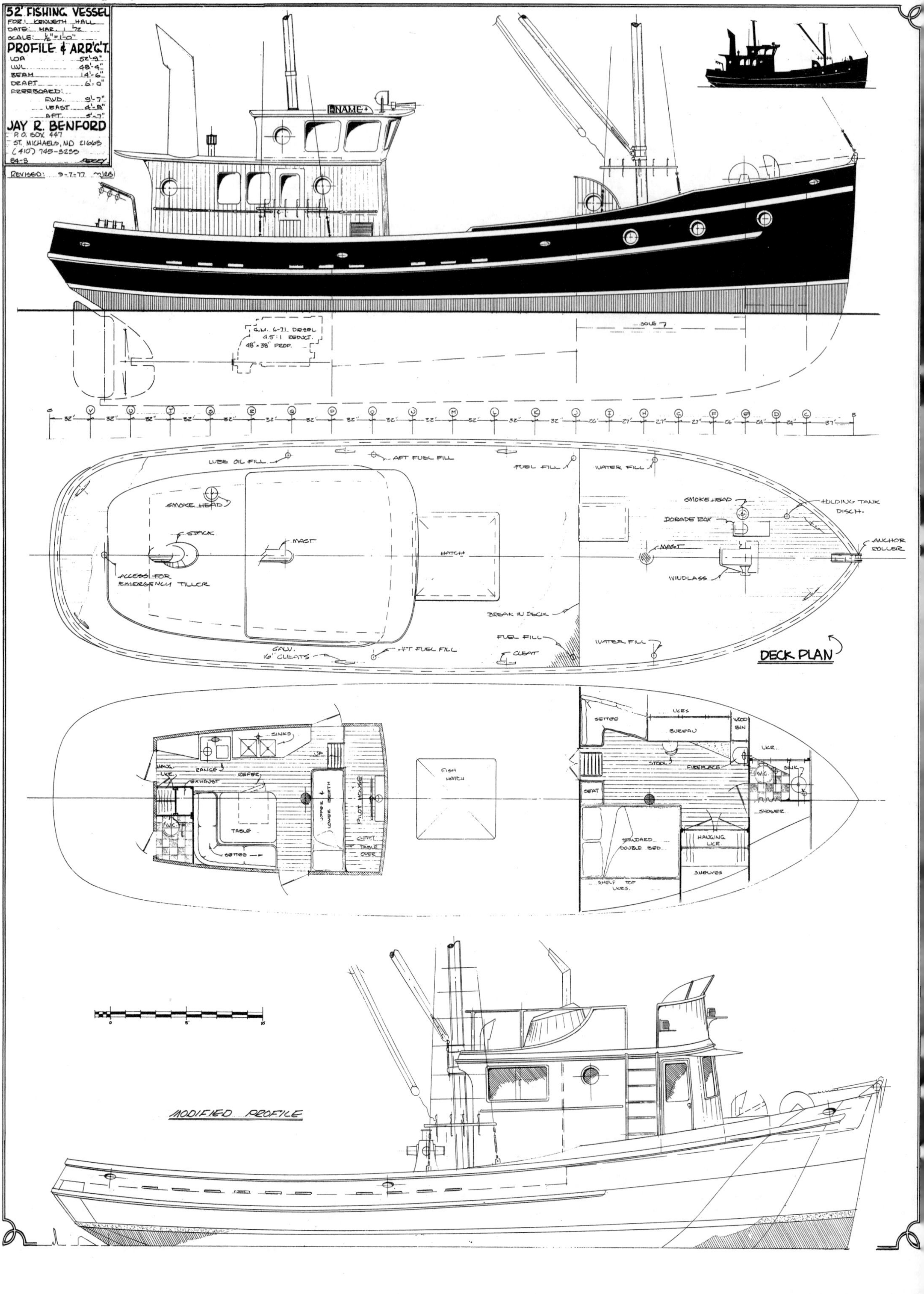

52' FISHING VESSEL
FOR: KENNETH HALL
DATE: MAR. 1 '72
SCALE: ½" = 1'-0"
PROFILE & ARR'G'T.
LOA 52'-9"
LWL 48'-4"
BEAM 14'-6"
DRAFT 6'-0"
FREEBOARD:
FWD. 9'-7"
LEAST 4'-8"
AFT 5'-7"

JAY R. BENFORD
P. O. BOX 447
ST. MICHAELS, MD 21663
(410) 745-3235

84-B

REVISED: 9-7-77 JRB

G.M. 6-71 DIESEL
4.5:1 REDUCT.
48" x 38" PROP.

SOLE

LUBE OIL FILL
AFT FUEL FILL
FUEL FILL
WATER FILL
SMOKE HEAD
STACK
MAST
HATCH
ACCESS FOR EMERGENCY TILLER
SMOKE HEAD
HOLDING TANK DISCH.
DORADE BOX
MAST
ANCHOR ROLLER
WINDLASS
BREAK IN DECK
FUEL FILL
CLEAT
WATER FILL
GALV. 16" CLEATS
AFT FUEL FILL

DECK PLAN

SINKS
RANGE
REFER.
EXHAUST
W.C.
TABLE
SETTEE
UPPER & LOWER BERTH
PILOT HOUSE
CHART TABLE OVER
FISH HATCH
SETTEE
LKRS
BUREAU
WOOD BIN
STOOL
FIREPLACE
LKR.
W.C.
SINK
SEAT
SHOWER
STANDARD DOUBLE BED
HANGING LKR
SHELVES
SHELF TOP LKRS.

MODIFIED PROFILE

52' FISHING VESSEL
FOR: KENNETH HALL
DATE: OCT 11, 75
SCALE: 1/2"=1'-0"

INBOARD PROFILES

LOA 52'-9"
LWL 48'-4"
BEAM 14'-6"
DRAFT 6'-0"
FREEBOARD
FWD 9'-7"
LEAST 4'-8"
AFT 5'-7"

JAY R. BENFORD
P.O. BOX 447
ST. MICHAELS, MD 21663
(410) 745-3235
84-8 GRANT

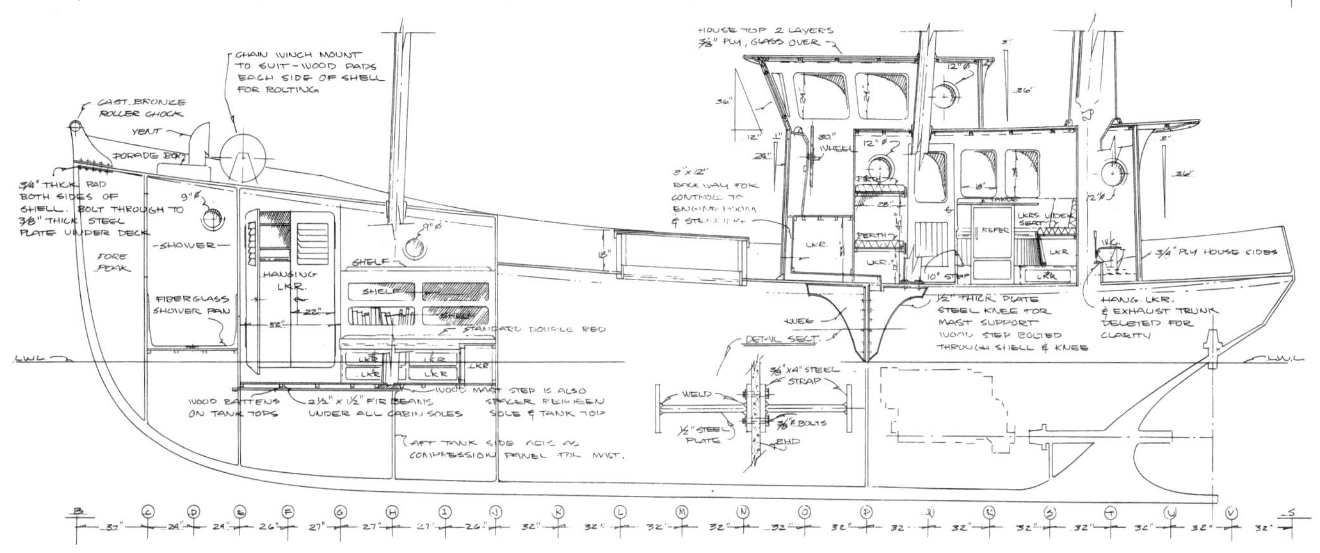

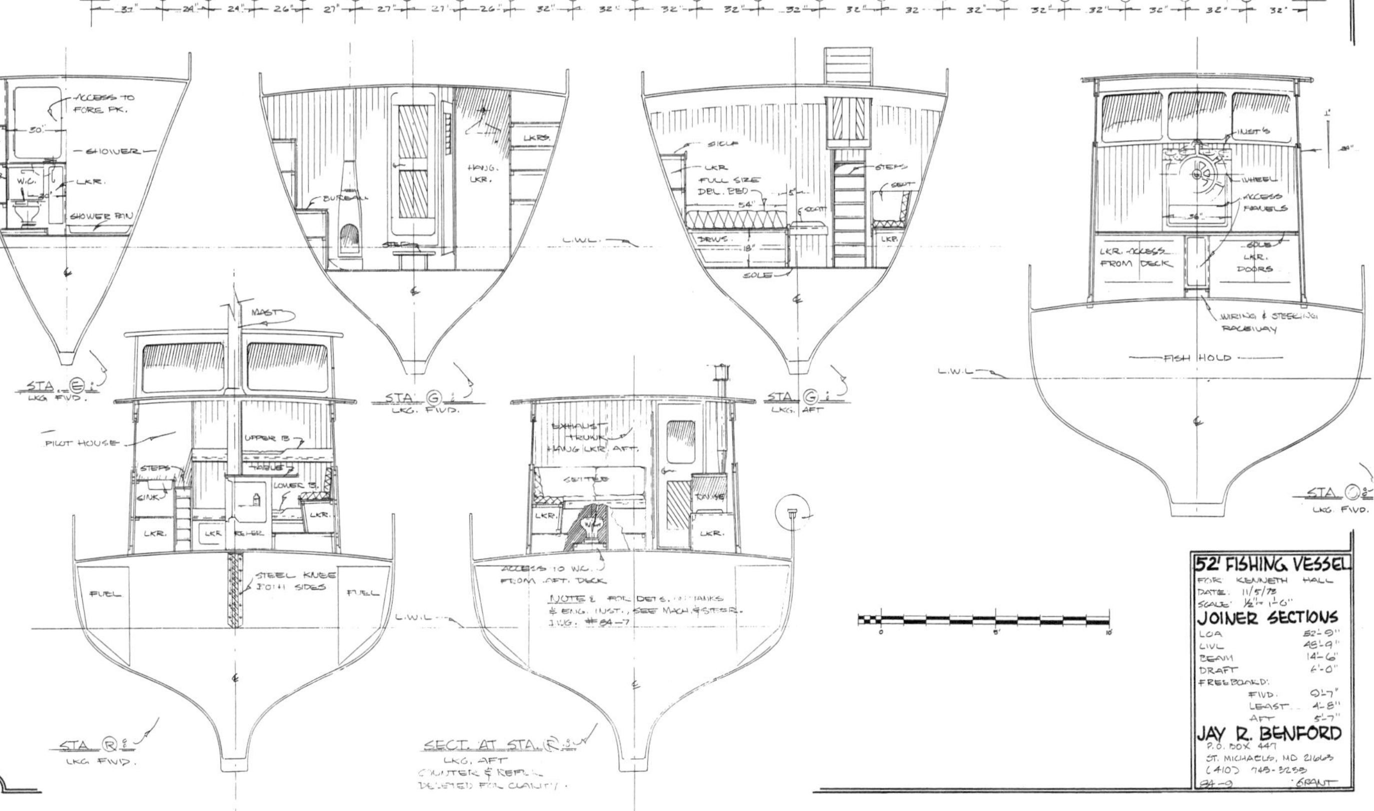

26' Deadrise Boat & Trawler Yacht

Design Number 4
1965

The first version of the 26' Deadrise Boat was done right after I'd been apprenticing with John Atkin. While there, I'd been immersed in the design work on a lot of nice, simple, straight-forward cruising boats. In the course of this, we'd also had a short cruise on a powerboat that John had earlier designed that was slightly smaller than this one. I liked the design and its charm. I took this as the inspiration in creating quite a different boat while keeping what I saw as the delights of its functionality.

On the Chesapeake Bay, Vee-bottomed boats are sometimes called *deadrise boats.* I adopted this name for this series of designs. These designs are not much like the lower freeboard and narrower working boats, but they do share a simplicity of concept and structure.

The 26' Deadrise Boat was the first design I had published in my own name. Ted Jones reviewed it in the November 1965 **Popular Boating** magazine's design section. This was back when magazines had real design review sections, something that seems to have fallen to the "bean counter's" relentless cost-cutting. It's a real shame, for the boating world could use more diversity of ideas in print. The major builders (advertisers, or major sources of operating revenues, in magazine parlance) are too busy copying each other and have lost track of what made boating interesting a few decades ago.

Without these design reviews, there are a lot fewer people exposed to the tremendous variety of choices they ***could*** have in boats. As it is, they have to really search out the few magazines that still have some coverage of custom designs in them. Anyone who's owned or even cruised on a classic or traditionally styled boat knows the amount of positive attention the boat receives. When talking to these onlookers I have found that most of them never realized that they too could own a boat like that. How sad it is that this diversity is lost in the rush of the production builders seeking to emulate the auto industry and create a mass market product.

The interest from Ted's design review led to creating some more versions, and these are also shown on the following couple pages. This is another area where the boating industry is hurt by the lack of design sections. Where else can a designer starting out on his or her own get their work published? Most cannot afford to carry the cost of paid advertising, which is one of the few choices left today. So, the door of opportunity is harder to open now than it was a generation ago and beginning designers are having an even more difficult time of finding those entry level positions.

The omission of any good interior passenger seating on the original (standard model) version was rectified in doing the designs of the other versions.

The Trawler Yacht (Brietz or Northwest Model) version has an aft cabin with galley and head in the after part and the dinette, plus helm and pilot seat in the raised pilothouse. The forward cabin has the sleeping accommodations. The midships cockpit will give the crew a good view of where you're heading. The wheel on the extension shaft through the front of the pilothouse lets the steering be handled from this cockpit. The mast carries a steadying sail and the ratlines let you go aloft to check the view further ahead. Flopper stopper stabilizers and poles could be fitted here for open water passages.

Our 30' Deadrise Boat, ***Petrel***, evolved from this one and led to the 32' ***Ladybug*** and the 35' ***Strumpet***, all of which turned out to be very nice boats. They all shared the pilothouse aft of a well deck layout, and the larger ones had the well deck continued around the house to the stern. While this could be done on the 26-footer, the resulting cabin would be somewhat smaller and thus less useful.

The engine sizes shown on these designs is larger than I would suggest today, unless you seriously wanted to try getting the boat running faster than displacement hull speeds. A little two or three cylinder diesel, well geared down, would be quite sufficient power and give good economy. Many of the new small diesels are quiet and smooth and seem to have been well refined.

Particulars:

Length overall	26'-0"
Length designed waterline	24'-0"
Beam	10'-0"
Draft	3'-0"
Freeboard:	
Forward	5'-6"
Least	3'-7"
Aft	4'-0"
Displacement, cruising trim*	8,400 lbs.
Displacement-length ratio	271
Sail area (riding sail)	67 sq. ft.
Water tankage	50 Gals.
Fuel tankage	110 Gals.
Headroom	4'-9" to 6'-5"

***CAUTION:** The displacement quoted here is for the boat in cruising trim. That is, with the fuel and water tanks filled, the crew on board, as well as the crews' gear and store in the locker. This should not be confused with the "shipping weight" often quoted as "displacement" by some manufacturers. This should be taken into account when comparing figures and ratios between this and other designs.

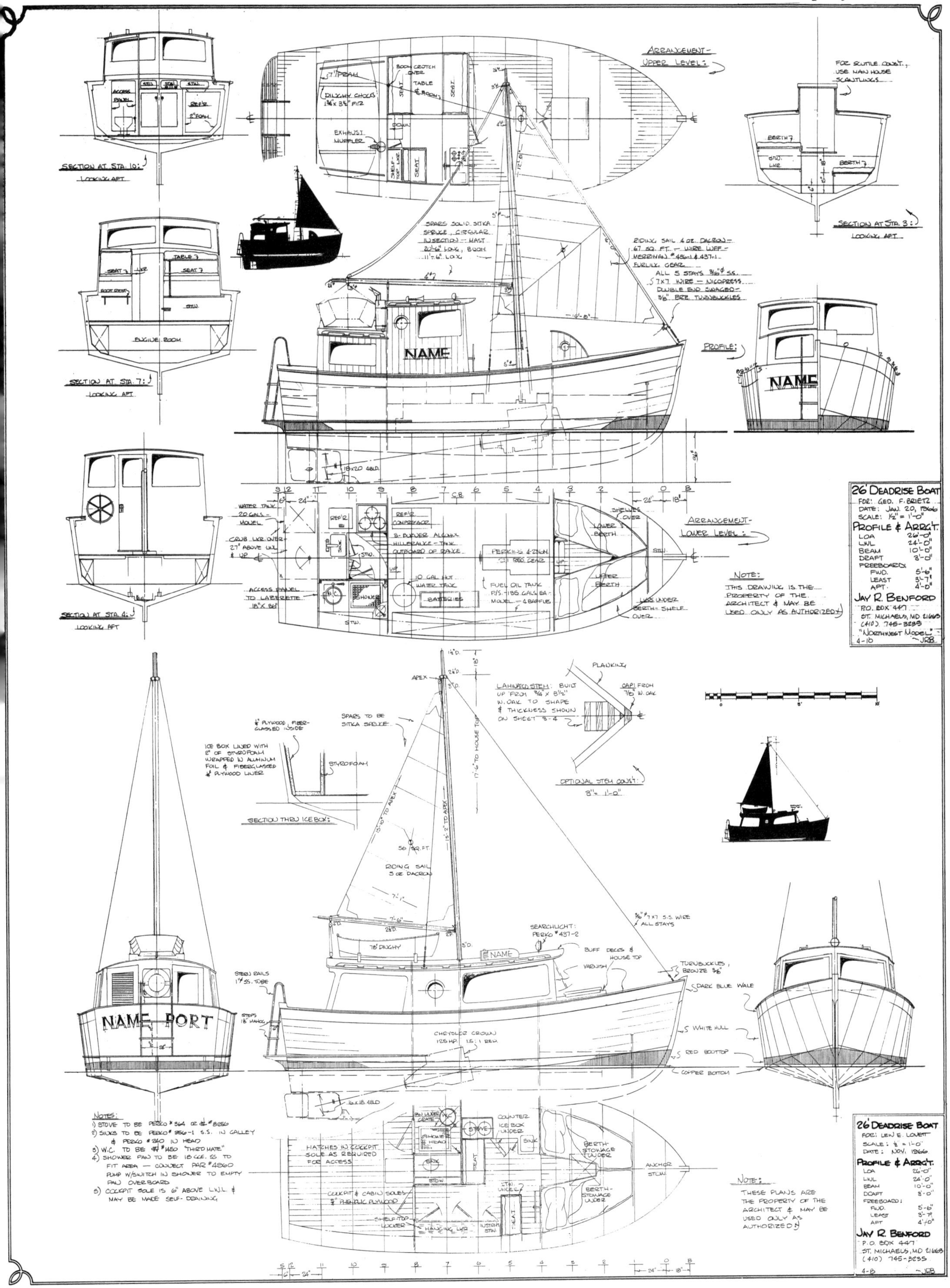

ARRANGEMENT-UPPER LEVEL:

7'-6" PRAM
DINGHY CHOCKS 1¾" x 3¾" FIR
BOOM CROTCH OVER
TABLE & BOOM
SEAT
EXHAUST MUFFLER
DOWN
SHELF-TOP LKR.
SEAT

FOR SCUTTLE CONST., USE MAIN HOUSE SCANTLINGS
BERTH
STW. LKR.
BERTH

SECTION AT STA. 10: LOOKING AFT

SECTION AT STA. 3: LOOKING AFT

SPARS SOLID SITKA SPRUCE, CIRCULAR IN SECTION - MAST 20'-6" LONG, BOOM 11'-6" LONG

RIDING SAIL 4 OZ. DACRON - 67 SQ. FT. - WIRE LUFF - MERRIMAN #486-1 & 437-1 FURLING GEAR
ALL 5 STAYS 3/16" ⌀ S.S. 7x7 WIRE - NICOPRESS DOUBLE END SWAGED - 3/8" BRZ. TURNBUCKLES

TABLE
SEAT
SEAT
FOOT REST
STW.
ENGINE ROOM

SECTION AT STA. 7: LOOKING AFT

NAME

PROFILE:

ARRANGEMENT-LOWER LEVEL:

WATER TANK 20 GALS. - MOYEL
CRUB LKR. OVER - 27" ABOVE LWL & UP
REF'R
REF'R COMPRESSOR
B-BORDER ALUMN. HILLERANGE TANK OUTBOARD OF RANGE
PERKINS 4-236 W/ RED. GEAR
10 GAL. HOT WATER TANK
BATTERIES
FUEL OIL TANK P/S - 135 GALS. EA. - MODEL - 4 BAFFLE
SHOWER
LOWER BERTH
SHELVES OVER
UPPER BERTH
LKRS. UNDER BERTH - SHELF OVER
ACCESS PANEL TO LAZERETTE 18" x 36"

SECTION AT STA. 4: LOOKING AFT

NOTE:
THIS DRAWING IS THE PROPERTY OF THE ARCHITECT & MAY BE USED ONLY AS AUTHORIZED.

26' DEADRISE BOAT
FOR: GEO. F. BRIETZ
DATE: JAN. 20, 1966
SCALE: ½" = 1'-0"

PROFILE & ARRG'T.

LOA	26'-0"
LWL	24'-0"
BEAM	10'-0"
DRAFT	3'-0"
FREEBOARD:	
FWD.	5'-6"
LEAST	3'-7"
APT.	4'-0"

JAY R. BENFORD
P.O. BOX 447
ST. MICHAELS, MD 21663
(410) 745-3235
"NORTHWEST MODEL"
4-10 ~JRB

LAMINATED STEM: BUILT UP FROM ¾ x 8½" W. OAK TO SHAPE & THICKNESS SHOWN ON SHEET 3-4
PLANKING
CAP: FROM 7/8" W. OAK
OPTIONAL STEM CONST.: 3" = 1'-0"

SPARS TO BE SITKA SPRUCE
½" PLYWOOD FIBERGLASSED INSIDE
ICE BOX LINED WITH 2" OF STYROFOAM WRAPPED IN ALUMINUM FOIL & FIBERGLASSED ¼" PLYWOOD LINER
STYROFOAM
SECTION THRU ICEBOX:

APEX
50 SQ. FT.
RIDING SAIL 5 OZ. DACRON
7'6" DINGHY
SEARCHLIGHT: PERKO #437-2
3/16" 7x7 S.S. WIRE ALL STAYS
BUFF DECKS & HOUSE TOP
VARNISH
TURNBUCKLES, BRONZE 3/8"
DARK BLUE WALE
STERN RAILS 1" S.S. TUBE
STEPS 18" MAHOG.
NAME PORT
CHRYSLER CROWN 125 HP. 1.5:1 RED.
WHITE HULL
RED BOOTTOP
COPPER BOTTOM

NOTES:
1) STOVE TO BE PERKO #364 OR ⊕ #8256
2) SINKS TO BE PERKO #356-1 S.S. IN GALLEY & PERKO #240 IN HEAD
3) W.C. TO BE ⊕ #1450 "THIRD MATE"
4) SHOWER PAN TO BE 18 GGE. SS TO FIT AREA — CONNECT PAR #4360 PUMP W/SWITCH IN SHOWER TO EMPTY PAN OVERBOARD
5) COCKPIT SOLE IS 6" ABOVE L.W.L. & MAY BE MADE SELF-DRAINING

HATCHES IN COCKPIT SOLE AS REQUIRED FOR ACCESS
COCKPIT & CABIN SOLES ¾" PHENOLIC PLYWOOD
SHELF-TOP LOCKER
HANGING LKR.
STOVE
SHOWER HEAD
SINK
COUNTER ICE BOX UNDER
BERTH-STOWAGE UNDER
ANCHOR STW.
SEAT

NOTE:
THESE PLANS ARE THE PROPERTY OF THE ARCHITECT & MAY BE USED ONLY AS AUTHORIZED.

26' DEADRISE BOAT
FOR: LEW E. LOVETT
SCALE: ½" = 1'-0"
DATE: NOV. 1966

PROFILE & ARRG'T.

LOA	26'-0"
LWL	24'-0"
BEAM	10'-0"
DRAFT	3'-0"
FREEBOARD:	
FWD.	5'-6"
LEAST	3'-7"
AFT	4'-0"

JAY R. BENFORD
P.O. BOX 447
ST. MICHAELS, MD 21663
(410) 745-3235
4-8 ~JRB

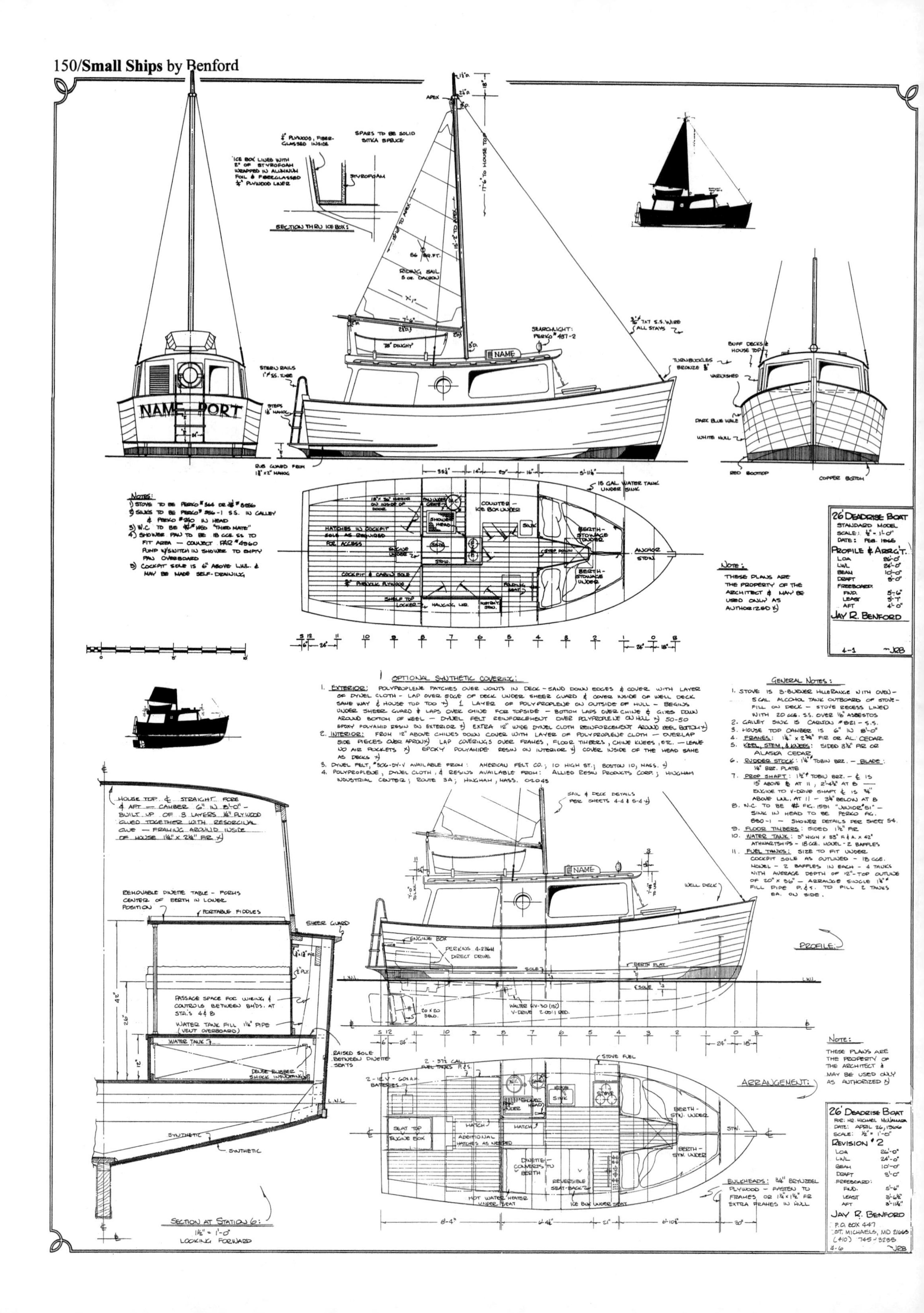

SPARS TO BE SOLID SITKA SPRUCE
SECTION THRU ICE BOX:
RIDING SAIL
SEARCHLIGHT: PERKO #487-2
NAME
NAME PORT
STERN RAILS 1" SS. TUBE
RUB GUARD FROM 1¾" x 2" MAHOG.
TURNBUCKLES BRONZE
BUFF DECKS & HOUSE TOP
VARNISHED
DARK BLUE WALE
WHITE HULL
RED BOOTTOP
COPPER BOTTOM
NOTES:
1) STOVE TO BE PERKO #344 OR #8256
2) SINKS TO BE PERKO #356-1 S.S. IN GALLEY & PERKO #240 IN HEAD
3) W.C. TO BE #1450 "THIRD MATE"
4) SHOWER PAN TO BE 18 GGE. SS TO FIT AREA — CONNECT PAR #4960 PUMP W/SWITCH IN SHOWER TO EMPTY PAN OVERBOARD
5) COCKPIT SOLE IS 6" ABOVE L.W.L. & MAY BE MADE SELF-DRAINING
18 GAL WATER TANK UNDER SINK
COUNTER — ICE BOX UNDER
HATCHES IN COCKPIT SOLE AS REQUIRED FOR ACCESS
BERTH — STOWAGE UNDER
ANCHOR STOW.
SHELF TOP LOCKER
HANGING LKR.
INSTR'M'T. STOW.
NOTE:
THESE PLANS ARE THE PROPERTY OF THE ARCHITECT & MAY BE USED ONLY AS AUTHORIZED
26' DEADRISE BOAT
STANDARD MODEL
SCALE: ½" = 1'-0"
DATE: FEB. 1966
PROFILE & ARRG'T.
LOA 26'-0"
LWL 24'-0"
BEAM 10'-0"
DRAFT 3'-0"
FREEBOARD:
FWD. 5'-6"
LEAST 3'-7"
AFT 4'-0"
JAY R. BENFORD
4-1 JRB
OPTIONAL SYNTHETIC COVERING:
1. EXTERIOR: POLYPROPLENE PATCHES OVER JOINTS IN DECK — SAND DOWN EDGES & COVER WITH LAYER OF DYNEL CLOTH — LAP OVER EDGE OF DECK UNDER SHEER GUARD & COVER INSIDE OF WELL DECK SAME WAY & HOUSE TOP TOO 1 LAYER OF POLYPROPLENE ON OUTSIDE OF HULL — BEGINS UNDER SHEER GUARD & LAPS OVER CHINE FOR TOPSIDE — BOTTOM LAPS OVER CHINE & GOES DOWN AROUND BOTTOM OF KEEL — DYNEL FELT REINFORCEMENT OVER POLYPROPLENE ON HULL 50-50 EPOXY POLYAMID RESIN ON EXTERIOR EXTRA 12" WIDE DYNEL CLOTH REINFORCEMENT AROUND KEEL BOTTOM
2. INTERIOR: FROM 12" ABOVE CHINES DOWN COVER WITH LAYER OF POLYPROPLENE CLOTH — OVERLAP SIDE PIECES OVER APRON LAP COVERINGS OVER FRAMES, FLOOR TIMBERS, CHINE KNEES, ETC. — LEAVE NO AIR POCKETS EPOXY POLYAMIDE RESIN ON INTERIOR COVER INSIDE OF THE HEAD SAME AS DECKS
3. DYNEL FELT, #306-DY-V AVAILABLE FROM: AMERICAN FELT CO.; 10 HIGH ST.; BOSTON 10, MASS.
4. POLYPROPLENE, DYNEL CLOTH, & RESINS AVAILABLE FROM: ALLIED RESIN PRODUCTS CORP.; HINGHAM INDUSTRIAL CENTER; ROUTE 3A; HINGHAM, MASS. 02043
GENERAL NOTES:
1. STOVE IS 3-BURNER HILLERANGE WITH OVEN — 5 GAL. ALCOHOL TANK OUTBOARD OF STOVE — FILL ON DECK — STOVE RECESS LINED WITH 20 GGE. S.S. OVER ⅛" ASBESTOS
2. GALLEY SINK IS CARLTON #821 — S.S.
3. HOUSE TOP CAMBER IS 6" IN 8'-0"
4. FRAMES: 1¼" x 2¾" FIR OR AL. CEDAR
5. KEEL, STEM, & KNEES: SIDED 3½" FIR OR ALASKA CEDAR
6. RUDDER STOCK: 1¼" TOBIN BRZ. — BLADE: ¼" BRZ. PLATE
7. PROP SHAFT: 1½" TOBIN BRZ. — ℄ IS 15" ABOVE B AT 11, 2'-4½" AT B — ENGINE TO V-DRIVE SHAFT ℄ IS ¾" ABOVE LWL AT 11 — 3¼" BELOW AT B
8. W.C. TO BE FIG. 1591 "JUNIOR 51" — SINK IN HEAD TO BE PERKO FIG. 880-1 — SHOWER DETAILS PER SHEET 54.
9. FLOOR TIMBERS: SIDED 1½" FIR
10. WATER TANK: 5" HIGH x 33" F & A. x 42" ATHWARTSHIPS — 18 GGE. MONEL — 2 BAFFLES
11. FUEL TANKS: SIZE TO FIT UNDER COCKPIT SOLE AS OUTLINED — 18 GGE. MONEL — 2 BAFFLES IN EACH — 4 TANKS WITH AVERAGE DEPTH OF 12" — TOP OUTLINE OF 20" x 36" — ARRANGE SINGLE 1¼" FILL PIPE P. & S. TO FILL 2 TANKS EA. ON SIDE.
SAIL & DECK DETAILS PER SHEETS 4-4 & 5-4
HOUSE TOP ℄ STRAIGHT FORE & AFT — CAMBER 6" IN 8'-0" — BUILT UP OF 3 LAYERS ⅜" PLYWOOD GLUED TOGETHER WITH RESORCINAL GLUE — FRAMING AROUND INSIDE OF HOUSE 1⅝" x 2¼" FIR
REMOVABLE DINETTE TABLE — FORMS CENTER OF BERTH IN LOWER POSITION
PORTABLE FIDDLES
SHEER GUARD
PASSAGE SPACE FOR WIRING & CONTROLS BETWEEN BH'DS. AT STA.'S 4 & 8
WATER TANK FILL 1¼" PIPE (VENT OVERBOARD)
WATER TANK
DENSE RUBBER SHOCK INSULATING
RAISED SOLE BETWEEN DINETTE SEATS
SYNTHETIC
SECTION AT STATION 6:
1½" = 1'-0"
LOOKING FORWARD
WELL DECK
ENGINE BOX
PERKINS 4-236M DIRECT DRIVE
BERTH FLAT
PROFILE:
STOVE FUEL
SINK
BERTH — STW. UNDER
STW.
SEAT TOP ENGINE BOX
ADDITIONAL HATCHES AS NEEDED
DINETTE — CONVERTS TO BERTH
REVERSIBLE SEAT-BACK
HOT WATER HEATER UNDER SEAT
ICE BOX UNDER SEAT
ARRANGEMENT:
NOTE:
THESE PLANS ARE THE PROPERTY OF THE ARCHITECT & MAY BE USED ONLY AS AUTHORIZED
BULKHEADS: ¾" BRYNZEEL PLYWOOD — FASTEN TO FRAMES OR 1¼" x 1½" FIR EXTRA FRAMES IN HULL
26' DEADRISE BOAT
FOR: MR. MICHAEL McNAMARA
DATE: APRIL 26, 1966
SCALE: ½" = 1'-0"
REVISION #2
LOA 26'-0"
LWL 24'-0"
BEAM 10'-0"
DRAFT 3'-0"
FREEBOARD:
FWD. 5'-6"
LEAST 3'-6½"
AFT 3'-11½"
JAY R. BENFORD
P.O. BOX 447
ST. MICHAELS, MD 21663
(410) 745-3235
4-6 JRB

27' Deadrise Boats

Design number 25

1966

Following the publication of the 26' Deadrise Boat came the commissions for these 27-footers. The first 27 drawn was the one on the following page. It is more of a day or short duration cruiser than the Bermuda Version which was done next.

The Bermuda Version came about from a request from our client to have a boat capable of more extended range cruising so that he could go from Bermuda to visit North America. Several boats have been built from this design, but I don't know if any of them have actually made this sort of voyage under power.

These fine cruisers will prove to be comfortable vessels, with their large cockpits for fishing and sunning, galleys in the deckhouses, enclosed heads with showers, and two permanent berths in foc'sles. A 20 to 30 h.p. diesel is all that is needed for up to six and a half knot service speed.

Photo courtesy Dr. Filippo Riva

Particulars:	**25**	**25A**
Length overall	27'-0"	27'-0"
Length designed waterline	25'-0"	25'-0"
Beam	10'-0"	10'-1"
Draft	3'-0"	3'-0"
Freeboard:		
Forward	5'-6"	5'-0"
Least	3'-7½"	2'-11"
Aft	4'-0"	3'-4"
Displacement, cruising trim*	9,100 lbs.	11,200 lbs.
Displacement-length ratio	260	320
Sail area (riding sail)	48 or 56 ft.²	48 sq. ft.
Water tankage	10 Gals.	50 Gals.
Fuel tankage	120 Gals.	250 Gals.
Headroom	5'-1" to 6'-7"	5'-4" to 6'-4"

***CAUTION:** The displacement quoted here is for the boat in cruising trim. That is, with the fuel and water tanks filled, the crew on board, as well as the crews' gear and store in the locker. This should not be confused with the "shipping weight" often quoted as "displacement" by some manufacturers. This should be taken into account when comparing figures and ratios between this and other designs.

OVER on our side of the Atlantic, we are only just entering the era of the American-type "powerboat". By this we mean the whole tribe of glass fibre, hard-chine planing boats with streamlined flying bridges and two large engines installed beneath a huge open cockpit.

The interesting thing is that on the other side of the Atlantic, where they have been exposed to this kind of boat for so much longer, a kind of backlash movement seems to have sprung up. Never have so many "character" boats been built than in the last few years, in the United States. Jaunty schooners and ketches from the boards of men like Atkins and Garden can be seen everywhere while even brigantines and full-rigged ships are not uncommon. In the field of pure motor boats the huge success of a boat like the Grand Banks is further evidence of the swing back to "boaty" boats.

Nothing could be more "boaty" than the little motor cruiser shown on this page. With her bold sheer, upright house and steadying sail she will always give pleasure to the eye of a seaman, whether she is bobbing gracefully at her mooring or puttering gravely along at her economical speed of about seven knots. She is being built for a Mr. Edward Barnes of Hamilton, Bermuda, who wants her for day-cruising and "occasional trips to Nova Scotia" which certainly seems an ambitious trip for a 27ft single-screw motor boat. To this end she has tanks for 250 gallons of fuel giving her the amazing range of 2,000 miles at 6 knots.

Fascinatingly, the engine chosen is a Brit, the "fisherman's friend" from Bridport, Dorset, though as a matter of fact the model 154 is Brit's conversion of an American diesel built by International Harvester. Its rated power of 40 b.h.p. is generous for this hull and the designer's power/speed graph shows that while only ten h.p. is required to drive her at six knots, 35 h.p. only pushes the speed up to a little over eight knots while at the same time fuel consumption is doubled. Still, the extra power is no doubt very welcome for use in strong headwinds or for the occasional towing job, aided by the 2 : 1 reduction and 18in propeller.

To provide accommodation suitable for both Bermuda and Nova Scotia cannot be easy in a 27ft boat, but what there is of it looks extremely sensible. For Bermuda, a largish cockpit and permanent bathing ladder; for Nova Scotia a cosy two-berth cabin with 6ft standing headroom. The w.c. compartment, which can be fitted with a shower, is entered from the cockpit which has the twin advantages of keeping smells out of the cabin and allowing a vastly improved galley arrangement. I shouldn't want to sleep so far forward on passage, but sleeping at sea is going to be pretty difficult anyway on a boat this size. The vee-bottomed hull-shape will make her rather stiff but the riding sail will do a lot to cut down the amount of roll. The foredeck has nine-inch deep bulwarks so as to provide a safe sitting area for children but one hopes it will not pick up too much water when this attractive little boat is butting her way through the Gulf Stream on one of those occasional trips to Nova Scotia.

Reprinted courtesy of **Yachting World** magazine from their September 1968 issue.

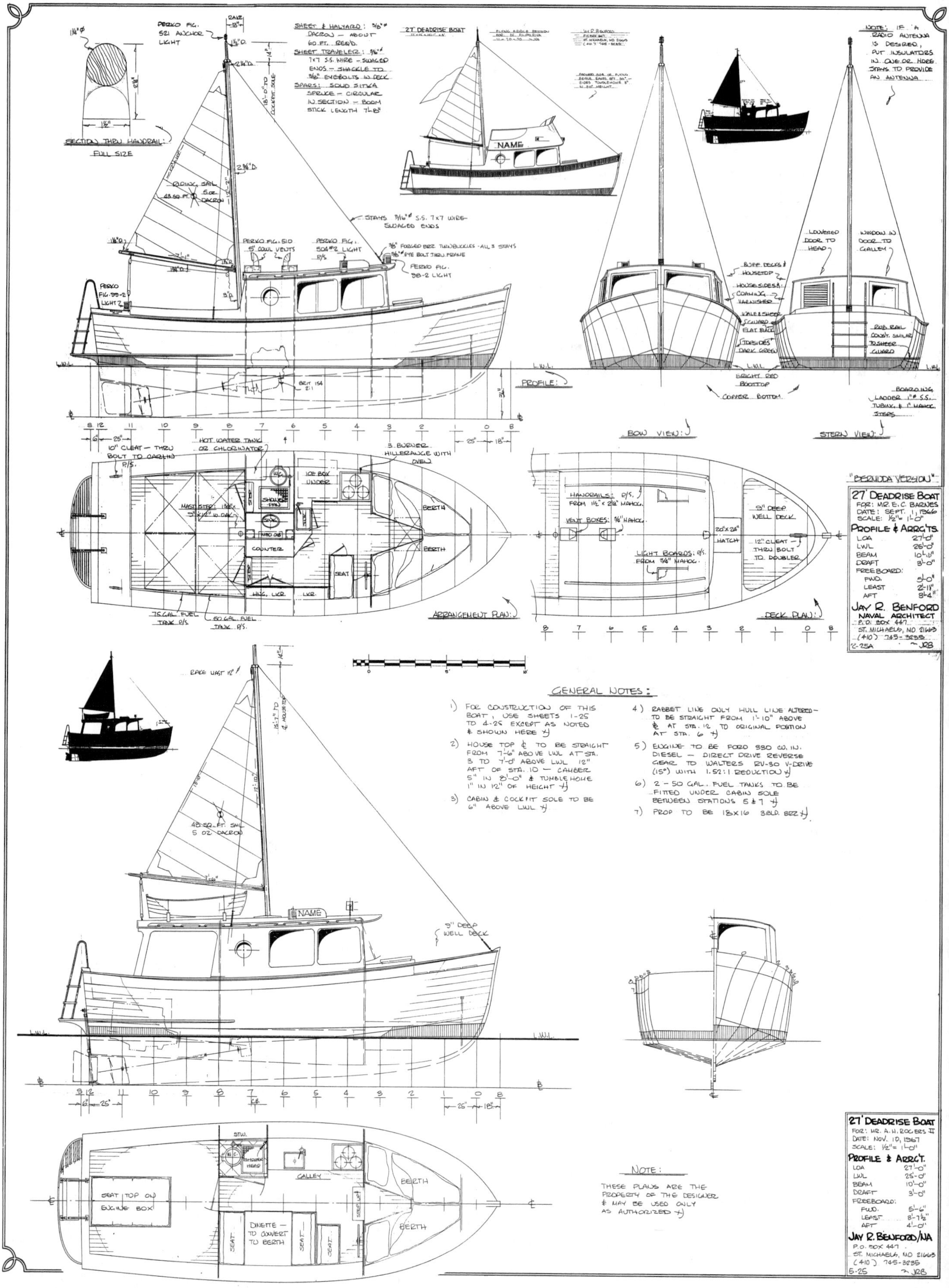

SHEET & HALYARD: 3/8"ø DACRON — ABOUT 60 FT. REQ'D.
SHEET TRAVELER: 3/16"ø 7x7 S.S. WIRE — SWAGED ENDS — SHACKLE TO 5/16" EYEBOLTS IN DECK
SPARS: SOLID SITKA SPRUCE — CIRCULAR IN SECTION — BOOM STICK LENGTH 7'-8"

PERKO FIG. 521 ANCHOR LIGHT

SECTION THRU HANDRAIL: FULL SIZE

STAYS 3/16"ø S.S. 7x7 WIRE — SWAGED ENDS

PERKO FIG. 510 5" COWL VENTS

PERKO FIG. 504 #2 LIGHT P/S

3/8" FORGED BRZ TURNBUCKLES - ALL 3 STAYS
3/8"ø EYE BOLT THRU FRAME

PERKO FIG. 58-2 LIGHT

PERKO FIG. 55-2 LIGHT

ROUND SAIL 43 SQ. FT. 5 OZ. DACRON

PROFILE

27' DEADRISE BOAT

NOTE: IF A RADIO ANTENNA IS DESIRED, PUT INSULATORS IN ONE OR MORE STAYS TO PROVIDE AN ANTENNA

LOUVERED DOOR TO HEAD

WINDOW IN DOOR TO GALLEY

BUFF DECKS & HOUSETOP
HOUSESIDES & COAMING VARNISHED
WALE & SHEER GUARD FLAT BLACK
TOPSIDES DARK GREEN
BRIGHT RED BOOTTOP
COPPER BOTTOM

RUB RAIL CONST. SIMILAR TO SHEER GUARD

BOARDING LADDER 1"ø S.S. TUBING & 1" MAHOG. STEPS

BOW VIEW

STERN VIEW

10" CLEAT — THRU BOLT TO CARLIN P/S.

HOT WATER TANK OR CHLORINATOR

3 BURNER HILLERANGE WITH OVEN

ICE BOX UNDER

MAST STEP: 1 3/4" x 5" x 12" W. OAK

SHOWER PAN

SINK

COUNTER

SEAT

BERTH

BERTH

HNG. LKR.

LKR.

75 GAL. FUEL TANK P/S

50 GAL. FUEL TANK P/S

ARRANGEMENT PLAN

HANDRAILS: P/S. FROM 1 1/2" x 2 1/4" MAHOG.

VENT BOXES: 3/4" MAHOG.

LIGHT BOARDS: P/S. FROM 3/4" MAHOG.

9" DEEP WELL DECK

20" x 24" HATCH

12" CLEAT — THRU BOLT TO DOUBLER

DECK PLAN

"BERMUDA VERSION"

27' DEADRISE BOAT
FOR: MR. E. C. BARNES
DATE: SEPT. 1, 1966
SCALE: 1/2" = 1'-0"

PROFILE & ARRG'TS.

LOA	27'-0"
LWL	25'-0"
BEAM	10'-1/2"
DRAFT	3'-0"
FREEBOARD:	
FWD.	5'-0"
LEAST	2'-11"
AFT	3'-4"

JAY R. BENFORD
NAVAL ARCHITECT
P.O. BOX 447
ST. MICHAELS, MD 21663
(410) 745-3235
2-25A

RAKE MAST 1/2"

48 SQ. FT. SAIL 5 OZ. DACRON

NAME

9" DEEP WELL DECK

GENERAL NOTES:

1) FOR CONSTRUCTION OF THIS BOAT, USE SHEETS 1-25 TO 4-25 EXCEPT AS NOTED & SHOWN HERE.
2) HOUSE TOP ℄ TO BE STRAIGHT FROM 7'-6" ABOVE LWL AT STA. 3 TO 7'-0" ABOVE LWL 12" AFT OF STA. 10 — CAMBER 5" IN 8'-0" & TUMBLEHOME 1" IN 12" OF HEIGHT.
3) CABIN & COCKPIT SOLE TO BE 6" ABOVE LWL.
4) RABBET LINE ONLY HULL LINE ALTERED — TO BE STRAIGHT FROM 1'-10" ABOVE ℄ AT STA. 12 TO ORIGINAL POSITION AT STA. 6.
5) ENGINE TO BE FORD 330 CU. IN. DIESEL — DIRECT DRIVE REVERSE GEAR TO WALTERS RV-30 V-DRIVE (15°) WITH 1.52:1 REDUCTION.
6) 2 — 50 GAL. FUEL TANKS TO BE FITTED UNDER CABIN SOLE BETWEEN STATIONS 5 & 7.
7) PROP TO BE 18x16 3BLD. BRZ.

SEAT TOP ON ENGINE BOX

GALLEY

HEAD

BERTH

BERTH

DINETTE — TO CONVERT TO BERTH

SEAT

SEAT

NOTE:
THESE PLANS ARE THE PROPERTY OF THE DESIGNER & MAY BE USED ONLY AS AUTHORIZED.

27' DEADRISE BOAT
FOR: MR. A. H. ROGERS II
DATE: NOV. 10, 1967
SCALE: 1/2" = 1'-0"

PROFILE & ARRG'T.

LOA	27'-0"
LWL	25'-0"
BEAM	10'-0"
DRAFT	3'-0"
FREEBOARD:	
FWD.	5'-6"
LEAST	3'-7 1/2"
AFT	4'-0"

JAY R. BENFORD/NA
P.O. BOX 447
ST. MICHAELS, MD 21663
(410) 745-3235
5-25

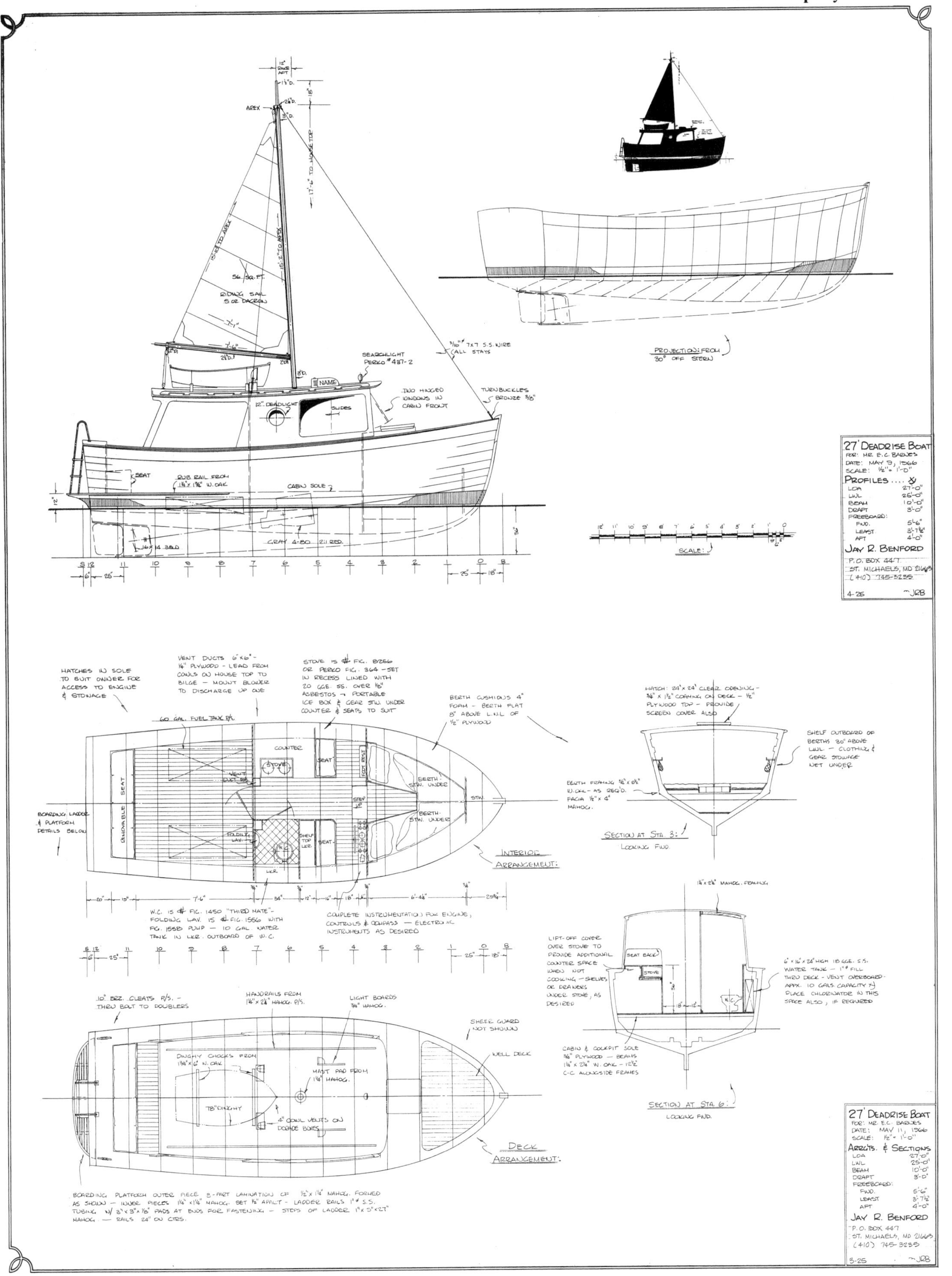

APEX
RIDING SAIL
5 OZ DACRON
SEARCHLIGHT
PERKO #487-2
3/16" ϕ 7X7 S.S. WIRE
ALL STAYS
TWO HINGED
WINDOWS IN
CABIN FRONT
TURNBUCKLES
BRONZE 3/8"
12" DEADLIGHT
SLIDES
NAME
SEAT
RUB RAIL FROM
1 1/4" X 1 7/8" W. OAK
CABIN SOLE
GRAY 4-80
PROJECTION: FROM
30° OFF STERN
27' DEADRISE BOAT
FOR: MR. E.C. BARNES
DATE: MAY 9, 1966
SCALE: 1/2" = 1'-0"
PROFILES
LOA 27'-0"
LWL 26'-0"
BEAM 10'-0"
DRAFT 3'-0"
FREEBOARD:
FWD. 5'-6"
LEAST 3'-7 1/2"
AFT 4'-0"
JAY R. BENFORD
P.O. BOX 447
ST. MICHAELS, MD 21663
(410) 745-3235
SCALE:
HATCHES IN SOLE
TO SUIT OWNER FOR
ACCESS TO ENGINE
& STOWAGE
VENT DUCTS 6"x6" -
1/4" PLYWOOD - LEAD FROM
COWLS ON HOUSE TOP TO
BILGE - MOUNT BLOWER
TO DISCHARGE UP ONE
STOVE IS # FIG. 8256
OR PERKO FIG. 364 - SET
IN RECESS LINED WITH
20 GGE. SS. OVER 1/8"
ASBESTOS - PORTABLE
ICE BOX & GEAR STW. UNDER
COUNTER & SEATS TO SUIT
BERTH CUSHIONS 4"
FOAM - BERTH FLAT
8" ABOVE L.W.L. OF
1/2" PLYWOOD
60 GAL. FUEL TANK P/S
COUNTER
STOVE
SEAT
BERTH
STW. UNDER
REMOVABLE SEAT
BOARDING LADDER
& PLATFORM
DETAILS BELOW
FOLDING LAV.
INTERIOR
ARRANGEMENT
W.C. IS # FIG. 1450 "THIRD MATE" -
FOLDING LAV. IS # FIG. 1556 WITH
FIG. 1558 PUMP - 10 GAL WATER
TANK IN LKR. OUTBOARD OF W.C.
COMPLETE INSTRUMENTATION FOR ENGINE,
CONTROLS & COMPASS - ELECTRONIC
INSTRUMENTS AS DESIRED
HATCH: 24" x 24" CLEAR OPENING -
3/4" x 1 1/2" COAMING ON DECK - 1/2"
PLYWOOD TOP - PROVIDE
SCREEN COVER ALSO
SHELF OUTBOARD OF
BERTHS 30" ABOVE
LWL - CLOTHING &
GEAR STOWAGE
NET UNDER
BERTH FRAMING
W. OAK - AS REQ'D.
FACIA 1/2" x 4"
MAHOG.
SECTION AT STA 3:
LOOKING FWD.
LIFT-OFF COVER
OVER STOVE TO
PROVIDE ADDITIONAL
COUNTER SPACE
WHEN NOT
COOKING - SHELVES
OR DRAWERS
UNDER STOVE, AS
DESIRED
SEAT BACK
STOVE
WATER TANK - 1" FILL
THRU DECK - VENT OVERBOARD -
APPX. 10 GALS. CAPACITY
PLACE CHLORINATOR IN THIS
SPACE ALSO, IF REQUIRED
CABIN & COCKPIT SOLE
3/4" PLYWOOD - BEAMS
C-C ALONGSIDE FRAMES
SECTION AT STA 6:
LOOKING FWD.
10" BRZ. CLEATS P/S. -
THRU BOLT TO DOUBLERS
HANDRAILS FROM
1 1/8" x 2 1/4" MAHOG. P/S.
LIGHT BOARDS
3/4" MAHOG.
SHEER GUARD
NOT SHOWN
WELL DECK
DINGHY CHOCKS FROM
W. OAK
MAST PAD FROM
1 1/8" MAHOG.
7'8" DINGHY
4" COWL VENTS ON
DORADE BOXES
DECK
ARRANGEMENT
BOARDING PLATFORM OUTER PIECE 3-PART LAMINATION OF 1/2" x 1 1/8" MAHOG. FORMED
AS SHOWN - INNER PIECES 1 1/4" x 1 1/8" MAHOG. SET 1/2" APART - LADDER RAILS 1" S.S.
TUBING W/ 3" x 3" x 1/8" PADS AT ENDS FOR FASTENING - STEPS OF LADDER 1" x 5" x 27"
MAHOG. - RAILS 24" ON CTRS.
27' DEADRISE BOAT
FOR: MR. E.C. BARNES
DATE: MAY 11, 1966
SCALE: 1/2" = 1'-0"
ARRGTS. & SECTIONS
LOA 27'-0"
LWL 25'-0"
BEAM 10'-0"
DRAFT 3'-0"
FREEBOARD:
FWD. 5'-6"
LEAST 3'-7 1/2"
AFT 4'-0"
JAY R. BENFORD
P.O. BOX 447
ST. MICHAELS, MD 21663
(410) 745-3235

30' Deadrise Boat *Petrel*

Design Number 42
1967

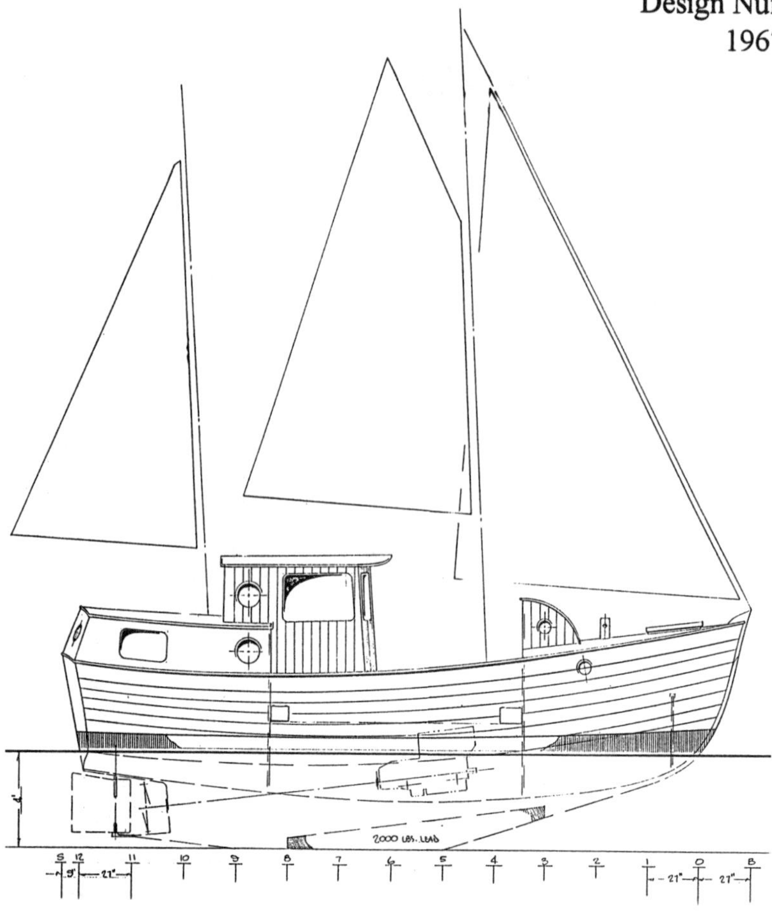

Our *Petrel* design is available in three versions, so far. The original has a well deck amidships and a cockpit aft, with a stateroom for two in the bow and the galley, head, dining area, and helm station in the after house.

The second version has eliminated the side decks and has a walk around deck around the after house for easier traffic fore and aft. The foc'sle shows an alternative for a double berth in place of the twin berths. This one also has the optional twin screw propulsion.

The third variation is a Trawler Yacht version with the forward well deck filled in with accommodations. The full width after cabin has a bigger space for having guests aboard, either for entertaining or along on a cruise. A couple other ideas for using this basic design and doing variations on it are shown in the sketches on this page.

We've also done some design work on making a longer (34'-6") version which would be built in steel also. This one, shown in profile below, would have the after cabin similar to the 30' Trawler Yacht with a private head added to the forward cabin.

Construction alternatives, to date, include carvel planked wood, plywood, or welded steel. As with any of our designs, the alternatives shown here have just scratched the surface of the wide range of possibilities of what might be — limited only by the size of the boat and one's imagination.

***CAUTION:** The displacement quoted here is for the boat in cruising trim. That is, with the fuel and water tanks filled, the crew on board, as well as the crews' gear and store in the locker. This should not be confused with the "shipping weight" often quoted as "displacement" by some manufacturers. This should be taken into account when comparing figures and ratios between this and other designs.

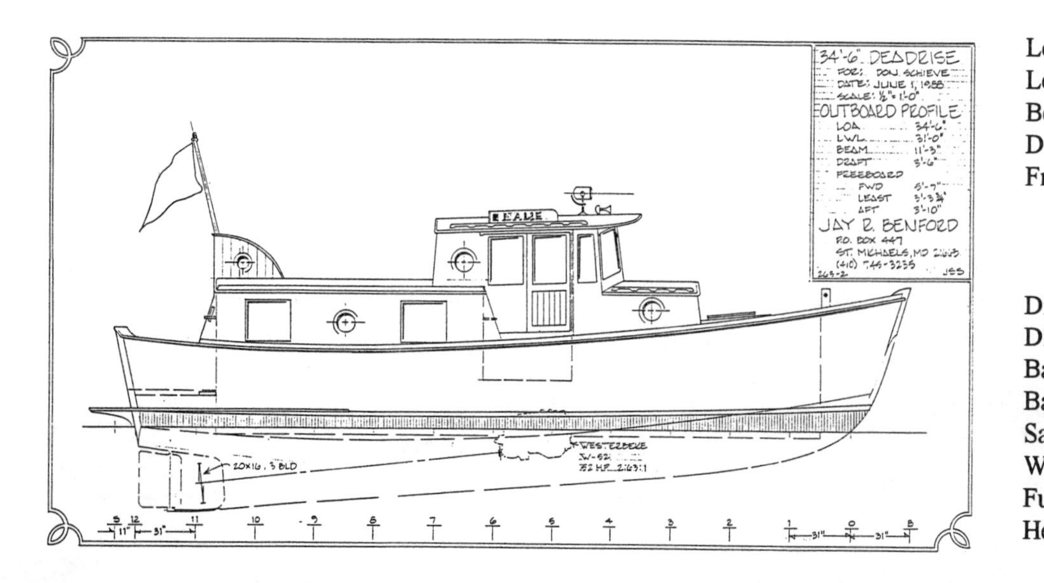

Particulars:

Length overall	30'-0"
Length designed waterline	27'-0"
Beam	11'-3"
Draft	3'-6"
Freeboard:	
Forward	5'-7"
Least	3'-3¾"
Aft	3'-10"
Displ., cruising trim*	13,800lbs
Displacement-length ratio	313
Ballast	935 lbs.
Ballast ratio	6.8%
Sail area	85sq.ft.
Water tankage	120gals
Fuel tankage	160gals
Headroom	5'-0" to 6'-7"

30' DEADRISE BOAT
FOR: DR. THOMAS C. JACOB
DATE: 12-18-74
SCALE: 1/2" = 1'-0"
PROFILE & ARRGT. #2

LOA	30'-0"
LWL	27'-0"
BEAM	11'-3"
DRAFT	3'-0"
FREEBOARD: (TO RAILCAP)	
FWD.	5'-7"
LEAST	3'-5¾"
AFT	3'-10"

JAY R. BENFORD
P.O. BOX 447
ST. MICHAELS, MD 21663
(410) 745-3235
42-B

NOTES:

1. JIB IS 85 SQ. FT. WITH WIRE LUFF FOR ROLLER FURLING — 9 OZ. TANBARK VIVATEX
2. JIB FURLING GEAR IS SIMPSON-LAWRENCE FIG. #1652 SIZE 2
3. ENGINES TO BE PISCES 27 HP (2 CYL. ISUZU BLOCK) DIESELS WITH 2:1 REDUCTION (OPPOSITE ROTATION — RH TO STBD — LH TO PORT)
4. PROPS 20"ø x 12" PITCH 3 BLADE
5. RUDDERS PERKO FIG. 818 NO. 3 SET JUST INBOARD OF PROP SHAFTS SO PROP SHAFTS CAN BE PULLED PAST RUDDERS. USE WAGNER HYDRAULIC STEERING MODEL N-85 w/24"ø DESTROYER WHEEL
6. 4" FOAM CUSHIONS ON SETTEES & BERTH — 2" SEATBACKS
7. KEEL PROFILE MODIFIED AS SHOWN FROM 6" ABOVE ℄ AT STA. 10 TO ORIGINAL LINE AT STA. 3 & AFT FROM 10 TO 30" ABOVE ℄ AT STERN.
8. WELL DECK FROM STA. 3½ TO STERN IS 27" BELOW & PARALLEL TO SHEER AT HULL SIDE — DECK IN WELL HAS 3" CAMBER IN MAX. BEAM
9. HEADSTAY ¼"ø GALV. 7x7 WIRE WITH SWAGED ENDS (NICOPRESS w/THIMBLE) & ½" GALV. (FORGED) TURNBUCKLE (JAW & JAW) ONTO ½" x 6" GALV. SHOULDER EYEBOLT (WILCOX-CRITTENDEN FIG. 2221) THRU STEMHEAD — SIMILAR EYEBOLT FOR FURLING GEAR
10. SHROUDS 5/16"ø GALV. 7x7 WIRE WITH SWAGED ENDS (NICOPRESS w/THIMBLE) & 5/8" GALV. (FORGED) TURNBUCKLE (JAW & JAW) CONNECTED ONTO 5/16" x 1" x 15" GALV. CHAINPLATE — LET CHAINPLATE IN FLUSH WITH HOUSE SIDE & THRU-BOLT TO DOOR FRAME
11. SHAFT ℄'S 20" ABOVE ℄ AT STA. 10 & 39" ABOVE ℄ AT STA. 6 — 24" OFF ℄ P. & S. — SHAFTS 1¼"ø ARMCO 17-4 PH OR EQUAL — USE SELF-ALIGNING STUFFING BOXES INBOARD & CUTLESS BEARINGS IN CAST BRONZE STRUTS — BOLT STRUTS & STUFFING BOX TO HUSKY BACKING PADS.

30' DEADRISE BOAT
FOR MR. DALLAS TILLING
DATE: NOVEMBER 1987
SCALE: 1/2" = 1'-0"
FRAMING

LOA	30'-0"
LWL	27'-0"
BEAM	11'-3"
DRAFT	3'-6"
FREEBOARD:	
FWD.	5'-7"
LEAST	3'-5¾"
AFT	3'-10"

JAY R. BENFORD/NA
P.O. BOX 447
ST. MICHAELS, MD 21663
(410) 745-3235
42-4

PARTICULARS:

DISPLACEMENT	6.16 L.T.
DISPL/LENGTH RATIO	313
PRISMATIC	0.647
C.B.	0.548 AFT

REVISED: 7-14-78

30' TRAWLER YACHT
FOR: MARINEX CORP.
DATE: 7-12-78
SCALE: 1/2" = 1'-0"
PROFILE & ARRGT.

LOA	30'-0"
LWL	27'-0"
BEAM	11'-3"
DRAFT	3'-6"
FREEBOARD:	
FWD.	5'-7"
LEAST	3'-5¾"
AFT	3'-10"

JAY R. BENFORD
P.O. BOX 447
ST. MICHAELS, MD. 21663
(410) 745-3235
42-12

Scale Model Helps Build a Deadrise Boat

Dallas Tinling's table top "shipyard" with scale size planking and the **Petrel** model under construction. Below, the finished **Petrel** model weighs anchor in a sea of grass.

What is a Deadrise Boat? Jay Benford began calling his simple, V-bottom cruisers "Deadrise Boats" to give them a formal name with a bit of distinctive character. If one takes the terminology to task when written without capital initial letters, the phrase is incomplete. For example an architect would say that a *boat* has *shallow deadrise* or *a deep deadrise*. The term *deadrise* in naval architecture denotes the rise of the bottom and is usually taken amidships. A horizontal line is drawn out from the keel at the point where the planking "cuts in." A vertical line extends down from the vessel's side. The amount of deadrise is from the chine to the base line, or a point on a round hull that is determined as being where side and bottom touch. Benford's applies "Deadrise Boat" in the same manner that a company might call its model the Sleekcraft Sportsfisherman and he is proper in using this privilege – with a capital D and B.

Petrel is the latest in Jay R. Benford's series of deadrise boat designs, of which there are four 26-foot versions and three 27-foot versions, plus this 30-footer designed for Dallas Tinling of Vista, Calif.

Tinling wanted a craft somewhat larger than the 26-foot design he had seen and wrote the Seattle naval architect for a complete set of study plans.

"One of his variations pleased me even more than the original," comments Tinling. "But while I had scaled down my desires, I still wanted a bit more boat than the stock plan. So I wrote him again and told him that I wanted to build the boat myself. He said he would design a boat to my desires, but he expected me to buy and use a good book on boat building. He also suggested that while he was working on the plans it might be a good idea for me to build the dinghy for the big boat, to get some experience. I learned that lofting is a lot more than just making a full size copy of the designer's plans. It includes a lot of details, rabbet lines, placement of fastenings, bevels, and joinery that are too extensive to be included on scale drawings.

"Then came the day when the plans for *Petrel* arrived. I had decided on a name for the boat before setting up the transom for the dinghy so that I could carve the name while I could still lay the wood flat on the work bench. I was anxious to get started building, but my workshop wasn't ready.

"I got to thinking it might be a good idea to loft the boat and build it as a scale model, as an outlet for my frustration at not being able to start the full size boat, and also to gain experience in lofting and to get an

Reprinted courtesy of Sea magazine from their September 1969 issue.

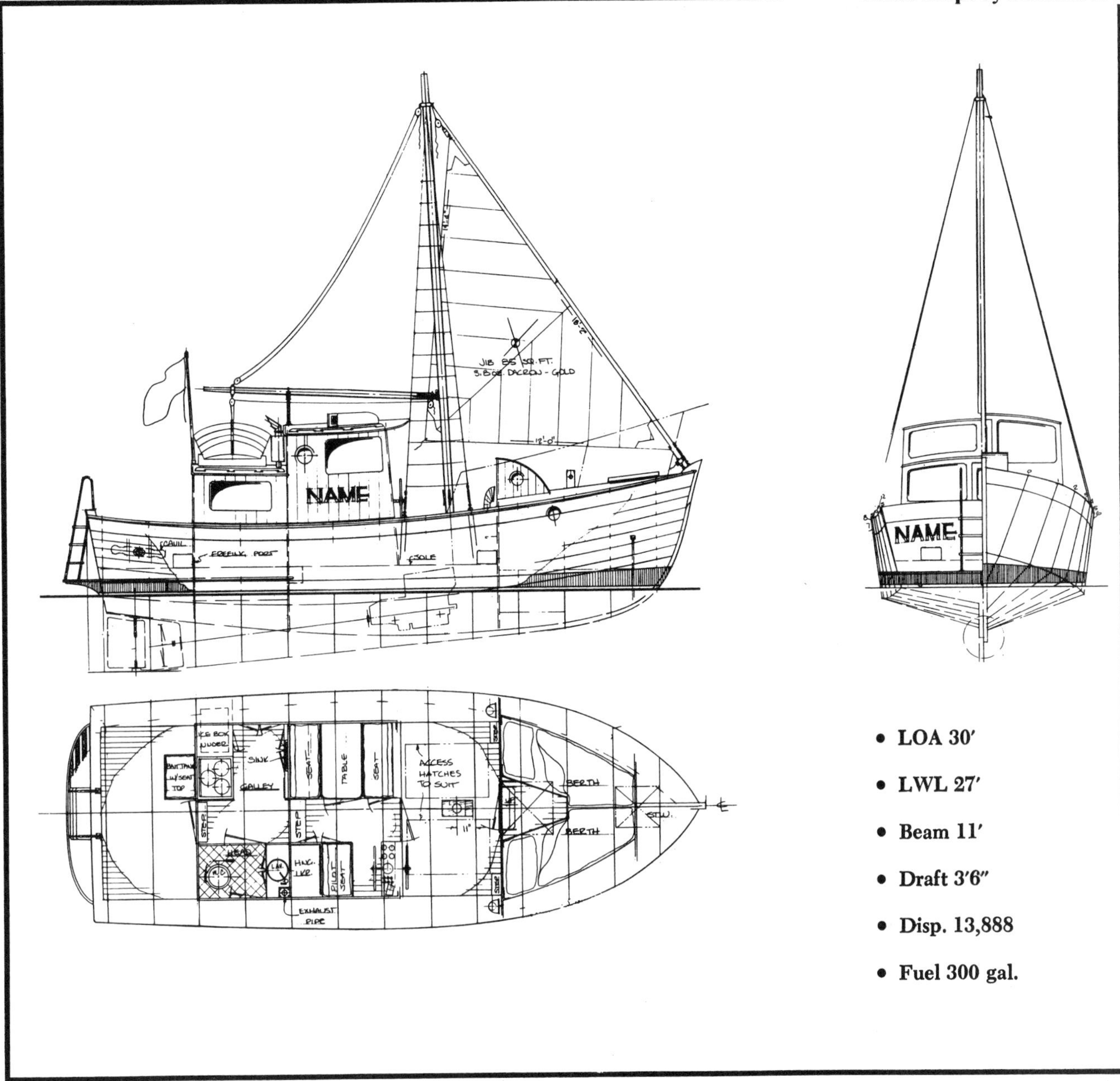

idea of what width of board would be required to shape the planks.

"I cut out a lot of scale size lumber, using scrap white pine for the framing and cedar scraps for the planking. The lofting and building took about three months of off-and-on evening work on a table set up in the living room.

"I did not go into the detail that characterizes fine ship models; the interior cabinetry and galley fixtures were just scale-size blocks of wood with pencil lines to indicate doors, drawers and burners.

"I got a great deal of pleasure from the project and a much greater appreciation of the proportions of the boat, especially the breadth and height relationships."

Petrel will be used for sportfishing and cruising in Southern California waters and the Gulf of California. She has two cockpits, midships and aft, which afford good protected deck space. She will be fitted with a Perkins 4-236 diesel and tankage for 300 gal. diesel fuel. This will give her a range of 1000 miles at 8 knots.

The arrangement provides fixed berths for two in the foc'sle and the dinette can be made into an extra double for guests—or they can sleep under the stars in the roomy cockpits. The dinette seats are raised to allow good vision out of the pilothouse windows and the aft seat is over the sink counter in the galley.

The steadying sail will help take the snap out of her motion in heavy weather, and works on sailboat jib roller furling gear for quick stowage. The boom is rigged to lift the skiff on and off and the ratlines on the shrouds provide access to the masthead lookout spot. A boarding platform and ladder will be provided aft.

She has straight frames throughout and construction complexity has been held down to facilitate amateur building, without any sacrifice in the basic qualities of the boat. Planking is carvel over sawn framing and with some minor drawing board changes, plywood sheathing could be used.

Benford has been working with Blackhawk Boat Company in British Columbia on construction of one or more of the 30-foot design on a semi-custom basis under his personal supervision. The design is being offered on a three to four-month delivery with teak decks, hot water pressure system with shower, diesel engine, galley and some electronics. Many options are offered on the semi-custom boat.

32' Cruiser

Design Number 48
1969 & 1972

The original version of this design was one of a series of preliminary studies I did in 1969 for a builder in Tacoma. We later had an opportunity to finish it off and it has proved to be a successful cruiser and liveaboard.

As a cozy retirement boat or as a vacation cruiser with lots of stowage and elbow room, this 32-footer has much to offer. In addition to the double berth, dresser, lounging seat and hanging locker in the foc'sle, the midships galley, dinette, and head with shower, she has a very large self-draining cockpit with comfortable lounging settee round the stern. Both socializing and fishing will prove most enjoyable here. Topside, on the boat deck, there's room for the ship's dinghy and a 100 sq. ft. steadying sail for rougher weather. This accommodation plan can be built on the chine hull of ***Ladybug*** or ***Ladybug***'s layout can be built on this round-bilged hull form.

Particulars:	**English**	**Metric**
Length overall	32'-0"	9.75 m
Length designed waterline	30'-0"	9.14 m
Beam	12'-0"	3.66 m
Draft	3'-6"	1.07 m
Freeboard:		
Forward	5'-11"	1.80 m
Least	3'-3"	0.99 m
Aft	3'-11"	1.19 m
Displacement, cruising trim*	21,275 lbs.	
Displacement-length ratio		
Ballast	1,500 lbs.	
Sail area (steadying sail)	100 sq. ft.	
Prismatic coefficient		
Pounds per inch immersion	1380	
Water tankage	130 Gals.	
Fuel tankage	130 Gals.	
Headroom	4'-10" to 7'-0"	

***CAUTION:** The displacement quoted here is for the boat in cruising trim. That is, with the fuel and water tanks filled, the crew on board, as well as the crews' gear and stores in the lockers. This should not be confused with the "shipping weight" often quoted as "displacement" by some manufacturers. This should be taken into account when comparing figures and ratios between this and other designs.

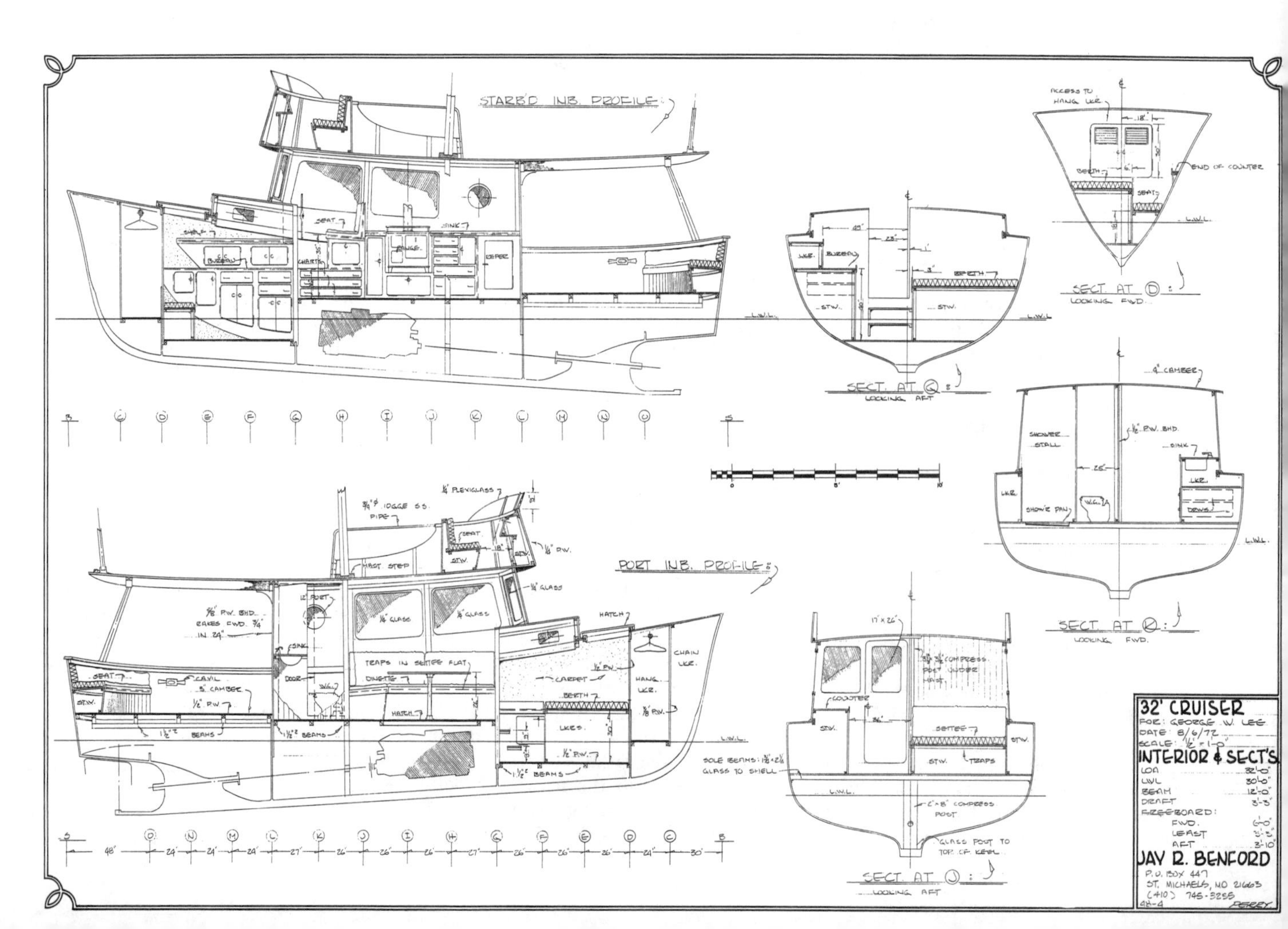

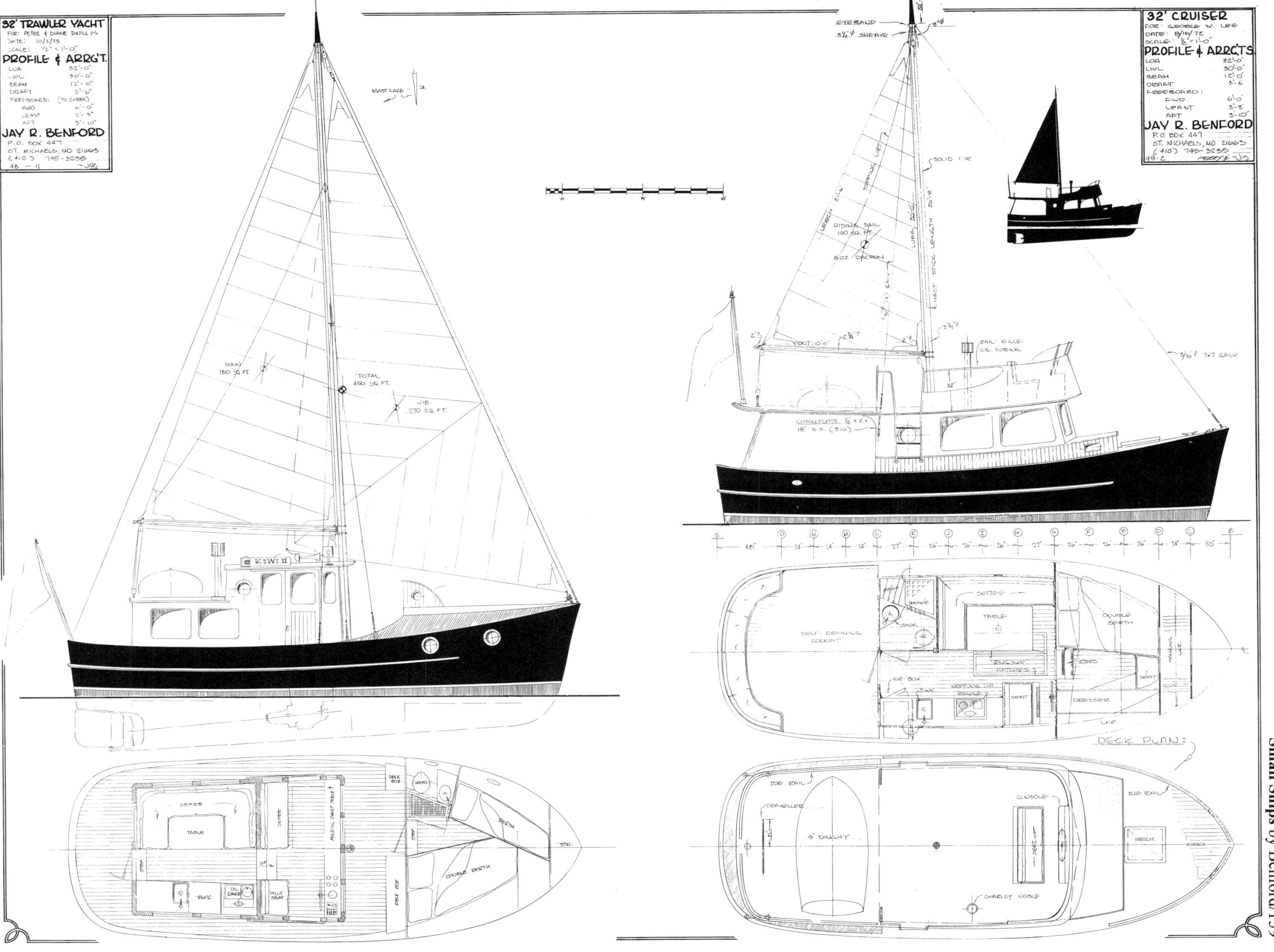
32' TRAWLER YACHT
FOR: PETER & DIANE DWILLES
DATE: 10/2/73
SCALE: 1/2" = 1'-0"
PROFILE & ARRG'T.
LOA 32'-0"
LWL 30'-0"
BEAM 12'-0"
DRAFT 3'-6"
FREEBOARD: (TO SHEER)
FWD 6'-0"
LEAST 3'-3"
AFT 3'-10"
JAY R. BENFORD
P.O. BOX 447
ST. MICHAELS, MD 21663
(410) 745-3235
48 — 11
MAST RAKE
MAIN 180 SQ. FT.
TOTAL 450 SQ. FT.
JIB 270 SQ. FT.
KIWI II
SETTEE
TABLE
FOLDING CHART TABLE
REFR.
OIL RANGE
PILOT SEAT
DECK BOX
STEP
BERTH
DOUBLE BERTH
32' CRUISER
FOR: GEORGE W. LEE
DATE: 8/10/72
SCALE: 1/2" = 1'-0"
PROFILE & ARRG'TS.
LOA 32'-0"
LWL 30'-0"
BEAM 12'-0"
DRAFT 3'-6
FREEBOARD:
FWD 6'-0
LEAST 3'-3
AFT 3'-10"
JAY R. BENFORD
P.O. BOX 447
ST. MICHAELS, MD 21663
(410) 745-3235
48-2
EYEBAND
SOLID FIR
RIDING SAIL 100 SQ. FT.
8 OZ. DACRON
TOPPING LIFT
LUFF 22'-0"
MAST STICK LENGTH 26'-8"
FOOT 10'-0"
RAIL: 10 GGE. S.S. TUBING
CHAINPLATE 1/4" x 2" x 18" S.S. (316)
3/16" 7x7 GALV.
SELF-DRAINING COCKPIT
SETTEE
TABLE
ENGINE HATCHES
ICE BOX
SINK
NEPTUNE 1-A RANGE
SEAT
STEPS
DOUBLE BERTH
DRESSER
DECK PLAN:
TOE RAIL
TRAVELLER
9' DINGHY
CONSOLE
SEAT
HATCH
CHARLEY NOBLE

32' Trawler Yacht — *Ladybug*

Design Number 48
1969

Designed in the fall of 1969, for Mr. Joseph Proulx of San Diego, the 32' ***Ladybug*** is an evolutionary step from our 30' Deadrise Boat, ***Petrel***, on the way to the creation of our 35' Trawler Yacht, ***Strumpet***. She's a good illustration of the several steps in the design process in which one boat evolves into another, in that she shows the definite individuality of her owner, yet the unmistakable relation to other sisters' basic premises.

She's a comfortable cruising home for one couple, with occasional visits from guests who wish to bunk on the convertible dinette. The foc'sle is completely self-contained and luxuriously appointed. A second head, with shower, is located aft to port. The main cabin contains the control station, dinette and a fireplace for comfort and warmth. Aft to starboard is the roomy galley.

One of the great security features of ***Ladybug*** is the deep well deck (about a foot above the waterline), with a minimum of 27" of solid bulwark, raising to 36" at the ends. The result is a feeling of really being ***in*** a little ship and not ***on*** a cockleshell, a wonderful feature when cruising with children as well. A boarding platform and ladder are provided at the stern.

The steadying sail rig helps reduce motion in mixed wave action, and also eases fuel consumption in reaching and running conditions.

Construction calls for strip planking over bulkheads with cold-molded veneers glued on diagonally over the strips. ***Ladybug*** herself was built over a 4½ year period by Joe Proulx, a professional yacht skipper and boatbuilder, known locally as "Joe Pro" for good reason. He lavished attention and love on her details, and we're all proud of the results. The photos of her don't begin to do her justice.

Another version (design number 49) was designed right after doing ***Ladybug***, this one with six inches more draft to encompass the larger prop swung by the engine chosen by her builder. This was the only significant difference between the two versions.

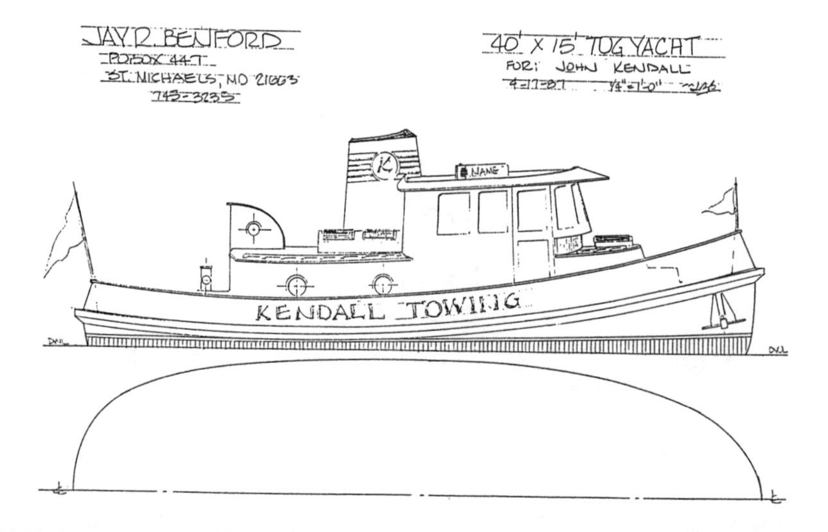

The next year, we designed a third version. This one had the cockpit aft, with a more conventional house layout, a flying bridge on top of the pilothouse, and a small rig for motorsailing or steadying the boat under power.

The 32' Cruiser on the previous pages shows a round-bilged hull form that can be used to build ***Ladybug*** for they are very similar in capacities and size.

Particulars of *Ladybug*

Item	English	Metric
Length overall	32'-0"	9.75 m
Length designed waterline	30'-0"	9.14 m
Beam	12'-0"	3.66 m
Draft	3'-6"	1.07 m
Freeboard:		
Forward	5'-11"	1.80 m
Least	3'-3"	0.99 m
Aft	3'-11"	1.19 m
Displacement, cruising trim*	15,100 lbs.	6,849 kg.
Displacement-length ratio	250	
Prismatic coefficient	0.695	
Sail area	274 sq. ft.	
Sail area-displacement ratio	7.18	
Prismatic coefficient	.695	
Pounds per inch immersion		
Water tankage	130 Gals.	
Fuel tankage	180 Gals.	
Headroom	5'-9" to 6'-9"	

***CAUTION:** The displacement quoted here is for the boat in cruising trim. That is, with the fuel and water tanks filled, the crew on board, as well as the crews' gear and stores in the lockers. This should not be confused with the "shipping weight" often quoted as "displacement" by some manufacturers. This should be taken into account when comparing figures and ratios between this and other designs.

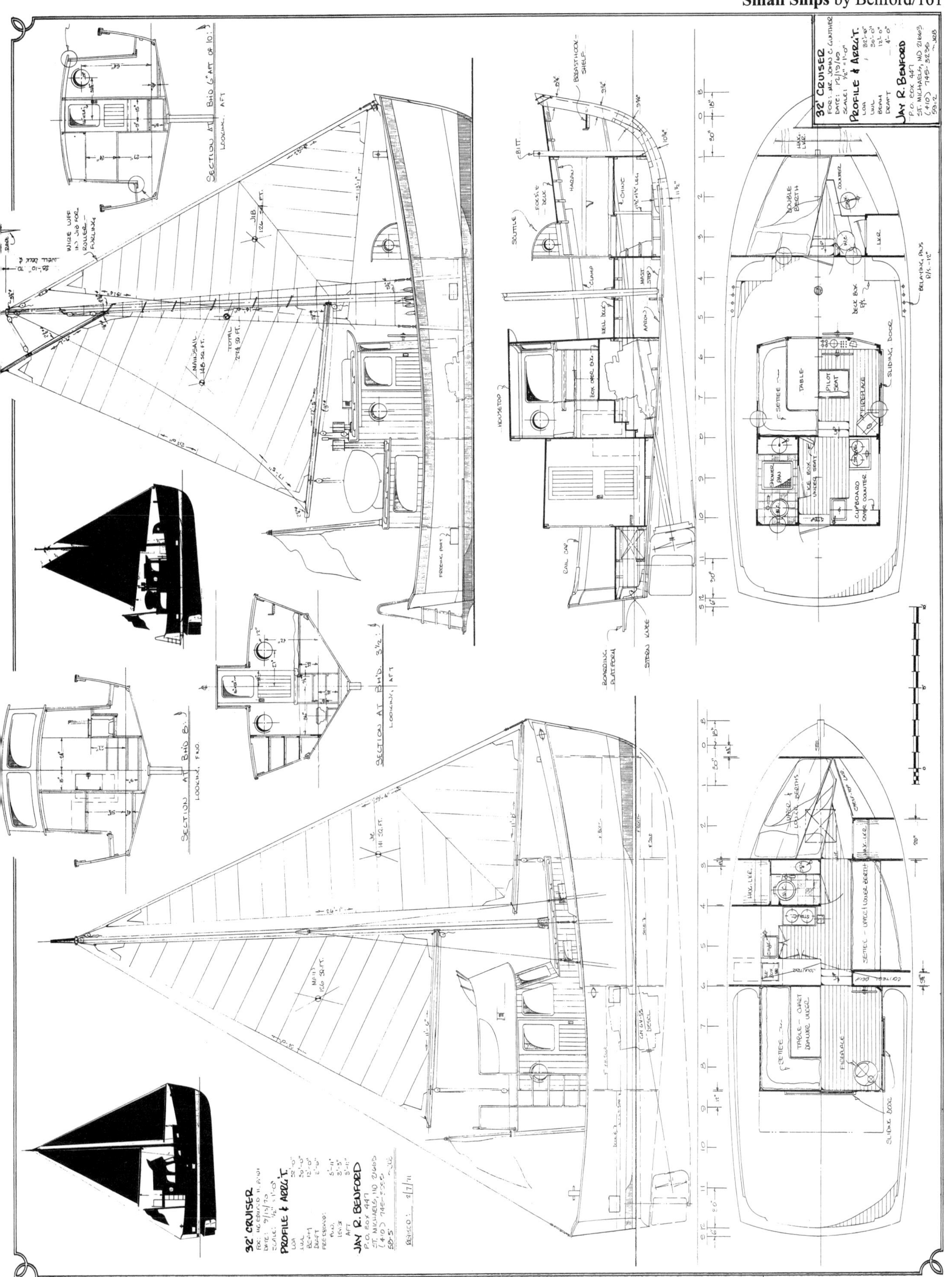
32' CRUISER
PROFILE & ARR'G.T.
JAY R. BENFORD
SECTION AT BHD 8
LOOKING FWD.
SECTION AT BHD 3½
LOOKING AFT
SECTION AT BHD 6" AFT OF 10
LOOKING AFT
MAINSAIL 148 SQ. FT.
JIB 126 SQ. FT.
TOTAL 274 SQ. FT.
DOUBLE BERTH
TABLE
SETTEE
PILOT SEAT
FIREPLACE
SLIDING DOOR
SHOWER PAN
ICE BOX UNDER SEAT
CUPBOARD OVER COUNTER
DECK BOX
BELAYING PINS
BOARDING PLATFORM
STERN KNEE
HOUSETOP
SCUTTLE
FOC'S'LE DECK
BREASTHOOK-SHELF
BITT
CHINE
CLAMP
MAST STEP
TABLE - CHART DRAWER UNDER
SETTEE
FIREPLACE
SLIDE DOOR

Ladybug

Benford Trawler Shaped By Skill And Affection

Ladybug is a TLC trawler, inspired by an article in SEA, designed by Seattle's Jay Benford, and brought to a festive San Diego launching in April by the energies of Joe and Phyllis Proulx.

Natty and sturdy aptly describe this ruggedly built 32-ft. cruiser since the high-gloss white hull is trimmed with bright red boottop and rail stripes.

Almost 4½ years of tender loving care was lavished by the Proulx' in the construction of *Ladybug* after they were inspired by the SEA article (Scale Model Helps Build a Deadrise Boat, SEA, Sept., 1969, Pg. 66) to contact Benford. The design discussed in the article was a 30-ft. trawler and the Proulx negotiated with Benford to provide plans for a 32-ft. version. From this point on, *Ladybug* dominated the Proulx' lives as she took shape in their backyard.

Varnished planking on the cabin sides, belaying pins, and ratlines up the 32-ft. mast give *Ladybug* a classic look. The teak swim step and transom boarding ladder add a modern touch without harming that vintage profile.

Appropriately, *Ladybug's* interior decor is red and black. This color scheme gives the forward owner's cabin almost a Victorian look—another plus in the craft's classic theme. The forward cabin is furnished with a double berth, head and wash basin. Access is through a foredeck scuttle.

The interior decor was Phyllis Proulx' responsibility and that included the actual installation of the nautical motif wallpaper in the wheelhouse.

The aft section of the main cabin houses the galley and a full lavatory compartment with shower. The galley equipment includes a Princess three-burner electric stove with oven, and a Kelvinator refrigerator-freezer combination unit. The latter was specially painted red to match the upholstery color.

The joiner work, hand-laid tile counters in galley, head and forward cabin, as well as the sound hull and cabin construction are ample evidence that Proulx, a professional yacht skipper and boatbuilder, exercised all of his nautical talents well in this building project.

He built the hull virtually single-handed with 1-1/8-in. mahogany planking and teak decks.

Power for *Ladybug* is a 453-cu-in. GM diesel with capacity aboard for 360 gal. of fuel divided into three tanks. She carries 85 gal. of water. Electricity afloat is provided by a 2½-kw Onan generator. *Ladybug* carries Wagner hydraulic steering.

The compact wheelhouse includes an L-shaped settee and dinette table as well as conical metal fireplace and a comfortable upholstered helmsman's seat. The switch control panel has a specially designed hinged transparent cover which prevents accidental tripping of the toggles without hiding the instrumentation.

Ladybug's mast isn't just for looks. She carries 280 sq.ft. of sail on a gaff rig which designer Benford is convinced will move the heavy hull if necessary.

The ship's dinghy, named *Baby Bug,* is carried atop the aft end of the main cabin.

There was one major problem which had to be solved before the first board was cut in this project: How to get *Ladybug* out of the Proulx' backyard when she was completed.

"There are 28 steps up to our front door," Phyllis Proulx pointed out. "We certainly couldn't get her out down that hill."

So from the beginning the Proulx' neighbor, Capt. James Conte, a Navy chaplain, has been a "partner" in the venture.

Ladybug "escaped" through Father Conte's backyard when it came time to launch. He also administered the blessing at the launching ceremonies.

In the 52 months it took to build *Ladybug,* Father Conte had time to do some designing of his own—on the fence that would have to be rebuilt after *Ladybug's* visit. Proulx is now trying to figure out if he can do as well building a wrought iron and stained glass fence as he did in building the sturdy little *Ladybug.*

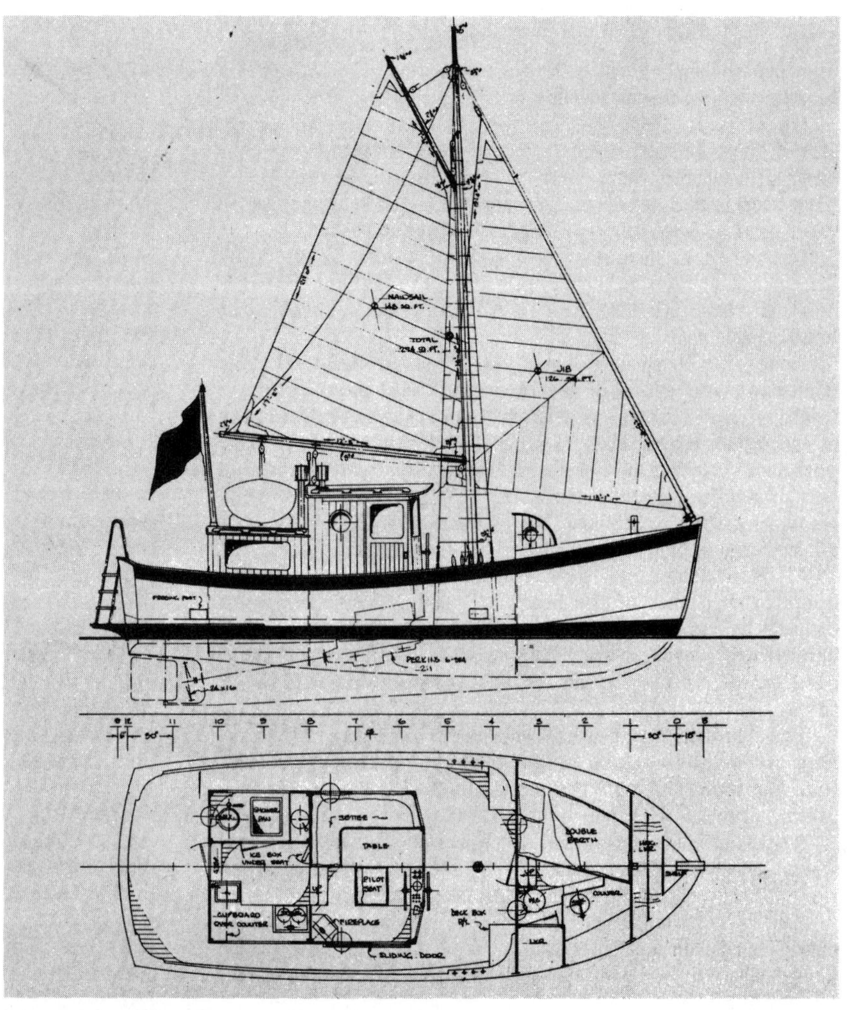

LOA 32 ft. LWL 30 ft. Beam 12 ft. 6 in. Draft 3 ft. 6 in. Displ. 16,000 lb.

Reprinted courtesy of Sea magazine from their July 1974 issue.

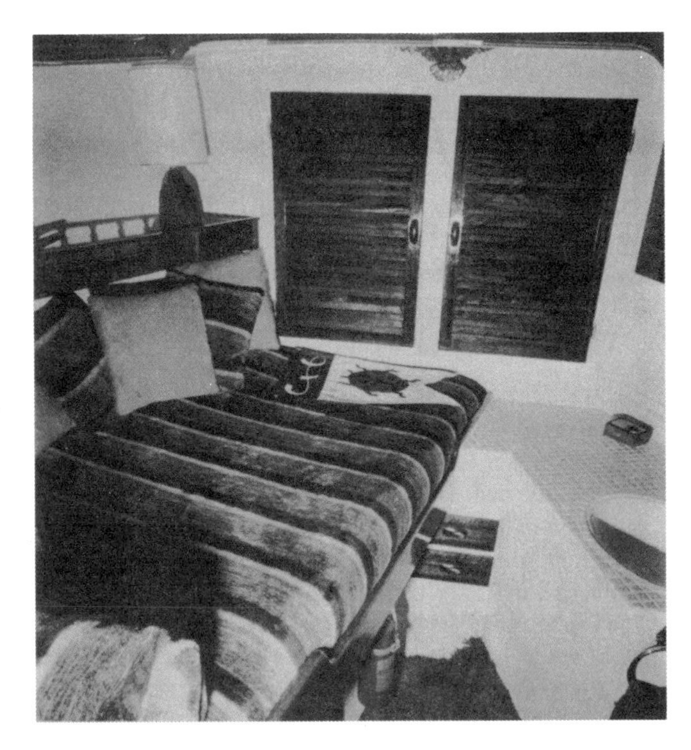

Sea Staff Photos

35' Trawler Yacht *Strumpet*

For: Ernest K. & Dodie Gann
Design Number 66
1970

Designers benefit from learning as much as possible about the personal cruising and living philosophy of their clients, so that they may create the most practical and efficient boat for them. The end result is that each custom designed boat makes a strong individual statement. The 35' Trawler Yacht, ***Strumpet***, is a very outspoken example of this phenomenon.

Looking at ***Strumpet*** we see strength, sturdiness, an appreciation for history, grace, purposefulness and privacy. It is no coincidence that these qualities are also outstanding characteristics of the people for whom she was created; Ernest K. and Dodie Gann. Knowing of their impressive background and experience — a dozen and a half commands, including the 117' brigantine, ***Albatros*** and the 60' ketch, ***Blackwatch***), and understanding some of the desires the Ganns had in mind for ***Strumpet***, where does a designer start?

The husky Scottish commercial fishing trawlers were the starting point. They were of appeal to the Ganns when they contacted Jay Benford after seeing his 30' Deadrise Boat, ***Petrel***, published in **National Fisherman.** The preliminaries for ***Strumpet*** thus started with a 30-footer on the boards, for this little packet was to be a personal boat, one about which Gann was later quoted as saying she "was laid out to drink six, eat four, and sleep two. And no more." But moving from a 60-footer to a smaller cruiser (even given the 22' Calkins Bartender in between) can be a difficult adjustment and the preliminaries for ***Strumpet*** soon grew to be 32'.

Completed plans were even drawn up at this point, and later stock plans for her were sold; that one became the lovely ***Ladybug***. However, her transom stern was soon changed to a handsome, canoe-sterned double-ender and she was drawn out to 35'. After five or six quite complete preliminary studies, she was finally built at that length.

In her 27" deep bulwarks, ***Strumpet*** really captures some of the big ship feeling of the Ganns' larger sirens. Because of these high bulwarks, being on deck gives one a secure feeling of being ***in*** as opposed to ***on*** a real boat. Since the layout of the vessel is focused on just two people, there is a good deal of space left for operating the boat, for moving about on deck.

An air of sturdiness is apparent from her design. It is backed up by the strength in her construction of Philippine mahogany and Alaskan cedar planking on steam bent white oak frames on 10" centers, over a stout backbone consisting of gumwood stem and sternpost, and fir keel, all of which are sided 5½". The six-cylinder Thornycroft Ford diesel, when geared down to a low revolution, hums along in her well insulated quarters. Access to her engine is sensible and workable. With a cruising speed of 1,000 to 1,200 rpm, she's guaranteed a long service life. Her easy pace also extends the enjoyment of those aboard.

Her design has a powerful sheer and simplicity of outfitting detail. But it doesn't stop there, for ***Strumpet*** is extremely stable, and unlike most cruisers she does not squat down aft, digging in her stern. Instead, she requires little effort to move forward, sliding through the water with pride, riding level, elegant and proud, not unlike a sailboat. The very small wake she leaves behind allows her to slip by with hardly any disturbance to those around her. Even at her top speed of 8.57 knots (higher than any displacement boats have a "right" to be), she maintains the same, steady, sea-worthy, graceful aura.

Strumpet's interior layout emphasizes the privacy, comfort and practicality that appeal to the Ganns. The steering station is sensibly laid out for maximum visibility, accessibility and ease of movement for the pilot. Sliding doors port and starboard, large windows and portlights all the way around, including a portlight in the aft end of the pilothouse, make the cabin not only airy but safe for the crew who wishes good visibility. The table nestles next to a settee which makes it comfortable for the crew to see and keep the captain company. The galley with its ample stowage space, is a couple of steps down going aft, but with pass-through bulkheads, and is open to communication as well as warmth from the diesel stove. (At a later point, the Ganns realized the fireplace in the main cabin was superfluous, and replaced it with a large chart table and stowage drawers.) The head fully aft is large and with ample hot water for a comfortable shower. It is convenient to a crew coming in the optional aft door, should wet raingear wish to be hung up first before being paraded through the living quarters. The foc'sle is cozily laid out for two, and was at one point modified to include a Tiny Tot heater and a double instead of single berths. A second head was not desired forward, so there is very ample room to move about in this stateroom.

Creating ***Strumpet*** was indeed a very satisfying series of evolutions in a design process. Experiencing her cruising was a pure delight! Eileen Crimmin commented in *Nautical Quarterly*'s summer, 1980, issue, "Seeing her at anchor or underway causes any knowledgeable sailor to sense immediately that, if this external view is the setting, then surely the rest of her is the jewel. . . .A jewel indeed."

32' *Strumpet* Preliminaries

There were also a couple other variations drawn up on the way to the 35-footer. These are shown on the pages following the drawings of ***Strumpet***. The chine 32-footer with rounded stern was the first of the preliminaries done to explore ideas with the Ganns. The round bilge one was a previous idea I'd generated.

The double-ended 32' cruiser is the last preliminary study we did for the Ganns before the final, round-bilged 35-footer. The builders, Jensen Shipyard in Friday Harbor suggested that there would be no real labor savings in building the chine hull as opposed to the round-bilged one. So we took the same concept and drew up the larger boat. It was decided that the small additional amount of material in building the three extra feet of length would be only a small change in cost since all the outfitting and equipment would be the same.

I've included this version here to show the evolution of ideas as well as to show an interesting boat. This design could be completed with only a little more work on our part. She could be done carvel planked, as designed, or in plywood, steel, or aluminum with a little revision to the lines.

The engine shown is the old Kelvin K3, rated 66 horsepower at 750 rpm. From its size it's evident that this would also function as ballast....

Particulars of *Strumpet*

Item	**English**	**Metric**
Length overall	35'-0"	10.67 m
Length designed waterline	32'-0"	9.75 m
Beam	12'-4½"	3.77 m
Draft	4'-6"	1.37 m
Freeboard:		
Forward	5'-9"	1.75 m
Least	3'-3"	0.99 m
Aft	4'-3"	1.30 m
Displacement, cruising trim*	24,600 lbs.	11,158 kg.
Ballast	3,000 lbs.	1,361 kg.
Displacement-length ratio	335	
Prismatic coefficient	0.628	
Pounds per inch immersion	1420	
Water	190 gals	719 liters
Fuel	285 gals.	1079 liters
Headroom	6'-2"	

***CAUTION:** The displacement quoted here is for the boat in cruising trim. That is, with the fuel and water tanks filled, the crew on board, as well as the crews' gear and stores in the lockers. This should not be confused with the "shipping weight" often quoted as "displacement" by some manufacturers. This should be taken into account when comparing figure and ratios between this and other designs.

Alternate 35' Trawler Version

The version shown below is a variation with a longer deckhouse and great cabin in the stern. This was evolved from the motorsailer version of ***Strumpet***, of which there are several versions. This one has the owner's stateroom in the stern along with the head and shower stall. The forward cabin has a couple berths and lockers with an optional second w.c. under a seat. The deckhouse has the galley along one side and an L-shaped settee opposite. This can have a dining table and be the social center. The mast and boom serve for lifting off the dinghy and can be the support for flopper-stoppers.

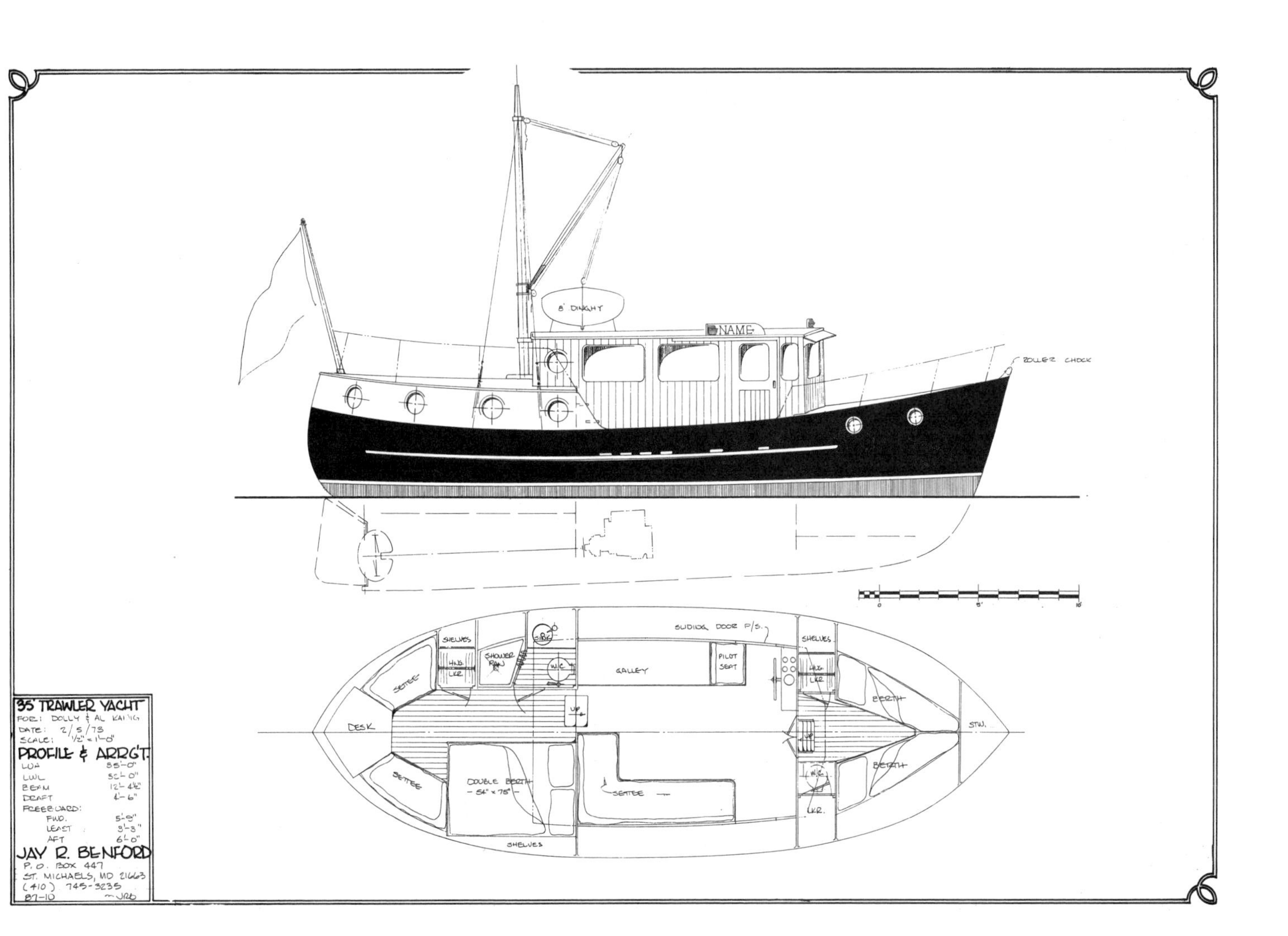

Above is STRUMPET's forward cabin and pilothouse. Below is her engine room, with good access both sides of the engine and well organized stores and equipment. Opposite above is a shot from her head looking forward through the galley into the pilothouse, with her Dickinson oil range along the port side of the galley. The lower photo, opposite, is of her head as seen from the pilothouse. Visible are the head and sink, and to the right behind the bulkhead is the shower stall.

Roy Montgomery Photos

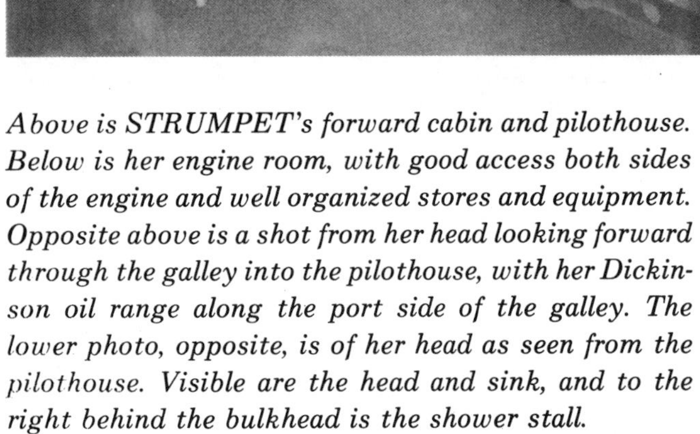

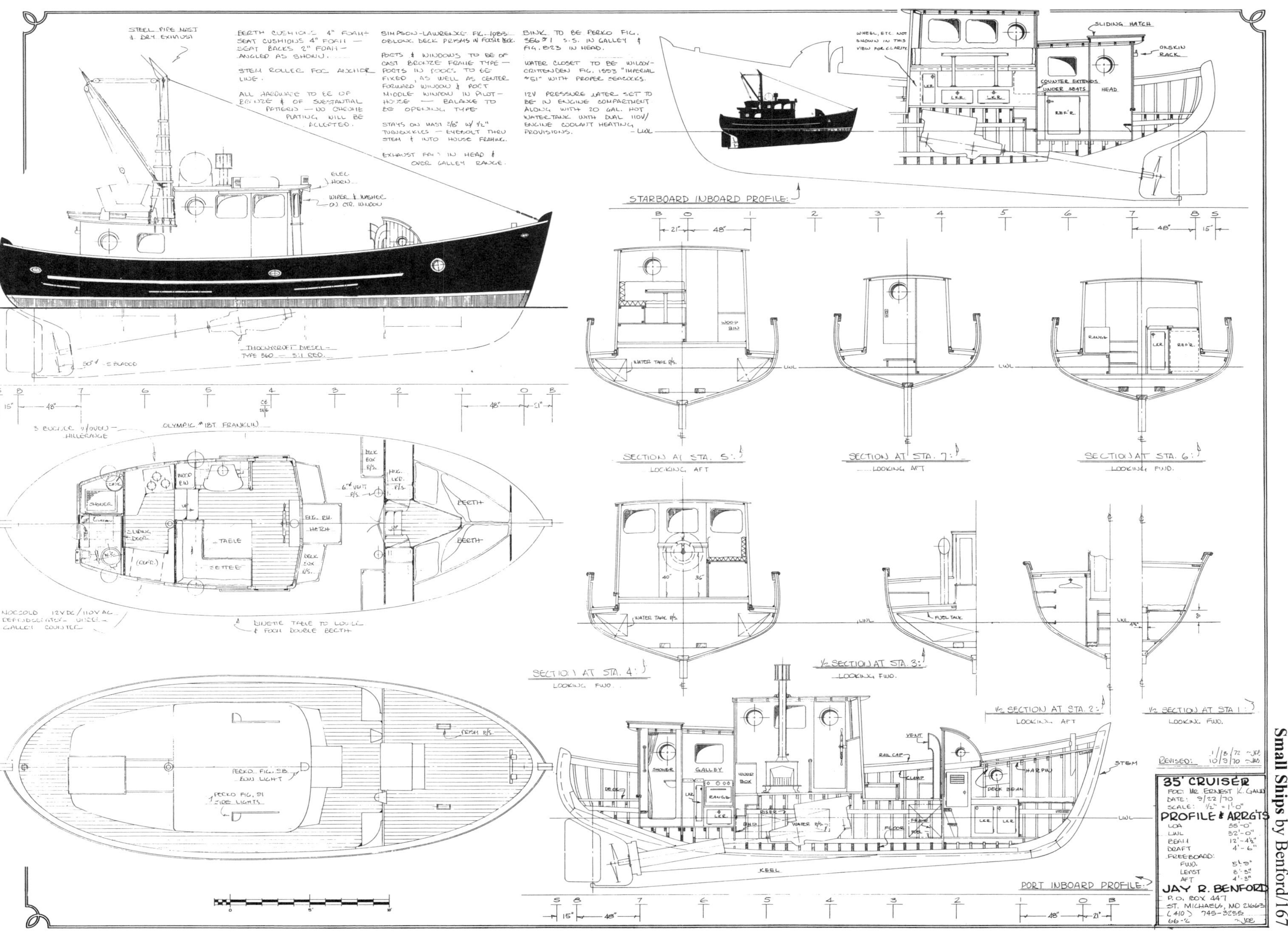

STEEL PIPE MAST & DRY EXHAUST
BERTH CUSHIONS 4" FOAM + SEAT CUSHIONS 4" FOAM — SEAT BACKS 2" FOAM — ANGLED AS SHOWN.
STEM ROLLER FOR ANCHOR LINE.
ALL HARDWARE TO BE OF BRONZE & OF SUBSTANTIAL PATTERNS — NO CHROME PLATING WILL BE ACCEPTED.
SIMPSON-LAWRENCE FIG. 1083 OBLONG DECK PRISMS IN FO'C'SLE DECK.
PORTS & WINDOWS TO BE OF CAST BRONZE FRAME TYPE — PORTS IN DOORS TO BE FIXED, AS WELL AS CENTER FORWARD WINDOW & PORT MIDDLE WINDOW IN PILOT-HOUSE — BALANCE TO BE OPENING TYPE
STAYS ON MAST 3/16" W/ 1/2" TURNBUCKLES — EYEBOLT THRU STEM & INTO HOUSE FRAMING.
EXHAUST FAN IN HEAD & OVER GALLEY RANGE.
SINK TO BE PERKO FIG. 366 #1 S.S. IN GALLEY & FIG. 823 IN HEAD.
WATER CLOSET TO BE WILCOX-CRITTENDEN FIG. 1503 "IMPERIAL #51" WITH PROPER SEACOCKS.
12V PRESSURE WATER SET TO BE IN ENGINE COMPARTMENT ALONG WITH 20 GAL. HOT WATER TANK WITH DUAL 110V/ ENGINE COOLANT HEATING PROVISIONS.
ELEC. HORN
WIPER & WASHER ON CTR. WINDOW
THORNYCROFT DIESEL — TYPE 360 — 3:1 RED.
OLYMPIC #18T FRANKLIN
3 BURNER W/OVEN — HILLERANGE
NORCOLD 12VDC/110VAC REFRIGERATOR UNDER GALLEY COUNTER
DINETTE TABLE TO LOWER & FORM DOUBLE BERTH
BERTH
TABLE
SETTEE
SHOWER
SLIDING DOOR
DECK BOX P/S.
6" VENT P/S.
ENG. RM. HATCH
PERKO FIG. 58 BOW LIGHT
PERKO FIG. 91 SIDE LIGHTS
FRESH P/S.
SLIDING HATCH
OILSKIN RACK
WHEEL, ETC. NOT SHOWN IN THIS VIEW FOR CLARITY.
COUNTER EXTENDS UNDER SEATS
HEAD
REFR.
LKR.
STARBOARD INBOARD PROFILE
WOOD BIN
WATER TANK P/S.
RANGE
LWL
SECTION AT STA. 5 — LOOKING AFT
SECTION AT STA. 7 — LOOKING AFT
SECTION AT STA. 6 — LOOKING FWD.
SECTION AT STA. 4 — LOOKING FWD.
FUEL TANK
1/2 SECTION AT STA. 3 — LOOKING FWD.
1/2 SECTION AT STA. 2 — LOOKING AFT
1/2 SECTION AT STA 1 — LOOKING FWD.
SHOWER
GALLEY
WOOD BOX
VENT
RAIL CAP
CLAMP
DECK BEAM
HARPIN
STEM
RISER
WATER P/S.
FLOOR
FRAME
KEEL
PORT INBOARD PROFILE
REVISED: 1/8/72 — JRB 10/9/70 — JRB
35' CRUISER
FOR: MR. ERNEST K. GANN
DATE: 9/22/70
SCALE: 1/2" = 1'-0"
PROFILE & ARRG'TS
LOA 35'-0"
LWL 32'-0"
BEAM 12'-4 1/2"
DRAFT 4'-6"
FREEBOARD:
FWD. 5'-5"
LEAST 3'-3"
AFT 4'-3"
JAY R. BENFORD
P.O. BOX 447
ST. MICHAELS, MD 21663
(410) 745-3235
66-2

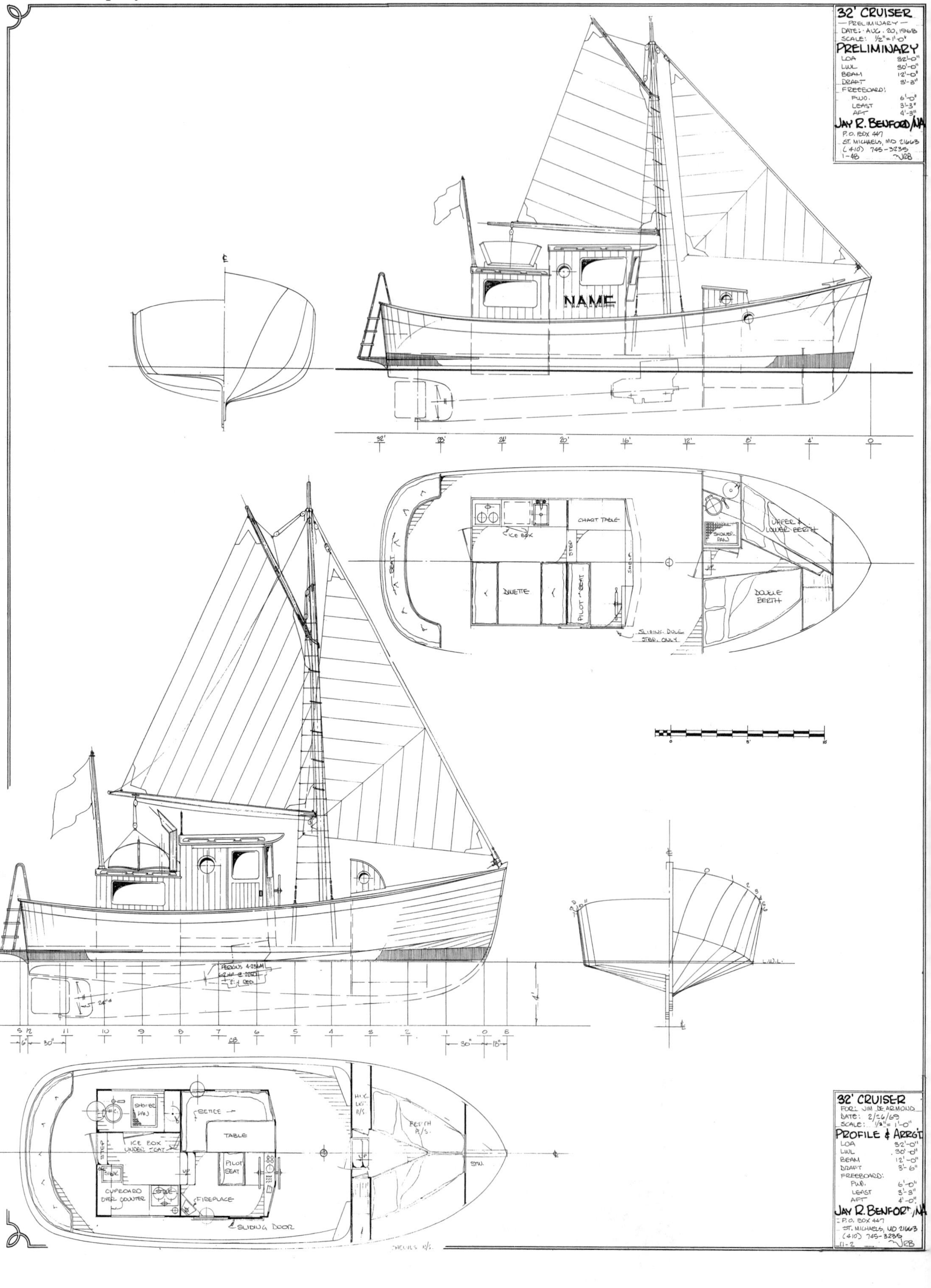

32' CRUISER
— PRELIMINARY —
DATE: AUG. 20, 1968
SCALE: 1/2" = 1'-0"

PRELIMINARY

LOA	32'-0"
LWL	30'-0"
BEAM	12'-0"
DRAFT	3'-8"
FREEBOARD:	
FWD.	6'-0"
LEAST	3'-3"
AFT	4'-3"

JAY R. BENFORD/NA
P.O. BOX 447
ST. MICHAELS, MD 21663
(410) 745-3235
1-48 JRB

NAME

32' 28' 24' 20' 16' 12' 8' 4' 0

CHART TABLE
ICE BOX
STEP
SHELF
DINETTE
PILOT SEAT
SEAT
SLIDING DOOR STBD. ONLY
SHOWER PAN
UPPER & LOWER BERTH
DOUBLE BERTH

0 5' 10'

PERKINS 4-236M
62HP @ 2250
2:1 RED.
24" ø
4'
S 12 11 10 9 8 7 6 5 4 3 2 1 0 B
6" 30" CB 30" 18"
L.W.L.

W.C.
SHOWER PAN
SETTEE
TABLE
ICE BOX UNDER SEAT
SINK
UP
PILOT SEAT
CUPBOARD OVER COUNTER
STOVE
FIREPLACE
SLIDING DOOR
H.V. LKR. P/S
UP
BERTH P/S
STW.
SHELVES P/S

32' CRUISER
FOR: JIM DE ARMOND
DATE: 2/26/69
SCALE: 1/2" = 1'-0"

PROFILE & ARRG'T

LOA	32'-0"
LWL	30'-0"
BEAM	12'-0"
DRAFT	3'-6"
FREEBOARD:	
FWD.	6'-0"
LEAST	3'-3"
AFT	4'-0"

JAY R. BENFORD/NA
P.O. BOX 447
ST. MICHAELS, MD 21663
(410) 745-3235
11-2 JRB

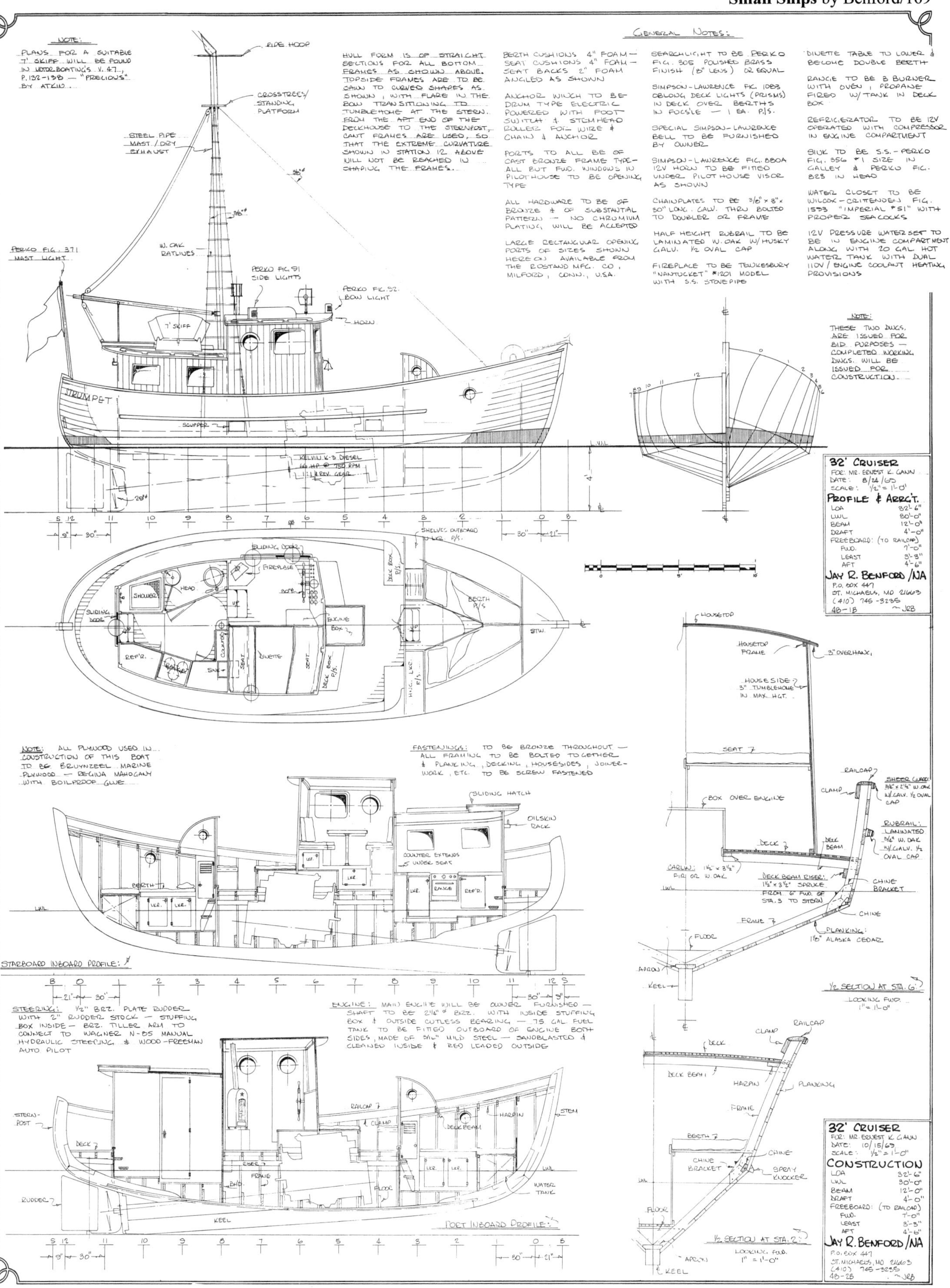

NOTE:
PLANS FOR A SUITABLE 7' SKIFF WILL BE FOUND IN MOTORBOATING'S V. 47, P. 132-138 — "PRECIOUS" BY ATKIN.
RIDE HOOP
CROSSTREES/ STANDING PLATFORM
STEEL PIPE MAST / DRY EXHAUST
W. OAK RATLINES
PERKO FIG. 371 MAST LIGHT
PERKO FIG. 91 SIDE LIGHTS
PERKO FIG. 52 BOW LIGHT
HORN
7' SKIFF
STRUMPET
SCUPPER
KELVIN K-3 DIESEL 66 HP @ 750 RPM 1:1 REV. GEAR
HULL FORM IS OF STRAIGHT SECTIONS FOR ALL BOTTOM FRAMES AS SHOWN ABOVE. TOPSIDE FRAMES ARE TO BE SAWN TO CURVED SHAPES AS SHOWN, WITH FLARE IN THE BOW TRANSITIONING TO TUMBLEHOME AT THE STERN. FROM THE AFT END OF THE DECKHOUSE TO THE STERNPOST, CANT FRAMES ARE USED, SO THAT THE EXTREME CURVATURE SHOWN IN STATION 12 ABOVE WILL NOT BE REACHED IN SHAPING THE FRAMES.
GENERAL NOTES:
BERTH CUSHIONS 4" FOAM — SEAT CUSHIONS 4" FOAM — SEAT BACKS 2" FOAM ANGLED AS SHOWN
ANCHOR WINCH TO BE DRUM TYPE ELECTRIC POWERED WITH FOOT SWITCH & STEMHEAD ROLLER FOR WIRE & CHAIN & ANCHOR
PORTS TO ALL BE OF CAST BRONZE FRAME TYPE — ALL BUT FWD. WINDOWS IN PILOTHOUSE TO BE OPENING TYPE
ALL HARDWARE TO BE OF BRONZE & OF SUBSTANTIAL PATTERN — NO CHROMIUM PLATING WILL BE ACCEPTED
LARGE RECTANGULAR OPENING PORTS OF SIZES SHOWN HEREON AVAILABLE FROM THE ROSTAND MFG. CO, MILFORD, CONN., U.S.A.
SEARCHLIGHT TO BE PERKO FIG. 305 POLISHED BRASS FINISH (8" LENS) OR EQUAL
SIMPSON-LAWRENCE FIG. 1083 OBLONG DECK LIGHTS (PRISMS) IN DECK OVER BERTHS IN FOC'S'LE — 1 EA. P/S.
SPECIAL SIMPSON-LAWRENCE BELL TO BE FURNISHED BY OWNER
SIMPSON-LAWRENCE FIG. 880A 12V HORN TO BE FITTED UNDER PILOTHOUSE VISOR AS SHOWN
CHAINPLATES TO BE 3/16" x 3" x 30" LONG. GALV. THRU BOLTED TO DOUBLER OR FRAME
HALF HEIGHT RUBRAIL TO BE LAMINATED W. OAK W/ HUSKY GALV. 1/2 OVAL CAP
FIREPLACE TO BE TEWKESBURY "NANTUCKET" #1201 MODEL WITH S.S. STOVEPIPE
DINETTE TABLE TO LOWER & BECOME DOUBLE BERTH
RANGE TO BE 3 BURNER WITH OVEN, PROPANE FIRED W/ TANK IN DECK BOX
REFRIGERATOR TO BE 12V OPERATED WITH COMPRESSOR IN ENGINE COMPARTMENT
SINK TO BE S.S. — PERKO FIG. 356 #1 SIZE IN GALLEY & PERKO FIG. 823 IN HEAD
WATER CLOSET TO BE WILCOX-CRITTENDEN FIG. 1533 "IMPERIAL #51" WITH PROPER SEACOCKS
12V PRESSURE WATER SET TO BE IN ENGINE COMPARTMENT ALONG WITH 20 GAL. HOT WATER TANK WITH DUAL 110V / ENGINE COOLANT HEATING PROVISIONS
NOTE: THESE TWO DWGS. ARE ISSUED FOR BID PURPOSES — COMPLETED WORKING DWGS. WILL BE ISSUED FOR CONSTRUCTION.
32' CRUISER
FOR: MR. ERNEST K. GANN
DATE: 8/24/65
SCALE: 1/2" = 1'-0"
PROFILE & ARRG'T.
LOA 32'-6"
LWL 30'-0"
BEAM 12'-0"
DRAFT 4'-0"
FREEBOARD: (TO RAILCAP) FWD. 7'-0" LEAST 3'-3" AFT 4'-6"
JAY R. BENFORD /NA
P.O. BOX 447 ST. MICHAELS, MD 21663 (410) 745-3235
4B-1B
SHELVES OUTBOARD P/S.
SLIDING DOOR
FIREPLACE
HEAD
SHOWER
SLIDING DOOR
REF'R.
SINK
SEAT
DINETTE
ENGINE BOX
DECK BOX P/S.
BERTH P/S
H.W.C. LKR. P/S
STW.
HOUSETOP
HOUSETOP FRAME
3" OVERHANG
HOUSESIDE 3" TUMBLEHOME IN MAX. HGT.
SEAT
BOX OVER ENGINE
DECK
RAILCAP
CLAMP
SHEER GUARD 3/4" x 2 1/4" W. OAK W/ GALV. 1/2 OVAL CAP
RUBRAIL: LAMINATED 3/4" W. OAK W/ GALV. 1/2 OVAL CAP
DECK BEAM
CARLIN: 1 3/4" x 3 1/2" FIR OR W. OAK
DECK BEAM RISER: 1 1/2" x 3 1/2" SPRUCE FROM 6" FWD. OF STA. 3 TO STERN
CHINE BRACKET
CHINE
FRAME
PLANKING: 1 1/8" ALASKA CEDAR
FLOOR
APRON
KEEL
1/2 SECTION AT STA. 6
LOOKING FWD. 1" = 1'-0"
NOTE: ALL PLYWOOD USED IN CONSTRUCTION OF THIS BOAT TO BE BRUYNZEEL MARINE PLYWOOD — REGINA MAHOGANY WITH BOILPROOF GLUE
FASTENINGS: TO BE BRONZE THROUGHOUT — ALL FRAMING TO BE BOLTED TOGETHER & PLANKING, DECKING, HOUSESIDES, JOINERWORK, ETC. TO BE SCREW FASTENED
SLIDING HATCH
OILSKIN RACK
COUNTER EXTENDS UNDER SEATS
BERTH
LKR.
RANGE
REF'R.
STARBOARD INBOARD PROFILE:
STEERING: 1/2" BRZ. PLATE RUDDER WITH 2" RUDDER STOCK — STUFFING BOX INSIDE — BRZ. TILLER ARM TO CONNECT TO WAGNER N-85 MANUAL HYDRAULIC STEERING & WOOD-FREEMAN AUTO PILOT
ENGINE: MAIN ENGINE WILL BE OWNER FURNISHED — SHAFT TO BE 2 1/4" Ø BRZ. WITH INSIDE STUFFING BOX & OUTSIDE CUTLESS BEARING — 75 GAL. FUEL TANK TO BE FITTED OUTBOARD OF ENGINE BOTH SIDES, MADE OF 3/16" MILD STEEL — SANDBLASTED & CLEANED INSIDE & RED LEADED OUTSIDE
STERNPOST
DECK
RAILCAP
CLAMP
DECK BEAM
HARPIN
STEM
RISER
FRAME
BHD.
FLOOR
WATER TANK
RUDDER
KEEL
PORT INBOARD PROFILE:
DECK BEAM
HARPIN
PLANKING
FRAME
BERTH
CHINE
CHINE BRACKET
SPRAY KNOCKER
FLOOR
APRON
KEEL
1/2 SECTION AT STA. 2
LOOKING FWD. 1" = 1'-0"
32' CRUISER
FOR: MR. ERNEST K. GANN
DATE: 10/15/65
SCALE: 1/2" = 1'-0"
CONSTRUCTION
LOA 32'-6"
LWL 30'-0"
BEAM 12'-0"
DRAFT 4'-0"
FREEBOARD: (TO RAILCAP) FWD. 7'-0" LEAST 3'-3" AFT 4'-6"
JAY R. BENFORD /NA
P.O. BOX 447 ST. MICHAELS, MD 21663 (410) 745-3235
4B-2B

Ernest K. Gann And The Strumpet

by Chris Caswell

Reprinted courtesy of Sea magazine from their September 1975 issue.

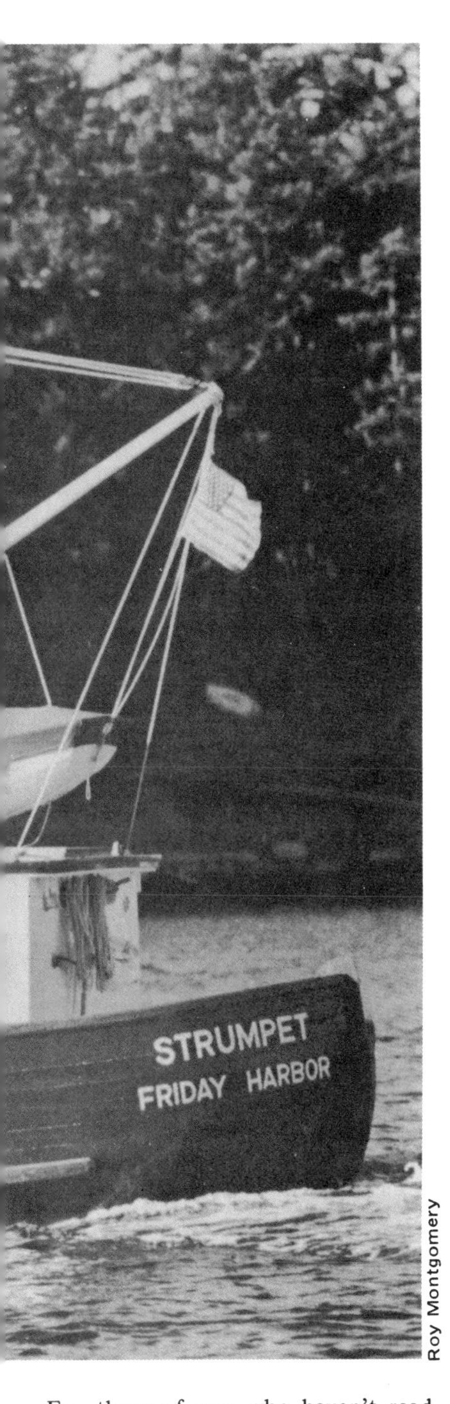

Roy Montgomery

Chris Caswell

For those of you who haven't read "The High And The Mighty," it is a story of an older airline pilot, forced by circumstance to fly as copilot with a younger and less experienced pilot on a TransPacific airliner. The older man, Dan Roman, is cynical, tough in a leathery way, and very, very remote. Plots and subplots involve most of the passengers aboard, but it is Roman, after engines fail and fires start, who manages to nurse the plane into a rainy San Francisco airport. Ever the private hero, Roman saunters into the mist as the operations manager bids him farewell, "So long, you old pelican."

It is a touching moment, and it is also appropriate for the story that follows.

For Ernest K. Gann, the author of "The High And The Mighty," is indeed a pelican. An accomplished pilot and the

author of eighteen books and innumerable articles, Gann is also a confirmed sailor. I hesitate to use the word "yachtsman" to describe him because of his attitudes toward the general yachting public, but he is definitely at home in both air and water.

Coming from an aristocratic Georgia family that felt his flying ambition was unseemly for a proper young man, Gann studied drama at Yale before departing for New York and employment at Radio City Music Hall. Giving up the production end of show business for the urge to fly, Gann joined a commercial airline, published his first book as a copilot of 29, and made captain at the surprisingly young age of 30. He continued to fly for nine years, including a stint with Air Transport Command during World War II. More books followed, including "Fate Is The Hunter," "Island In The Sky," and "Twilight For The Gods."

Since those early flying days, Gann has become a modern-day Antoine de Saint-Exupery and has flown all manner of aircraft from F-111 supersonic fighters to rickety WW I Sopwith Camels, including such unlikely craft as the Goodyear blimp.

Wedged in between these accomplishments like the scribbled notes in a pilot's logbook are eighteen different vessels that Gann has owned and cruised on many of the world's oceans. This lifelong infatuation with the sea started on the freshwater lakes of Minnesota: "...as a boy, with an old wooden cement trough as hull, my mother's

sheets, and sailing away on the great adventure. Almost drowned along with three other kids when we capsized a long way from shore. . .then there was just one boat after another."

Although his flying life has led him to call a variety of places home, he has now settled on San Juan Island off the Washington coast. It is there that this encounter takes place, on a bright blue afternoon aboard his boat. Gann carefully schedules every morning for writing, so the only time available is during a brief session of slapping varnish on the stained brightwork of his cruiser, punctuated by shouts of greeting to friends.

Gann's present boat is the *Strumpet,* a traditional 35-foot cruiser designed by Jay Benford of Seattle. "I saw some of Jay's designs and liked his attitude." Benford has been called a romantic because of the tradition of his designs, but he combines this with a practicality about boats and life.

Strumpet was designed specifically for Gann and his wife Dodie, a former Olympic skier and once Gann's secretary. "Yeah, the boat was very specially designed. After having some really big boats, *Strumpet* was laid out to drink six, eat four, and sleep two. And no more. We got tired of running a guest ranch for people. On long voyages, when you're paying the bill, it's really something else. Most people who can afford to go and help out can't get away. Those who can get away can't afford it. So *Strumpet* is for two.

"I told Jay what I wanted and he went away and drew some ideas and came back and we talked some more and he went away. . .hell, I forget how many times. . .five or six. This went on for about a year. Hi, Molly," he calls to a passerby.

Strumpet was built in the Jensen Shipyard in Friday Harbor, a one-man operation with a handful of employees. "This boat is one of the last survivors of some pride—not too much—but some pride of workmanship. It's an almost forgotten emotion, apparently."

His passing reference to bigger boats and floating guest ranches is a mild understatement when some of his previous boats are listed. There is *Black Watch,* just before *Strumpet.* Now named *Mara,* she's a 60-foot schooner Gann had built in Denmark. "I never brought her to the States, though. We kept her in Europe and sailed her everywhere. She was unbelievable. . .a little boy's dream. The after cabin wasn't one of those junky things on fiberglass boats. . .it was right out of Captain Kidd's diary."

There was *Albatross,* a one-time Dutch training ship of 117 feet overall and 92 tons which Gann acquired in Holland, converted into a brigantine, cruised across the Atlantic, into the South Seas, back to the Mediterranean and through the Greek Islands. Shortly after Gann sold her, she was hit by a white squall in the Gulf of Mexico, capsized, and sank taking five people with her.

There was *Raccoon,* a small daysailer that came apart under Gann near Cape Cod, and *Thetis,* a lovely 28-foot-yawl. For a time, the sea became Gann's livelihood as a commercial salmon and albacore fisherman out of San Francisco. Working up to owning first one vessel, the *Fred Holmes,* and then a second, the *Mike,* the venture alternately prospered and starved until Gann and his partner passed the two vessels on to other dreamers.

Uncle Sam was a 40-foot Friendship sloop that was lost to a hurricane in Connecticut during World War II when Gann was on a flight, and *Restless,* a 30-foot classic dispatch launch, was returned to steam power on San Francisco Bay because Gann thought he might like a steam yacht.

Others left their mark although Gann never owned them outright: the *Don Quixote* in Spain (a lateen rigged fishboat), the *Butterfly* in Brazil (a balsa catamaran owned by a waiter), *Li Po* (a junk in Asian waters), and *Henrietta* (a Caribbean trading sloop).

Content, for now at least, with *Strumpet,* Gann finds his time laden. "That's the hitch. I'm so involved in so many other things that we don't seem to get away too often. When we do, it's just for a night or two. Given a chance, we run over to Canoe Cove in Canada to see friends or down to Seattle to work on the boat. Or we just go somewhere in the San Juans and *sit!*

"We do a little fishing—I've got long line gear aboard—and we set some crab pots. I'm working on books now, and movies. I'm just finishing the screenplay for 'Band of Brothers.' But I think the best book I've written was 'Song Of The Sirens.' Most of my books I don't like, so that'll give you some idea. But that one I do. I don't know, I've written so many books and I'm so tired of it. . . what I'd like to do is go away on *Strumpet* for a couple of weeks or even longer. . ."

Strumpet is ready to go when Gann is. Built of a combination of Philippine mahogany and Alaskan cedar planking over oak frames, her lines are clean and traditional. The wheelhouse sports a ship's telegraph that has been converted into the engine gearshift, and Gann delights in ringing for full astern as he docks *Strumpet.* A dinette and wood-burning fireplace share the wheelhouse and, two steps aft, is the compact galley with its diesel-fired stove. In the stern is

a head with a stall shower. The for has two berths and storage spac personal gear. A removable ca shades the midship deck, and a din stored on the wheelhouse roof.

"The only mistake I made wa getting a slow-speed engine li wanted in the first place. I didn't the guts to go ahead and spen money and I wish I had now. The p is a six-cylinder Ford diesel—it's a engine, but it's not very romantic. Jay I wanted a heavy and slow e but the cost was not to be believe the weight. . .my God, the boat w have gone to the bottom. We did tests on Lake Washington and she 2 1/2 gallons an hour at eight kn that's plenty fast enough for me.

Settled aboard his romantic cr Gann takes pleasure in the p crowds on the docks. "It's unb able. . .there are half a million cruisers up and down this dock people always come as if draw magnet to *Strumpet.* Even sailors, w is a very pleasant thing. People say God, that *looks* like a boat.' I'v nice people come up to *Strumpe* say, 'Where can I buy one of thes

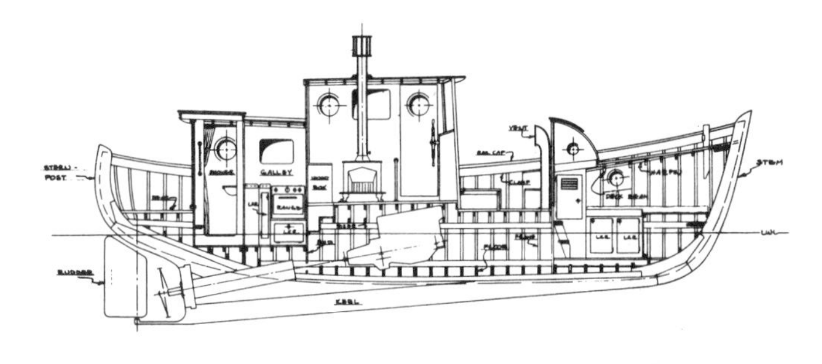

hough they have them on counters in narine stores.

"A lot of these (his gesture includes a lock of recent fiberglass yachts) are for ale because there's no work to be done on them. When you work on a boat, 'ou give something to it and when you give something you begin to love it. A boat that has no work involved, you just lon't love for very long. It's just like a voman. If she's easy. . .the hell with her vhen it's over. But if you give to her, you'll fall in love. And it's ridiculous to ay a boat is no work." Gann snorts in lisgust. "The price of boats has gone. . . vell, like everything else. When a man gets eighteen dollars an hour to pound nails–not always straight–something's very wrong."

A group of men walk past *Strumpet* carrying several large fish. "Must be kindivers, damn 'em. You used to be ble to catch fish easily, now they're near gone. About a month ago I set a leep line on a buoy by Orcas Island, went away about a mile and had lunch. could see the buoy with my binoculars out when I came back, somebody'd cut he line and stolen the catch. It never

Ernest K. Gann

would've happened five years ago.

"It's a different boating public now. . .you could be drowning and they'd go right on by. Those big ocean pounders will go right by a little row-boat at 20 knots!

"But after people run around fast for a while–that's the main thing, how fast is it, not how good is it–all of a sudden they get tired and the payments are heavy and that's it. Then they sell cheap. But have you ever tried to buy a good boat? No way. There's a little coasting schooner in Santa Barbara I've been after for fifteen years and she's still not for sale. Of course he's not going to sell–I wouldn't either."

Gann's gesture encompasses the dock again, "But you can buy all you want of those or those. Buy 'em by the hundreds. That ought to prove something to somebody. If you try to buy a good wood cruiser, if you're mighty lucky you might get one. But you're going to search a long time."

Gann steps into the wheelhouse and points out the autopilot. "Now here's an outfit for you–Wood-Freeman. They still believe in pride of workmanship. When *Strumpet* was new, their autopilot wandered, so I called them. The man said try this and try that and call me back. So it still didn't work and I called back and he said he'd send me a kit and to let him know. So I called back and told him the same thing and asked if he could come up to the island. He did and he said he'd had some trouble with that system and they weren't proud of it at all. Said they were changing and he'd bring up a whole new unit. They replaced it and didn't charge a cent. Boy, I'll sing their praises to the sky. You know, I just put two $6000 engines in my plane. After a hundred hours, I find that the manufacturer had put the wrong cylinders in the block. It's a damn wonder it didn't blow up both engines someplace. That's modern workmanship for you.

"You know, I had to pay *extra* to get a brass wheel for *Strumpet.* They even charge you *not* to chrome it. . .can you believe it? I keep saying chrome is for cars."

The plane Gann mentions is his Cessna 310 which he keeps at the Friday Harbor airport. "I travel away from the island as little as possible. Once in a while Dodie and I will fly down to San Francisco to go out for dinner because the restaurants in the San Juans are probably the worst in the entire world, bar none. I'll put our greasy spoons up against any greasy spoon. But anyway, I rarely fly by airline any more–I get so tired of being treated like cattle."

Gann has lived in the San Juans for nine years and still enjoys the solitude. He writes daily, is an ardent conservationist (his efforts to curb air pollution on Puget Sound earned him an award), and dabbles in painting (one of his paintings hangs in a local pub and another is aboard *Strumpet*).

As one of his friends notes, "Ernie really likes to keep his life separate from his readers. That farm is as close to a fort as you'll find." Gann's Red Mill Farm is located roughly in the center of San Juan Island, halfway between his plane at Friday Harbor and *Strumpet* in Roche Harbor. Red Mill was originally 600 acres on an unmarked dirt lane, although some has been sold in smaller parcels. The farm sprawls over rolling hills and includes several buildings, a barn, cattle, horses, rabbits and, of course, the red mill.

Back down at Roche Harbor, the stream of dock walkers continue to stop and gaze at *Strumpet*. Few are aware that it belongs to *the* Ernest Gann, but they recognize that it is a world apart from their own high-speed cruisers.

Gann finishes his boat chores, packs the paint cans into a tired canvas bag not unlike the scarred leather flight case carried by Dan Roman and strolls off into the bright sunlight of summer.

So long, you old pelican.

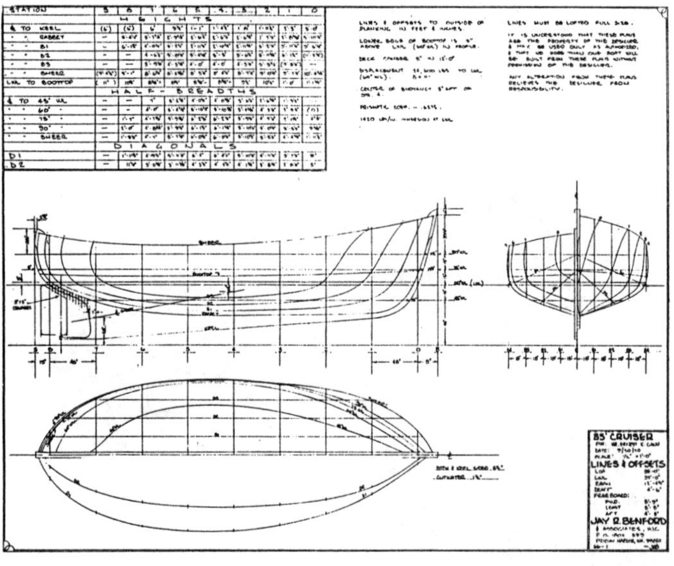

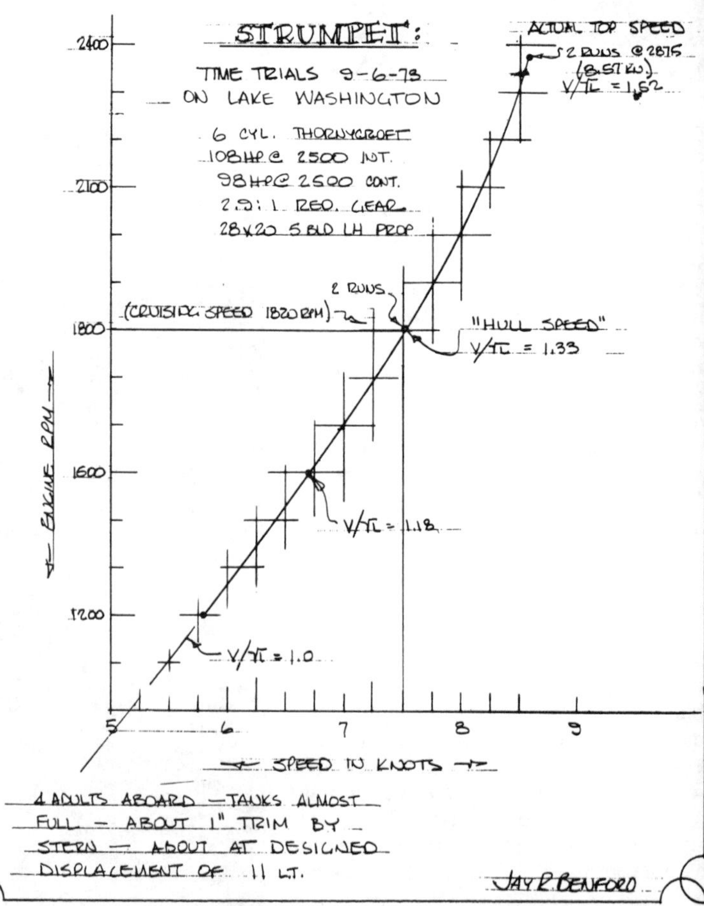

Most of the literature on designing displacement boats (or vessels that do not rise up and plane on the lift generated by their speed) indicates that they should not be able to go beyond the speed-length ratio of 1.33 times the square root of their waterline length, often referred to as "hull speed". ***Strumpet*** proved to be able to run comfortably beyond this speed thought of as an outer limit for displacement hulls. This speed curve shows the timed trials we ran on her and the resulting speeds.

27½' Trawler Yacht

Design number 71

1971 & 1978

This design was started as a preliminary design done in 1971. The original client's plans changed and he did not proceed with building the boat. Later, in 1977 and 1978 we had others who wanted to build the boat and the plans were then completed. The preliminary version is shown below and the later version on the next page.

The foc'sle is home to a double berth and shelf-top locker. The self-draining cockpit amidships has room for comfortably sitting out in deck chairs. Under the cockpit is the engine room, tanks, machinery and storage hold. The pilothouse is a couple steps up from the gracious aft cabin, and thus a social part of the activities. The steering station is to port. I prefer this location, for the danger zone is to starboard and thus the helmsman will be looking in that direction while talking with the crew on the starboard pilothouse seat instead of having to turn his back on the crew or guests. The galley is to starboard in the aft cabin, partially recessed under the pilothouse seating. A large round settee in the stern looks on the galley forward to starboard. The head has room for a shower and the oilskins hang outboard and drain into the shower sump. The steadying sail will help the motion in a seaway and the ratlines will let you go aloft for a better view over the horizon. Poles for trailing flopper stopper stabilizers can be added. The boom can be used to lift the dinghy on and off, or any salvage or treasure that might need to be loaded into the hold....

Particulars:	**English**	**Metric**
Length overall	27'-6"	8.38m
Length designed waterline	25'-6"	7.77m
Beam	10'-6"	3.20m
Draft	3'-0"	0.91m
Freeboard:		
Forward	5'-6"	1.68m
Least	3'-9"	1.14m
Aft	4'-3"	1.30m
Displacement, cruising trim	14,170 lbs.	6,426kg.
Displacement-length ratio	382	
Prismatic coefficient	.583	
Water tankage	150 Gals.	568 litres
Fuel tankage	150 Gals.	568 litres
Headroom	6'-2" to 6'-7"	1.88 to 2.01m

***CAUTION:** The displacement quoted here is for the boat in cruising trim. That is, with the fuel and water tanks filled, the crew on board, as well as the crews' gear and store in the locker. This should not be confused with the "shipping weight" often quoted as "displacement" by some manufacturers. This should be taken into account when comparing figures and ratios between this and other designs.

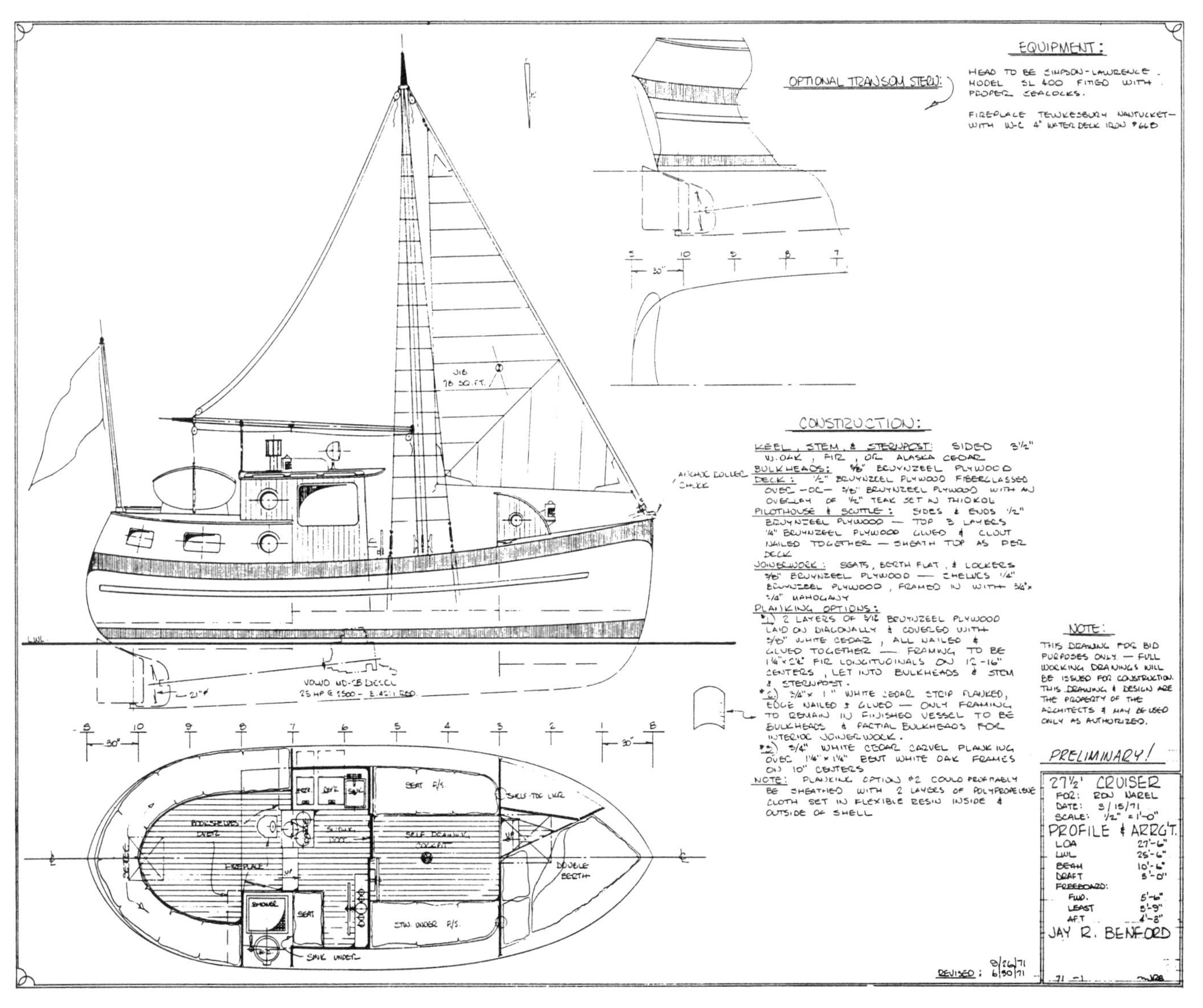

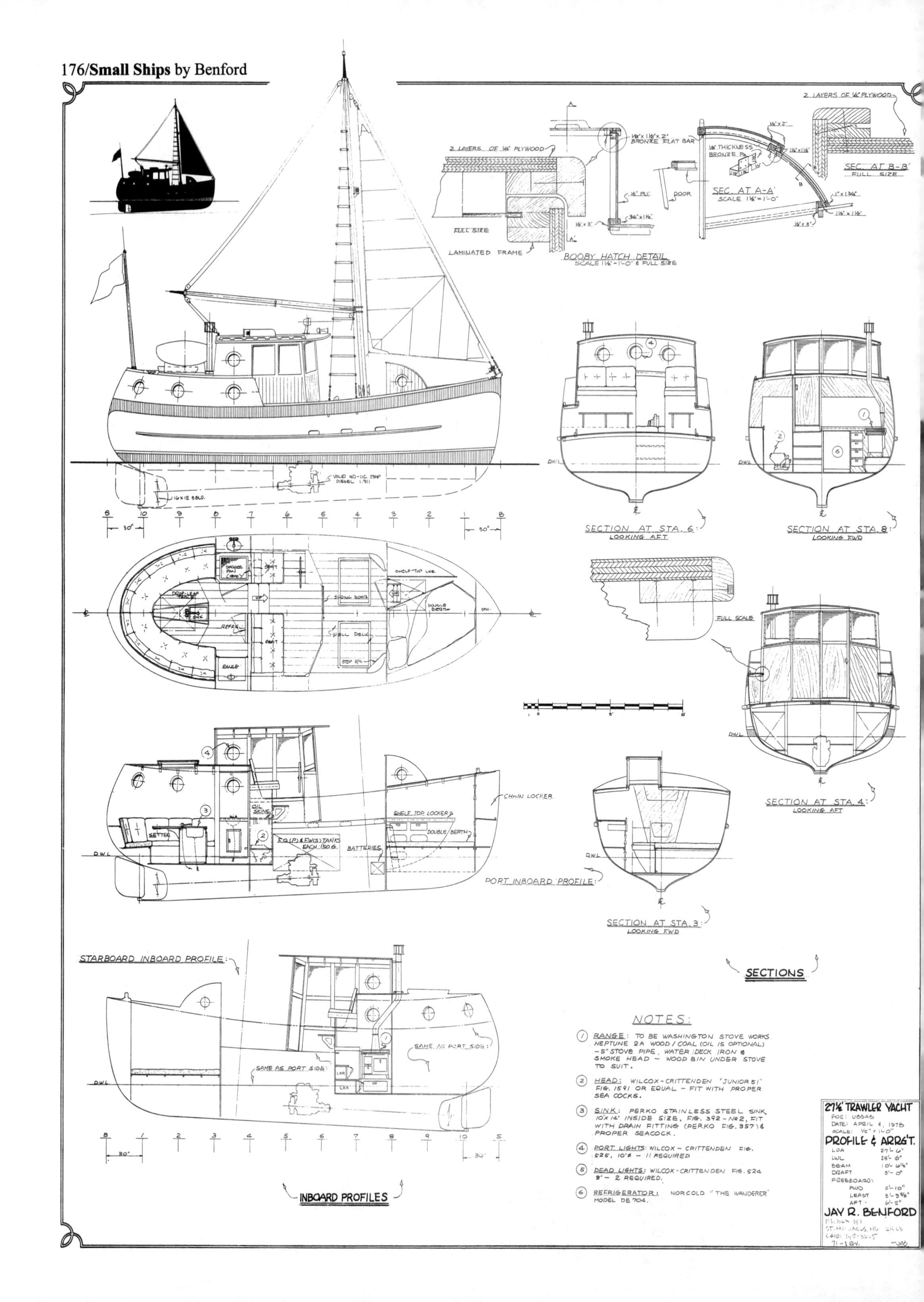

2 LAYERS OF 1/4" PLYWOOD
FULL SIZE
LAMINATED FRAME
BOOBY HATCH DETAIL
SCALE 1½"=1'-0" & FULL SIZE
SEC. AT A-A'
SCALE 1½"=1'-0"
SEC. AT B-B'
FULL SIZE
DOOR
SECTION AT STA. 6
LOOKING AFT
SECTION AT STA. 8
LOOKING FWD
FULL SCALE
SECTION AT STA. 4
LOOKING AFT
SECTION AT STA. 3
LOOKING FWD
SECTIONS
CHAIN LOCKER
SHELF TOP LOCKER
DOUBLE BERTH
BATTERIES
SETTEE
PORT INBOARD PROFILE
STARBOARD INBOARD PROFILE
SAME AS PORT SIDE
INBOARD PROFILES
DWL
NOTES:
1 RANGE: TO BE WASHINGTON STOVE WORKS NEPTUNE 2A WOOD / COAL (OIL IS OPTIONAL) – 5" STOVE PIPE, WATER DECK IRON & SMOKE HEAD – WOOD BIN UNDER STOVE TO SUIT.
2 HEAD: WILCOX-CRITTENDEN "JUNIOR 51" FIG. 1591 OR EQUAL – FIT WITH PROPER SEA COCKS.
3 SINK: PERKO STAINLESS STEEL SINK 10"x14" INSIDE SIZE, FIG. 392 – No 2, FIT WITH DRAIN FITTING (PERKO FIG. 357) & PROPER SEACOCK.
4 PORT LIGHTS: WILCOX – CRITTENDEN FIG. 525, 10"Ø – 11 REQUIRED
5 DEAD LIGHTS: WILCOX-CRITTENDEN FIG. 524 8" – 2 REQUIRED.
6 REFRIGERATOR: NORCOLD "THE WANDERER" MODEL DE 704.
27½' TRAWLER YACHT
FOR: UBBAB
DATE: APRIL 4, 1978
SCALE: ½" = 1'-0"
PROFILE & ARRG'T.
LOA 27'-6"
LWL 25'-6"
BEAM 10'-6½"
DRAFT 3'-0"
FREEBOARD:
FWD 5'-10"
LEAST 3'-9⅝"
AFT 6'-2"
JAY R. BENFORD

30' Beaching Cruiser

Design number 23

1966

This design commission came about as a referral from Bill Garden, who'd done a 22-footer along these lines. He was busy with larger boat commissions and kindly passed on my name to the client. Every young designer very much appreciates getting this sort of opportunity and I've tried to do the same from time to time.

We first did the preliminary study for the 28-footer, shown below, on the way to creating the final 30-footer shown on the next page, which evolved over a number of years. The development of the ideas from one to the other is evident and the 30-footer benefits from her extra length.

This stylish and sporty little cruiser is designed for lightweight aluminum construction. She has surprising accommodations for her low profile. There are settee-berths port and starboard around the portable table. The galley and the enclosed head are amidships, handy to the cockpit and in the area of least motion. The cockpit, with U-shaped settee and pilotseat is all the way aft, with a protective windshield for the helmsman.

Her modest size diesel engine will move her along at a good pace and with good economy.

Particulars:

Length overall	30'-0"
Length designed waterline	26'-0"
Beam	10'-0"
Draft	1'-8"
Freeboard:	
Forward	4'-6"
Least	2'-8"
Aft	3'-0"
Displacement, cruising trim	7,850 lbs.
Water tankage	50 Gals.
Fuel tankage	50 Gals.
Headroom	6'-0"

***CAUTION:** The displacement quoted here is for the boat in cruising trim. That is, with the fuel and water tanks filled, the crew on board, as well as the crews' gear and store in the locker. This should not be confused with the "shipping weight" often quoted as "displacement" by some manufacturers. This should be taken into account when comparing figures and ratios between this and other designs.

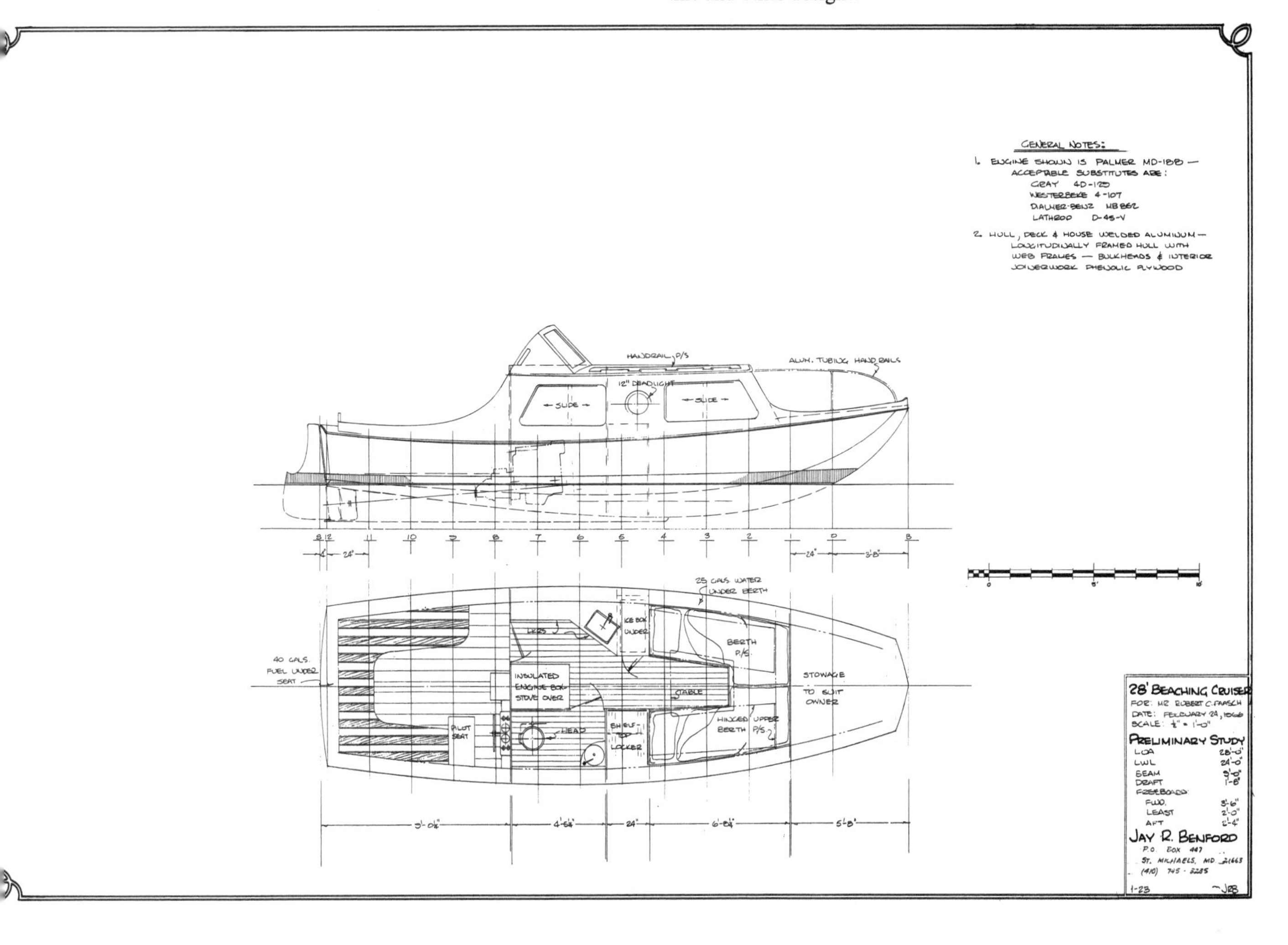

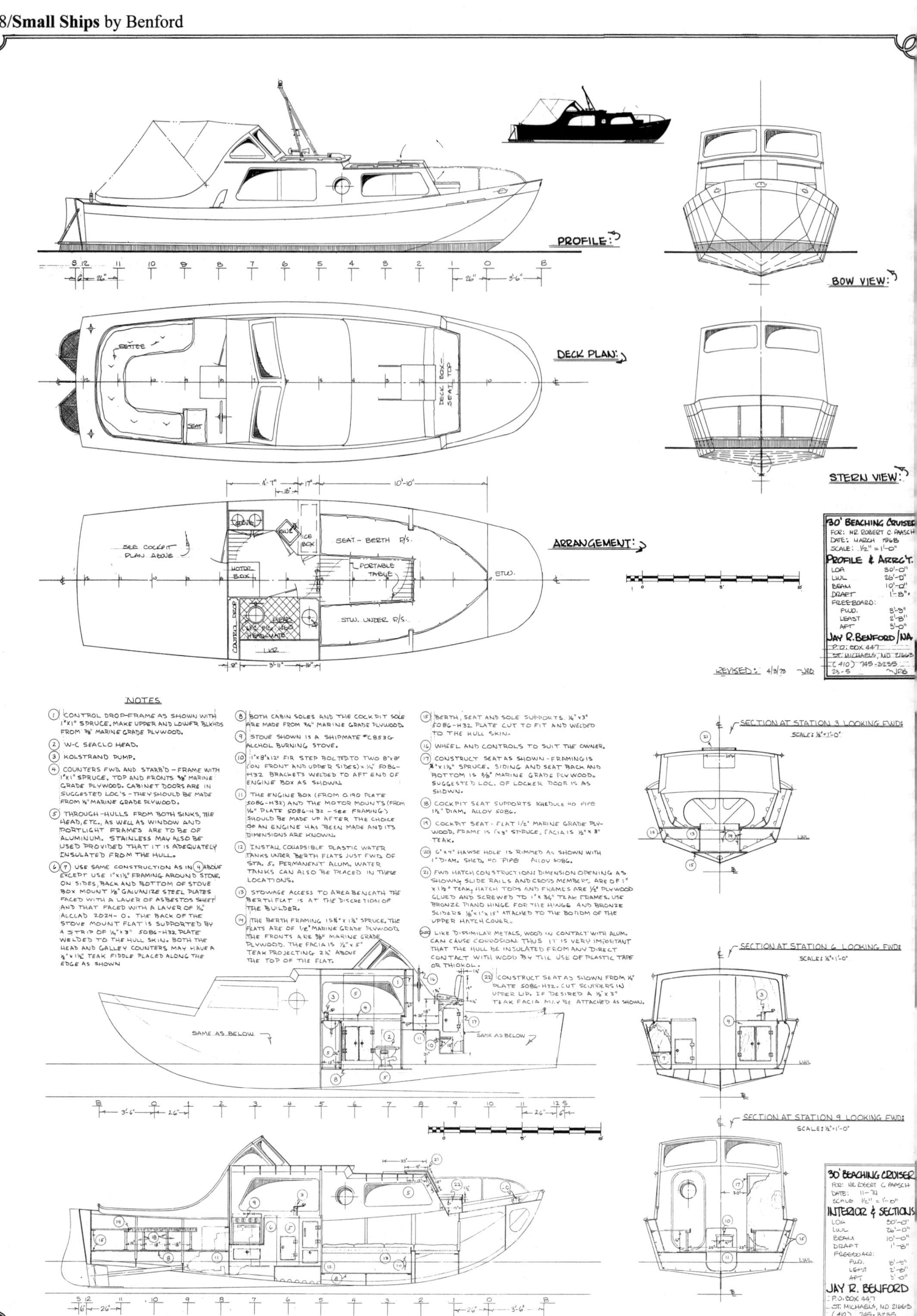
PROFILE:
BOW VIEW:
DECK PLAN:
SETTEE
SEAT
DECK BOX–SEAT TOP
STERN VIEW:
ARRANGEMENT:
SEE COCKPIT PLAN ABOVE
STOVE
ICE BOX
SEAT – BERTH P/S.
MOTOR BOX
PORTABLE TABLE
STW.
STW. UNDER P/S.
HEAD-MATE
LKR.
30' BEACHING CRUISER
FOR: MR. ROBERT C. PAASCH
DATE: MARCH 1968
SCALE: 1/2" = 1'-0"
PROFILE & ARRG'T.
LOA 30'-0"
LWL 26'-0"
BEAM 10'-0"
DRAFT 1'-8"
FREEBOARD:
FWD. 3'-9"
LEAST 2'-8"
APT 3'-0"
JAY R. BENFORD/NA
P.O. BOX 447
ST. MICHAELS, MD 21663
(410) 745-3235
23-5 ~JRB
REVISED: 4/3/73 ~JRB
NOTES
(1) CONTROL DROP-FRAME AS SHOWN WITH 1"X1" SPRUCE. MAKE UPPER AND LOWER BLKHDS FROM 3/8" MARINE GRADE PLYWOOD.
(2) W-C SEACLO HEAD.
(3) KOLSTRAND PUMP.
(4) COUNTERS FWD. AND STARB'D – FRAME WITH 1"X1" SPRUCE. TOP AND FRONTS 3/8" MARINE GRADE PLYWOOD. CABINET DOORS ARE IN SUGGESTED LOC'S – THEY SHOULD BE MADE FROM 1/2" MARINE GRADE PLYWOOD.
(5) THROUGH-HULLS FROM BOTH SINKS, THE HEAD, ETC., AS WELL AS WINDOW AND PORTLIGHT FRAMES ARE TO BE OF ALUMINUM. STAINLESS MAY ALSO BE USED PROVIDED THAT IT IS ADEQUATELY INSULATED FROM THE HULL.
(6)(7) USE SAME CONSTRUCTION AS IN (4) ABOVE EXCEPT USE 1"X1 1/2" FRAMING AROUND STOVE. ON SIDES, BACK AND BOTTOM OF STOVE BOX MOUNT 1/8" GALVANIZE STEEL PLATES FACED WITH A LAYER OF ASBESTOS SHEET AND THAT FACED WITH A LAYER OF 1/16" ALCLAD 2024-0. THE BACK OF THE STOVE MOUNT FLAT IS SUPPORTED BY A STRIP OF 1/4"X3" 5086-H32 PLATE WELDED TO THE HULL SKIN. BOTH THE HEAD AND GALLEY COUNTERS MAY HAVE A 1/2"X1 1/2" TEAK FIDDLE PLACED ALONG THE EDGE AS SHOWN
(8) BOTH CABIN SOLES AND THE COCKPIT SOLE ARE MADE FROM 3/4" MARINE GRADE PLYWOOD.
(9) STOVE SHOWN IS A SHIPMATE #C853G ALCHOL BURNING STOVE.
(10) 1"X8"X12" FIR STEP BOLTED TO TWO 8"X8" (ON FRONT AND UPPER SIDES) X 1/4" 5086-H32 BRACKETS WELDED TO AFT END OF ENGINE BOX AS SHOWN.
(11) THE ENGINE BOX (FROM 0.190 PLATE 5086-H32) AND THE MOTOR MOUNTS (FROM 1/4" PLATE 5086-H32 – SEE FRAMING) SHOULD BE MADE UP AFTER THE CHOICE OF AN ENGINE HAS BEEN MADE AND ITS DIMENSIONS ARE KNOWN.
(12) INSTALL COLLAPSIBLE PLASTIC WATER TANKS UNDER BERTH FLATS JUST FWD. OF STA. 5. PERMANENT ALUM. WATER TANKS CAN ALSO BE PLACED IN THESE LOCATIONS.
(13) STOWAGE ACCESS TO AREA BENEATH THE BERTH FLAT IS AT THE DISCRETION OF THE BUILDER.
(14) THE BERTH FRAMING IS 3/4"X1 1/2" SPRUCE. THE FLATS ARE OF 1/2" MARINE GRADE PLYWOOD. THE FRONTS ARE 3/8" MARINE GRADE PLYWOOD. THE FACIA IS 1/2"X5" TEAK PROJECTING 2 1/2" ABOVE THE TOP OF THE FLAT.
(15) BERTH, SEAT AND SOLE SUPPORTS. 1/4"X3" 5086-H32 PLATE CUT TO FIT AND WELDED TO THE HULL SKIN.
(16) WHEEL AND CONTROLS TO SUIT THE OWNER.
(17) CONSTRUCT SEAT AS SHOWN – FRAMING IS 3/4"X1 1/2" SPRUCE. SIDING AND SEAT BACK AND BOTTOM IS 3/8" MARINE GRADE PLYWOOD. SUGGESTED LOC. OF LOCKER DOOR IS AS SHOWN.
(18) COCKPIT SEAT SUPPORTS SCHEDULE 40 PIPE 1 1/2" DIAM. ALLOY 5086.
(19) COCKPIT SEAT – FLAT 1/2" MARINE GRADE PLYWOOD. FRAME IS 1"X3" SPRUCE, FACIA IS 1/2"X3" TEAK.
(20) 6"X4" HAWSE HOLE IS RIMMED AS SHOWN WITH 1" DIAM. SHED. 40 PIPE ALLOY 5086.
(21) FWD HATCH CONSTRUCTION: DIMENSION OPENING AS SHOWN, SLIDE RAILS AND CROSS MEMBERS ARE OF 1"X1 1/2" TEAK, HATCH TOPS AND FRAMES ARE 1/2" PLYWOOD GLUED AND SCREWED TO 1"X3/4" TEAK FRAMES. USE BRONZE PIANO HINGE FOR THE HINGE AND BRONZE SLIDERS 1/8"X1"X15" ATTACHED TO THE BOTTOM OF THE UPPER HATCH COVER.
(INFO) LIKE DISSIMILAR METALS, WOOD IN CONTACT WITH ALUM. CAN CAUSE CORROSION. THUS IT IS VERY IMPORTANT THAT THE HULL BE INSULATED FROM ANY DIRECT CONTACT WITH WOOD BY THE USE OF PLASTIC TAPE OR THIOKOL.
(22) CONSTRUCT SEAT AS SHOWN FROM 1/4" PLATE 5086-H32. CUT SCUPPERS IN UPPER LIP. IF DESIRED A 1/2"X3" TEAK FACIA MAY BE ATTACHED AS SHOWN.
SAME AS BELOW
SECTION AT STATION 3 LOOKING FWD:
SCALE: 1/2"=1'-0"
SECTION AT STATION 6 LOOKING FWD:
SCALE: 1/2"=1'-0"
SECTION AT STATION 9 LOOKING FWD:
SCALE: 1/2"=1'-0"
LWL
30' BEACHING CRUISER
FOR: MR. ROBERT C. PAASCH
DATE: 11-74
SCALE 1/2" = 1'-0"
INTERIOR & SECTIONS
LOA 30'-0"
LWL 26'-0"
BEAM 10'-0"
DRAFT 1'-8"
FREEBOARD:
FWD. 3'-9"
LEAST 2'-8"
APT 3'-0"
JAY R. BENFORD
P.O. BOX 447
ST. MICHAELS, MD 21663
(410) 745-3235
23-7

Benford 30

Designs Number 52, 101, & 173
1967 through 1978

Over a period of years we've designed a couple dozen varied boats, all based on this family of hull forms. These have ranged from the power boats shown on the following pages through ocean crossing sailboats.

The selection of power versions here gives some idea of the possibilities of a variation on a theme. Hull number 52, the first of this series was round bilged and built in both 'glass and wood. Hull 101 is the variation that was molded in fiberglass as the Benford 30, as well as being built in wood. Hull 173 is the multi-chine version drawn up to be built in steel. It could be the basis for building a stitch and glue version in plywood and epoxy.

The Polaris Mk II on the next page is one of my all-time favorites. While the accommodations are a bit Spartan for long-term living aboard, they are fine for casual or vacation cruising. The well deck in the waist is a great place for fishing, sunning, and dining alfresco. The great cabin has an enclosed head with shower sharing the space, the galley and an elegant elliptical settee and drop-leaf table. There's room for having up to six for dinner or socializing.

The Mk IV version on the same page is a deeper keel version built on the molded fiberglass hull. A later version of this is shown on the fourth page of these designs.

The second page has a couple quasi-motorsailer versions. The Polaris version has a 3'-6" draft keel variation. The other one has less sail area with no mainsail shown, but this could be added. The layout has a three sided walk-around on the main deck, with a deck at the sheer level for the foc'sle. The head is in the aft port corner of the cabin, with entry into the main cabin. It could also have a door onto the after deck for ease in cleaning up and direct access for swimmers.

The third page has the Polaris Mk I, the first of these well deck variations. It's got a lower profile and no raised deck for the great cabin. The stack functions as a headroom space for the aft cabin. The Orion version is a smaller version of a lot of the cruisers found in the Pacific Northwest. She has room for a family of four or two couples to go cruising, with the dinette making for two separate sleeping spaces.

The fourth page has a couple versions drawn for the 'glass hulls. One is an updating of the Polaris theme within the volume of the B30 molded hull. The other is a cool-climate trawler version, with the interior volumes maximized. It's interior is much like some of the sailing versions we've built, with an enclosed pilothouse in the place of the sailing cockpit. The foc'sle has a double berth and the settees in the great cabin would be supplemental berthing for occasional guests. The head has room for a shower and the galley has plenty of room for the cook to work out of the way of the fore and aft traffic. This one would be great for making that passage from Seattle to Skagway. She's shown in artist Don Kotts' drawings below.

Particulars:	**52**	**101**	**173**
Length overall	30'-0"	30'-0"	30'-0"
Length designed waterline	23'-0"	23'-6"	23'-4"
Beam	10'-7"	10'-6"	10'-8"
Draft	2'-6" shoal to 4'-6" keel		
Freeboard:			
Forward	4'-9"		
Least	3'-0"		
Aft	4'-5¾"		
Displacement, cruising trim	9,500 to 10,975 lbs.		
Displacement-length ratio	349 to 377		
Ballast	up to 3,300 lbs.		
Ballast ratio	up to 34%		
Prismatic coefficient	.55		
Pounds per inch immersion	807		
Water tankage	100 Gals.		
Fuel tankage	100 Gals.		
Headroom	4'-10½" to 6'-10½"		
GM	2.8' to 4'		

***CAUTION:** The displacement quoted here is for the boat in cruising trim. That is, with the fuel and water tanks filled, the crew on board, as well as the crews' gear and stores in the lockers. This should not be confused with the "shipping weight" often quoted as "displacement" by some manufacturers. This should be taken into account when comparing figures and ratios between this and other designs.

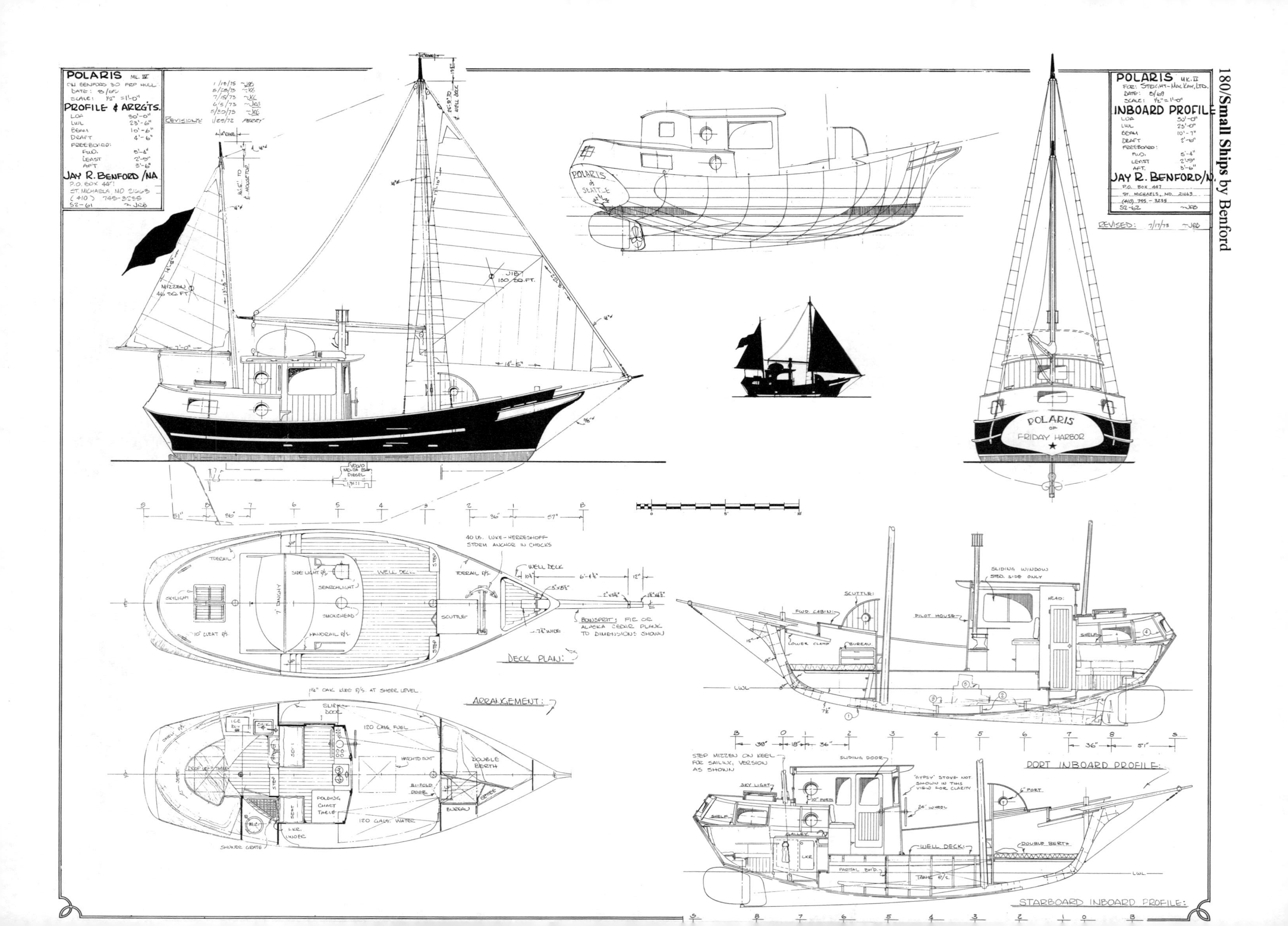
POLARIS MK. IV
ON BENFORD 30 FRP HULL
PROFILE & ARRGTS.
LOA 30'-0"
LWL 23'-6"
BEAM 10'-6"
DRAFT 4'-6"
FREEBOARD:
FWD. 5'-4"
LEAST 2'-9"
AFT 3'-6"
JAY R. BENFORD /NA
P.O. BOX 447
ST. MICHAELS, MD 21663
REVISIONS:
MIZZEN 46 SQ. FT.
JIB 180 SQ. FT.
POLARIS MK. II
FOR: STEIGHT-MACKAY, LTD.
INBOARD PROFILE
LOA 30'-0"
LWL 23'-0"
BEAM 10'-7"
DRAFT 2'-0"
FREEBOARD:
FWD. 5'-4"
LEAST 2'-5"
AFT. 3'-6"
JAY R. BENFORD/N.
P.O. BOX 447
ST. MICHAELS, MD. 21663
REVISED: 7/17/73
POLARIS OF SEATTLE
POLARIS OF FRIDAY HARBOR
40 LB. LUKE-HERRESHOFF STORM ANCHOR IN CHOCKS
WELL DECK
BOWSPRIT: FIR OR ALASKA CEDAR PLANK TO DIMENSIONS SHOWN
TOERAIL
SKYLIGHT
7' DINGHY
SIDE LIGHT P/S
SEARCHLIGHT
SMOKEHEAD
HANDRAIL P/S
SCUTTLE
10" CLEAT P/S
DECK PLAN:
ARRANGEMENT:
SLIDING DOOR
120 GALS. FUEL
120 GALS. WATER
FOLDING CHART TABLE
DOUBLE BERTH
BUREAU
BI-FOLD DOOR
SHOWER GRATE
SCUTTLE
FWD. CABIN
PILOT HOUSE
SLIDING WINDOW STBD. SIDE ONLY
HEAD
LOWER CLAMP
SHELF
LWL
PORT INBOARD PROFILE:
STEP MIZZEN ON KEEL FOR SAILING VERSION AS SHOWN
SLIDING DOOR
'GYPSY' STOVE NOT SHOWN IN THIS VIEW FOR CLARITY
24" WHEEL
6" PORT
SKY LIGHT
GALLEY
WELL DECK
DOUBLE BERTH
PARTIAL BH'D.
TANK P/S.
STARBOARD INBOARD PROFILE:

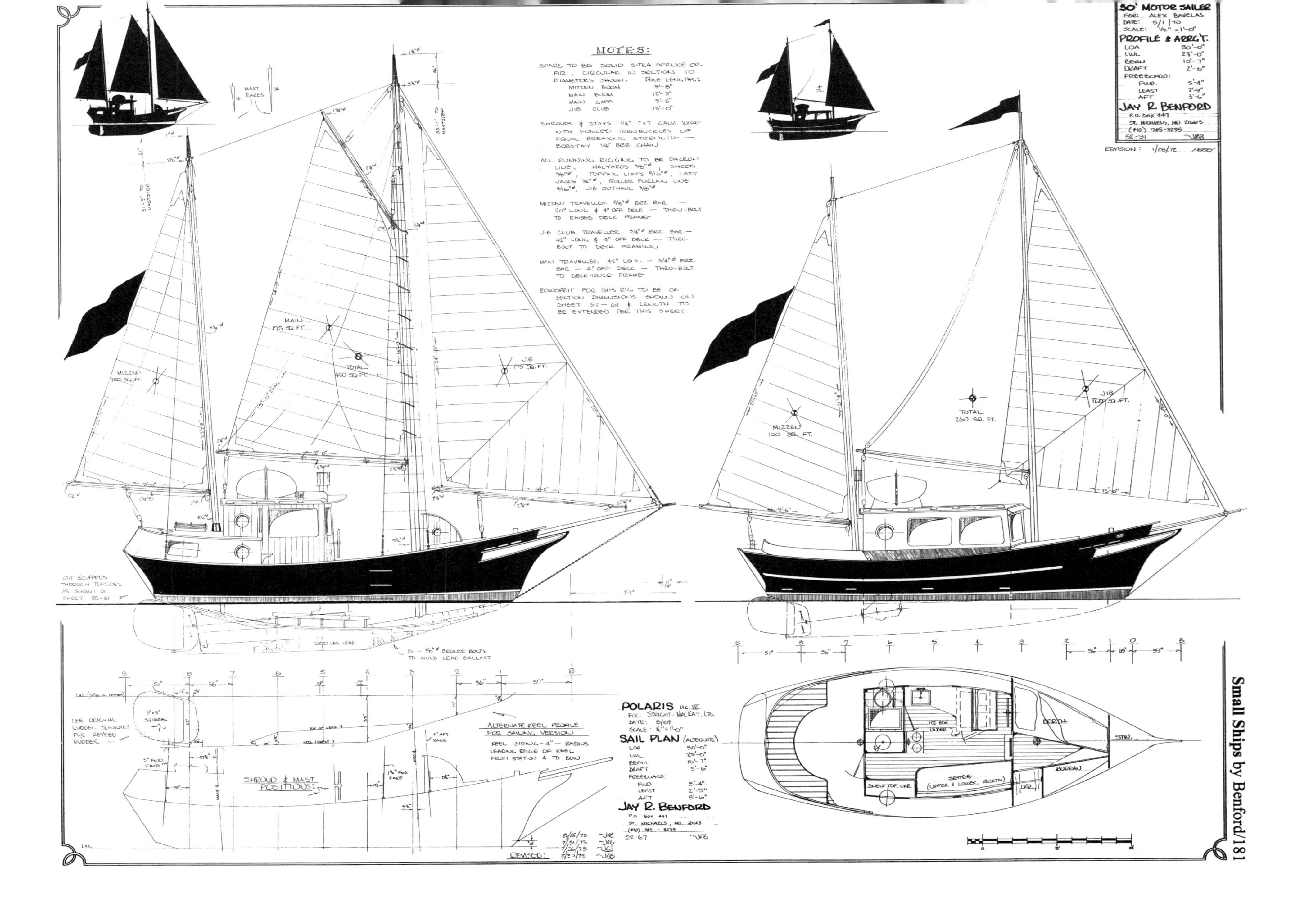
30' MOTOR SAILER
FOR: ALEX BAVELAS
DATE: 5/1/70
SCALE: 1/2" = 1'-0"
PROFILE & ARRG'T.
LOA 30'-0"
LWL 23'-0"
BEAM 10'-7"
DRAFT 2'-6"
FREEBOARD:
FWD. 5'-4"
LEAST 2'-9"
AFT 3'-6"
JAY R. BENFORD
P.O. BOX 447
ST. MICHAELS, MD 21663
(410) 745-3235
52-91
REVISION: 1/28/72
NOTES:
SPARS TO BE SOLID SITKA SPRUCE OR FIR, CIRCULAR IN SECTION TO DIAMETERS SHOWN. POLE LENGTHS:
MIZZEN BOOM 9'-8"
MAIN BOOM 12'-9"
MAIN GAFF 9'-5"
JIB CLUB 13'-0"
SHROUDS & STAYS 1/4" 7x7 GALV. WIRE WITH FORGED TURNBUCKLES OF EQUAL BREAKING STRENGTH — BOBSTAY 1/4" BBB CHAIN
ALL RUNNING RIGGING TO BE DACRON LINE. HALYARDS 3/8"ø, SHEETS 3/8"ø, TOPPING LIFTS 5/16"ø, LAZY JACKS 1/4"ø, ROLLER FURLING LINE 3/16"ø, JIB OUTHAUL 3/8"ø
MIZZEN TRAVELLER 5/8"ø BRZ. BAR — 20" LONG & 4" OFF DECK — THRU-BOLT TO RAISED DECK FRAME
JIB CLUB TRAVELLER 3/4"ø BRZ. BAR — 42" LONG & 4" OFF DECK — THRU-BOLT TO DECK FRAMING
MAIN TRAVELLER 42" LONG — 3/4"ø BRZ. BAR — 4" OFF DECK — THRU-BOLT TO DECKHOUSE FRAME
BOWSPRIT FOR THIS RIG TO BE OF SECTION DIMENSIONS SHOWN ON SHEET 52-61 & LENGTH TO BE EXTENDED PER THIS SHEET
MAST RAKES
MAIN 175 SQ. FT.
TOTAL 450 SQ. FT.
JIB 175 SQ. FT.
MIZZEN 100 SQ. FT.
TOTAL 260 SQ. FT.
JIB 160 SQ. FT.
MIZZEN 100 SQ. FT.
CUT SCUPPERS THROUGH TOPSIDES AS SHOWN ON SHEET 52-61
2000 LBS. LEAD
5 - 5/8"ø BRONZE BOLTS TO HOLD LEAD BALLAST
USE ORIGINAL RUDDER SCANTLINGS FOR REVISED RUDDER
ALTERNATE KEEL PROFILE FOR SAILING VERSION
KEEL SIDING - 4" - RADIUS LEADING EDGE OF KEEL FROM STATION 4 TO BOW
SHROUD & MAST POSITIONS
POLARIS MK. III
FOR: STRIGHT-MACKAY, LTD.
DATE: 8/69
SCALE: 1/2" = 1'-0"
SAIL PLAN (ALTERNATE)
LOA 30'-0"
LWL 23'-0"
BEAM 10'-7"
DRAFT 5'-6"
FREEBOARD:
FWD. 5'-4"
LEAST 2'-9"
AFT 3'-6"
JAY R. BENFORD
P.O. BOX 447
ST. MICHAELS, MD. 21663
(410) 745 - 3235
52-67
REVISED:
BERTH
STW.
BUREAU
ICE BOX UNDER
SETTEE (UPPER & LOWER BERTH)
SHELF TOP LKR.
LKR.

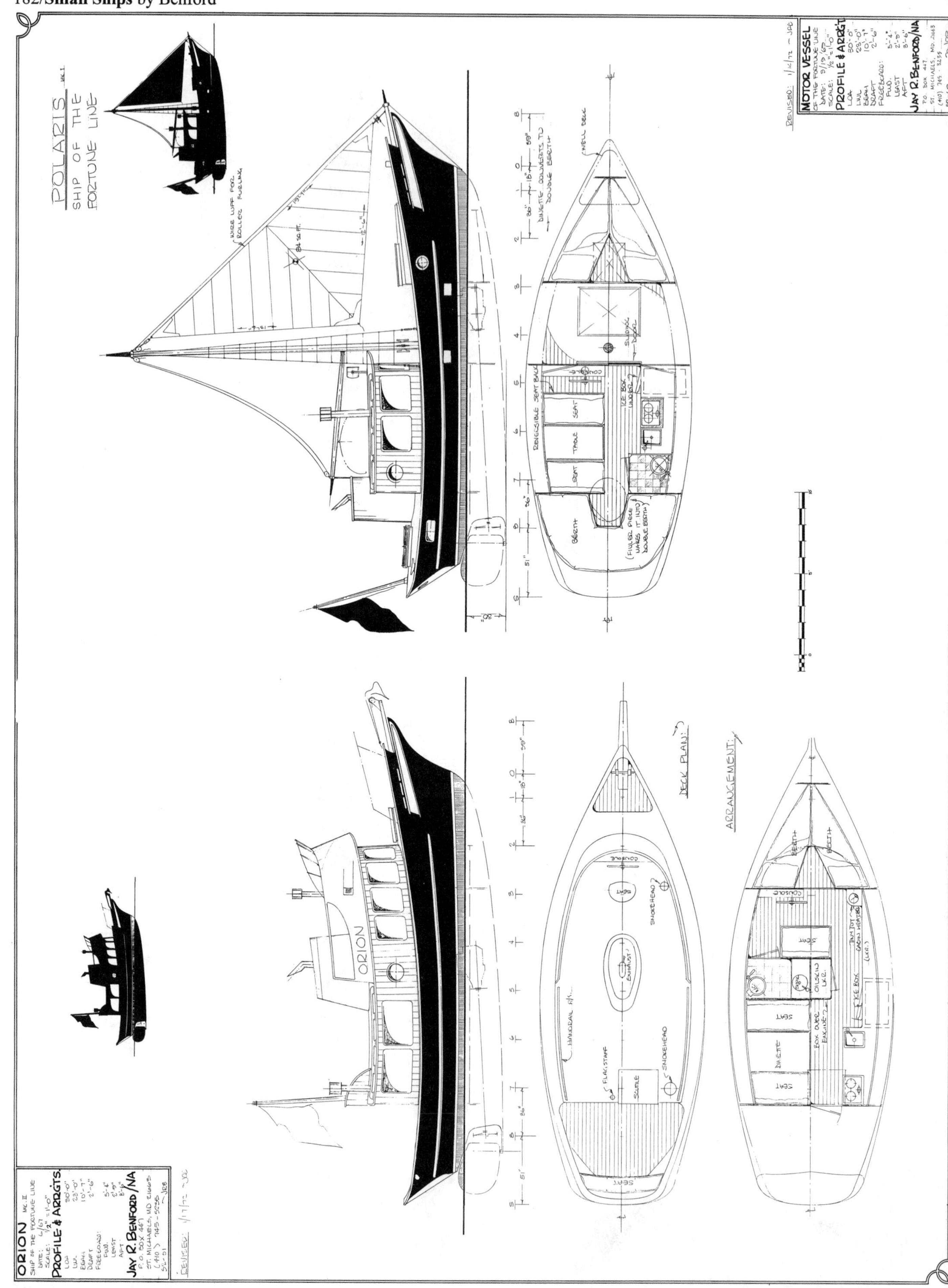
POLARIS MK.I
SHIP OF THE FORTUNE LINE
WIRE LUFF FOR ROLLER FURLING
REVISED: 1/12/72 ~ JRB
MOTOR VESSEL
OF THE FORTUNE LINE
DATE: 9/19/67
SCALE: 1/2" = 1'-0"
PROFILE & ARRG'T
LOA 30'-0"
LWL 23'-0"
BEAM 10'-7"
DRAFT 2'-6"
FREEBOARD:
FWD. 5'-4"
LEAST 2'-9"
AFT 3'-6"
JAY R. BENFORD/NA
P.O. BOX 447
ST. MICHAELS, MD. 21663
(410) 745 - 3235
52-60
~JRB
DINETTE CONVERTS TO DOUBLE BERTH
WELL DECK
SLIDING DOOR
CONSOLE
REVERSIBLE SEAT BACK
SEAT
TABLE
ICE BOX UNDER
BERTH
FILLER PIECE MAKES IT INTO DOUBLE BERTH
DECK PLAN
ARRANGEMENT
CONSOLE
SEAT
SMOKEHEAD
EXHAUST
HANDRAIL P/S
FLAGSTAFF
SCUTTLE
BERTH
BERTH
SEAT
OILSKIN LKR.
ICE BOX
BOX OVER ENGINE
DINETTE
ORION
ORION MK. II
SHIP OF THE FORTUNE LINE
DATE: 6/67
SCALE: 1/2" = 1'-0"
PROFILE & ARRGTS.
LOA 30'-0"
LWL 23'-0"
BEAM 10'-7"
DRAFT 2'-6"
FREEBOARD:
FWD. 5'-4"
LEAST 2'-9"
AFT 3'-6"
JAY R. BENFORD/NA
P.O. BOX 447
ST. MICHAELS, MD 21663
(410) 745-3235
52-51
~JRB
REVISED: 1/17/72 ~ JRB

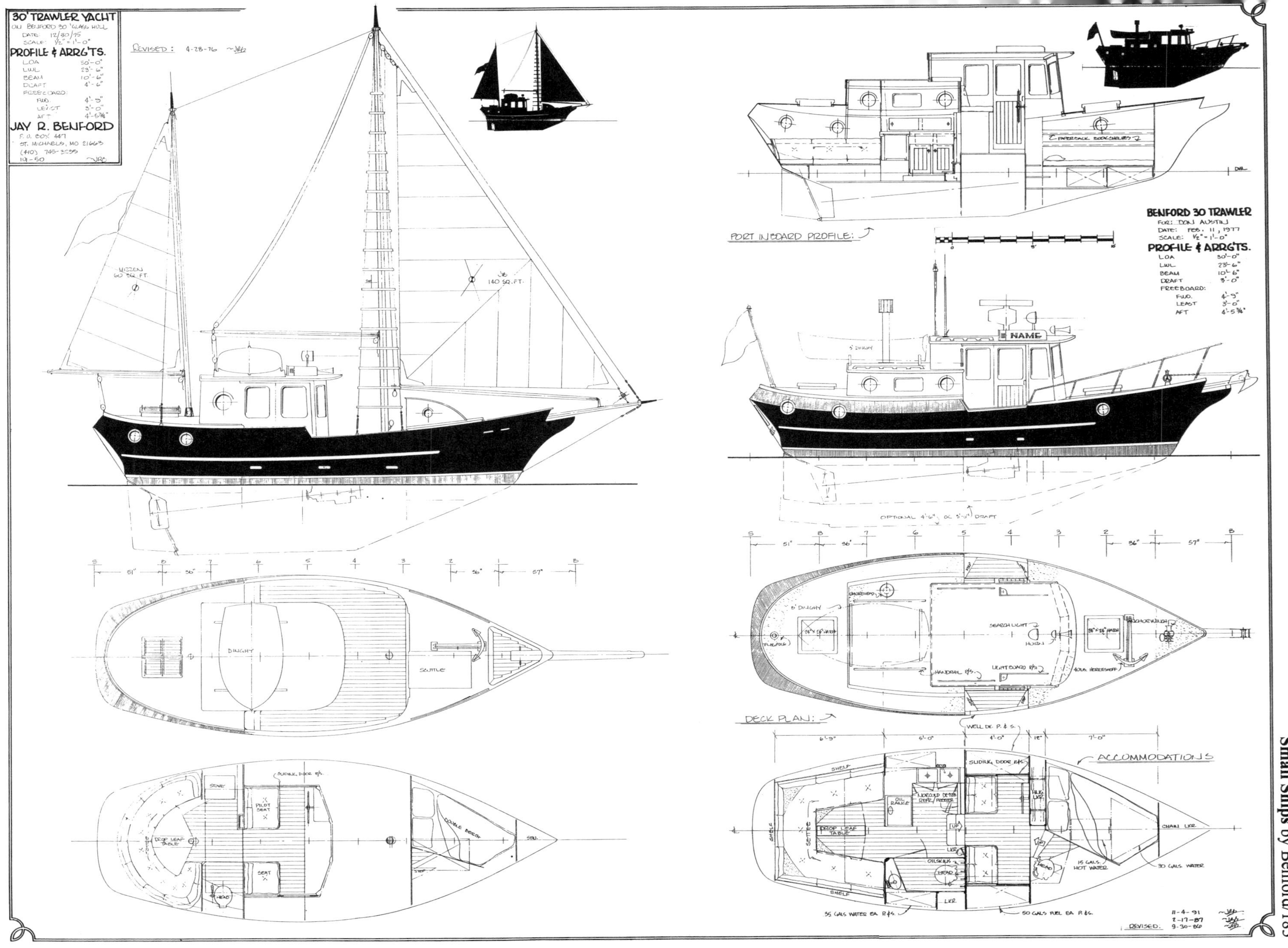
30' TRAWLER YACHT
ON BENFORD 30' GLASS HULL
DATE: 12/30/75
SCALE: 1/2" = 1'-0"
PROFILE & ARRG'TS.
LOA 30'-0"
LWL 23'-6"
BEAM 10'-6"
DRAFT 4'-6"
FREEBOARD:
FWD. 4'-9"
LEAST 3'-0"
AFT 4'-5 3/4"
JAY R. BENFORD
P.O. BOX 447
ST. MICHAELS, MD 21663
(410) 745-3235
101-50
REVISED: 4-28-76
MIZZEN 60 SQ. FT.
JIB 140 SQ. FT.
DINGHY
SCUTTLE
STOVE
PILOT SEAT
SEAT
DROP LEAF TABLE
HEAD
DOUBLE BERTH
PORT INBOARD PROFILE:
BENFORD 30 TRAWLER
FOR: DON AUSTIN
DATE: FEB. 11, 1977
SCALE: 1/2" = 1'-0"
PROFILE & ARRG'TS.
LOA 30'-0"
LWL 23'-6"
BEAM 10'-6"
DRAFT 3'-0"
FREEBOARD:
FWD. 4'-9"
LEAST 3'-0"
AFT 4'-5 3/4"
NAME
8' DINGHY
OPTIONAL 4'-6" OR 5'-0" DRAFT
SEARCH LIGHT
HORN
HANDRAIL P/S.
LIGHTBOARD P/S
DECK PLAN:
WELL DK. P. & S.
ACCOMMODATIONS
SLIDING DOOR P/S
DROP LEAF TABLE
SETTEE
SHELF
OIL RANGE
OILSKINS
HEAD
LKR
CHAIN LKR
15 GALS. HOT WATER
30 GALS WATER
35 GALS WATER EA. P/S.
50 GALS FUEL EA. P/S.
REVISED: 11-4-91 2-17-87 9-30-80

36 & 39' Cruisers

Designs number 73, 130 & 178
1971, 1975 & 1978

This handsome 36' Cruiser (#73) has a practical, no-nonsense air about her. Suitable as a home afloat, she's capable of extended coastwise cruising. She has a double berth in the foc'sle with plenty of shelf-top locker space port and starboard of the berth, and full length hanging lockers each side aft of the berth. Three steps up and aft find one in the pilothouse, which has a large pilot seat to port, passenger seating to starboard of that, which faces a fireplace abaft of the pilot's seat. Down a couple steps amidships is another hanging locker, with head and shower opposite. A dinette to starboard and U-shaped galley to port end the aft cabin. In the stern is a self-draining cockpit. There's stowage on deck for an eight foot skiff and she has a large flying bridge. This type of layout could be done on the hull plans of the second 36-footer.

Another 36' Cruiser, design number 130, is laid out for construction in plywood. This one shows a different approach to styling and layout. The forward stateroom is down a couple steps from the rest of the cabin, which is all on one level. The cabin sole and side decks are all one level, and there is a walk-around deck under the overhang of the boatdeck. Both sides of the house have built-in ladders to the boatdeck, making access to and from the flying bridge and dinghy storage quick and easy.

The 39-footer (#178) evolved from a request we had for a modified version of the 36-footer with a second enclosed head in the forward stateroom instead of the head in the cabin in the 36-footer. We also made the beam a bit wider and made access around the deckhouse easier.

Particulars:	**#73**	**#130**	**#178**
Length overall	36'-6"	36'-6"	39'-0"
Length designed waterline	35'-0"	35'-0"	35'-0"
Beam	12'-0"	13'-6"	14'-0"
Draft	3'-9"	3'-6"	3'-6"
Freeboard:			
Forward	6'-6"	6'-6"	6'-5"
Least	3'-8"	3'-6"	3'-6"
Aft	4'-0"	4'-3"	4'-3"
Displacement, cruising trim*		17,250	17,250
Water tankage (gallons)		120	300
Fuel tankage (gallons)		300	300
Headroom (varies)		6'-3" to 6'-7"	

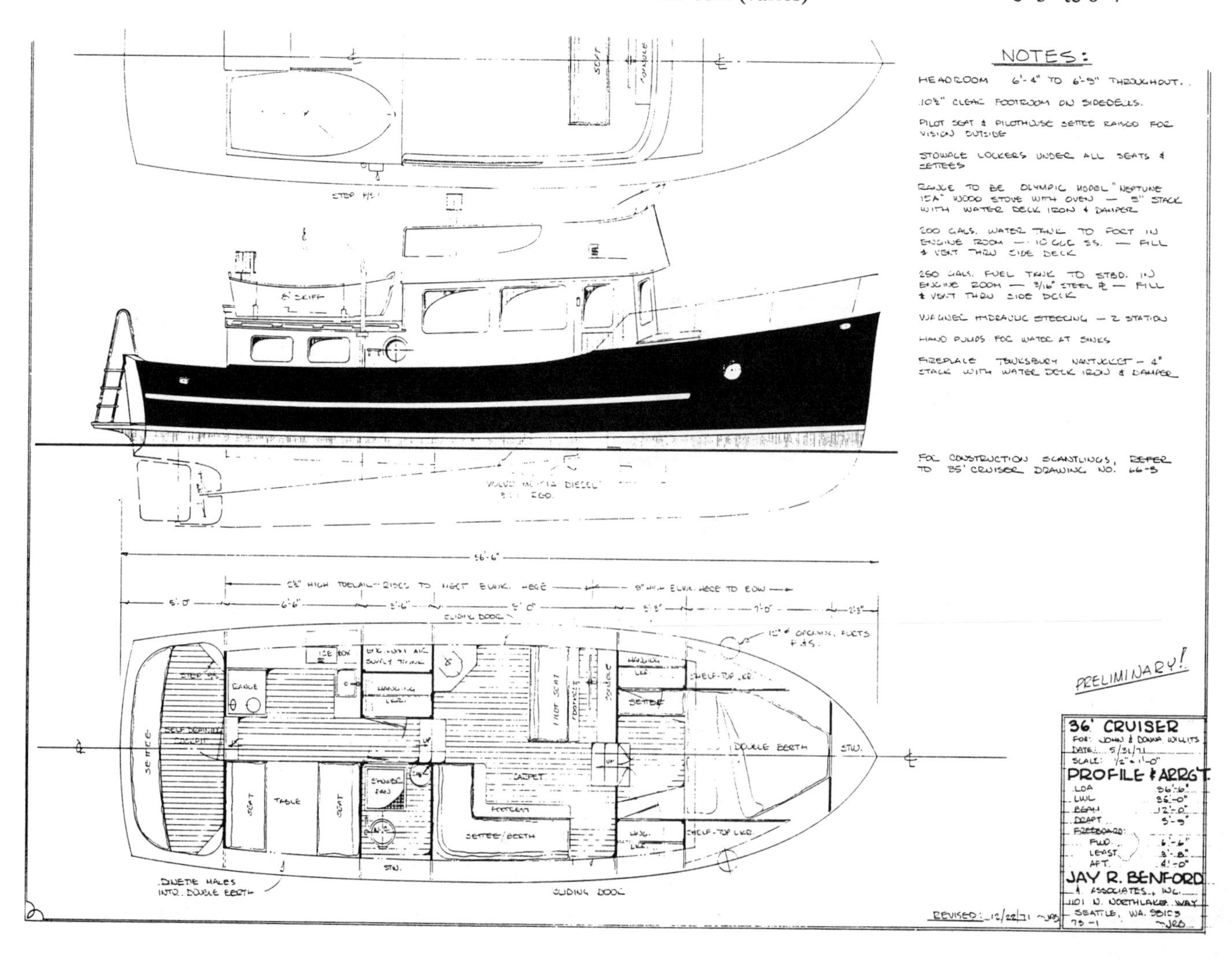

36' CRUISER
FOR: MR. HUGH E. ELWOOD
DATE: 8/17/75
SCALE: 1/2" = 1'-0"

PROFILE & ARRGT.

LOA	36'-7 1/2"
LWL	35'-0"
BEAM	13'-6"
DRAFT	3'-6"
FREEBOARD:	
FWD	6'-6"
LEAST	3'-6"
AFT	4'-9"

JAY R. BENFORD
P.O. BOX 447
ST. MICHAELS, MD 21663
(410) 745-3235
130-2

NOTES:

1. BERTH & SEAT CUSHIONS 4" FOAM — SEATBACKS 2" FOAM — COVERS BY OWNER TO SUIT.
2. WINDOWS & PORTS TO BE 1/4" LEXAN SET IN PROPER FRAMES — SIZES AS SHOWN.
3. 12" x 18" LOUVRE OVER DUCT FOR AIR TO ENGINE ROOM.
4. LOUVRE OVER EXHAUST FAN FROM HEAD.
5. SETTEE MAKES INTO UPPER & LOWER BERTH — SEAT BACK HINGES UP & HOOKS IN RAISED POSITION.
6. SETTEE IS EXTENSION BERTH — MAKES DOUBLE & SEAT BACK TO HINGE UP FOR SINGLE UPPER.
7. WILCOX-CRITTENDEN "SKIPPER" HEAD — FIT WITH PROPER SEACOCKS ON INLET & OUTLET.
8. WILCOX-CRITTENDEN "JUNIOR 51" HEAD — FIT WITH PROPER SEACOCKS ON INLET & OUTLET.
9. SHOWER GRATE OVER PAN — DRAIN TO BILGE PUMP PICK-UP.
10. POLAR SINK — 10 1/2" x 13 1/4" OVAL 5" DEEP — DRAIN THROUGH SEACOCK.

INBOARD PROFILES

SAME AS POST SIDE

FOR ADDITIONAL FRAMING DETAILS SEE SHEET 130-4

DECK PLAN
STARBOARD SIDE SIMILAR

TANK LAYOUT

~130 GALS. WATER
10 GGE. GALV.

MOUNT BATTERIES OPPOSITE FOR WEIGHT BALANCE

20 GAL. HOT WATER

~150 GALS. FUEL OIL EA. P & S.
10 GGE. BLACK IRON

DOUBLE BERTH
DRESSER
TABLE
SETTEE
REF'R
DICKINSON PACIFIC
REF'R UNDER COUNTER

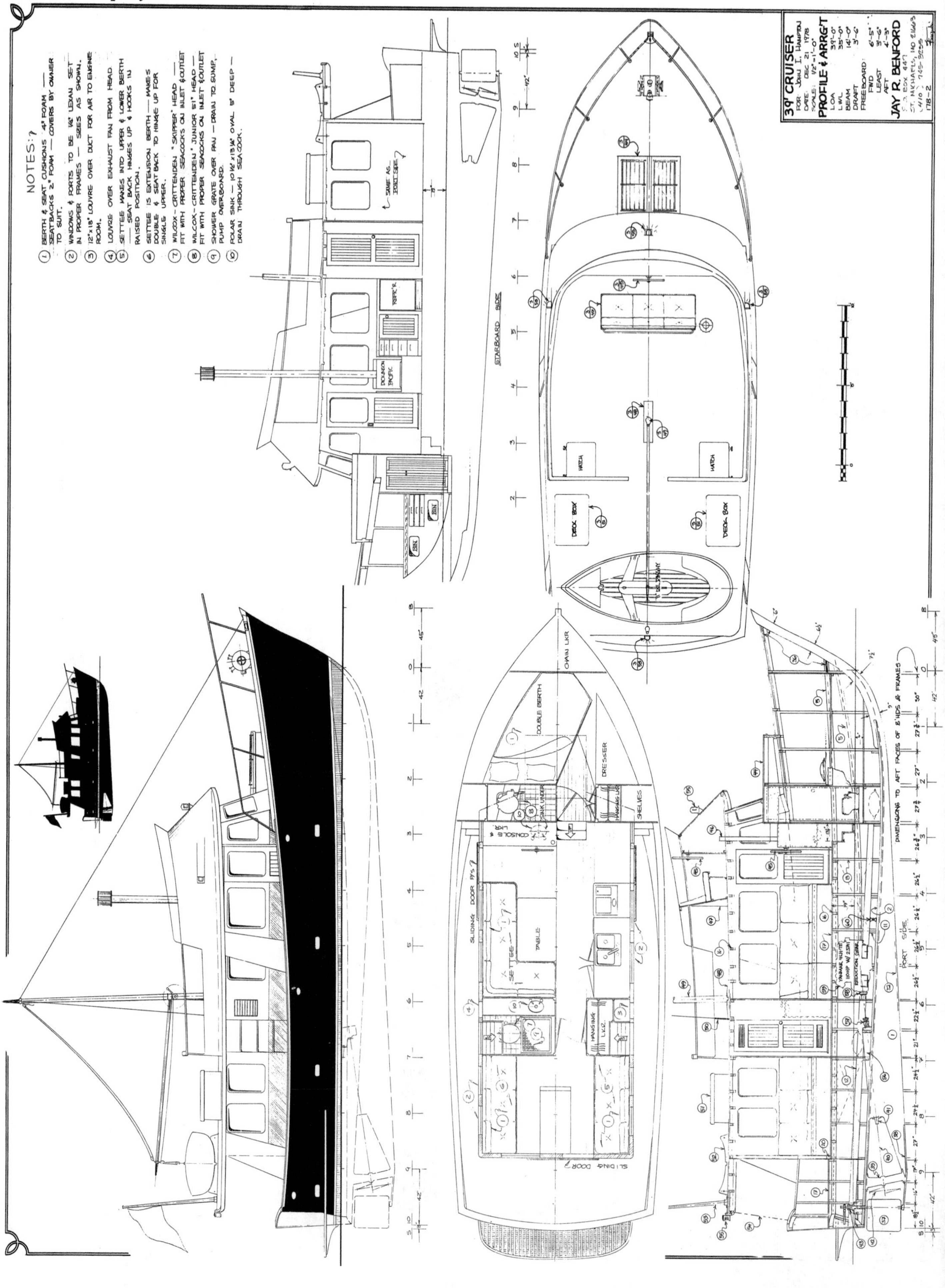

NOTES:

1. BERTH & SEAT CUSHIONS 4" FOAM — SEATBACKS 2" FOAM — COVERS BY OWNER TO SUIT.
2. WINDOWS & PORTS TO BE 1/4" LEXAN SET IN PROPER FRAMES — SIZES AS SHOWN.
3. 12"x18" LOUVRE OVER DUCT FOR AIR TO ENGINE ROOM.
4. LOUVRE OVER EXHAUST FAN FROM HEAD.
5. SETTEE MAKES INTO UPPER & LOWER BERTH — SEAT BACK HINGES UP & HOOKS IN RAISED POSITION.
6. SETTEE IS EXTENSION BERTH — MAKES DOUBLE & SEAT BACK TO HINGE UP FOR SINGLE UPPER.
7. WILCOX-CRITTENDEN "SKIPPER" HEAD — FIT WITH PROPER SEACOCKS ON INLET & OUTLET.
8. WILCOX-CRITTENDEN "JUNIOR 51" HEAD — FIT WITH PROPER SEACOCKS ON INLET & OUTLET.
9. SHOWER GRATE OVER PAN — DRAIN TO SUMP, PUMP OVERBOARD.
10. POLAR SINK — 10 1/2 x 13 1/4 OVAL 5" DEEP — DRAIN THROUGH SEACOCK.

SLIDING DOOR P/S
SETTEE
TABLE
CONSOLE & LKR.
SINK UNDER
DOUBLE BERTH
CHAIN LKR
DRESSER
HANGING LKR
SHELVES
HANGING LKR.
SLIDING DOOR

DICKINSON PACIFIC
REFR'R
SAME AS PORT SIDE
BIN
BIN
STARBOARD SIDE

DECK BOX
HATCH
DECK BOX
HATCH
DINGHY

YANMAR 4LH-TE 100HP W/ 2.5:1 REDUCTION GEAR
PORT SIDE
DIMENSIONS TO AFT FACES OF B'HDS & FRAMES

39' CRUISER
FOR: JOHN I. HAMPTON
DATE: DEC 21 1978
SCALE: 1/2"=1'-0"

PROFILE & ARRG'T

LOA	39'-0"
LWL	35'-0"
BEAM	14'-0"
DRAFT	3'-6"
FREEBOARD:	
FWD	6'-5"
LEAST	3'-6"
AFT	4'-3"

JAY R. BENFORD
P.O. BOX 447
ST. MICHAELS, MD 21663
(410) 745-3235
178-2

40' Trawler Yachts

Designs Number 324 & 329
1992 & 1994

These two 40' Trawler Yachts are a good study in diversity. The different hull forms and particularly the beams make for very different spaces in the boats.

The double ender (pages 186B & C) is a molded fiberglass hull from Cascade Yachts. We originally designed the hull as an offshore sailing yacht. For this version, we've reduced the draft and raised the freeboard to make a very capable motor yacht. She'll be outfitted at Higgins Yacht Yard in St. Michaels, just a few blocks from our office. She has a seakindly hull form and will have a comfortable motion underway. For extended offshore work, we'd add flopper-stoppers or an active stabilizer.

This vessel has two staterooms; the larger one forward as the owner's quarters and the smaller one under the pilothouse as the guest cabin. The head has a separate shower stall and good elbow room. There is plenty of room in the galley for people to get past the chef and/or bring in a couple chairs from the aft deck to sit around the saloon table in cooler or rainy weather. The after deck is generously sized and will be a popular place while aboard.

Like many of the plans shown in this book, the pages of drawings here have more than our usual study plans on them. While the study plans are larger, careful study of these drawings will yield a great deal of information.

The wider, transom stern one (page 186D) has more emphasis on longer term living aboard, with things like a washer and dryer fitted. She's intended for coastwise cruises and could make passages to the Bahamas.

Her accommodations, except for the pilothouse which doubles as a guest cabin, are all on one level. The master stateroom is forward, with a large hanging locker, dressers and washer and dryer in it. Next aft is the head with a shower stall to starboard and a good-sized galley to port. The saloon has half-height book cases across both aft bulkheads and the forward port bulkhead. The galley can have an opening over the shelves to pass food and drinks in and out. There is an inside staircase up to the pilothouse; much easier to negotiate than climbing a steep ladder and crossing an open deck — particularly in poor weather.

The power curve here is for the double ender and shows how much power is needed in calm conditions versus what's needed in heavy weather. My own thoughts would be towards using a 50 hp diesel to give more than enough power for light-air full-speed cruising (8.2 to 8.3 knots on the sailing versions) and the ability to maintain about half speed in a gale. Increasing the power to 75 doesn't increase the cruising speed, but will extend the speed while powering in heavy weather. The curves also show how little power is really needed for slightly more modest cruising speeds. This same relationship of power to reasonable cruising speeds applies to the other boats in this book, with the power required changed to suit the displacement.

Particulars:		**324**	**329**
Length overall		40'-0"	40'-0"
Length designed waterline		40'-0"	34'-0"
Beam		15'-0"	12'-4"
Draft		3'-6"	3'-6"
Freeboard:	Forward	8'-0"	8'-0"
	Least		3'-4"
	Aft	1'-3"	4'-4"
Displacement, cruising trim*		40,000 lbs.	21,750
Displacement-length ratio		279	247
Prismatic coefficient		.64	.56
Water tankage		330 Gals.	300 Gals.
Fuel tankage		315 Gals.	300 Gals.
Headroom		6'-7"	6'-1" to 6'-9"

***CAUTION:** The displacement quoted here is for the boat in cruising trim. That is, with the fuel and water tanks filled, the crew on board, as well as the crews' gear and stores in the lockers. This should not be confused with the "shipping weight" often quoted as "displacement" by some manufacturers. This should be taken into account when comparing figures and ratios between this and other designs.

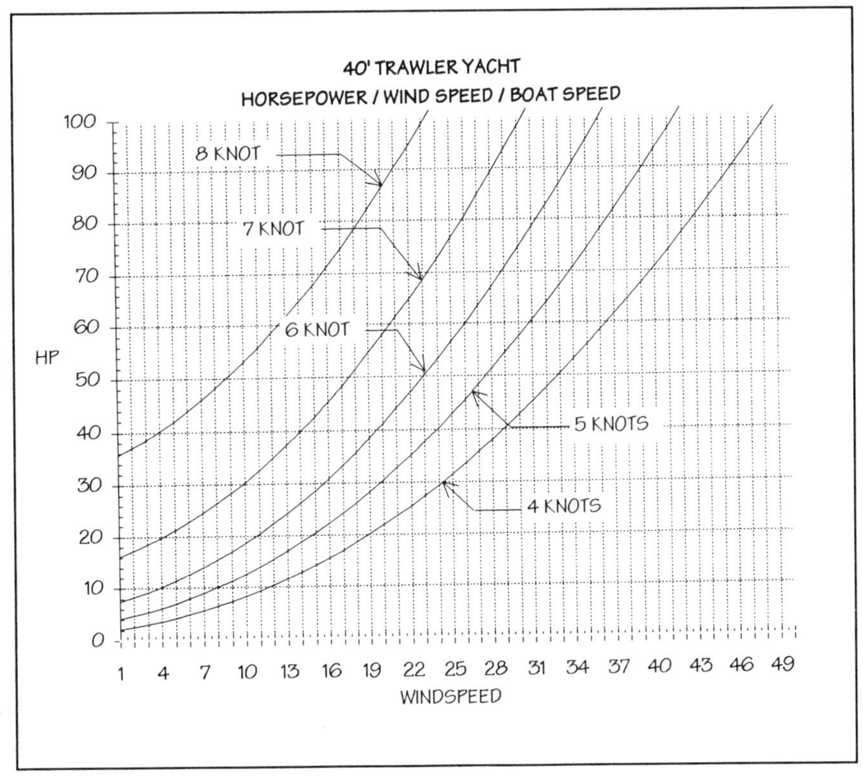

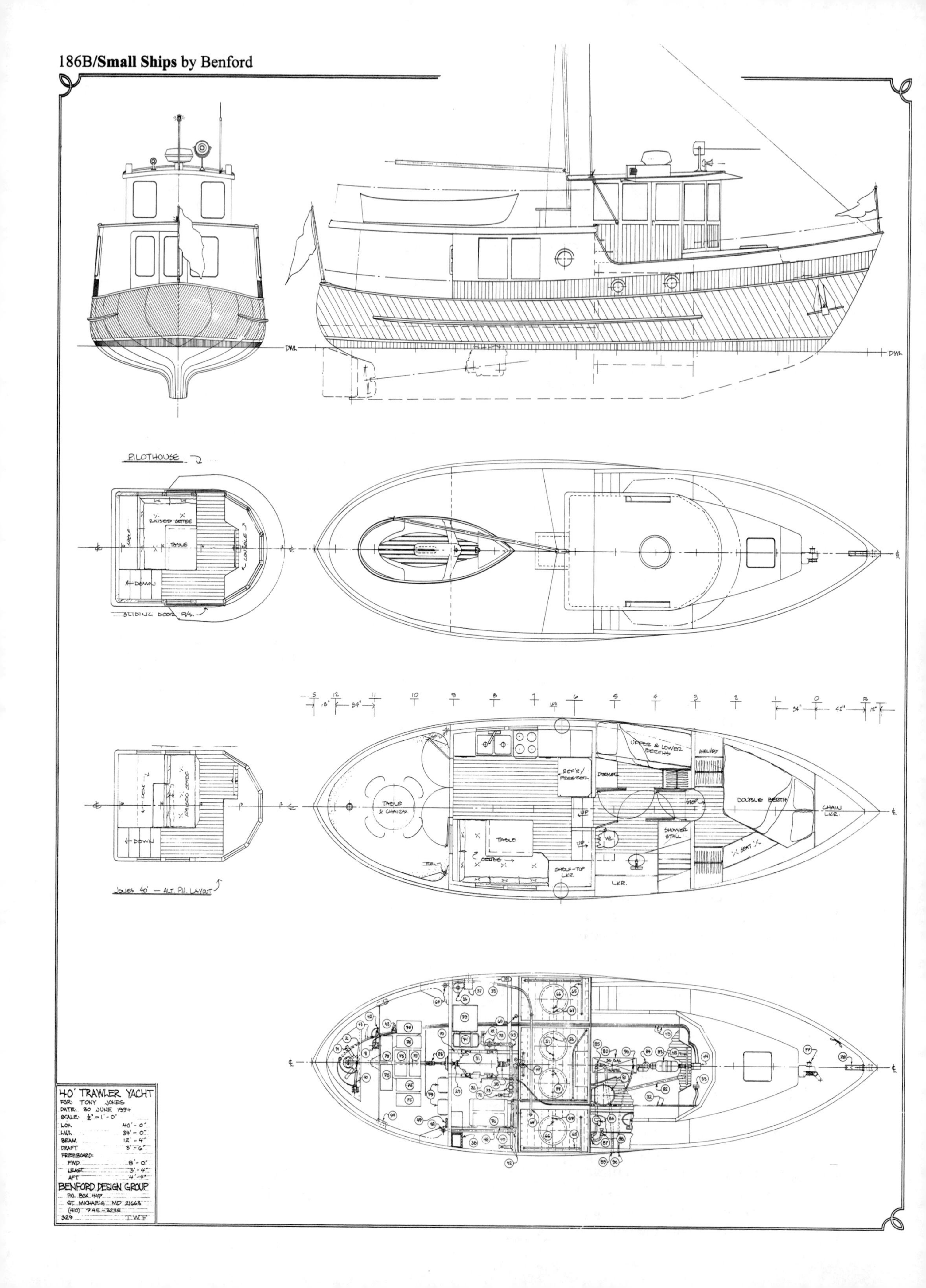

DWL
PILOTHOUSE
RAISED SETTEE
TABLE
CONSOLE
DOWN
SLIDING DOOR P/S.
UPPER & LOWER BERTHS
SHELVES
REFR/ FREEZER
DRESSER
UP
TABLE & CHAIRS
TABLE
SETTEE
SHELF-TOP LKR.
LKR.
WC
SHOWER STALL
STEP
DOUBLE BERTH
SEAT
CHAIN LKR.
DOWN
Jones 40' — ALT. P.H. LAYOUT
40' TRAWLER YACHT
FOR: TONY JONES
DATE: 30 JUNE 1994
SCALE: ½" = 1'-0"
LOA 40'-0"
LWL 34'-0"
BEAM 12'-4"
DRAFT 3'-6"
FREEBOARD:
FWD 8'-0"
LEAST 3'-4"
AFT 4'-4"
BENFORD DESIGN GROUP
P.O. BOX 447
ST. MICHAELS MD 21663
(410) 745-3235
329 I.W.F

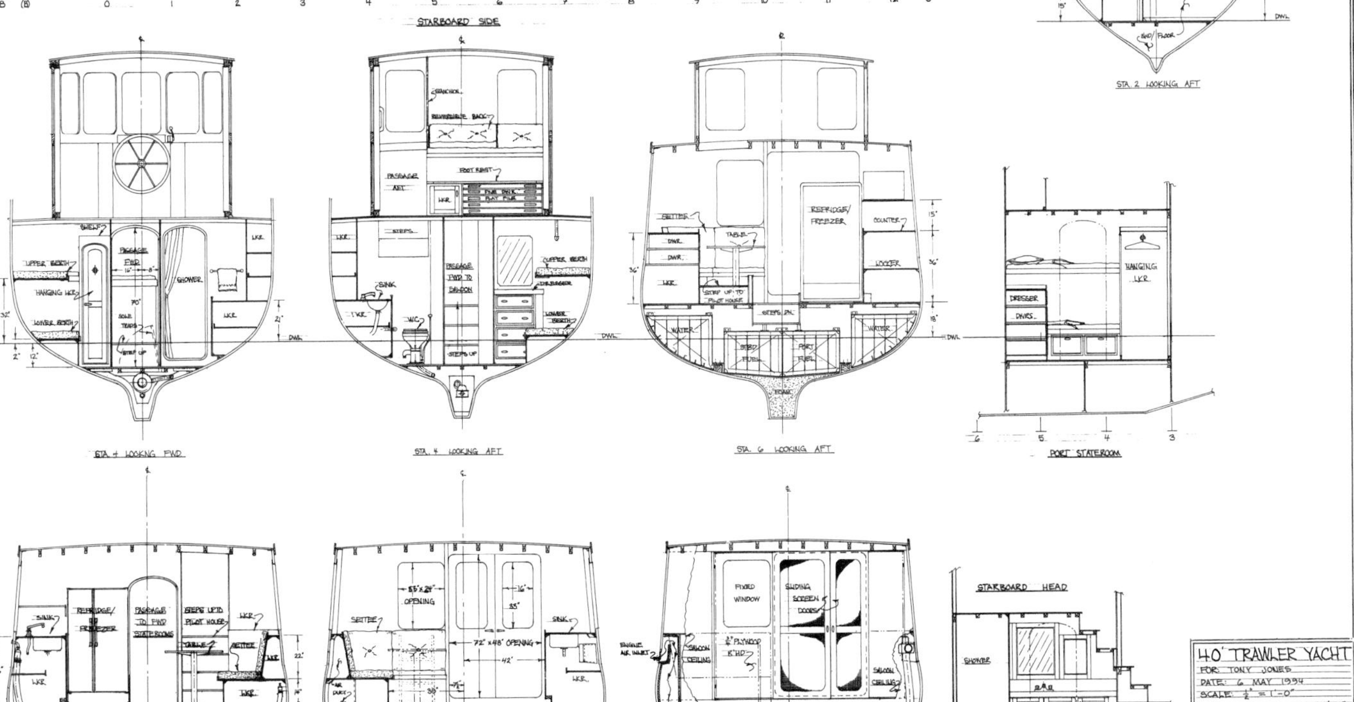

40' TRAWLER YACHT
FOR: TONY JONES
DATE: 6 MAY 1994
SCALE: ½" = 1'-0"

LOA	40'-0"
LWL	34'-0"
BEAM	12'-4"
DRAFT	3'-6"
FREEBOARD:	
FWD	8'-0"
LEAST	3'-4"
AFT	4'-4"

BENFORD DESIGN GROUP
P.O. BOX 447
ST. MICHAELS MD 21663
(410) 745-3235
329 T.W.F.

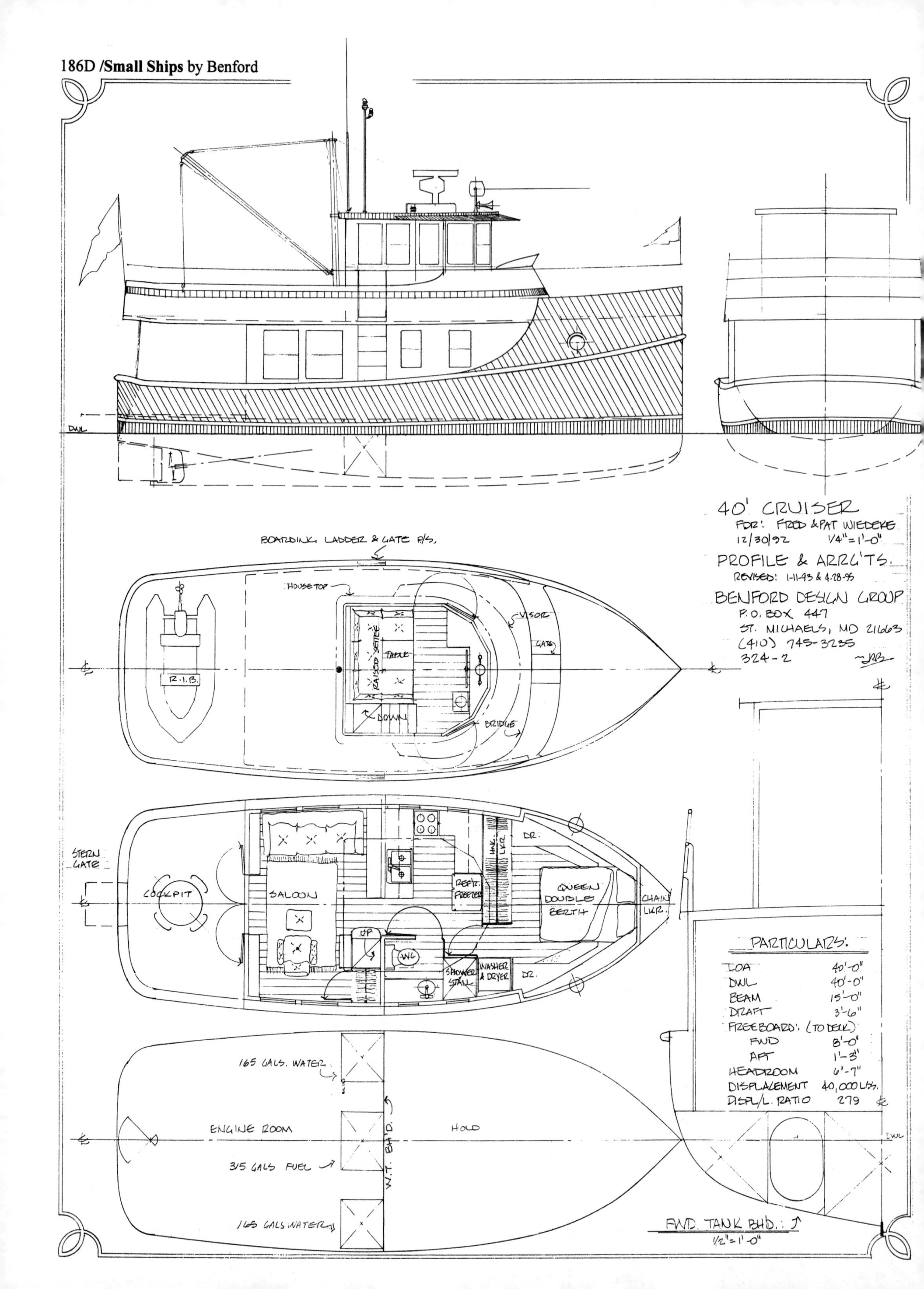

40' CRUISER
FOR: FRED & PAT WIEDEKE
12/30/92 1/4"=1'-0"
PROFILE & ARRG'TS.
REVISED: 1-11-93 & 4-28-95
BENFORD DESIGN GROUP
P.O. BOX 447
ST. MICHAELS, MD 21663
(410) 745-3235
324-2
DWL
BOARDING LADDER & GATE P/S.
HOUSETOP
VISOR
GATE
RAISED SETTEE
TABLE
DOWN
BRIDGE
R.I.B.
STERN GATE
COCKPIT
SALOON
UP
WC
SHOWER STALL
WASHER & DRYER
REFR. FREEZER
HNG. LKR.
DR.
QUEEN DOUBLE BERTH
CHAIN LKR.
PARTICULARS:
LOA 40'-0"
DWL 40'-0"
BEAM 15'-0"
DRAFT 3'-6"
FREEBOARD: (TO DECK)
FWD 8'-0"
AFT 1'-3"
HEADROOM 6'-7"
DISPLACEMENT 40,000 LBS.
DISPL/L. RATIO 279
165 GALS. WATER
ENGINE ROOM
315 GALS FUEL
W.T. BHD.
HOLD
165 GALS WATER
FWD. TANK BHD.
1/2"=1'-0"

44 & 55' Trawler Yachts

Designs Number 79 & 95
1971 & 1972

These two designs are of a type that became popular first in the Pacific Northwest, in large measure through the design work and innovation of Bill Garden. I lived in the Puget Sound area for eighteen years and had a lot of exposure to his designs and work, plus several delightful visits with him on his island office. He has an enviable situation and has earned his place in the history of yacht design.

44' Trawler Yacht

This design, along with our 35-footer ***Strumpet***, was featured in Bob Beebe's **Voyaging Under Power**. I was honored to be chosen to have a couple of our designs included in this pioneering work. She's a comfortable, husky, and seaworthy cruiser designed for long passages.

There are two alternate engine positions. The original has the engine in a trunk which has four opening sides for access, including one into the guest stateroom. The alternative has the engine slid far enough aft to be under the saloon, allowing for turning the former trunk into some locker space. I'd expect an economical cruising speed to be in the range of seven to eight knots. One builder reported getting 10 knots in a lightly fitted out version, but I'd not expect this speed with any consistency without changing the hull form to flatten the run.

Both staterooms could have private access to the head by repositioning the master stateroom door to the after end of the passageway by the head and adding a separate door into the head from the guest stateroom.

She has a spacious layout, with berths for five, washer and dryer, plenty of stowage space. With fuel economy in mind, this handsome craft will provide her owner with an economical and capable home afloat.

A couple of the preliminary versions of the 44 are also included, to give some other ideas of how this size vessel could be laid out. These were done in 1968 and 1969 for a company in Tacoma to help show the type of boats they would like to build.

55' Trawler Yacht *Lady K*

The commission for this design came from a couple who had seen our 44 and wanted a bigger version for their cruising. The stretching of the boat, while giving her a flatter run and fuller stern, makes running into semi-displacement speeds a bit easier. Also, having a longer and more slender hull form will make for nice motion in a seaway.

The layout of the accommodation plan is similar with the spaces expanded. A workbench has been added along the passageway, close to the engine room entrance.

A cruising speed of 10 knots will give a 700-mile range on the 1,000 gallon fuel tanks. Main saloon and galley is aft within the walk-around deck. The huge head with a double-sink, shower and tub is amidships opposite the workbench, pantry, washer and dryer, and adjacent to the stateroom with upper and lower berths. The master stateroom, with a double berth, two hanging lockers and two dressers is forward.

A potential rearrangement of the accommodations would be to move the guest stateroom aft and the head forward, giving room to provide a private head for each of the staterooms.

Particulars:	**44 (#79)**	**55 (#95)**
Length overall	44'-0"	55'-0"
Length designed waterline	42'-0"	53'-0"
Beam	15'-1"	15'-0"
Draft	5'-0"	5'-0"
Freeboard:		
Forward	7'-6½"	7'-6"
Least	4'-0½"	4'-0"
Aft	4'-6"	4'-6"
Displacement, cruising trim*	45,400 lbs.	56,000 lbs.
Displacement-length ratio	274	168
Sail area (riding sail)	150 sq. ft.	179 sq. ft.
Prismatic coefficient	0.58	0.64
Pounds per inch immersion		3,060
Water tankage	325 Gals.	180 Gals.
Fuel tankage	1,000 Gals.	1,000 Gals.
Headroom	6'-3" to 6'-5"	6'-0" to 6'-3"
GM		6'

***CAUTION:** The displacement quoted here is for the boat in cruising trim. That is, with the fuel and water tanks filled, the crew on board, as well as the crews' gear and stores in the lockers. This should not be confused with the "shipping weight" often quoted as "displacement" by some manufacturers. This should be taken into account when comparing figure and ratios between this and other designs.

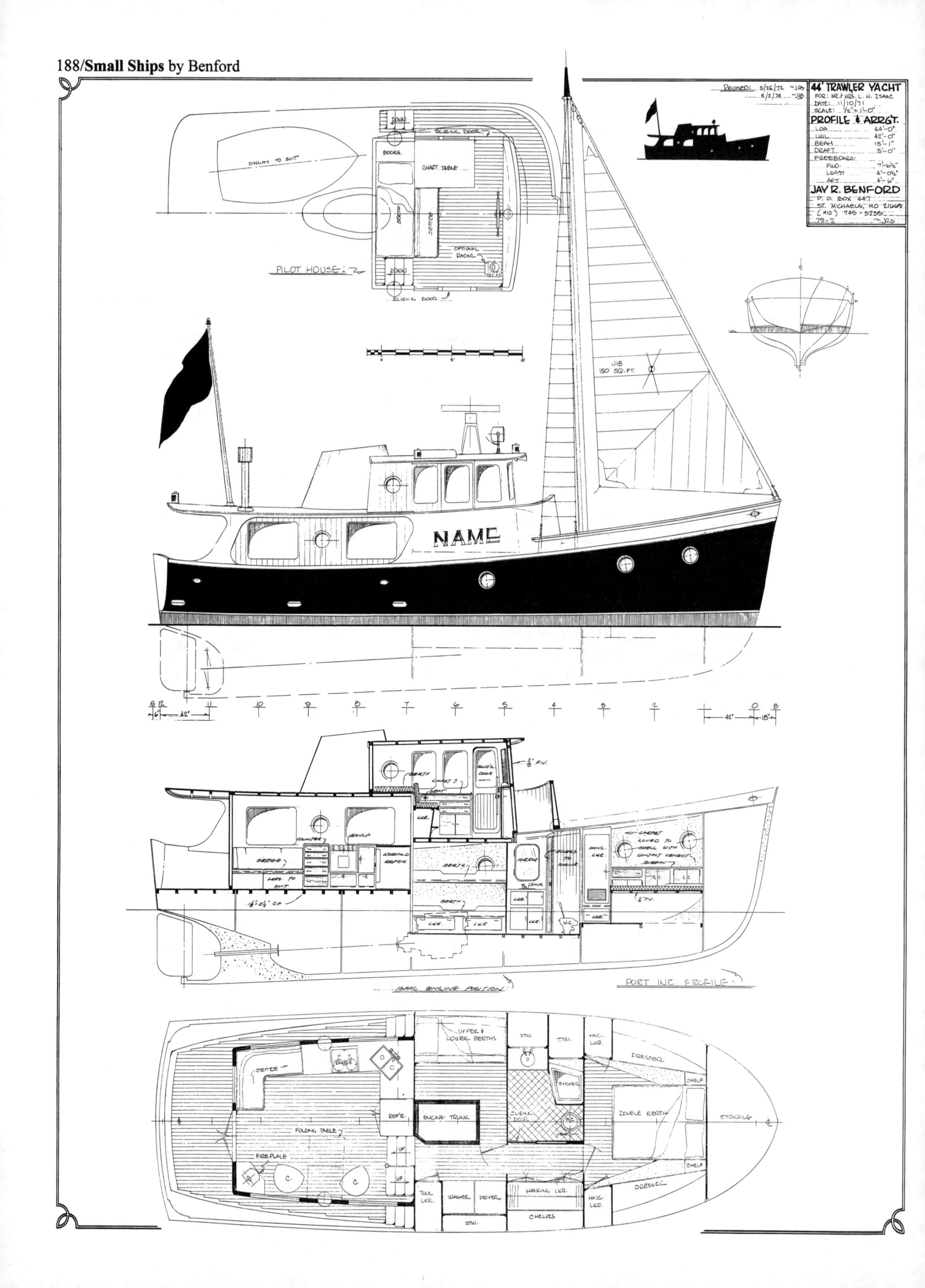
REVISED: 5/25/72 ~JRB
5/2/74 ~JRB
44' TRAWLER YACHT
FOR: MR. & MRS. L. H. ISAAC
DATE: 11/10/71
SCALE: 1/2" = 1'-0"
PROFILE & ARRGT.
LOA 44'-0"
LWL 42'-0"
BEAM 15'-1"
DRAFT 5'-0"
FREEBOARD:
FWD. 7'-6½"
LEAST 4'-0½"
AFT 4'-6"
JAY R. BENFORD
P. O. BOX 447
ST. MICHAELS, MD 21663
(410) 745-3235
79-2 ~JRB
DINGHY TO SUIT
DOWN
SLIDING DOOR
BOOKS
CHART TABLE
BERTH
SETTEE
OPTIONAL RADAR
PILOT HOUSE
JIB
150 SQ. FT.
NAME
ENGINE TRUNK
FOLDING TABLE
FIREPLACE
SETTEE
REF'R.
UP
TOOL LKR.
WASHER
DRYER
STW.
HANGING LKR.
SHELVES
HNG. LKR.
UPPER & LOWER BERTHS
SLIDING DOOR
SHOWER
DOUBLE BERTH
DRESSER
SHELF
STOWAGE
PORT INB. PROFILE
ISAAC ENGINE POSITION

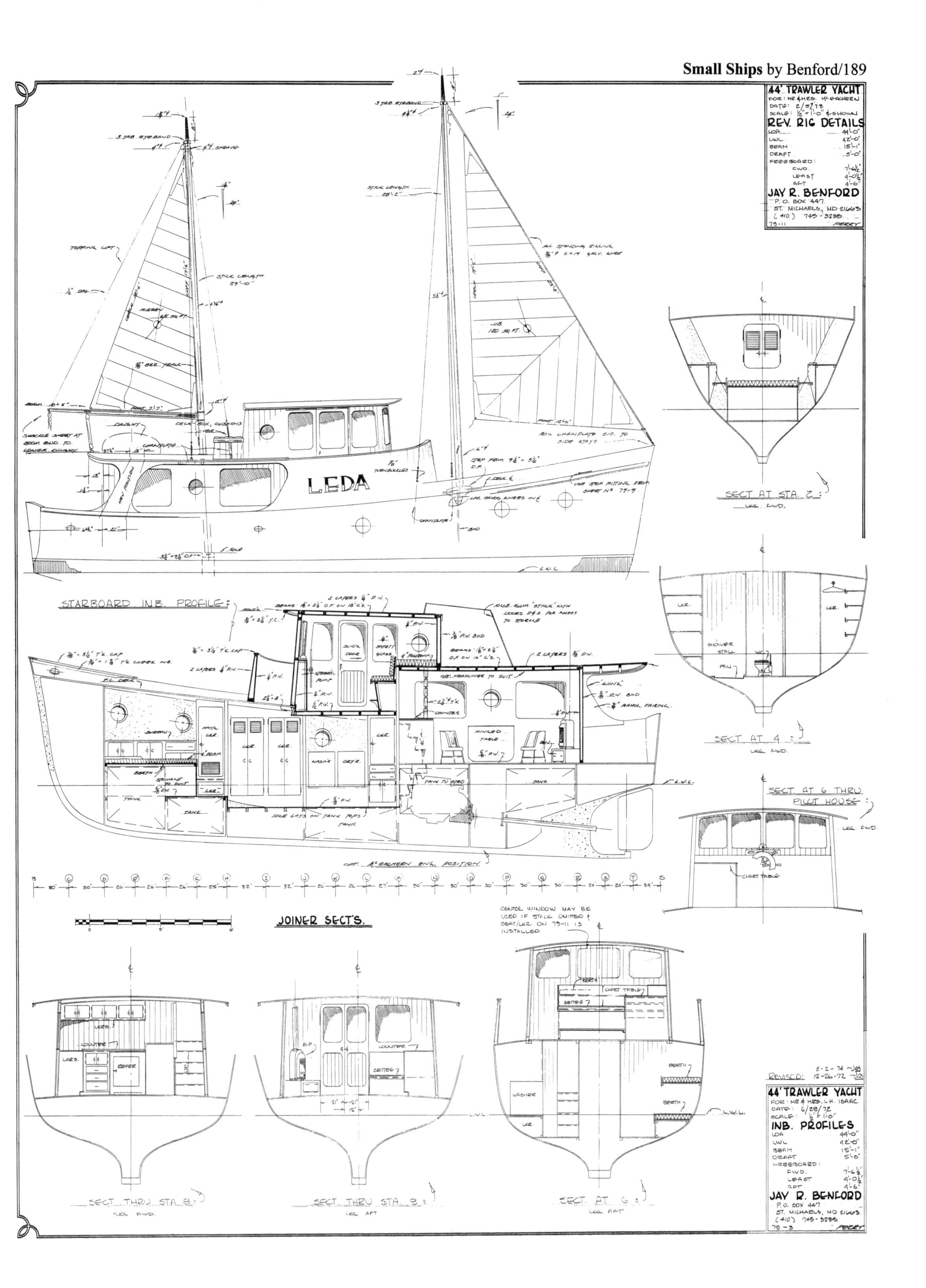
44' TRAWLER YACHT
FOR: MR & MRS. McEACHERN
DATE: 2/5/73
SCALE: 1/2"=1'-0" & SHOWN
REV. RIG DETAILS
LOA 44'-0"
LWL 42'-0"
BEAM 15'-1"
DRAFT 5'-0"
FREEBOARD:
FWD. 7'-6 1/2"
LEAST 4'-0 1/2"
AFT 4'-6"
JAY R. BENFORD
P. O. BOX 447
ST. MICHAELS, MD 21663
(410) 745-3235
73-11
LEDA
MIZZEN
JIB
120 SQ. FT.
TOPPING LIFT
STICK LENGTH
DINGHY
CHAINPLATE
SECT AT STA. 2
LKG. FWD.
SECT AT 4
LKG. FWD.
SECT AT 6 THRU
PILOT HOUSE
LKG. FWD.
CHART TABLE
STARBOARD INB. PROFILE
OPT. McEACHERN ENG. POSITION
TANK
JOINER SECT'S.
CENTER WINDOW MAY BE
USED IF STACK OMITTED &
SEAT/LKR ON 73-11 IS
INSTALLED
SECT THRU STA 8
LKG. FWD.
SECT THRU STA 3
LKG. AFT
SECT AT 6
LKG. AFT
BERTH
WASHER
REVISED: 5-2-74
12-26-72
44' TRAWLER YACHT
FOR: MR & MRS. L.H. ISAAC
DATE: 6/28/72
SCALE: 1/2"=1'-0"
INB. PROFILES
LOA 44'-0"
LWL 42'-0"
BEAM 15'-1"
DRAFT 5'-0"
FREEBOARD:
FWD. 7'-6 1/2"
LEAST 4'-0 1/2"
AFT 4'-6"
JAY R. BENFORD
P.O. BOX 447
ST. MICHAELS, MD 21663
(410) 745-3235
73-3

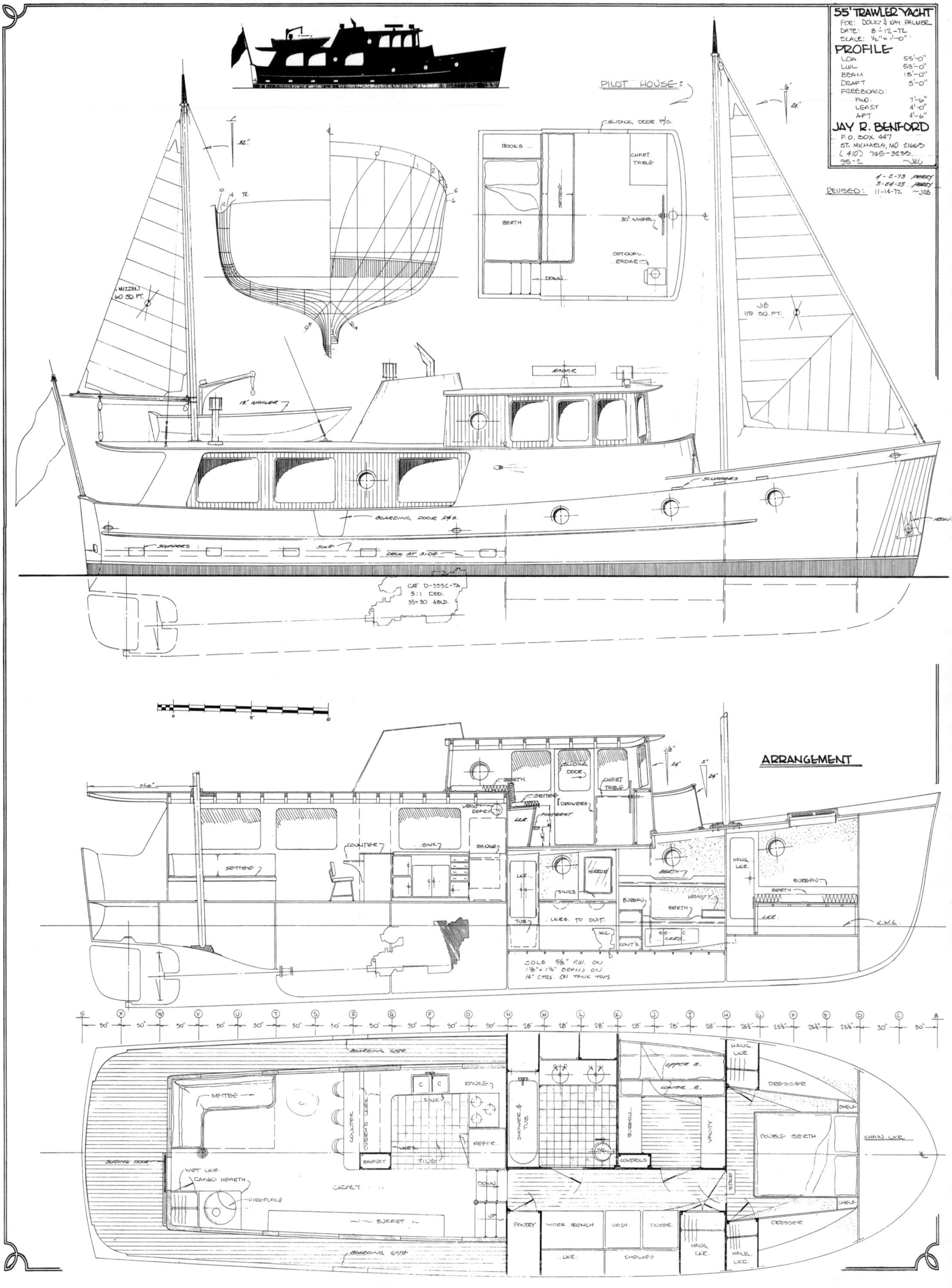

55' TRAWLER YACHT
FOR: DOUG & KAY PALMER
DATE: 8-12-72
SCALE: 1/2" = 1'-0"

PROFILE

LOA	55'-0"
LWL	53'-0"
BEAM	15'-0"
DRAFT	5'-0"
FREEBOARD:	
FWD.	7'-6"
LEAST	4'-0"
AFT	4'-6"

JAY R. BENFORD
P.O. BOX 447
ST. MICHAELS, MD 21663
(410) 745-3235
95-2 ~JRB

REVISED: 11-14-72 ~JRB
1-2-73 PERRY
3-24-73 PERRY

PILOT HOUSE

SLIDING DOOR P/S
BOOKS
CHART TABLE
SETTEE
BERTH
30" WHEEL
OPTIONAL RADAR
DOWN

MIZZEN 60 SQ. FT.
JIB 119 SQ. FT.
13' WHALER
RADAR
SCUPPERS
BOARDING DOOR P/S
SOLE
DECK AT SIDE

CAT D-353C-TA
3:1 RED.
35 x 30 4BLD.

ARRANGEMENT

BERTH
SETTEE
DRAWERS
FOOTREST
CHART TABLE
SLIDING DOOR
VENT REFER
COUNTER
SINK
RANGE
SETTEE
LKR.
MIRROR
SINKS
TUB
LKRS. TO SUIT
W.C.
CONT'S
BUREAU
BERTH
VANITY
HANG LKR.
BUREAU
BERTH
LKR.
L.W.L.

SOLE 5/8" P.W. ON
1 1/8" x 1 1/2" BEAMS ON
14" CTRS. ON TANK TOPS

BOARDING GATE
SETTEE
COUNTER
OVERHD LKRS.
RANGE
SINK
REFER
EXHAUST
TILE
LKRS.
SHOWER & TUB
BUREAU
VANITY
UPPER B.
LOWER B.
HANG LKR.
DRESSER
SHELF
DOUBLE BERTH
CHAIN LKR.
CONTROLS
STEP
SLIDING DOOR
WET LKR.
CAKED HEARTH
FIREPLACE
CARPET
DOWN
UP
BUFFET
PANTRY
WORK BENCH
WASH
DRYER
HANG LKR.
SHELF
DRESSER
HANG LKR.
LKR.
SHELVES
BOARDING GATE

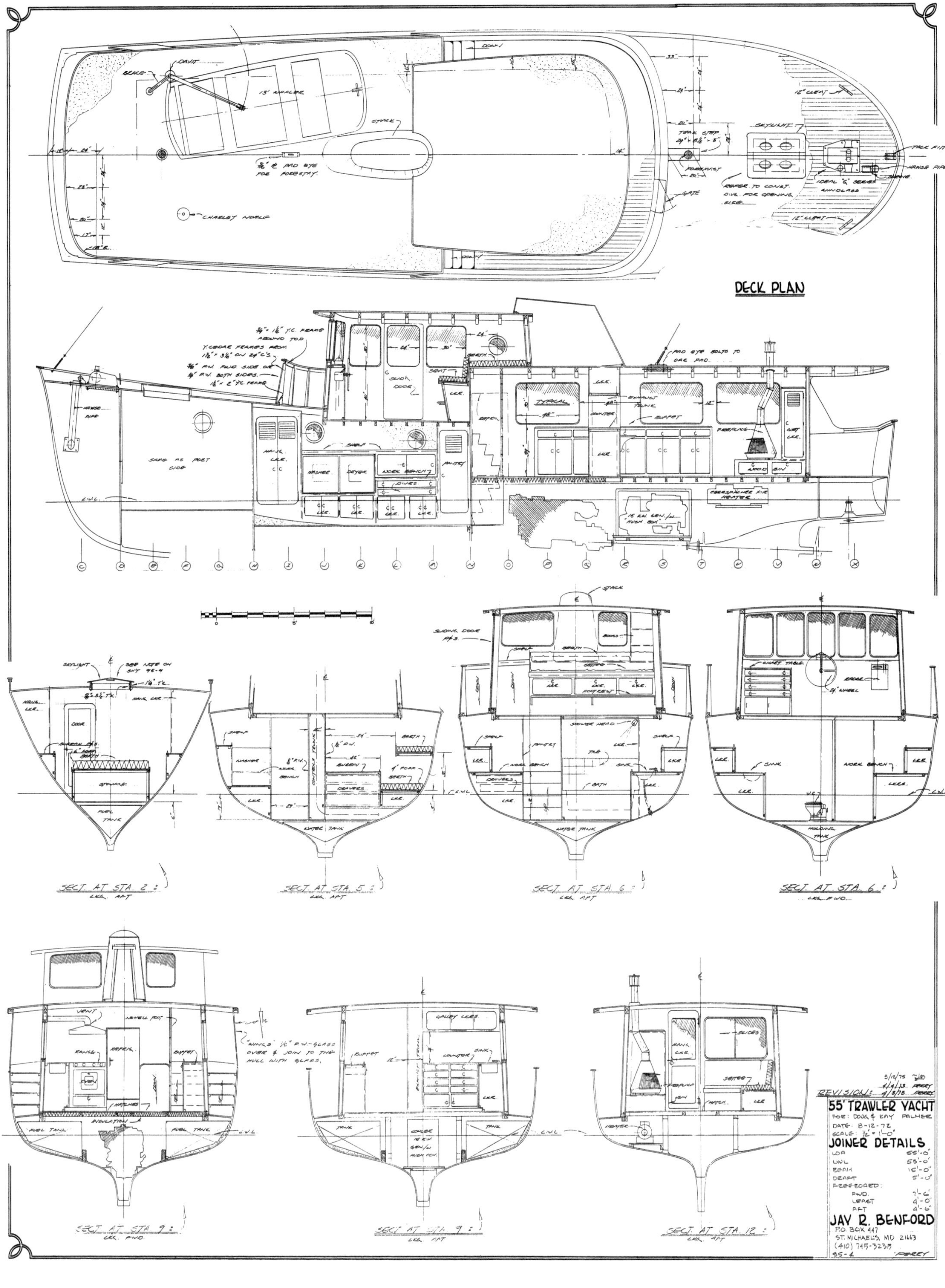

DECK PLAN
13' WHALER
DAVIT
CHARLEY NOBLE
SKYLIGHT
WINDLASS
HAWSE PIPE
SAME AS PORT SIDE
WASHER
DRYER
WORK BENCH
PANTRY
FIREPLACE
WOOD BIN
SECT. AT STA. 2: LKG. AFT
SECT. AT STA. 5: LKG. AFT
SECT. AT STA. 6: LKG. AFT
SECT. AT STA. 6: LKG. FWD.
SECT. AT STA. 9: LKG. FWD.
SECT. AT STA. 9: LKG. AFT
SECT. AT STA. 12: LKG. AFT
REVISION:
55' TRAWLER YACHT
FOR: DOUG & KAY PALMER
DATE: 8-12-72
SCALE: 1/2" = 1'-0"
JOINER DETAILS
LOA 55'-0"
LWL 53'-0"
BEAM 16'-0"
DRAFT 5'-0"
FREEBOARD:
FWD. 7'-6"
LEAST 4'-0"
AFT 4'-6"
JAY R. BENFORD
P.O. BOX 447
ST. MICHAELS, MD 21663
(410) 745-3235
55-4

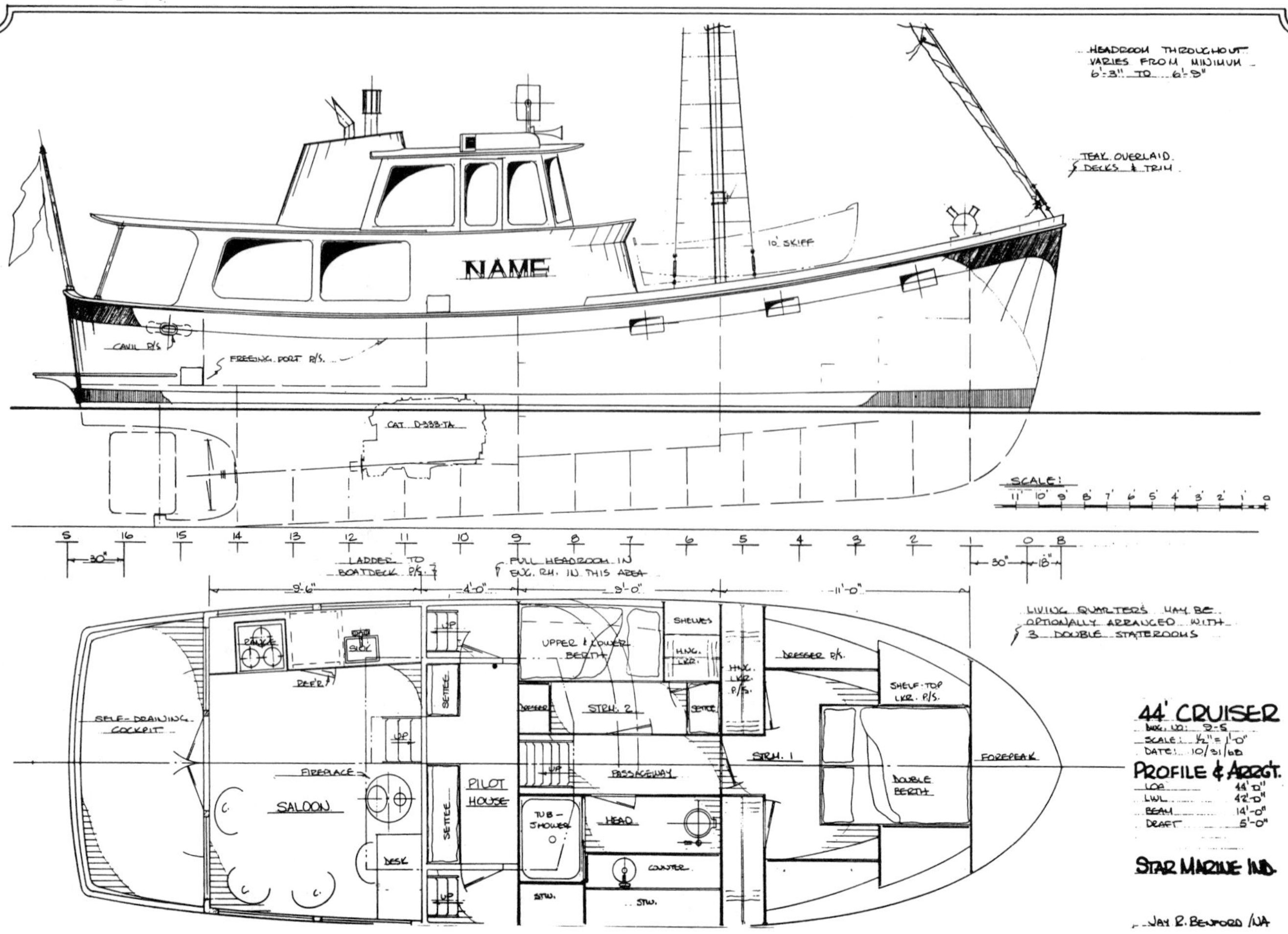

HEADROOM THROUGHOUT VARIES FROM MINIMUM 6'-3" TO 6'-9"

TEAK OVERLAID DECKS & TRIM

NAME

10' SKIFF

CAVIL P/S.

FREEING PORT P/S.

CAT. D-333-TA

SCALE: 11' 10' 9' 8' 7' 6' 5' 4' 3' 2' 1' 0

S 16 15 14 13 12 11 10 9 8 7 6 5 4 3 2 1 0 B

30" 18"

LADDER TO BOATDECK P/S.

FULL HEADROOM IN ENG. RM. IN THIS AREA

9'-6" 4'-0" 5'-0" 11'-0"

LIVING QUARTERS MAY BE OPTIONALLY ARRANGED WITH 3 DOUBLE STATEROOMS

SELF-DRAINING COCKPIT

RANGE SINK REF'R. UP SETTEE FIREPLACE SALOON DESK PILOT HOUSE UP

UPPER & LOWER BERTH SHELVES HNG. LKR. DRESSER STRM. 2 SETTEE PASSAGEWAY TUB-SHOWER HEAD COUNTER STW.

HNG. LKR. P/S. DRESSER P/S. SHELF-TOP LKR. P/S. STRM. 1 DOUBLE BERTH FOREPEAK

44' CRUISER
DWG. NO: S-5
SCALE: 1/2" = 1'-0"
DATE: 10/31/68
PROFILE & ARRG'T.
LOA 44'-0"
LWL 42'-0"
BEAM 14'-0"
DRAFT 5'-0"

STAR MARINE IND.

JAY R. BENFORD /NA

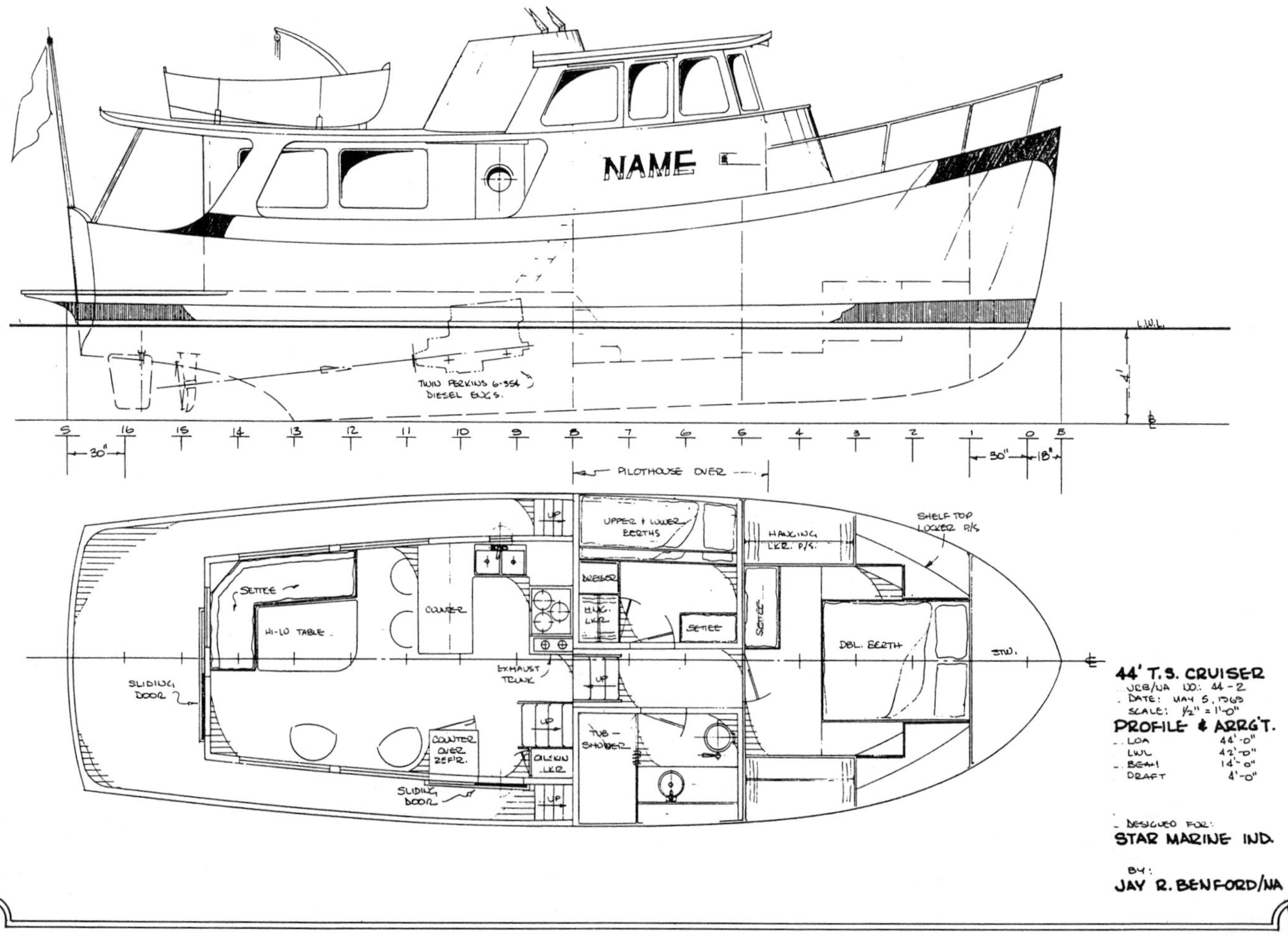

NAME

L.W.L.

TWIN PERKINS 6-354 DIESEL ENGS.

4'

S 16 15 14 13 12 11 10 9 8 7 6 5 4 3 2 1 0 B

30" 18"

PILOTHOUSE OVER

SETTEE HI-LO TABLE COUNTER SLIDING DOOR EXHAUST TRUNK COUNTER OVER REFR. OILSKIN LKR. SLIDING DOOR UP

UPPER & LOWER BERTHS DRESSER HNG. LKR. SETTEE TUB-SHOWER HANGING LKR. P/S. SETTEE SHELF TOP LOCKER P/S. DBL. BERTH STW.

44' T.S. CRUISER
JRB/NA NO: 44-2
DATE: MAY 5, 1969
SCALE: 1/2" = 1'-0"
PROFILE & ARRG'T.
LOA 44'-0"
LWL 42'-0"
BEAM 14'-0"
DRAFT 4'-0"

DESIGNED FOR:
STAR MARINE IND.

BY:
JAY R. BENFORD/NA

Little Sindbad

45' Motoryacht For Offshore Passagemaking

Design Number 246

1986

Most of our custom design commissions come from experienced yachtsmen. This one is a good example. The client had owned an 80-foot steel motoryacht for twenty-five years prior to commissioning this design. It had been built in Scandinavia, to his order, along the lines of a fishing trawler. During the course of his ownership, he'd cruised the boat a good deal and had a chance to thoroughly sort out the boat.

That experience had given him plenty of time to think about what his next boat should be. He had concluded that the same sort of boat in a smaller scale would be desirable. This would give him a boat that could be operated without the three paid crew aboard that the 80-footer had, and let him continue to cruise with his wife and/or friends aboard.

The styling of the boat went through a long evolution, with the very pleasing results shown on the drawings on the next page. She has all the character and charm of the larger boat, with the practicality and handiness of being able to be single-handed.

This design was one of the first ones on which we used our Fast Yacht computer design software. We had just recently purchased this and an HP engineering workstation and this was fortuitous timing. The computer let us keep making evolutionary changes to the size and shape of the boat as the concept evolved and grew to accommodate not only the cabins but the enlarged stores and tank capacities that were ultimately included.

The fuel tankage for the boat gives an estimated range sufficient for transoceanic cruising as shown on the curves below. The construction is very heavy, with one-quarter inch steel plating on the hull. The deckhouses and great cabin are aluminum construction.

The forward mast has ratlines and a crow's nest. This will provide a great place for getting better visibility over the horizon or close aboard in getting into a strange harbor or anchorage. A jib could be carried on the forestay, perhaps with a roller-furling gear, to aid as a steadying sail.

The after mast is an extended dry stack, similar to the ones I first designed in the '60s while doing tug designs for Foss Launch & Tug in Seattle. These proved excellent for getting the noise away from the crew and letting whatever soot might come out be blown away more readily. It is well braced and stayed, to provide supports for the poles for the flopper-stopper stabilizers. There is a bracket for holding the radar well elevated for better long range work and supports to hold up an awning over the after deck.

For those who want to cruise in the French canals, the stack can be built with a detachment joint to allow passage through the section with the lowest clearances.

Particulars:

Length overall		45'-6"
Length designed waterline		41'-7½"
Beam		14'-4"
Draft		5'-0"
Freeboard:	Forward	7'-0"
	Least	2'-8"
	Aft	7'-6"
Displacement, cruising trim		65,600 lbs.
Displacement-length ratio		406
Pounds per inch immersion		2,216
Water tankage		530 Gals.
Fuel tankage		1,475 Gals.
Headroom		6'-4" to 6'-8"
Ballast		6,000 lbs.
GM		3.5'

***CAUTION:** The displacement quoted here is for the boat in cruising trim. That is, with the fuel and water tanks filled, the crew on board, as well as the crews' gear and stores in the lockers. This should not be confused with the "shipping weight" often quoted as "displacement" by some manufacturers. This should be taken into account when comparing figures and ratios between this and other designs.

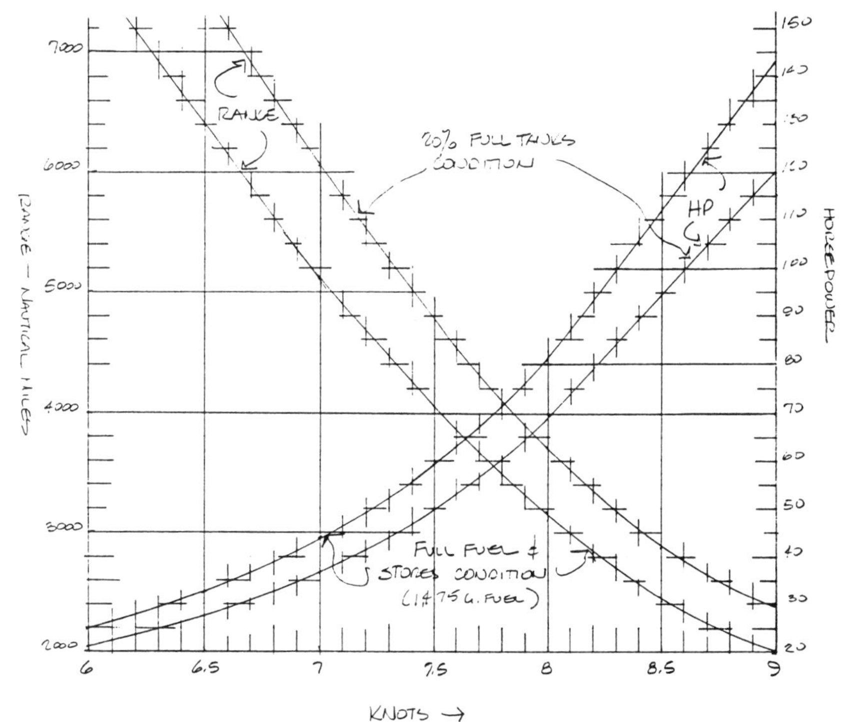

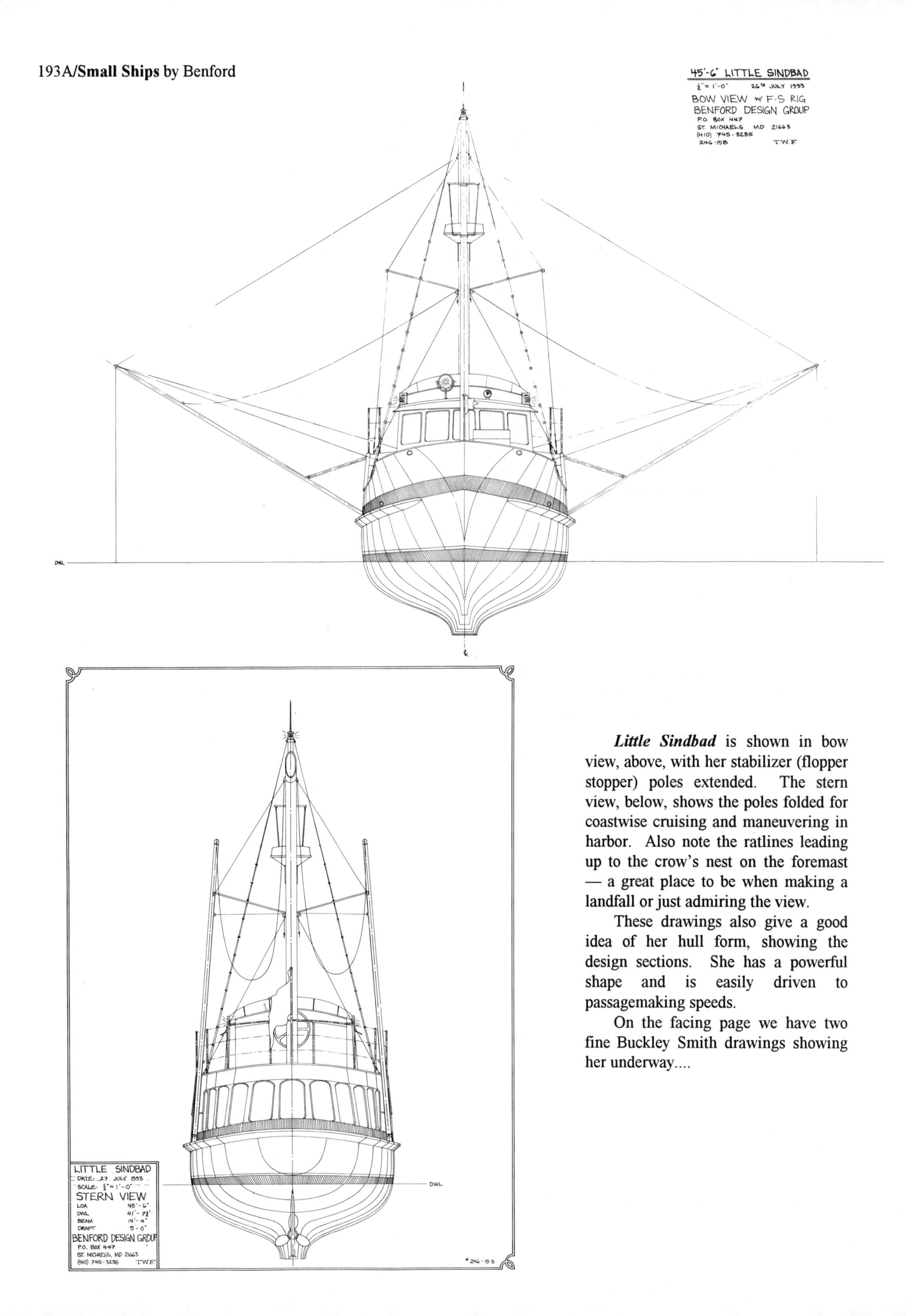

Little Sindbad is shown in bow view, above, with her stabilizer (flopper stopper) poles extended. The stern view, below, shows the poles folded for coastwise cruising and maneuvering in harbor. Also note the ratlines leading up to the crow's nest on the foremast — a great place to be when making a landfall or just admiring the view.

These drawings also give a good idea of her hull form, showing the design sections. She has a powerful shape and is easily driven to passagemaking speeds.

On the facing page we have two fine Buckley Smith drawings showing her underway....

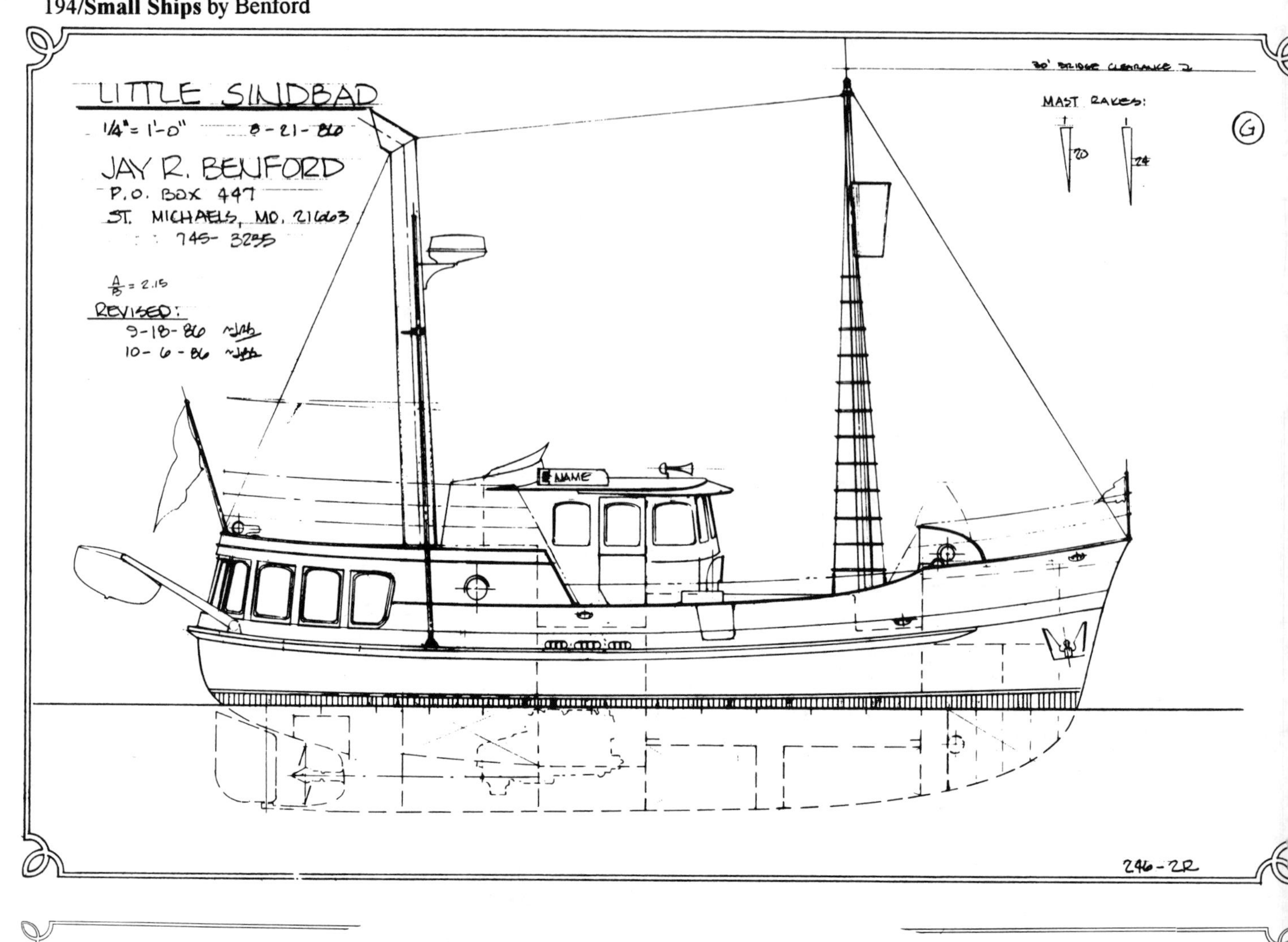
LITTLE SINDBAD
1/4" = 1'-0"
8-21-86
JAY R. BENFORD
P.O. BOX 447
ST. MICHAELS, MD. 21663
745-3235
A/B = 2.15
REVISED:
9-18-86
10-6-86
MAST RAKES:
G
NAME
246-2R

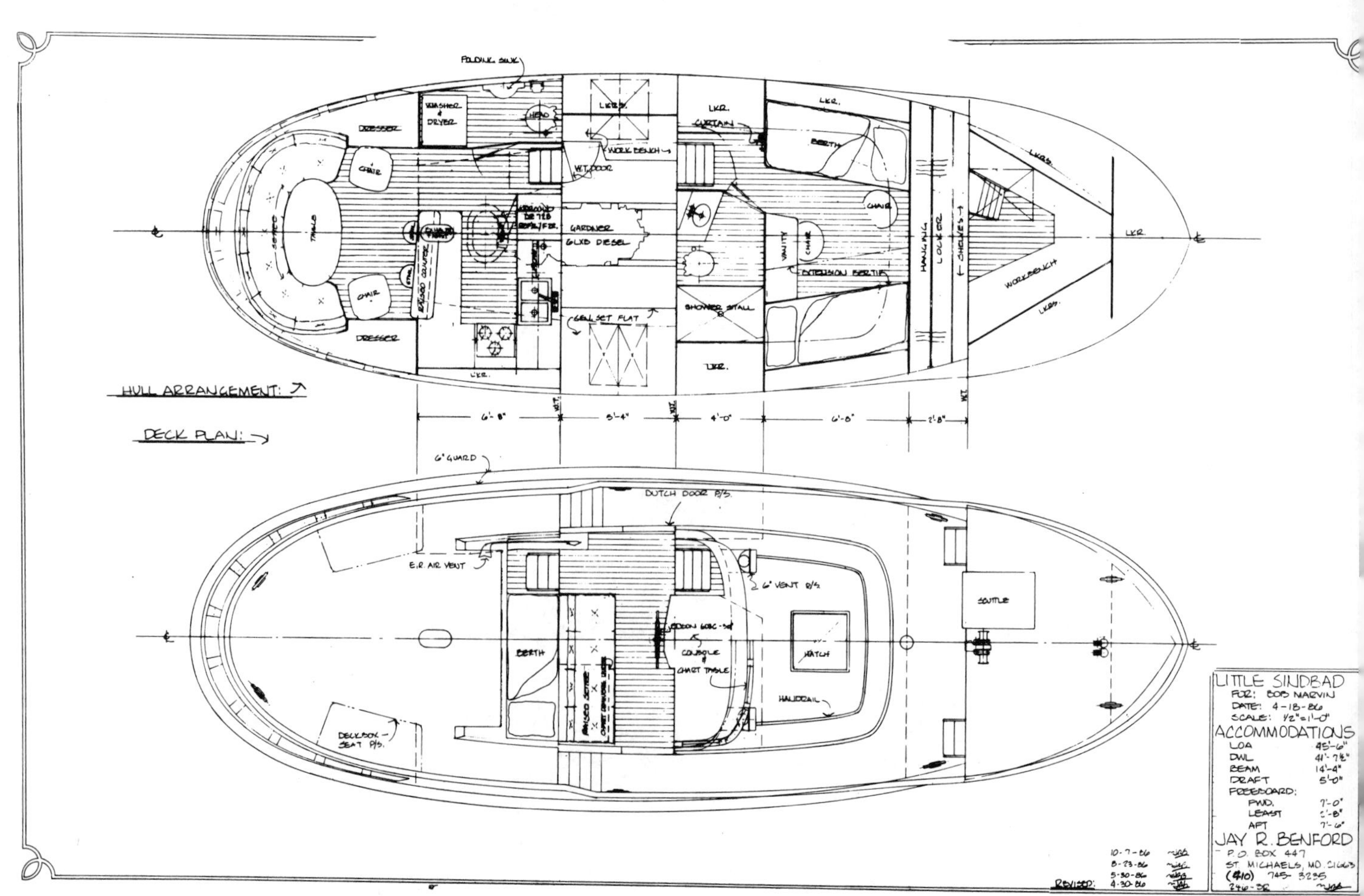
FOLDING SINK
WASHER & DRYER
DRESSER
HEAD
LKR.
WORK BENCH
W.T. DOOR
CHAIR
TABLE
GARDNER
6LXB DIESEL
GENSET FLAT
CURTAIN
BERTH
VANITY
EXTENSION BERTHS
SHOWER STALL
HANGING LOCKER
SHELVES
WORKBENCH
HULL ARRANGEMENT:
DECK PLAN:
6'-8"
5'-4"
4'-0"
6'-8"
2'-8"
6" GUARD
DUTCH DOOR P/S
E.R. AIR VENT
6" VENT P/S
SCUTTLE
BERTH
CONSOLE
CHART TABLE
HATCH
HANDRAIL
DECKBOX - SEAT P/S
LITTLE SINDBAD
FOR: BOB MARVIN
DATE: 4-18-86
SCALE: 1/2"=1'-0"
ACCOMMODATIONS
LOA 45'-6"
DWL 41'-7½"
BEAM 14'-4"
DRAFT 5'-0"
FREEBOARD:
FWD. 7'-0"
AFT 7'-6"
JAY R. BENFORD
P.O. BOX 447
ST. MICHAELS, MD. 21663
(410) 745-3235
REVISED:
10-7-86
8-23-86
5-30-86
4-30-86

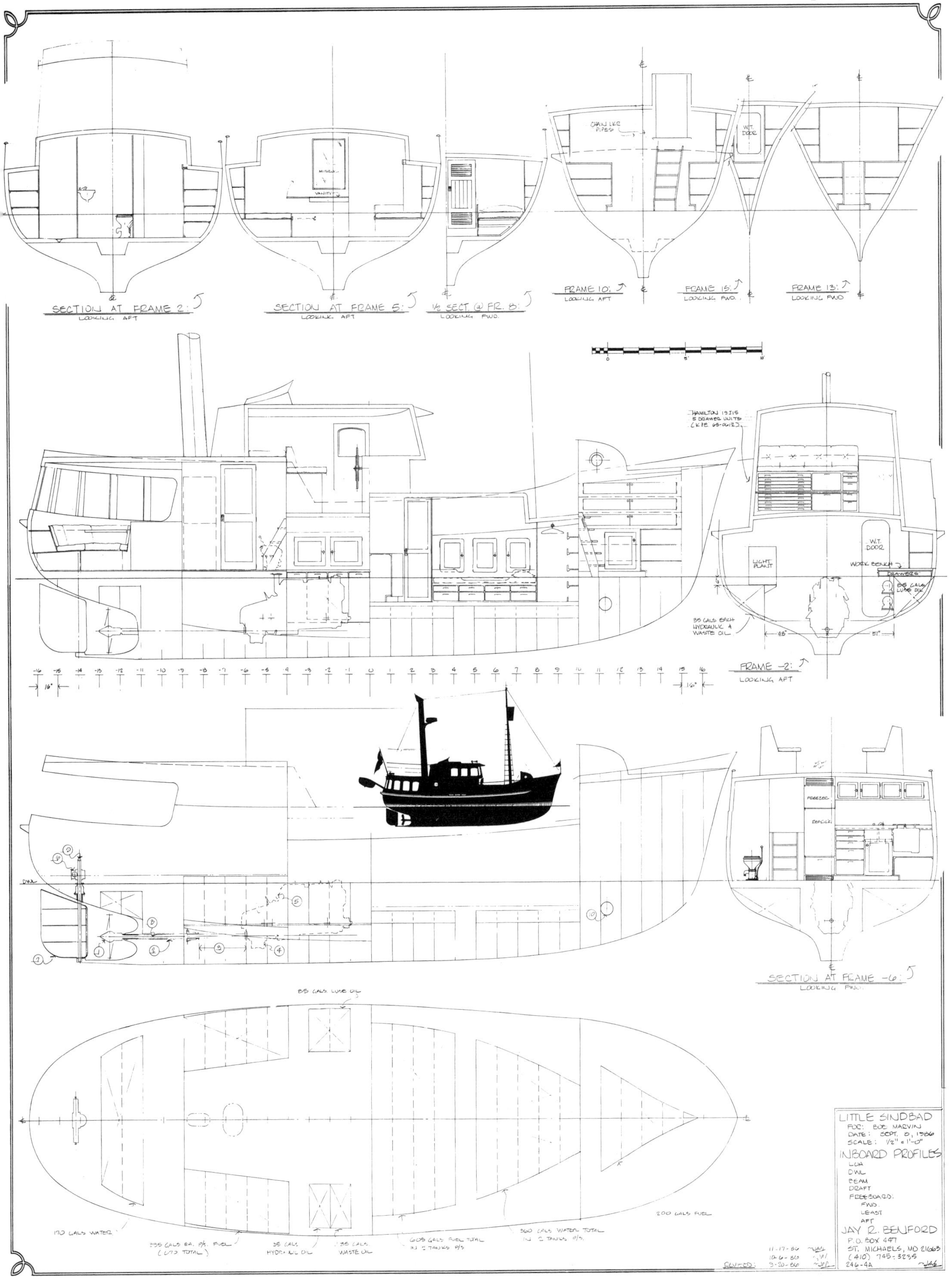

SECTION AT FRAME 2:
LOOKING AFT

SECTION AT FRAME 5:
LOOKING AFT

½ SECT. @ FR. 8:
LOOKING FWD.

FRAME 10:
LOOKING AFT

FRAME 15:
LOOKING FWD.

FRAME 13:
LOOKING FWD

CHAIN LKR PIPES

W.T. DOOR

HAMILTON 19515 5 DRAWER UNITS (KTE 65-0612)

W.T. DOOR

LIGHT PLANT

WORK BENCH

DRAWERS

85 GALS LUBE OIL

85 GALS EACH HYDRAULIC & WASTE OIL

FRAME -2:
LOOKING AFT

DWL

FREEZER

REFRIG.

SECTION AT FRAME -6:
LOOKING FWD.

85 GALS LUBE OIL

170 GALS WATER

335 GALS EA. P/S FUEL (670 TOTAL)

35 GALS HYDRAULIC OIL

35 GALS WASTE OIL

605 GALS FUEL TOTAL IN 2 TANKS P/S

360 GALS WATER TOTAL IN 2 TANKS P/S

200 GALS FUEL

REVISED: 11-17-86, 10-6-86, 9-20-86

LITTLE SINDBAD
FOR: BOB MARVIN
DATE: SEPT. 8, 1986
SCALE: ½" = 1'-0"

INBOARD PROFILES

LOA
DWL
BEAM
DRAFT
FREEBOARD:
FWD.
LEAST
AFT

JAY R. BENFORD
P.O. BOX 447
ST. MICHAELS, MD 21663
(410) 745-3235
246-4A

47' Motor Yacht

Design Number 247
1986

Shortly after doing the 45' ***Little Sindbad*** design, we did this conceptual design for a 47' Motor Yacht. It is an evolutionary development of ***Little Sindbad***'s hull, with a quite different approach to the design of the houses and accommodations.

The owner's cabin is forward with its own private head and good sized lockers. Under the pilothouse is the guest stateroom, with a head opposite on the passageway and the washer and dryer alongside. Up in the saloon, the galley is U-shaped to keep the cook out of the traffic.

The pilothouse has a long settee and room for a pilot berth on the shelf aft of it. The helm is well elevated and will have good forward visibility. The overhang on the front of the pilothouse will provide a place for chartwork. The forward trunk cabin top has enough length for a little sailing dinghy and larger ones can be carried on top of the saloon. The after hatch on the housetop is over a ladder for access to the housetop and small craft carried there.

The engine room is under the saloon, with comfortable sitting headroom. The integral tankage is outboard and the genset can be located to hook onto the main prop shaft to provide "get-home" capability.

The design was also done with the option of having a motorsailing rig, and this can be accommodated if desired. It's an option to adding stabilizers for passage making work and can help stretch the useful range of the fuel carried. This is a very capable motor yacht and would be a handsome addition to any harbor and make a comfortable offshore passage maker.

Particulars:

Length overall	47'-3"
Length designed waterline	43'-1½"
Beam	14'-11
Draft	5'-0"
Freeboard (to decks):	
Forward	5'-6"
Least	2'-2"
Aft	2'-9"
Displacement, cruising trim*	58,600 lbs.
Displacement-length ratio	326
Water tankage	600 Gals.
Fuel tankage	1,000 Gals.
Headroom	6'-7"
GM	4.6'

***CAUTION:** The displacement quoted here is for the boat in cruising trim. That is, with the fuel and water tanks filled, the crew on board, as well as the crews' gear and stores in the lockers. This should not be confused with the "shipping weight" often quoted as "displacement" by some manufacturers. This should be taken into account when comparing figures and ratios between this and other designs.

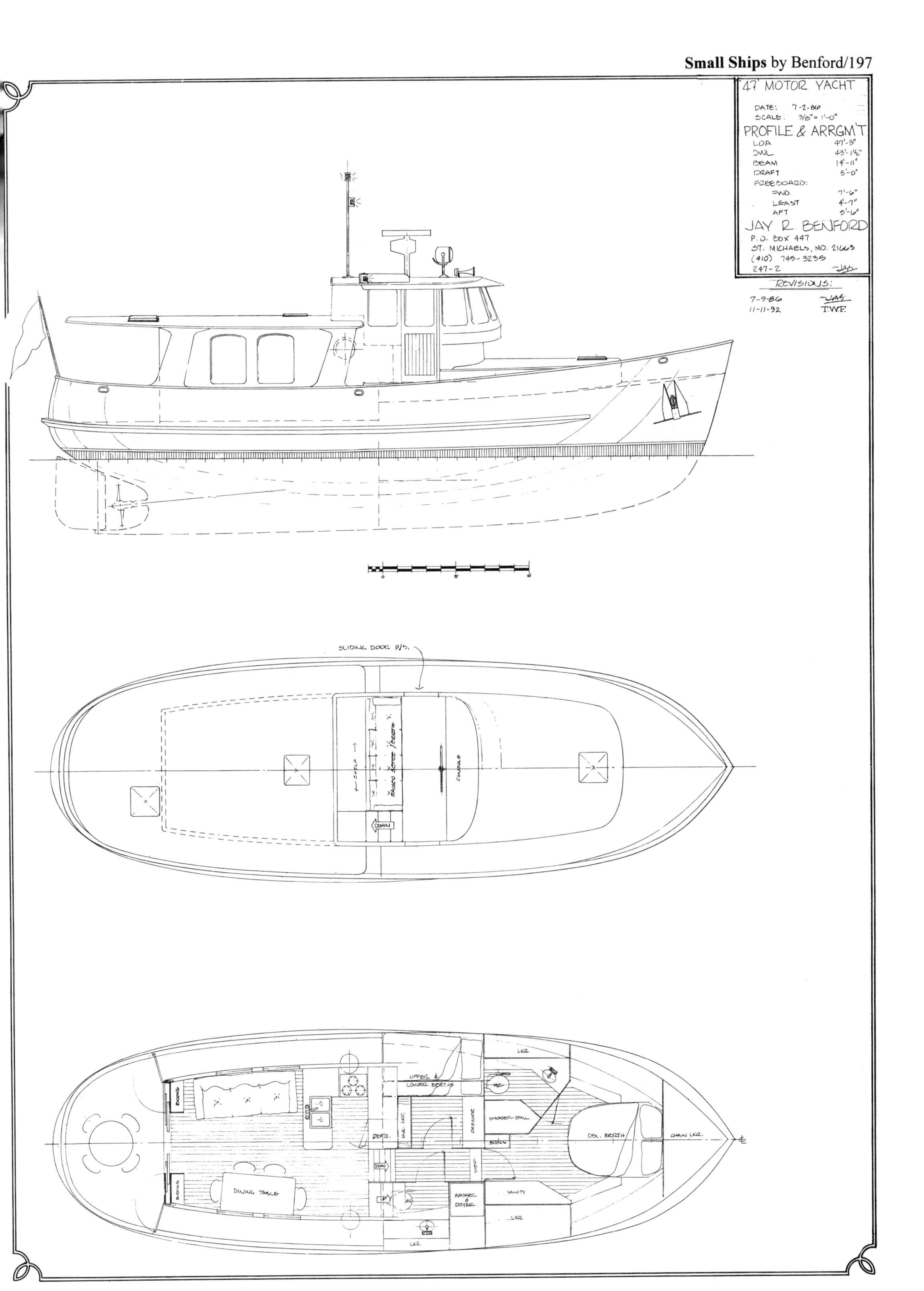

47' MOTOR YACHT
DATE: 7-2-86
SCALE: 3/8" = 1'-0"
PROFILE & ARRGM'T
LOA 47'-3"
DWL 43'-1½"
BEAM 14'-11"
DRAFT 5'-0"
FREEBOARD:
FWD. 7'-6"
LEAST 4'-7"
AFT 5'-6"
JAY R. BENFORD
P.O. BOX 447
ST. MICHAELS, MD. 21663
(410) 745-3235
247-2
REVISIONS:
7-9-86
11-11-92
T.W.F.
SLIDING DOOR P/S.
SHELF
RAISED SETTEE/BERTH
CONSOLE
DOWN
BOOKS
DINING TABLE
REFR.
UPPER & LOWER BERTHS
DRESSER
SHOWER STALL
DBL. BERTH
CHAIN LKR.
LKR.
VANITY
WASHER & DRYER
STEP

48' Trawler Yacht

Design Number 115

1974

This design draws on the fishing vessels of the Pacific Northwest for the inspiration for her design. These trollers have been going to sea for decades and been well proven. This version has ballast, like the commercial boats, and adds extensive accommodations instead of a cargo hold.

The design was conceived for a partnership that intended to mold the boats in Airex®-cored fiberglass with a rather heavy laminate schedule. We got as far as working up the lines and offsets and the laminate schedule when their partnership dissolved. We were sorry not to have the opportunity to complete the design. We'd be interested in talking about finishing her off with anyone else who might like the idea or a variation of her. She could easily be worked out for building in wood or steel if that would be the preferred method.

The saloon has the galley aft with the cook in an area out of the fore and aft traffic in the boat. There is a free-standing fireplace opposite, near the stairs to the master stateroom and some bookshelves. Forward in the saloon is the dining table and chairs, a couple of sofas and the helm. At the very forward end are the stairs leading to the forward staterooms and head. In the bow is a stateroom which could be used for a paid crew, some kids, or the overflow of guests. Just aft of that is the double stateroom for the guests with a head opposite. The head has a separate shower stall.

Amidships, under the saloon, is the engine room. There is a generous amount of room here for the establishment of a real workshop. There is also room to make her a twin screw and/or add a second generator. Access to the engine room is from the aft passageway which also leads to the ship's storeroom, master stateroom and head.

This sort of deep-sea vessel could easily carry a good-sized steadying sail rig and/or flopper-stoppers for easing the motion. She not only has good initial stability but the long range of stability needed to recover from a knockdown.

She can carry her small craft on the housetop, using the mast and boom to launch and retrieve them. The flying bridge can be used in nice weather and when a bit higher line of sight is wanted for making landfalls. Or, just for having a nice place to sit and enjoy being out in the sun.

On the main deck, the sheltered walk-around side decks will be good for shelter from rain, and give shade on sunny days. The large open after deck has plenty of room for a collection of deck chairs.

The speed and range curve will give an idea of what she requires in the way of an engine and how far she can go on her tankage. Here's real ocean-crossing potential....

Particulars:

Length overall	48'-0"
Length designed waterline	43'-4"
Beam	16'-2"
Draft	6'-6"
Freeboard:	
Forward	8'-9"
Least	6'-1"
Aft	6'-9"
Displacement, cruising trim	80,250 pounds
Displacement-length ratio	440
Ballast	16,000 pounds
Ballast ratio	20%
GM	3.5'
Prismatic coefficient	.62
Pounds per inch immersion	2,600 pounds
Water tankage	600 Gals.
Fuel tankage	1,500 Gals.
Headroom	6'-4" to 8'-0"

***CAUTION:** The displacement quoted here is for the boat in cruising trim. That is, with the fuel and water tanks filled, the crew on board, as well as the crews' gear and stores in the lockers. This should not be confused with the "shipping weight" often quoted as "displacement" by some manufacturers. This should be taken into account when comparing figures and ratios between this and other designs.

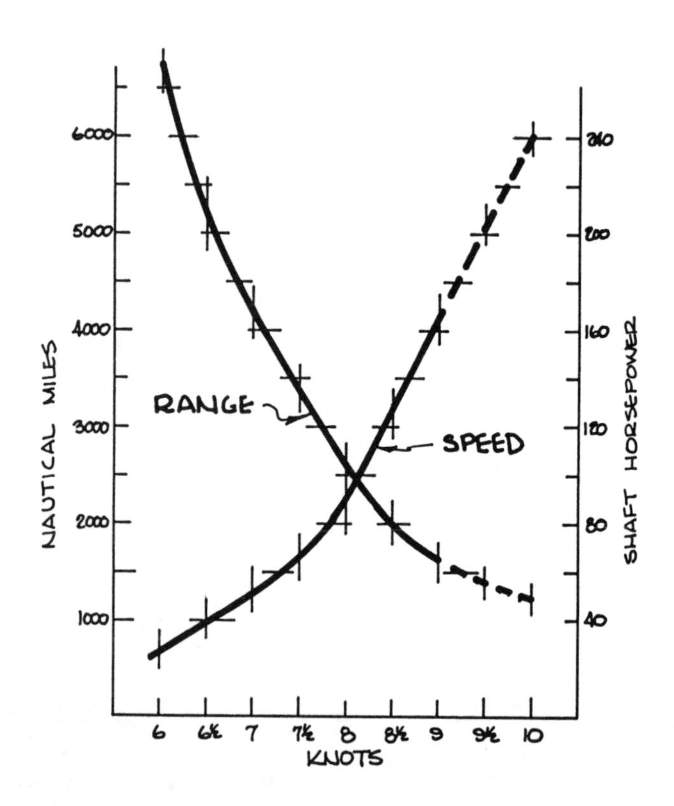

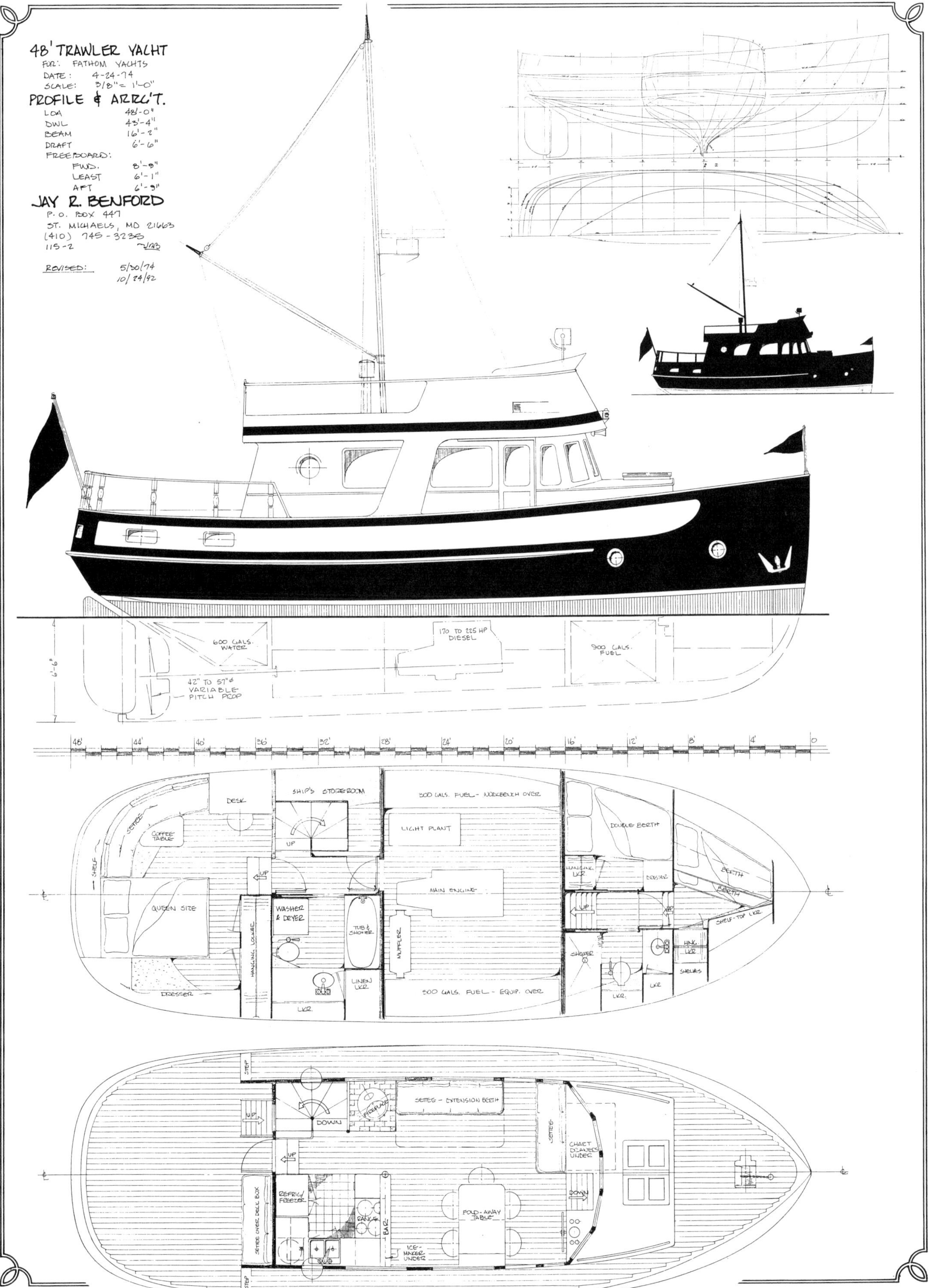
48' TRAWLER YACHT
FOR: FATHOM YACHTS
DATE: 4-24-74
SCALE: 3/8" = 1'-0"
PROFILE & ARR'G'T.
LOA 48'-0"
DWL 43'-4"
BEAM 16'-2"
DRAFT 6'-6"
FREEBOARD:
FWD. 8'-9"
LEAST 6'-1"
AFT 6'-9"
JAY R. BENFORD
P. O. BOX 447
ST. MICHAELS, MD 21663
(410) 745-3235
115-2
JRB
REVISED: 5/30/74
10/24/92
600 GALS. WATER
170 TO 225 HP DIESEL
900 GALS. FUEL
42" TO 57" Ø VARIABLE PITCH PROP
6'-6"
48' 44' 40' 36' 32' 28' 24' 20' 16' 12' 8' 4' 0
DESK
SHIP'S STOREROOM
300 GALS. FUEL - WORKBENCH OVER
LIGHT PLANT
DOUBLE BERTH
SETTEE
COFFEE TABLE
SHELF
UP
UP
HANG. LKR.
DRESSER
BERTH
BERTH
MAIN ENGINE
QUEEN SIZE
WASHER & DRYER
TUB & SHOWER
MUFFLER
SHELF-TOP LKR.
UP
SHOWER
HNG. LKR.
HANGING LOCKER
LINEN LKR.
SHELVES
DRESSER
300 GALS. FUEL - EQUIP. OVER
LKR.
LKR.
LKR.
STEP
UP
DOWN
FIREPLACE
SETTEE - EXTENSION BERTH
SETTEE
UP
CHART DRAWERS UNDER
REFRIG./FREEZER
DOWN
SETTEE OVER DECK BOX
RANGE
BAR
FOLD-AWAY TABLE
ICE-MAKER UNDER
STEP

54' Trawler Yacht

Design Number 302

1990

The 54' Trawler Yacht design was commissioned by a boatbuilding company for a prospective client who wanted a trawler yacht type that would operate in the 10-15 knot speed range. In looking at it again for inclusion in this book, I was reminded of my desire at the time to drop her freeboard by at least six inches to reduce the apparent bulk of her appearance. With a little work in revamping tank locations, we could probably cut a bit more height off her.

Instead of having a projecting swimming or boarding platform, we've built a recess into the transom. This makes for an area that can be used for these functions and still have the structural integrity of the main hull, with the lower guard rail wrapping all around the stern at this level. If there is any concern about being hit from astern, mooring stern-to, or keeping the maximum waterline length relative to overall length, this is a good solution. We've used the space underneath for a fuel tank. By letting this tank or the forward ones be used or filled, you can keep the boat in trim.

The cockpit has a set of stairs leading up to the after deck, for easy boarding. There is also a door and access into the master stateroom, which can be an escape hatch or a private exit for late night skinny dipping.

The cover over the cockpit is at table height, and has a folding drop-leaf for dining on the after deck. There is room for several chairs to be pulled up to the table, or it can be used without the leaf as a buffet.

The master stateroom has a queen double, large dressers, a full sized washer and dryer stacked in a closet, a walk-in closet and a private head with a two person Jacuzzi tub/shower.

Next forward is a passageway between the engine rooms for the twin engines. With the tanks in the bilge and forward of here, these spaces are open and less crowded with all the additional machinery and workbench that will be fitted.

The galley is below the pilothouse, and has a dinette. This space is more closed in than on the Coasters and not as well lit or ventilated. While larger windows could be fitted, they would not be in character with the rest of the boat.

The forward head is accessed off the passageway, for use both by the forward stateroom and anyone in the galley area. The forward stateroom has shown a twin berth layout, but there's room to have a double berth as an alternative.

The tall stacks alongside the pilothouse are actually twin stacks for the engine exhausts. There will also be room in them for vents for the hot air to rise out of the engine room. This stack is a solution to getting the exhaust high enough to let any soot be blown away. It also means that the noise is well above deck level and it won't be as noisy aboard as with a lower outlet. I first used this on some of the Foss Tugs in the middle 1960's and the operators liked the results.

The boat deck has rails around the forward part of it to keep the passengers safe. The crane/davit for the RIB is socketed into one of the aft corners of the deckhouse, to give it a rigid mounting place.

The Portuguese bridge wraps around the pilothouse, giving shelter to the decks there. It would have a gate in the center for access to the foredeck. The pilothouse has a notch in the after port corner for stairs to the boat deck. Inside the pilothouse, the raised settee has room for the crew to sit up where they can see where they're headed and to socialize in harbor or at anchor.

The saloon has a very open layout, with room for a variety of accommodation plans. In addition to the settees shown as built-ins, with storage underneath, several chairs would be in order. The aft bulkhead has twin sliding doors and screens to make the space usable all year. There are also wing doors on both sides, to close off the side decks from the after deck. These would be handy if the aft deck were to be enclosed for off season use and in foul weather to keep the rain or spray out.

Particulars:

Length overall	54'-0"
Length designed waterline	51'-3"
Beam	17'-0"
Draft	4'-6"
Freeboard:	
Forward	8'-9"
Amidships	6'-6"
Aft cockpit	1'-6"
Displacement, cruising trim*	97,000 lbs.
Displacement-length ratio	322
Water tankage, Gals.	660
Fuel tankage, Gals.	1,060
Headroom	6'-7"
GM	4.1'

***CAUTION:** The displacement quoted here is for the boat in cruising trim. That is, with the fuel and water tanks filled, the crew on board, as well as the crews' gear and stores in the lockers. This should not be confused with the "shipping weight" often quoted as "displacement" by some manufacturers. This should be taken into account when comparing figures and ratios between this and other designs.

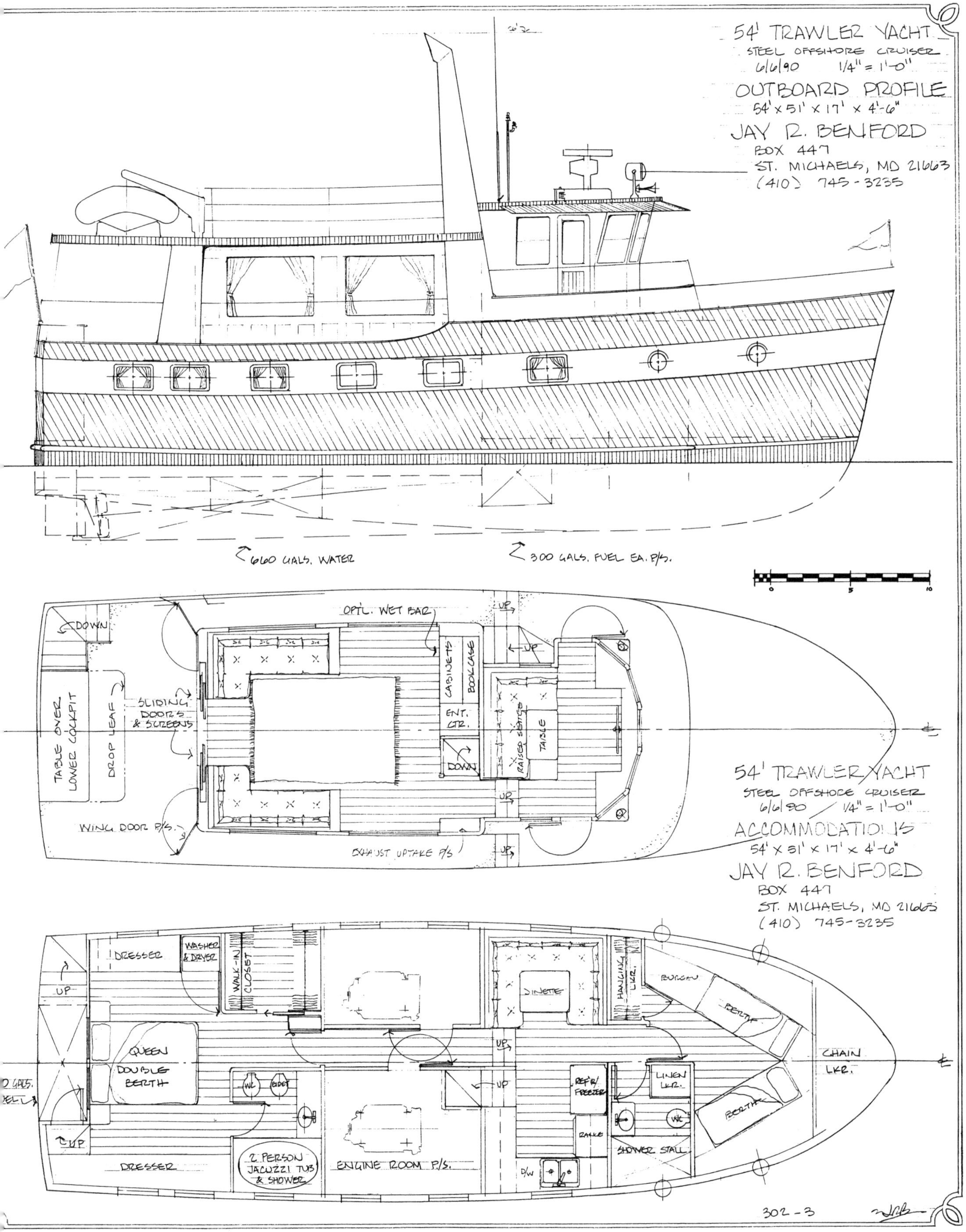

54' TRAWLER YACHT
STEEL OFFSHORE CRUISER
6/6/90 1/4" = 1'-0"
OUTBOARD PROFILE
54' x 51' x 17' x 4'-6"
JAY R. BENFORD
BOX 447
ST. MICHAELS, MD 21663
(410) 745-3235

660 GALS. WATER
300 GALS. FUEL EA. P/S.

0 5 10

DOWN
OPT'L. WET BAR
UP
CABINETS
BOOKCASE
ENT. CTR.
DOWN
RAISED SETTEE
TABLE
TABLE OVER LOWER COCKPIT
DROP LEAF
SLIDING DOORS & SCREENS
WING DOOR P/S.
EXHAUST UPTAKE P/S

54' TRAWLER YACHT
STEEL OFFSHORE CRUISER
6/6/90 1/4" = 1'-0"
ACCOMMODATIONS
54' x 51' x 17' x 4'-6"
JAY R. BENFORD
BOX 447
ST. MICHAELS, MD 21663
(410) 745-3235

DRESSER
WASHER & DRYER
WALK-IN CLOSET
UP
DINETTE
HANGING LKR.
BUREAU
BERTH
QUEEN DOUBLE BERTH
WC
BIDET
UP
CHAIN LKR.
REFR/FREEZER
LINEN LKR.
WC
BERTH
RANGE
SHOWER STALL
UP
DRESSER
2 PERSON JACUZZI TUB & SHOWER
ENGINE ROOM P/S.
D/W
GALS.
FUEL

302-3

56' & 64' TRAWLER YACHTS

Design Number 327

1994

This design started out as a mid-50-footer. In the first round of preliminary revisions we stretched it out to 64'. We developed that version for a while creating an interesting and salty little ship. Then we started looking for ways to simplify the boat and make it more affordable in the short run, without losing the practicality of the boat nor giving up the accommodations for the owners and guests. We hoped this would lead to the owner being able to start construction sooner. So far, though, neither version has proceeded to the building contract — yet....

The 64 is shown on page 201C with the layout for the engine room and tankage at the top of page 201B. She has twin engine rooms with a workshop and storage area forward under the cargo hatch in the well deck.

The 64 has twin 13' Whalers over the well deck with a crane to lift them off and a 10' Avon RIB on the boat deck with a separate davit. The 56 has a 13' Whaler and 12' Jet boat on the boat deck which will share a common davit or small crane.

The major changes that were made in downsizing the 64 to make the 56 were to eliminate the well deck, shorten the foc'sle, move the crew quarters down and aft under the galley and dining area and create a small aft deck for boarding.

The 56 is shown on page 201D and her lower deck and engine room and tankage layout is shown on the bottom of 201B. The two perspectives shown on 201B are for the 56-footer showing her exteriors and how the traffic on the stairs would flow on one of them.

One of the features that is retained in both versions is a dumb waiter running from the "basement" to the boat deck. This allows for sending an ice chest full of cold drinks to where it is needed or moving hot meals up and down. You will find it as the core of the central staircase in both vessels. We put it there since the stairs were also an element that served from top to bottom and wrapped the stairs around the dumb waiter trunk, providing openings forward, aft or sideways as was needed.

These designs have the saloon up, over the owner's party staterooms. This puts it at a level to look over most other boats in a marina and to enjoy more free-flowing air for ventilation. It also allows for a large, covered after deck for sitting out and socializing or dining.

The pilothouse has an elevated seating area for the guests and a large console and chart table for serious navigation work. The helmsman has a raised seat with footrest for comfort on long watches. The bridge wraps around the front of the pilothouse and will provide a secure place for the watchstanders to move about and take sights.

Both vessels have three staterooms aft, each with its own private head. The master (furthest aft) has a small cockpit aft of it on the 56 for boarding and line handling. There is also an escape door/hatch in the aft bulkhead.

The matching guest staterooms have two twin beds that can be combined to make a queen sized double. This is done by moving the nightstand and outboard bed and rebolting them in reversed positions. We've kept the inboard beds fixed, to allow for access to the linen storage drawers under them from the passageway. This makes for more flexibility in who uses the rooms and how they are accommodated.

The crew quarters are forward in the 64 and moved down and aft on the 56. We are planning on further modifying the 56 to have the starboard crew cabin become a ship's office with a desk, files and computer facilities.

These designs take advantage of the simplified structural design philosophy we evolved with the development of the Coasters. The hull forms have flared bows that are much finer than the original Coasters, being intended for more extended offshore work. This also makes them a bit easier on their power requirements for the cruising speeds and will help stretch the fuel out on a long passage. They also have large water tanks that will make living aboard and doing dishes, laundry and taking showers all a regular occurrence. This is just one of the many features that make these great cruising yachts and homes afloat.

Particulars:	**56'**	**64'**
Length overall	56'-0"	64'-0"
Length designed waterline	53'-6"	61'-0"
Beam	20'-0"	20'-0"
Draft	5'-0"	5'-0"
Freeboard:		
Forward	8'-0"	8'-6"
Waist	4'-0"	4'-0"
Aft	1'-9"	7'-6"
Displacement, cruising trim*	135,000 lbs.	145,500
Displacement-length ratio	394	286
Prismatic coefficient	.55	.55
Water tankage	1,250	1,260 Gals.
Fuel tankage	1,800	2,550 Gals.
Headroom	6'-7"	6'-7"

***CAUTION:** The displacement quoted here is for the boat in cruising trim. That is, with the fuel and water tanks filled, the crew on board, as well as the crews' gear and stores in the lockers. This should not be confused with the "shipping weight" often quoted as "displacement" by some manufacturers. This should be taken into account when comparing figures and ratios between this and other designs.

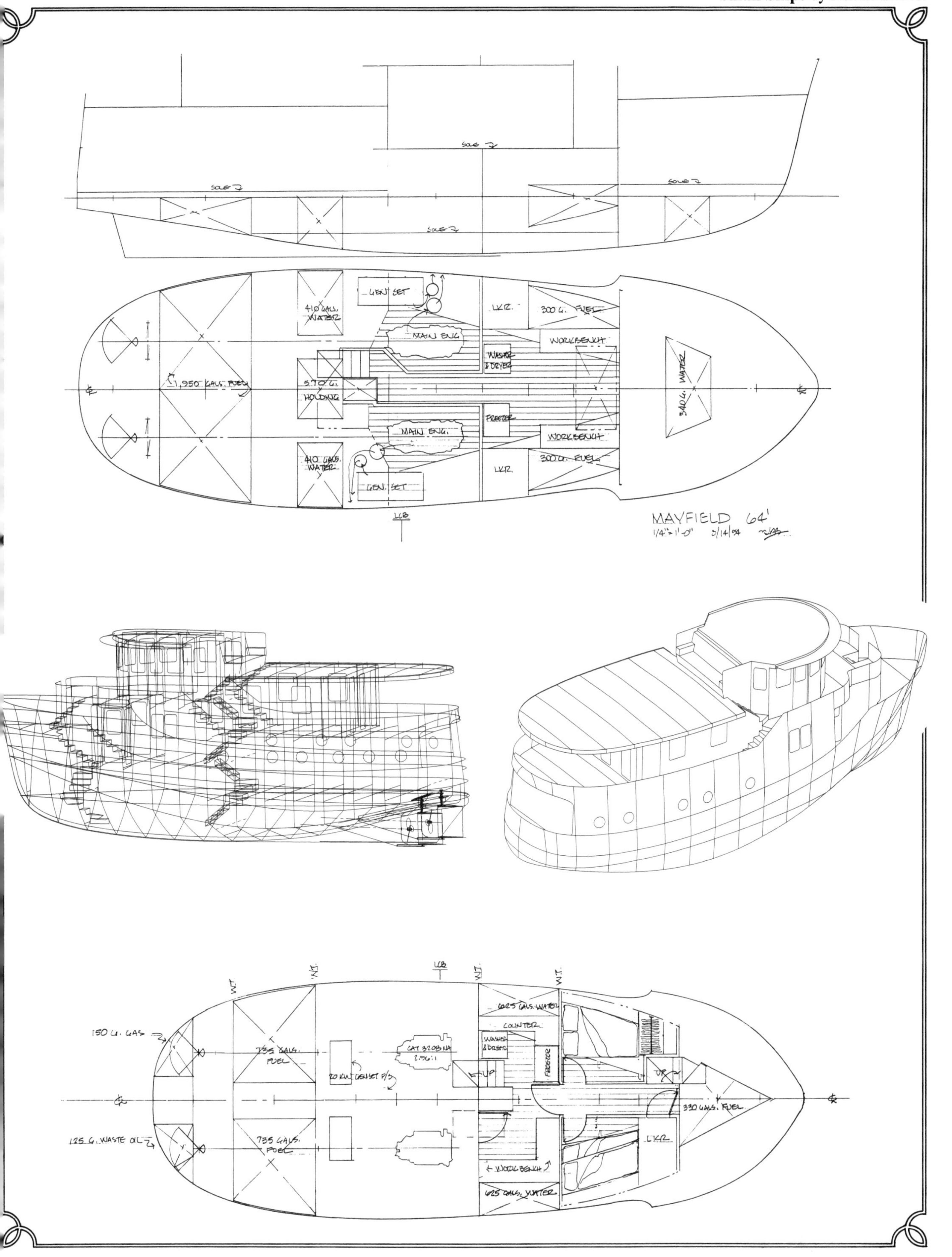

SOLE
410 GALS. WATER
GEN SET
MAIN ENG.
LKR.
300 G. FUEL
WORKBENCH
WASHER & DRYER
1,950 GALS. FUEL
570 G. HOLDING
FREEZER
340 G. WATER
LCB
MAYFIELD 64'
1/4"=1'-0"
150 G. GAS
125 G. WASTE OIL
735 GALS. FUEL
CAT 3208 NA
20 KW GEN SET P/S
625 GALS. WATER
COUNTER
UP
330 GALS. FUEL
WORKBENCH
W.T.

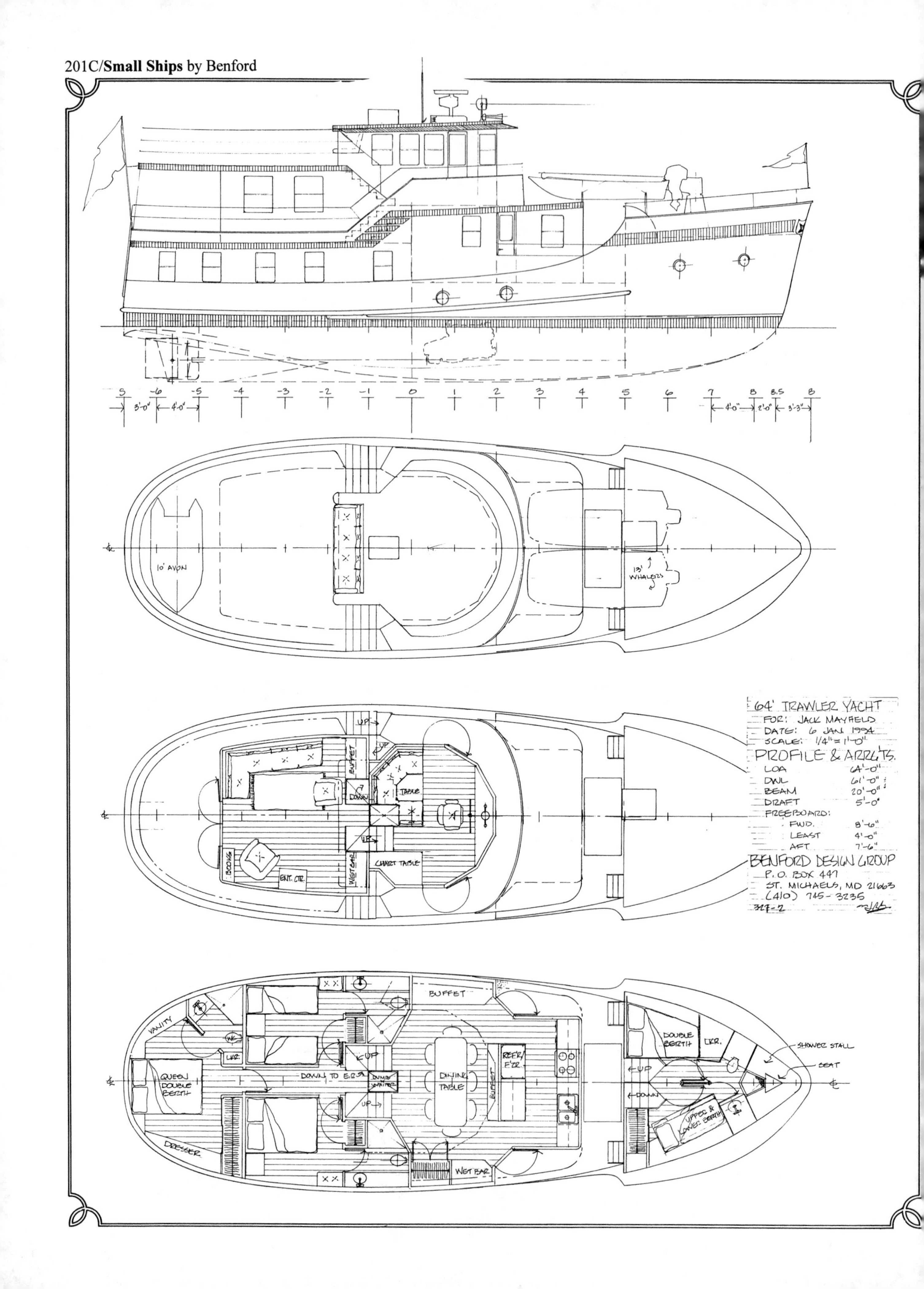

10' AVON
13' WHALERS
UP
DOWN
BUFFET
TABLE
BOOKS
ENT. CTR.
WET BAR
CHART TABLE
64' TRAWLER YACHT
FOR: JACK MAYFIELD
DATE: 6 JAN. 1994
SCALE: 1/4" = 1'-0"
PROFILE & ARR'G'TS.
LOA 64'-0"
DWL 61'-0"
BEAM 20'-0"
DRAFT 5'-0"
FREEBOARD:
FWD. 8'-6"
LEAST 4'-0"
AFT 7'-6"
BENFORD DESIGN GROUP
P.O. BOX 447
ST. MICHAELS, MD 21663
(410) 745-3235
317-2
VANITY
QUEEN DOUBLE BERTH
LKR.
DRESSER
DOWN TO E.R.
DUMB WAITER
DINING TABLE
REFR/FZR.
WET BAR
DOUBLE BERTH
SHOWER STALL
SEAT
UPPER & LOWER BERTH

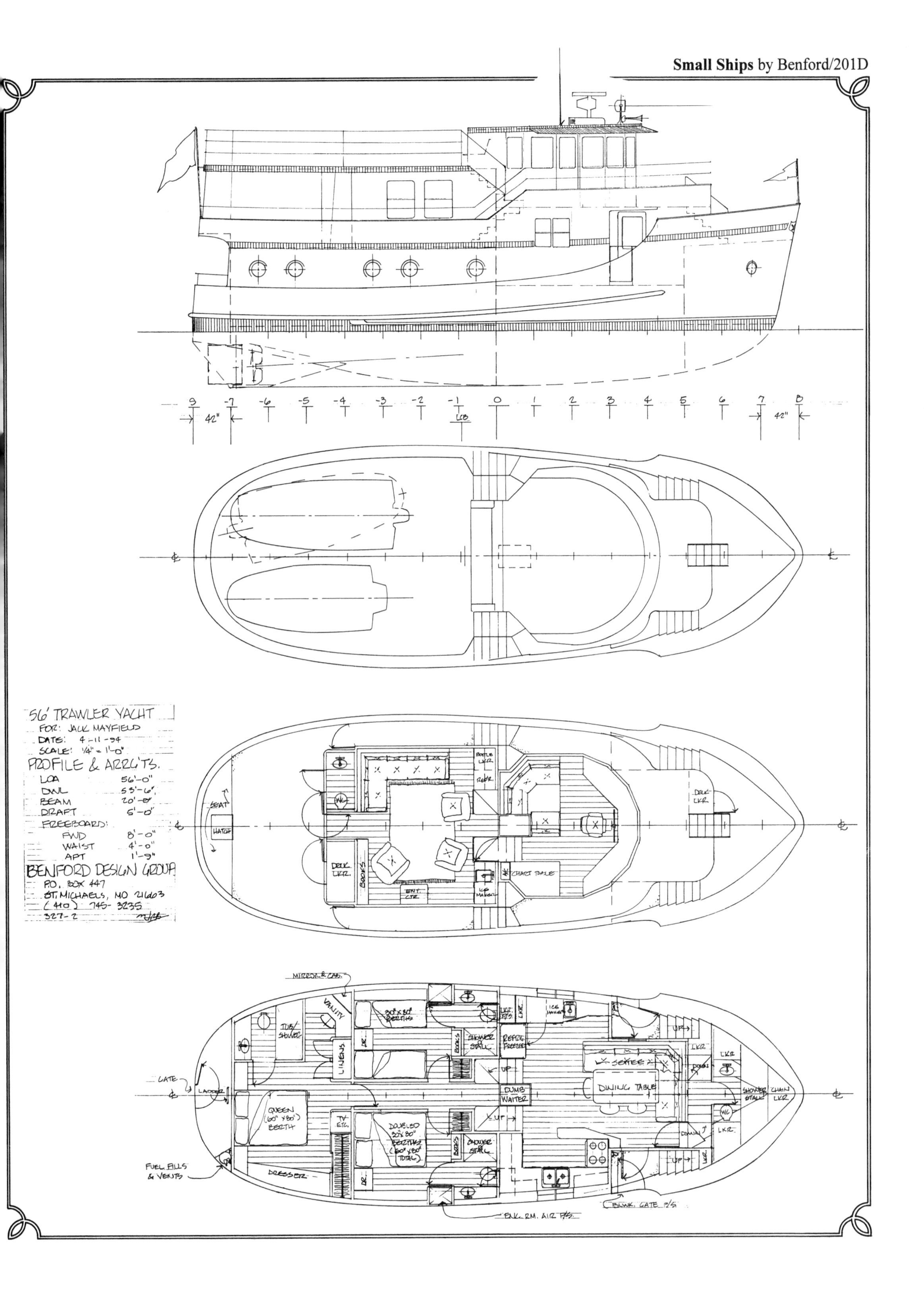

56' TRAWLER YACHT
FOR: JACK MAYFIELD
DATE: 4-11-94
SCALE: 1/4" = 1'-0"
PROFILE & ARRG'TS.
LOA 56'-0"
DWL 53'-6"
BEAM 20'-0"
DRAFT 5'-0"
FREEBOARD:
FWD 8'-0"
WAIST 4'-0"
APT 1'-9"
BENFORD DESIGN GROUP
P.O. BOX 447
ST. MICHAELS, MD 21663
(410) 745-3235
327-2
42"
LCB
SEAT
HATCH
DECK LKR.
BOOKS
ENT. CTR.
ICE MAKER
CHART TABLE
REFR.
BOTTLE LKR.
MIRROR & CAB.
VANITY
TUB/SHOWER
LINENS
DR.
30"x80" BERTHS
BOOKS
SHOWER STALL
REFRIG. FREEZER
ICE MAKER
UP
SETTEE
DINING TABLE
LKR
SHOWER STALL
CHAIN LKR
GATE
LADDER
QUEEN (60"x80") BERTH
TV ETC.
DOUBLED 30"x80" BERTHS (60"x80" TOTAL)
DUMB WAITER
DOWN
FUEL FILLS & VENTS
DRESSER
ENG. RM. AIR P/S
BLWK. GATE P/S

58 & 61' Aircraft Carriers

Design Number 320

1992

The concept of carrying an airplane with floats on a yacht has a lot of attraction to people who like to fly. It lets a businessman with a busy schedule go cruising and yet be able to return to his work if some need arises.

It also lets one go exploring in a different way when off cruising, using the plane to check out in advance proposed cruising routes and see if the areas to be traveled have obstructions or fixed bridges that are too low to get under. It could be used by an avid fisherman to scout out the areas in which there is activity so the fishing has a higher chance of being productive.

In order to carry an airplane on deck, the boat needs to have good stability, to ensure that the weight and windage of the airplane will not endanger the boat. The structure where the plane will be stowed needs to be designed to be up to the task of supporting it. And, the launching crane needs to be installed in the structure in a way that will not overload the local structure.

In the course of designing this boat we had the drawings pretty well finished and the boat was under construction at 58-feet. Then, during one of our "what if" discussions with our clients, they asked if they could have the stem tilted forward another three feet to give it a more rakish appearance. We concurred that we could revise the design with little change to the existing plans to make this work. So, in short order, the boat grew three feet and is built that way.

The accommodations are laid out to provide for operating with a paid crew, if desired, with the crew having their own quarters in the bow. Aft there are three double staterooms, each with their own enclosed head and shower. The furthest aft one is considered the owner's quarters and thus is slightly larger than the others, and enjoys access to the aft deck.

Amidships is the entry into the galley and dining area, with an office space in one corner. Below this is the workshop and engine room aft and storage hold forward, both of which have standing headroom.

The saloon is on the upper aft deck, with a wet bar on the aft deck. There is also the more formal dining area at the forward end of this space.

The pilothouse is well elevated, at the height of most other boats' flying bridges, so there will be good visibility and the passengers will enjoy the raised seating so they can also see where they're going. There is a stairway leading up from inside to the hatch in the pilothouse top for access to the flight deck.

Outside the pilothouse, on the wrap-around bridge, there is an option for centerline stairs to the foredeck. These would be handy for quick access to the bow for line handling and checking on the anchor rode.

Included are drawings for both the 58 and 61-foot versions. We've also picked variations in the styling to show how the character of the boat can change with some little changes to the shell.

The sketch below is the one I made for the meeting at which the clients decided to proceed with the design. From it one can see how much some design projects evolve over the course of their development.

Also included is a profile variation idea I had for how to do a freighter yacht with this sort of more stylized exterior shape. All these are offered up as ideas — the 61-footer is the one that was the final choice by the clients and is being built.

Particulars:

Length overall	58'-0" or 61'-0"
Length designed waterline	55'-0"
Beam	20'-0"
Draft	4'-6"
Freeboard:	
Forward	8'-0"
Waist	4'-5"
Aft	1'-8½"
Displacement, cruising trim	119,000 lbs.
Displacement-length ratio	319
Pounds per inch immersion	4253 pounds
Water tankage	1,080 Gals.
Fuel tankage	1,385 Gals.
Holding tankage	500 Gals.
Headroom	6'-7"

***CAUTION:** The displacement quoted here is for the boat in cruising trim. That is, with the fuel and water tanks filled, the crew on board, as well as the crews' gear and stores in the lockers. This should not be confused with the "shipping weight" often quoted as "displacement" by some manufacturers. This should be taken into account when comparing figures and ratios between this and other designs.

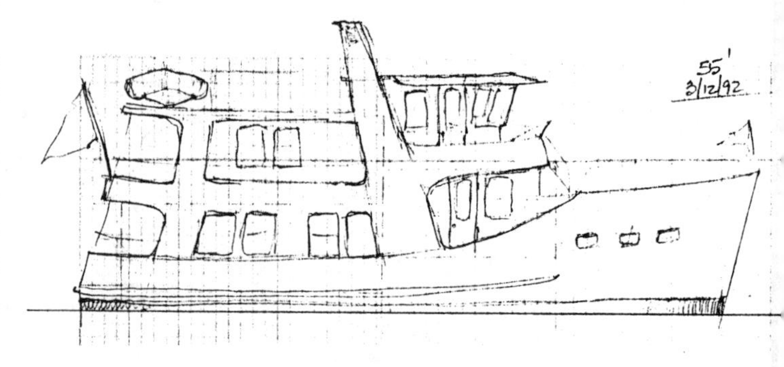

The preliminary sketch (above) that preceded the agreement to proceed with this design shows another styling variation and an alternative radar arch that would have engine room vents built into it.

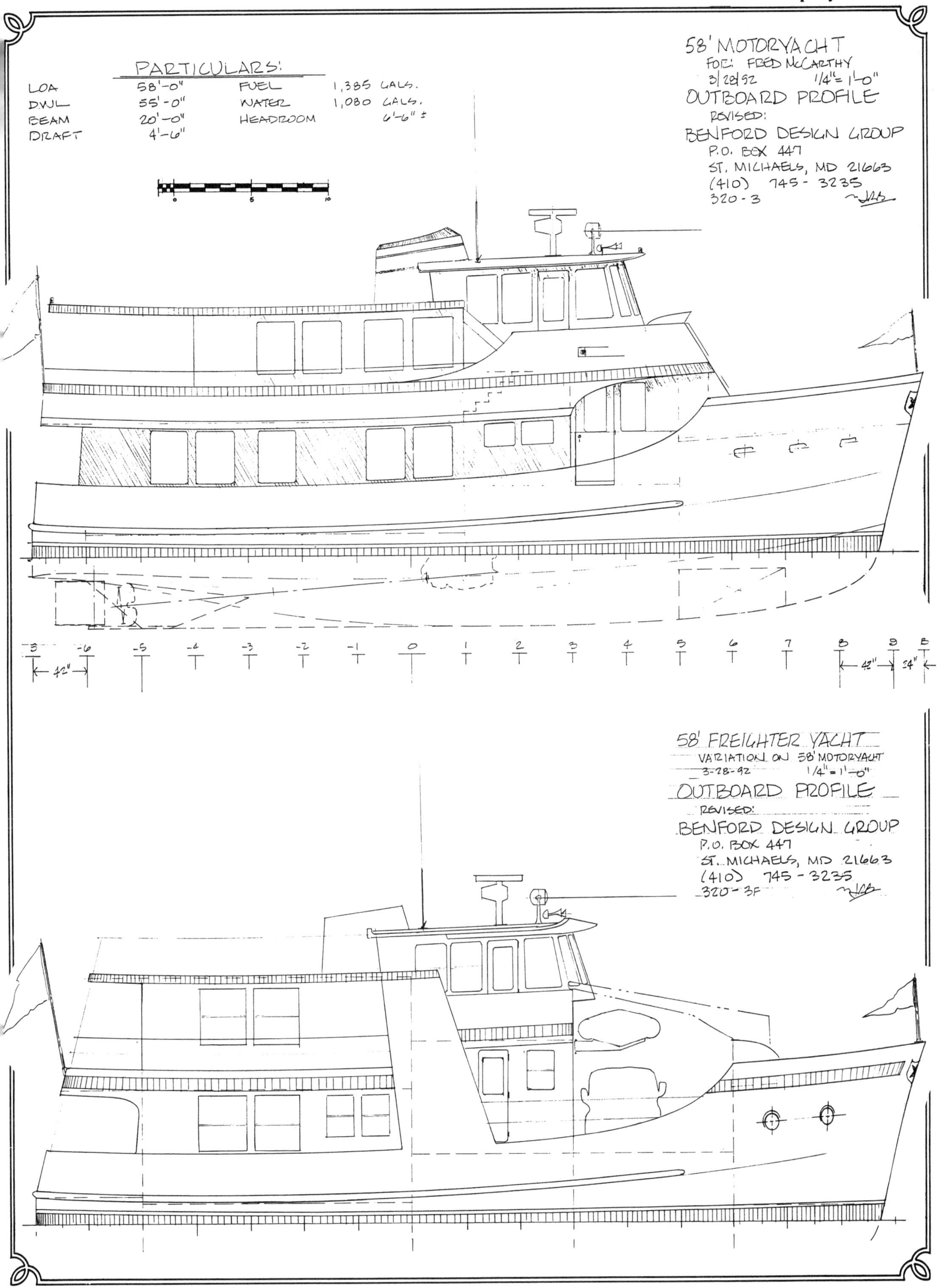
PARTICULARS:
LOA 58'-0"
DWL 55'-0"
BEAM 20'-0"
DRAFT 4'-6"
FUEL 1,385 GALS.
WATER 1,080 GALS.
HEADROOM 6'-6" ±
58' MOTORYACHT
FOR: FRED McCARTHY
3/28/92 1/4" = 1'-0"
OUTBOARD PROFILE
REVISED:
BENFORD DESIGN GROUP
P.O. BOX 447
ST. MICHAELS, MD 21663
(410) 745 - 3235
320 - 3
58' FREIGHTER YACHT
VARIATION ON 58' MOTORYACHT
3-28-92 1/4" = 1'-0"
OUTBOARD PROFILE
REVISED:
BENFORD DESIGN GROUP
P.O. BOX 447
ST. MICHAELS, MD 21663
(410) 745 - 3235
320 - 3F

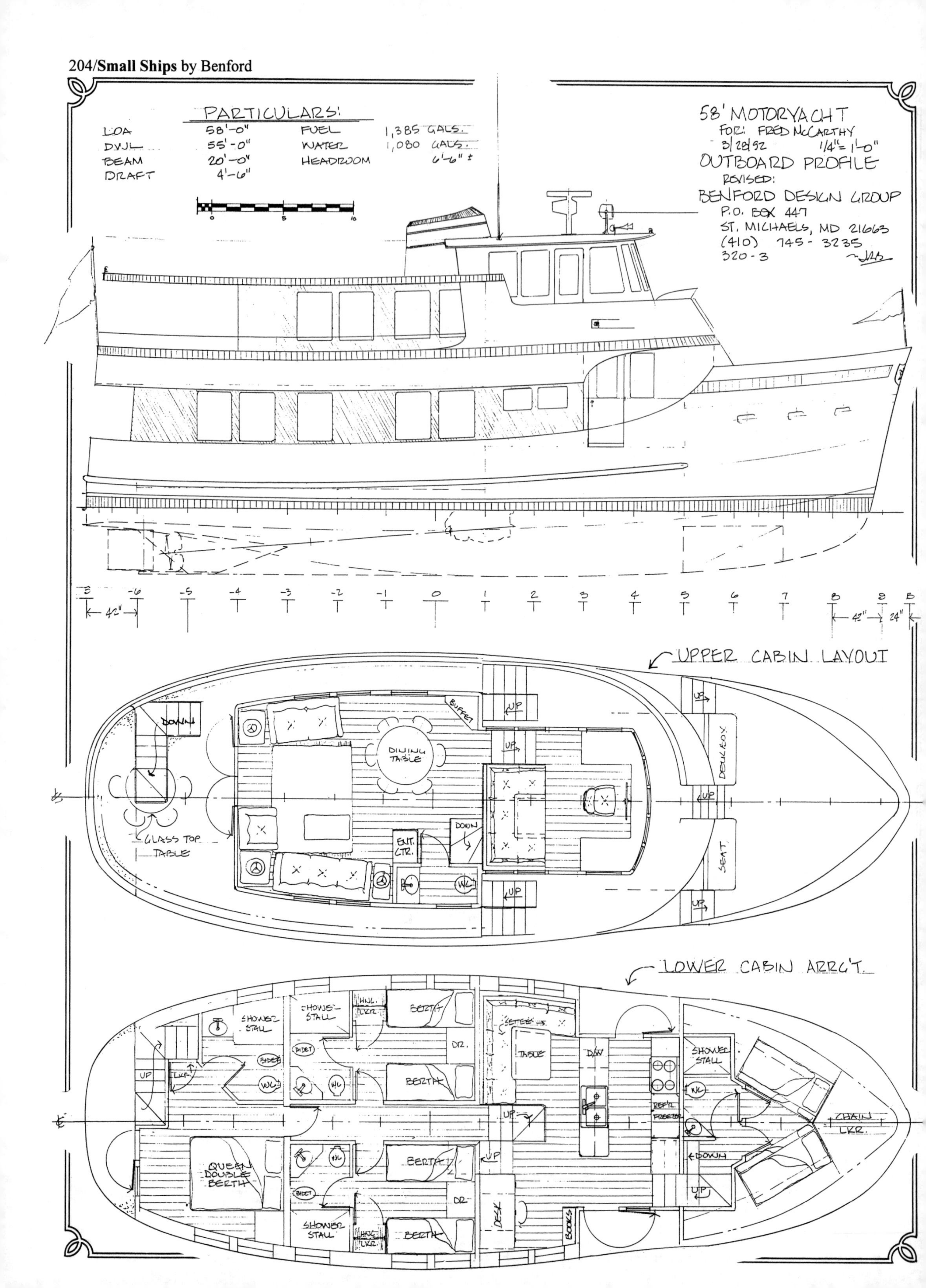

PARTICULARS:
LOA 58'-0"
DWL 55'-0"
BEAM 20'-0"
DRAFT 4'-6"
FUEL 1,385 GALS.
WATER 1,080 GALS.
HEADROOM 6'-6" ±
58' MOTORYACHT
FOR: FRED McCARTHY
3/28/92 1/4" = 1'-0"
OUTBOARD PROFILE
REVISED:
BENFORD DESIGN GROUP
P.O. BOX 447
ST. MICHAELS, MD 21663
(410) 745-3235
320-3
UPPER CABIN LAYOUT
DOWN
GLASS TOP TABLE
BUFFET
DINING TABLE
UP
DOWN
ENT. CTR.
WC
DECK BOX
SEAT
LOWER CABIN ARR'G'T.
SHOWER STALL
BIDET
WC
LKR
UP
HNG. LKR.
BERTH
DR.
QUEEN DOUBLE BERTH
SETTEE
TABLE
D/W
REF'R. FREEZER
DESK
BOOKS
CHAIN LKR.

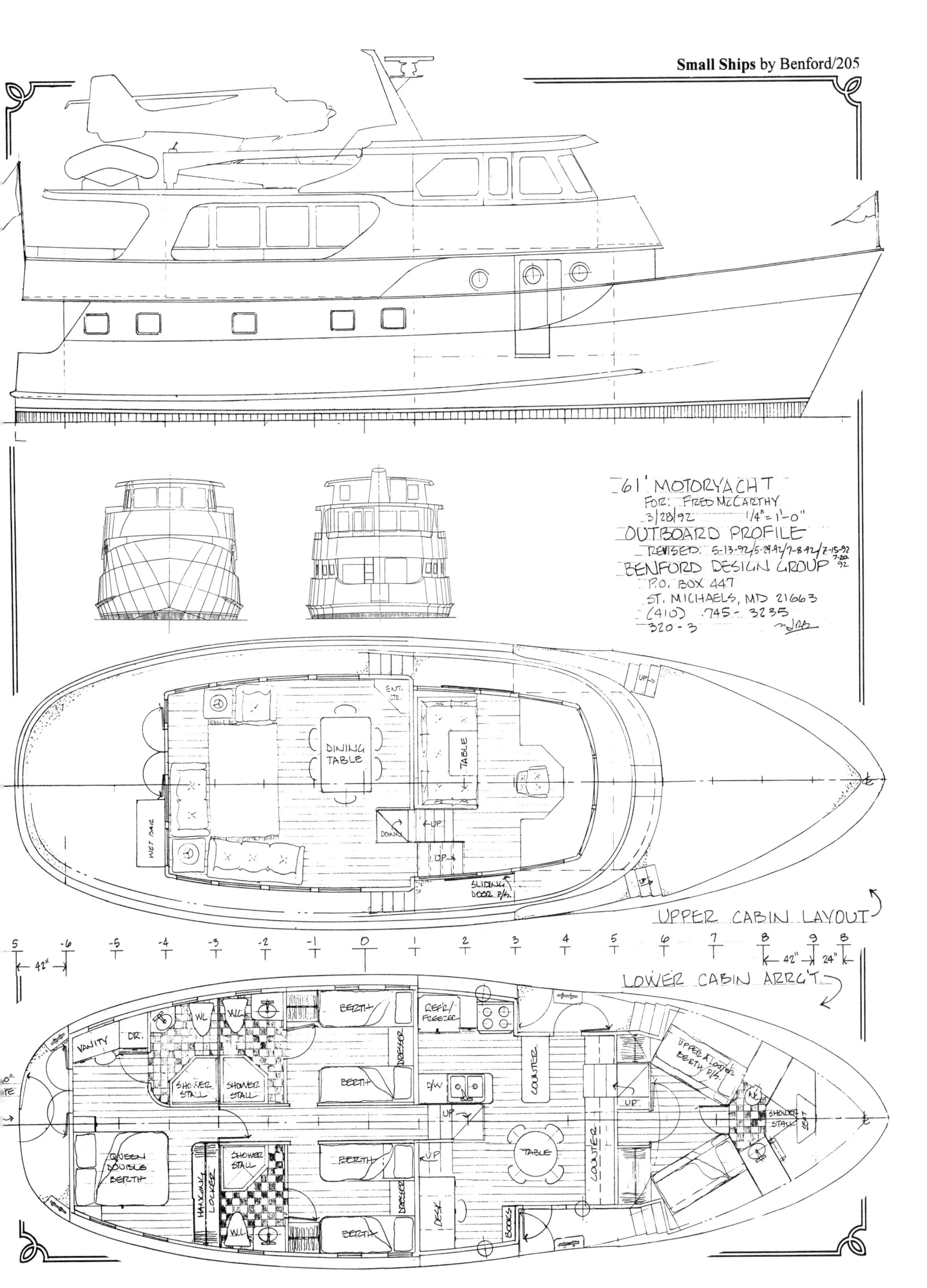
61' MOTORYACHT
FOR: FRED McCARTHY
3/28/92 1/4" = 1'-0"
OUTBOARD PROFILE
REVISED: 5-13-92/5-29-92/7-8-92/7-15-92 7-20-92
BENFORD DESIGN GROUP
P.O. BOX 447
ST. MICHAELS, MD 21663
(410) 745-3235
320-3
UPPER CABIN LAYOUT
DINING TABLE
TABLE
WET BAR
ENT. CTR.
DOWN
UP
SLIDING DOOR P/S.
LOWER CABIN ARR'G'T.
VANITY
DR.
W.C.
SHOWER STALL
BERTH
DRESSER
REFR/FREEZER
D/W
COUNTER
UPPER & LOWER BERTH P/S.
QUEEN DOUBLE BERTH
HANGING LOCKER
DESK
BOOKS
TABLE
SEAT
42"
24"

62' Trawler Yacht *Indenture*

Design Number 76
1972

This 62' Cruiser was designed for a nautically oriented dentist who also wanted to efficiently service his clients in the otherwise inaccessible coastal regions of Alaska. Designed for comfortable living and working aboard, the pilothouse is separate from the main deck living quarters and lower deck working quarters. The dental working area is below decks amidships and forward. 3,000 gallons of fuel and 560 of water give her a good cruising range. She's powered with a slow turning diesel and carries a steadying sail rig and is intended for making long passages. I take full blame for naming the design....

Particulars:

Length overall	62'-0"
Length designed waterline	57'-0"
Beam	18'-0"
Draft	8'-0"
Freeboard:	
Forward	7'-3½"
Least	3'-3"
Aft	3'-9"
Displacement, cruising trim	148,000 lbs.
Displacement-length ratio	357
Ballast	15,676 lbs.
Ballast ratio	11%
Sail area (riding sail)	482 sq. ft.
Sail area-displacement ratio	2.76
Prismatic coefficient	.62
Pounds per inch immersion	4,050
Water tankage	560 Gals.
Fuel tankage	3,000 Gals.
Headroom	6'-2" to 6'-4"

***CAUTION:** The displacement quoted here is for the boat in cruising trim. That is, with the fuel and water tanks filled, the crew on board, as well as the crews' gear and stores in the lockers. This should not be confused with the "shipping weight" often quoted as "displacement" by some manufacturers. This should be taken into account when comparing figures and ratios between this and other designs.

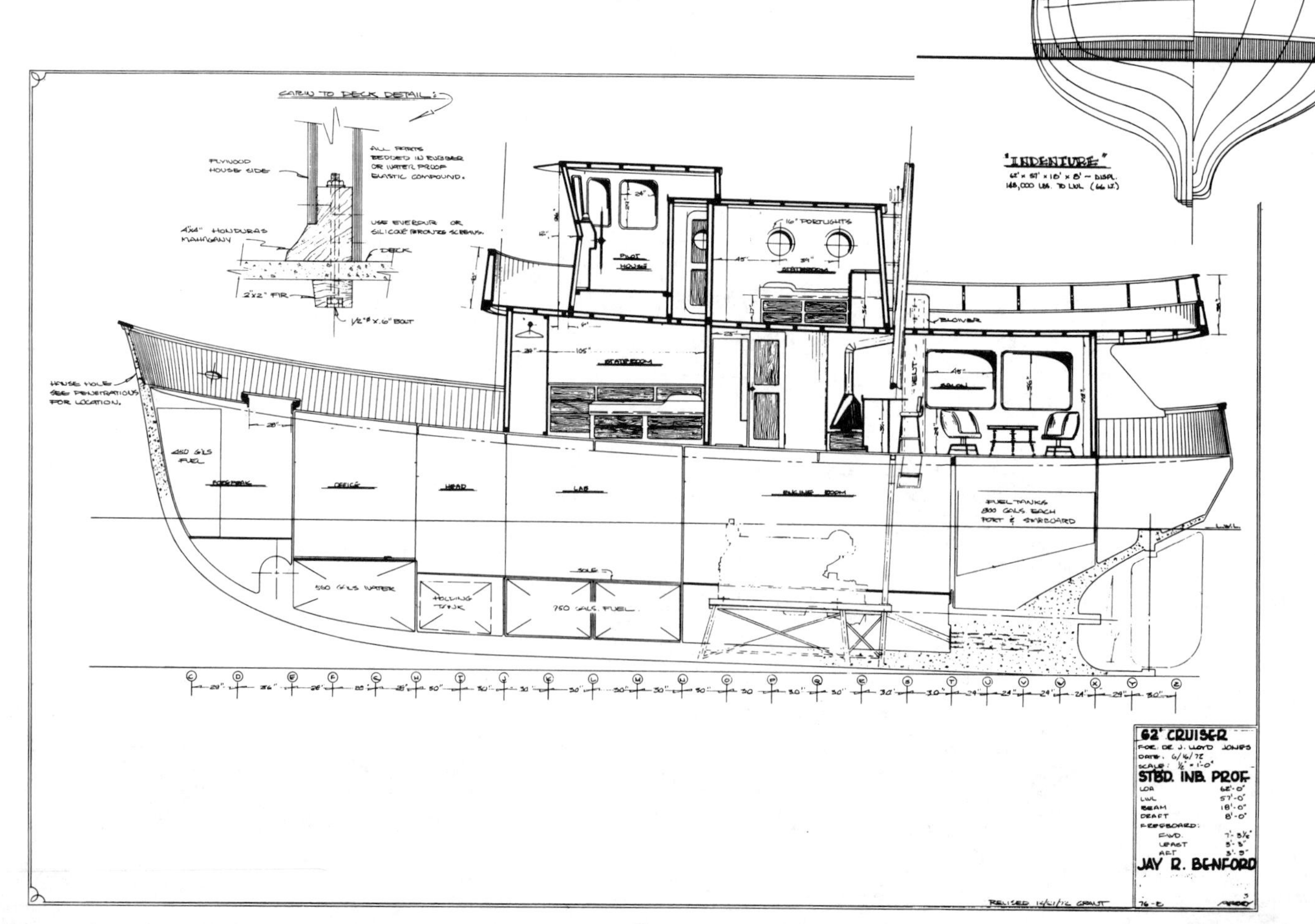

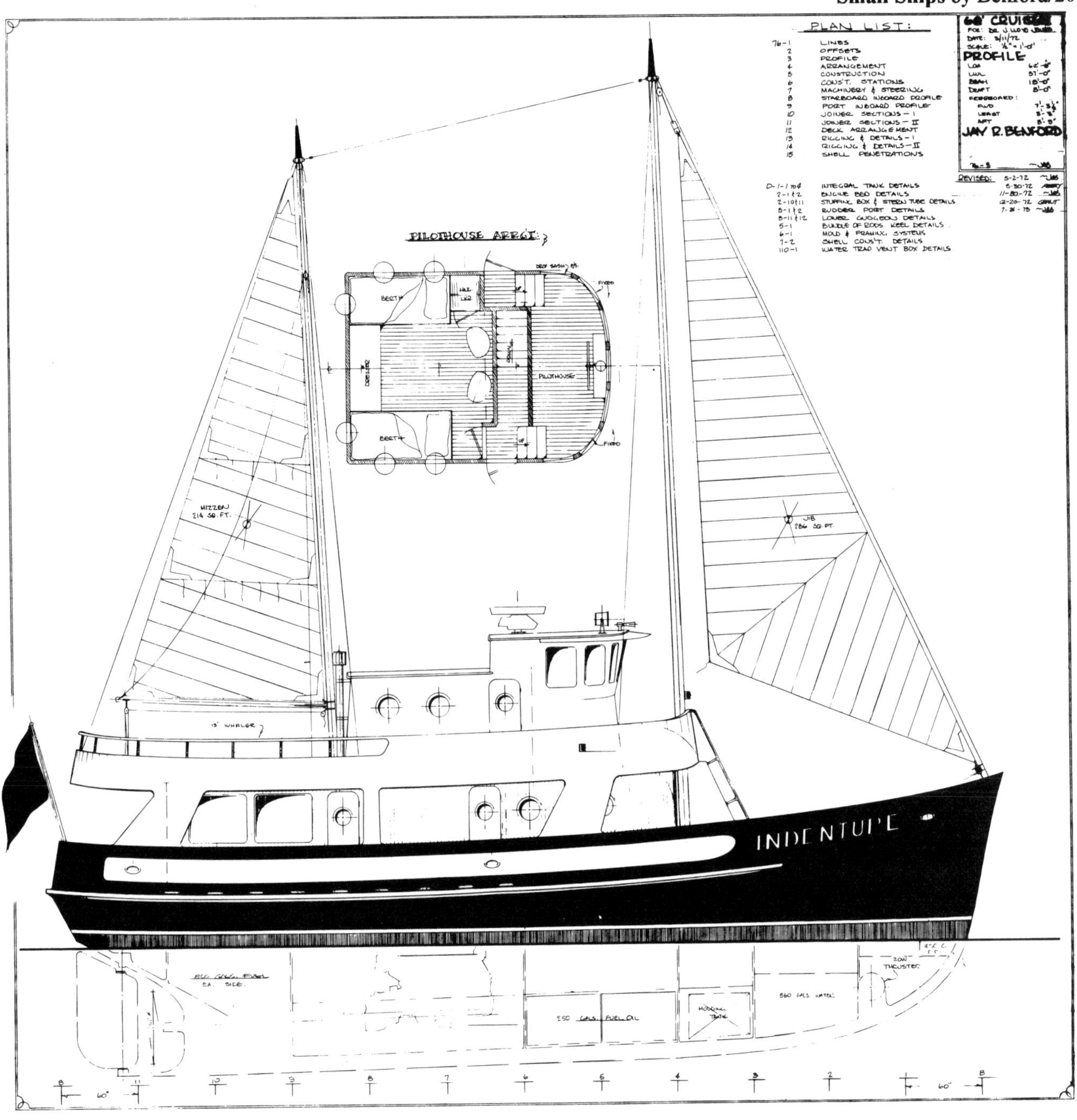
PLAN LIST:
76-1 LINES
2 OFFSETS
3 PROFILE
4 ARRANGEMENT
5 CONSTRUCTION
6 CONS'T. STATIONS
7 MACHINERY & STEERING
8 STARBOARD INBOARD PROFILE
9 PORT INBOARD PROFILE
10 JOINER SECTIONS – I
11 JOINER SECTIONS – II
12 DECK ARRANGEMENT
13 RIGGING & DETAILS – I
14 RIGGING & DETAILS – II
15 SHELL PENETRATIONS
D-1-1 to 4 INTEGRAL TANK DETAILS
2-1 & 2 ENGINE BED DETAILS
2-10 & 11 STUFFING BOX & STERN TUBE DETAILS
3-1 & 2 RUDDER PORT DETAILS
3-11 & 12 LOWER GUDGEON DETAILS
5-1 BUNDLE OF RODS KEEL DETAILS
6-1 MOLD & FRAMING SYSTEMS
7-2 SHELL CONS'T. DETAILS
110-1 WATER TRAP VENT BOX DETAILS
PROFILE
LOA
LWL 57'-0"
BEAM 18'-0"
DRAFT 8'-0"
FREEBOARD:
FWD
LEAST
AFT
JAY R. BENFORD
REVISED:
PILOTHOUSE ARR'G'T.
BERTH
PILOTHOUSE
MIZZEN 214 SQ. FT.
JIB 286 SQ. FT.
13' WHALER
INDENTURE
550 GALS. FUEL OIL
560 GALS. WATER
THRUSTER
60"

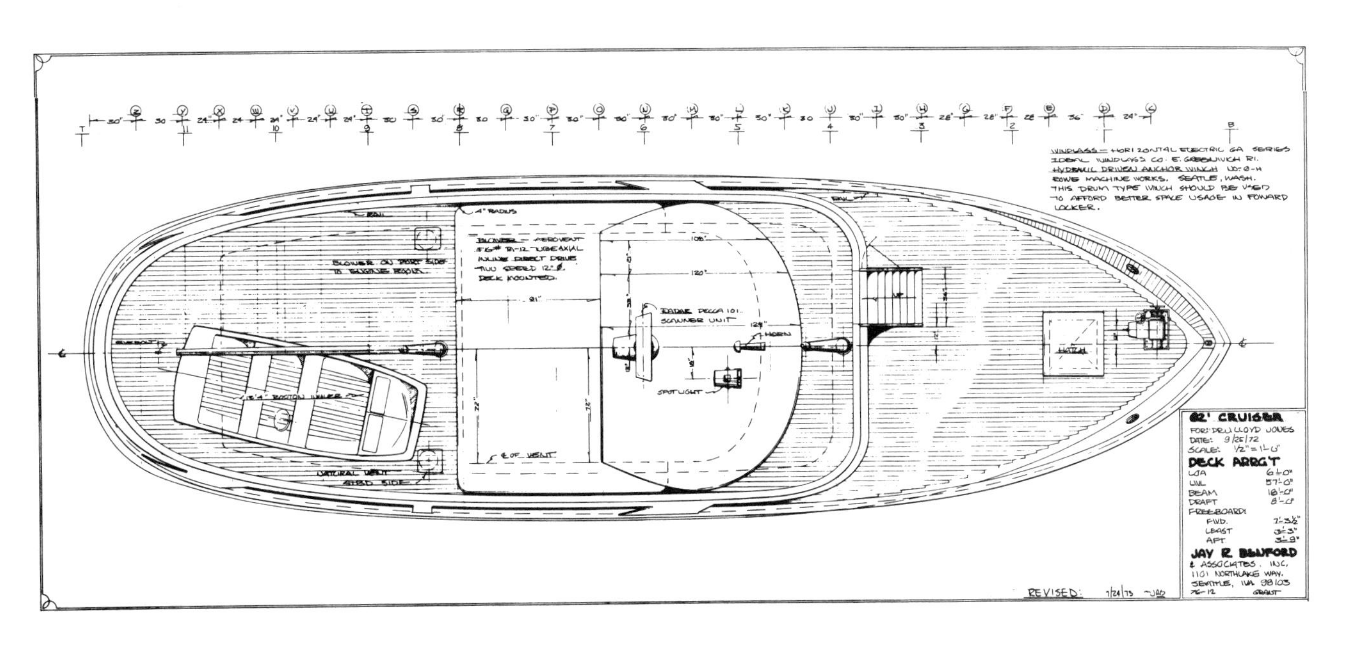
WINDLASS – HORIZONTAL ELECTRIC GA SERIES
IDEAL WINDLASS CO. E. GREENWICH RI.
HYDRAULIC DRIVEN ANCHOR WINCH NO. 0-H
ROWE MACHINE WORKS. SEATTLE, WASH.
THIS DRUM TYPE WINCH SHOULD BE USED
TO AFFORD BETTER SPACE USAGE IN FORWARD
LOCKER.
BLOWER ON PORT SIDE
SCANNER UNIT
SPOTLIGHT
NATURAL VENT
62' CRUISER
FOR: DR. J. LLOYD JONES
DATE: 9/25/72
SCALE: 1/2" = 1'-0"
DECK ARRG'T
LWL 57'-0"
BEAM 18'-0"
DRAFT 8'-0"
FREEBOARD:
FWD. 7'-3½"
LEAST 3'-3"
AFT 3'-9"
JAY R. BENFORD
& ASSOCIATES, INC.
1101 NORTHLAKE WAY.
SEATTLE, WA 98103
76-12
REVISED:

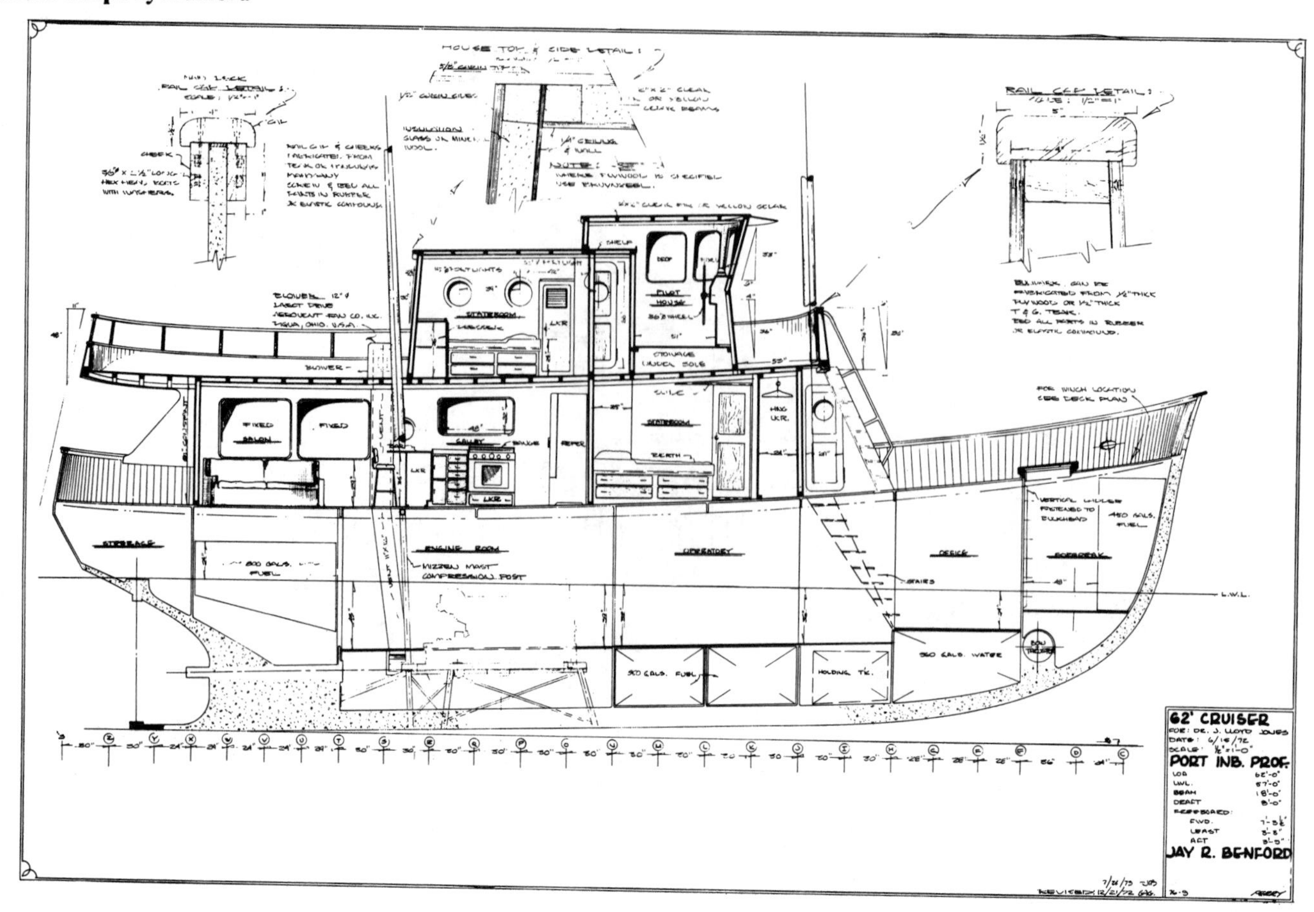
HOUSE TOP & SIDE DETAIL:
RAIL CAP DETAIL:
STATEROOM
PILOT HOUSE
SALON
GALLEY
FIXED
BERTH
LKR
ENGINE ROOM
MIZZEN MAST COMPRESSION POST
OPERATORY
DECK
FOREPEAK
STAIRS
460 GALS. FUEL
800 GALS. FUEL
350 GALS. WATER
HOLDING TK.
L.W.L.
62' CRUISER
FOR: DR. J. LLOYD JONES
PORT INB. PROF.
LOA 62'-0"
LWL 57'-0"
BEAM 18'-0"
DRAFT 8'-0"
FREEBOARD:
FWD.
LEAST
AFT
JAY R. BENFORD

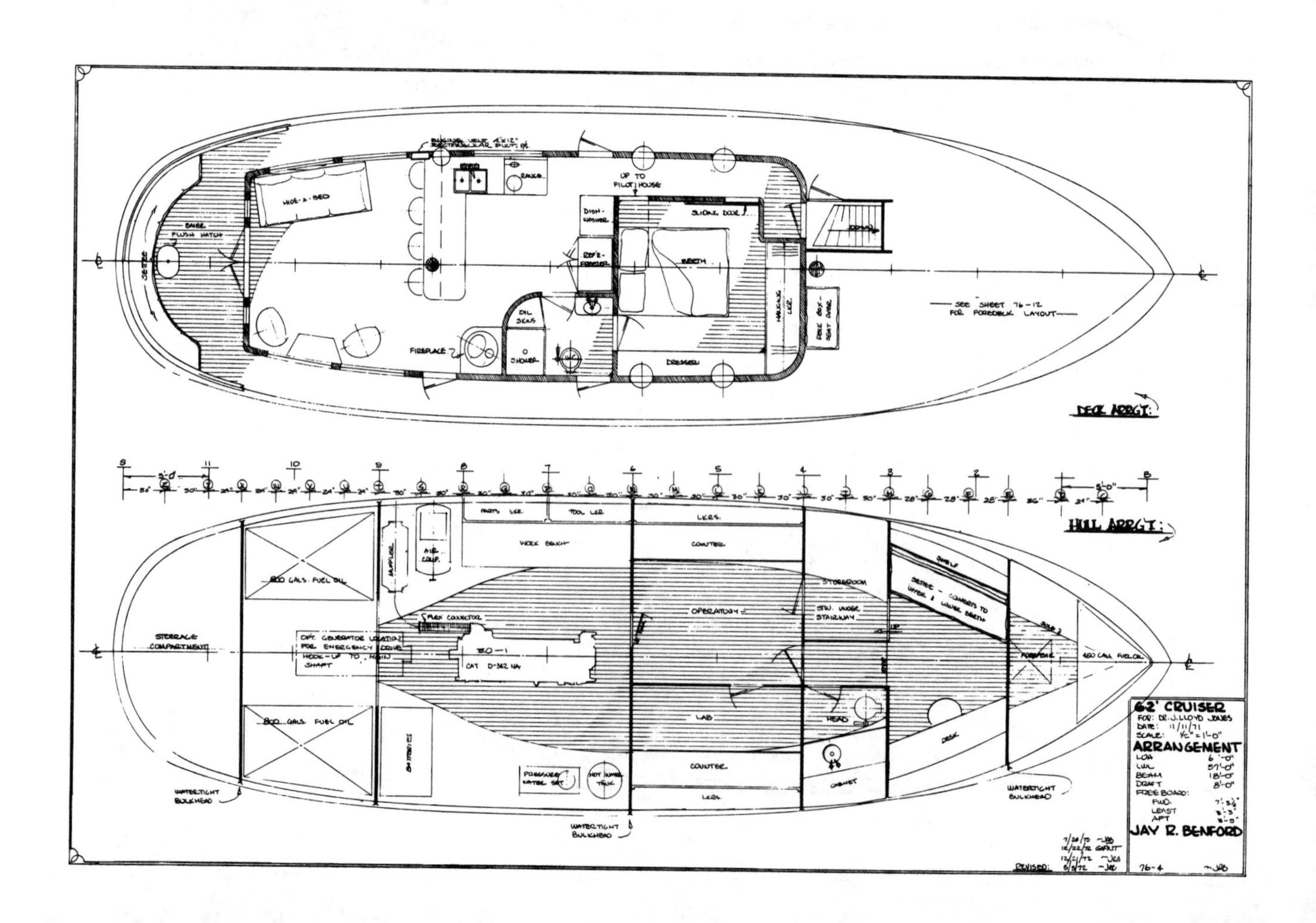
UP TO PILOT HOUSE
HIDE-A-BED
DISH-WASHER
REF-FREEZER
BERTH
DRESSER
FIREPLACE
SHOWER
SEE SHEET 76-12 FOR FOREDECK LAYOUT
DECK ARRG'T:
HULL ARRG'T:
800 GALS. FUEL OIL
STERRAGE COMPARTMENT
CAT. D-342 NA
WORK BENCH
COUNTER
OPERATORY
LAB
LKRS.
HEAD
STOREROOM
FOREPEAK
460 GALS. FUEL OIL
WATERTIGHT BULKHEAD
62' CRUISER
FOR: DR. J. LLOYD JONES
ARRANGEMENT
LOA 62'-0"
LWL 57'-0"
BEAM 18'-0"
DRAFT 8'-0"
FREEBOARD:
FWD.
LEAST
AFT
JAY R. BENFORD

65' Trawler Yacht

Design Number 278

1988

This trawler yacht is an extension of our design philosophy of simplicity and rugged construction. We've applied it to this trawler yacht as we have to the variety of freighter yacht designs in this book. The result is a rugged little ship, with all the comforts and features usually found on trawler yachts.

We did a number of studies on the profile for the boat, and some of these are shown below. Some of the profile variations require moving the foc'sle accommodations aft a bit, to take into account the finer bow shape and the overhanging stem. While reducing the size of the hold, this finer bow would let her be driven at higher speed in rough going. This is a mixed blessing, for a slower speed is usually more comfortable in rough weather.

The profile shown on the following page has a crane on the foredeck, like ***Key Largo***, and can load and offload the 17' skiff and/or any weighty items to be carried in the cargo hold.

The saloon shows two different sides, with and without side decks. I prefer the walkaround decks for greater practicality in working the ship, but some services require maximizing the interior volume.

The interior is set up to function well as a three couple ("six-pack") charter boat operation, with the crew having their quarters in the bow. They have fore and aft access by going through the watertight doors in the hold and coming up the stairs that lead into the galley area.

The accommodations in the hull also show a couple variations for the middle staterooms. They could be converted to small meeting rooms, or opened across the centerline into one good sized room, for a meeting, a larger party, or a big playroom for the kids.

One of the key design elements to making a multi-level design like this function well is to have the traffic flow well sorted out. In doing this we have provided for stairs and passageways that are wider than found in most yachts. The amidships stairs from the side decks to the saloon level and up to the pilothouse level and on to the deck over the saloon give port and starboard alternatives to this route.

Other designs of ours can be restyled like this for those who like the livability and generous sizing of the accommodations and want an alternative look and appearance to the boat.

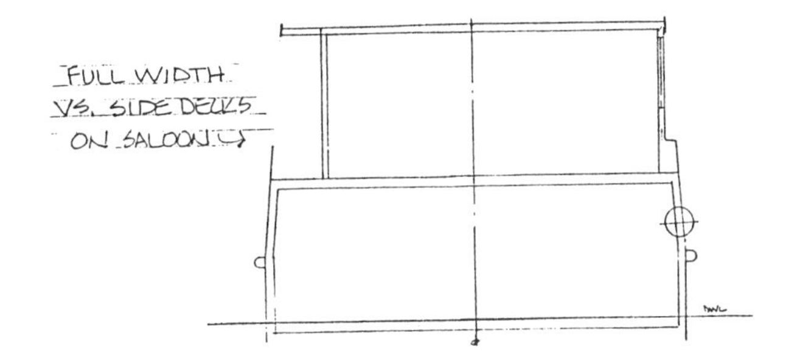

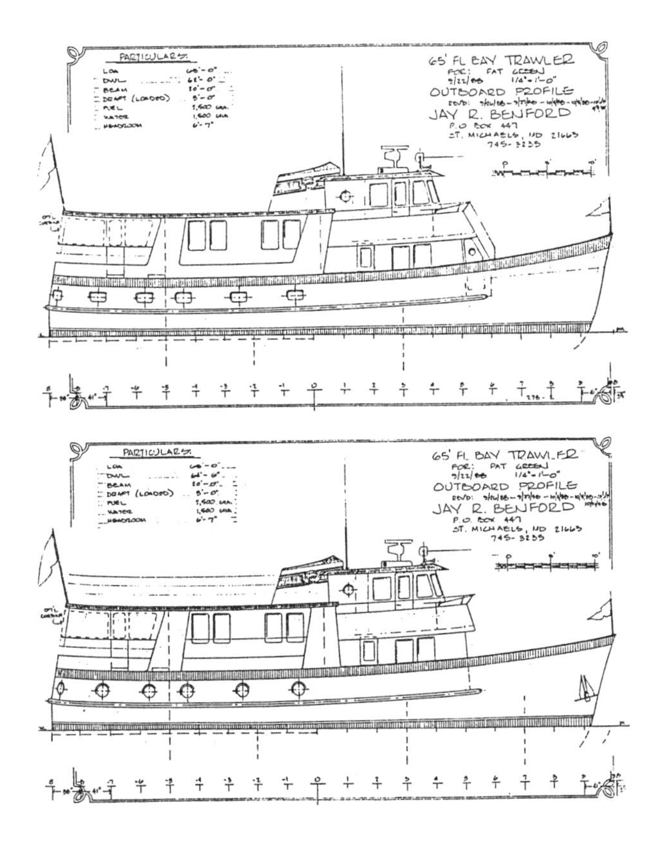

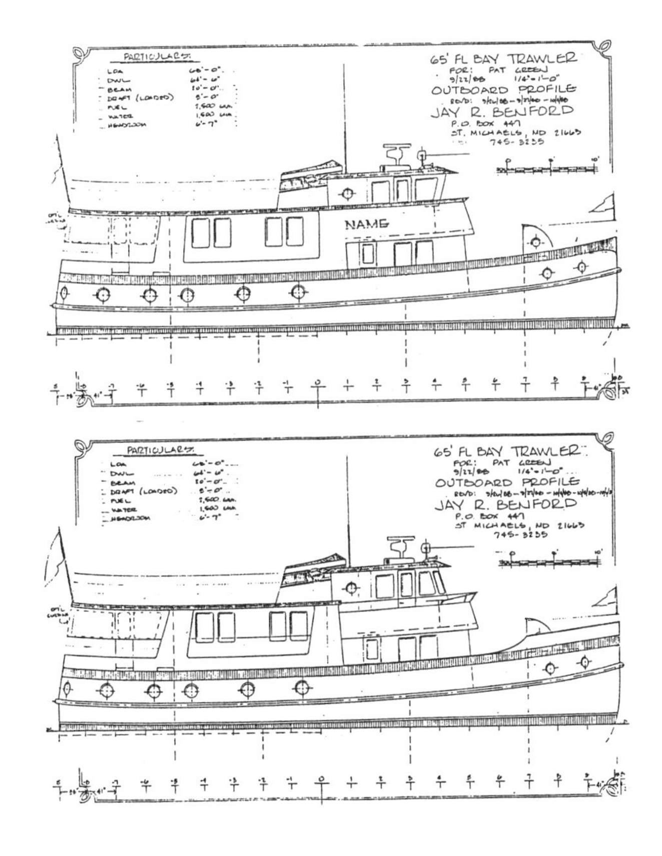

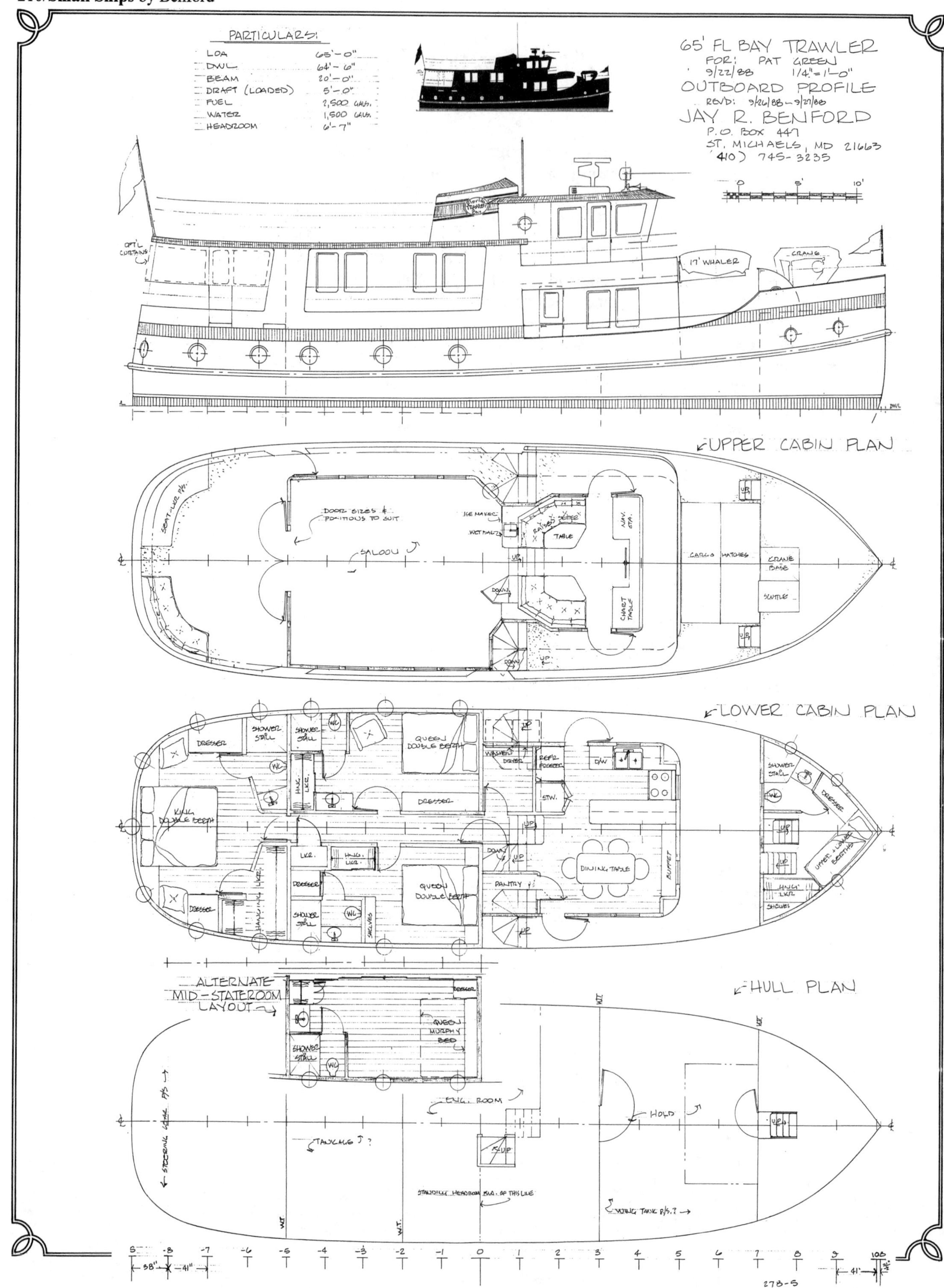
PARTICULARS:
LOA 65'-0"
DWL 64'-6"
BEAM 20'-0"
DRAFT (LOADED) 5'-0"
FUEL 2,500 GALS.
WATER 1,500 GALS.
HEADROOM 6'-7"
65' FL BAY TRAWLER
FOR: PAT GREEN
9/22/88 1/4"=1'-0"
OUTBOARD PROFILE
REV'D: 9/26/88 - 9/27/88
JAY R. BENFORD
P.O. BOX 447
ST. MICHAELS, MD 21663
(410) 745-3235
OPT'L CURTAINS
17' WHALER
CRANE
UPPER CABIN PLAN
DOOR SIZES & POSITIONS TO SUIT
SALOON
ICE MAKER
WET BAR
RAISED SETTEE
TABLE
NAV. STA.
CHART TABLE
CARGO HATCHES
CRANE BASE
SCUTTLE
LOWER CABIN PLAN
SHOWER STALL
DRESSER
KING DOUBLE BERTH
HANGING LKR.
QUEEN DOUBLE BERTH
WASHER DRYER
REF'R FREEZER
STW.
DW
DINING TABLE
BUFFET
PANTRY
LKR.
SHELVES
UPPER + LOWER BERTHS
ALTERNATE MID-STATEROOM LAYOUT
QUEEN MURPHY BED
HULL PLAN
ENG. ROOM
HOLD
TANKAGE ?
STEERING GEAR P/S
STANDING HEADROOM FWD. OF THIS LINE
WING TANK P/S ?
278-5

135' Canal Barge

Design Number 331
1995

Many of these ex-working barges in Europe have been converted for yacht use. This one's planned to be converted in the Netherlands for family use cruising on the thousands of miles of European rivers and canals. Later, she may also be brought over to the US and used for coastal and river cruising.

In interests of simplicity, we're maintaining the existing machinery, crew quarters and pilothouse aft. The owner correctly notes that this will keep the paid crew working at the after end of the boat, leaving the forward end undisturbed for the owner's family and guests.

The original 95' long cargo hold is being greatly shortened with the addition of the commodious living spaces in the after two-thirds of it. The forward end of it is being used as the "garage" for the 22' power boat, the 13' Whaler stacked over the Land Rover, and other smaller water toys. A crane is being fitted in the hold to lift these things in and out. We've shown a pair of windows on each side to let in light.

Moving aft on the house-top are found some skylights for letting light and air into the master stateroom and the dining saloon. There is also a large cockpit cut into the house-top.

This 'midships cockpit has a fixed top over it to give rain and sun shade. The forward half has seating built-in to give lounging space. The after half has room for a dining table and chairs. This is likely to be the most used space on the boat. It has a good view of where you're going and recessing it into the house-top gives it deep and secure surroundings for keeping kids of all ages safely aboard.

Just forward of the pilothouse is the viewing saloon, with deeper windows so seated passengers can see the world slide by. It also has port and starboard doors for entry and exit.

Under the viewing saloon is the galley and breakfast bar. There is generous storage for provisions in the lockers and freezers. This area is also home to the washer and dryer. Two deadlights have been added to each side to let in some natural light.

Particulars:		**Metric**	**English**
Length overall		41 m	135'-0"
Beam		5.2 m	17'-1"
Draft		1.54 m	4'-11"
Freeboard:	Forward	1.9 m	6'-3"
	Least	.95 m	3'-0"
	Aft	1.65 m	4'-11"
Water tankage			8,580 Gals.
Fuel tankage			3,800 Gals.
Headroom		2.0 m	6'-7"

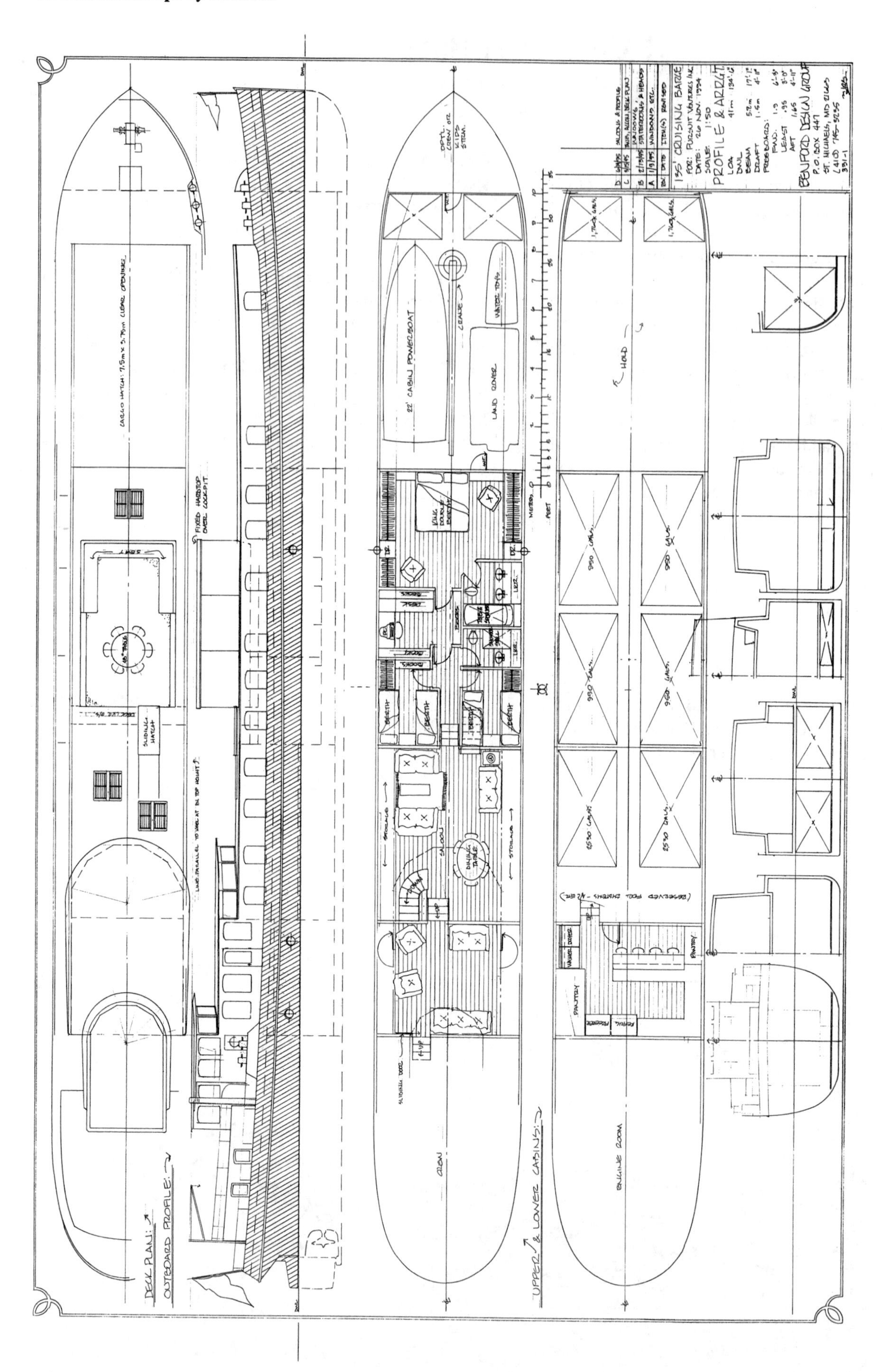

34' & 38' Fantail Cruisers

Designs Number 206 & 214
1982 & 1983

In 1969 I purchased a 1926, 50' fantail stern cruiser, the ***Kiyi***. I lived aboard for a very pleasant year (1970), and made the mistake of selling her. I thought that the expanding design business required moving into larger quarters. I should have just rented office space and kept on living aboard....

These vessels were designed with the memories of a vessel I'd owned and loved years ago still fresh in my mind. With this thought and because she was such a fine echo of the practical vessels of years ago, I called the 34-footer the ***Memory***. Built and outfitted in the spirit of her design, she will create many more fine memories for her owners and crew. Her light, easily driven hull makes for low construction costs, and low fuel consumption permitting long summer cruises to be a thing of reality again. She was originally built cold-molded, and a fiberglass mold was taken off one of the cold-molded hulls. The cold-molded scantlings are quite conservative, providing for robust structure.

In today's world, boatbuilding costs are calculated on a dollars-per-pound basis. This means that a boat that is a bit lighter should be a bit less expensive to build. This is because most material is sold by the pound, and the lighter boat will require a smaller engine and lighter gear.

Memory is a serious cruiser of lighter displacement which opens up cruising to those who previously couldn't afford a vessel this size. Economics and aesthetics can go hand in hand with her elegant fantail stern bringing in a classic appearance that will always be appealing.

In style, ***Memory*** is a smaller version of the ***Kiyi***, with much the same appearance in a boat two-thirds as long. The ***Kiyi*** had the same beam as ***Memory***, and was quite easily driven. Everywhere I went, the ***Kiyi*** was admired by others who agreed with me that she was very lovely. Boats built from this design are likely to have the same response.

Before the days of the horsepower race and before cruising became so crowded that everyone felt they had to rush to the next marina to get a slip for the night, there existed a much more economical style of cruiser. They were easily driven at modest speeds by small engines. My thoughts on this are expressed in more detail in The Height of Fuelishness chapter.

My comments on 10 knots being desirable, as quoted in the interview in that chapter notwithstanding, I'd suggest an even slower speed as being desirable and worth considering. This is particularly true with smaller vessels, such as the 34-footer shown in this chapter. A slower speed allows the shorter vessel to operate at a lower speed-length ratio which is even more conducive to good economy of operation.

This 34' fantail cruiser is intended for speeds in the 6 to 8 knot range, with 6 to 7 being quite economical. We've specified a 22½ horsepower diesel, even though she needs only about 12 to 15 horsepower to reach her top speed. This will allow a margin for adverse weather conditions as well as towing another vessel, if aid is needed. She can be slightly over wheeled so the engine doesn't have to operate at full rpm's.

Particulars:	**34**	**38**
Length overall	34'-0"	38'-0"
Length designed waterline	31'-0"	34'-6"
Beam	10'-6"	12'-0"
Draft	2'-6"	2'-9"
Freeboard		
Forward	5'-3"	6'-3"
Least	2'-9"	3'-6¾"
Aft	4'-0"	4'-6"
Displacement, cruising trim*	9,200 lbs.	11,765 lbs.
Displacement-length ratio	138	128
Prismatic coefficient	0.586	0.575
Pounds per inch immersion		1,155
Water tankage	80 Gals.	100 Gals.
Fuel tankage	50 Gals.	60 Gals.
Headroom	6'-3"	6'-6"

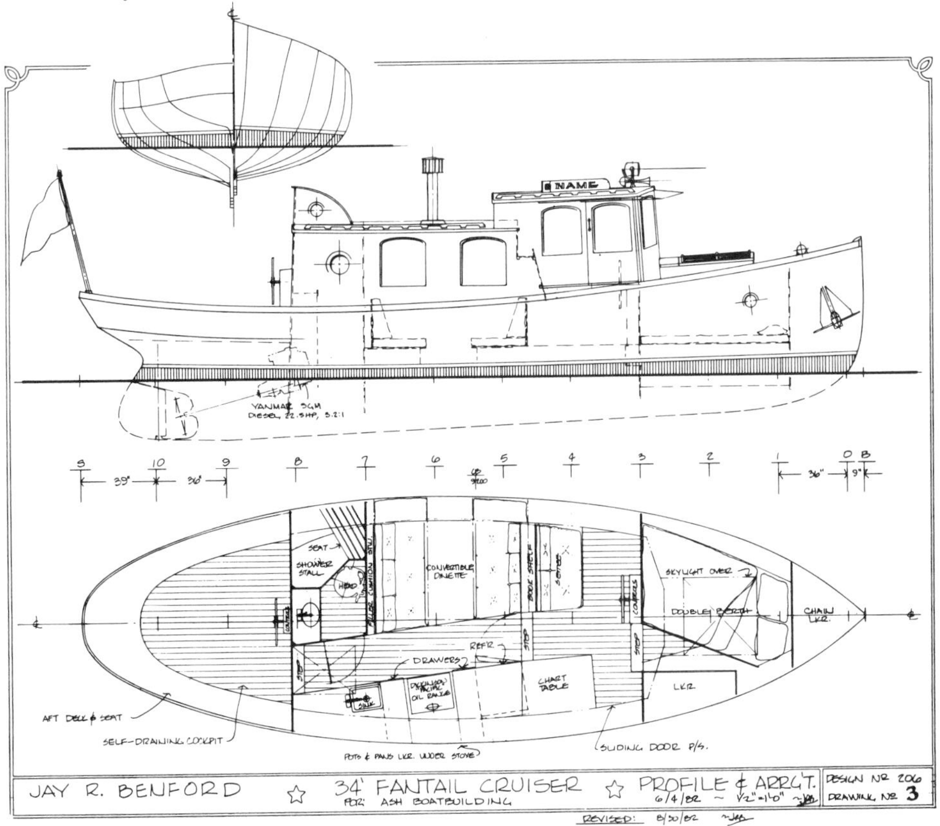

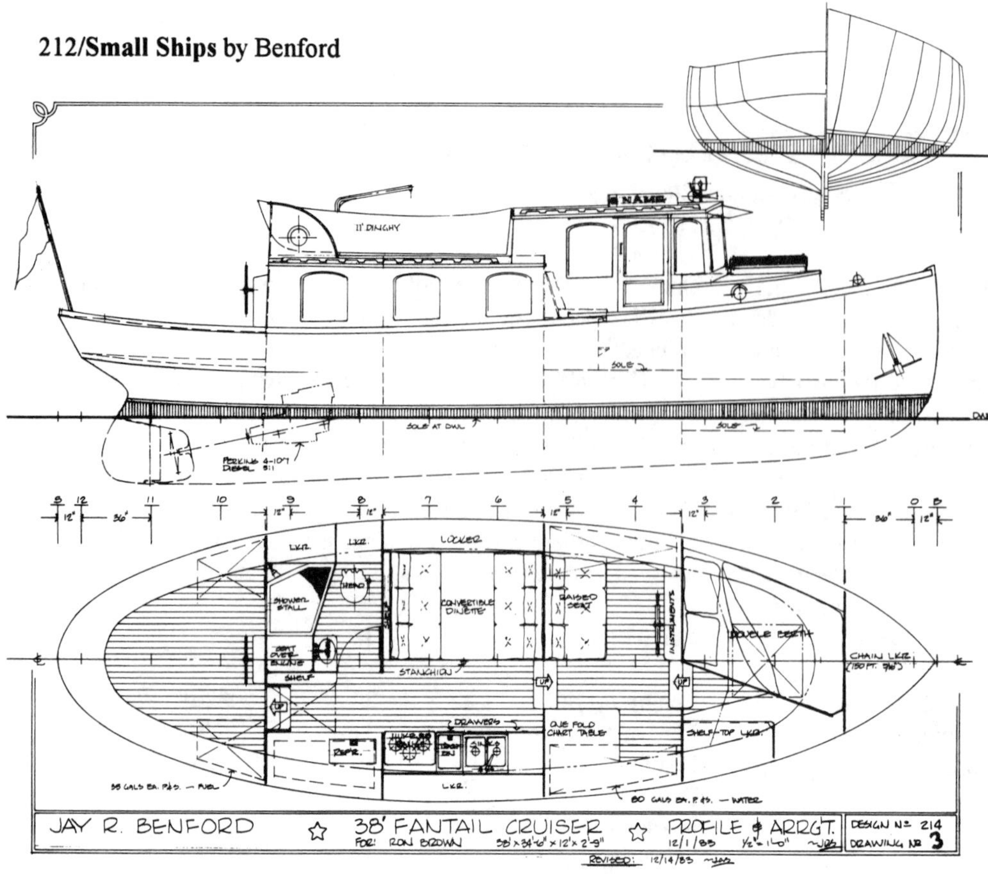

The pilothouse has a two person seat, raised for good visibility. The chart table is close to the helm, and is a one-fold chart size. The sliding door port and starboard makes it easy for the helmsman to look out and check docking situations and gives good ventilation. Her forward stateroom has a comfortable double berth, with a large skylight overhead. This provides a good source of ventilation and makes the cabin well lit, as well as adding to the headroom and sense of spaciousness there, and lets the crew see what sort of day they have ahead of them before they get up. There is good stowage under the berth and in the locker to starboard.

The aft cabin has a convertible dinette, for guests to stay aboard. The galley opposite follows the slanted passageway to the aft door. The passage is offset to clear the engine, which is on centerline under the head sink. Keeping the engine aft, we've shortened the drive line, and kept the engine away from the pilothouse, so the pilothouse will be quieter. The enclosed head has a separate shower stall with a seat, so the whole head compartment won't be soaked every time someone showers.

The aft deck is at comfortable sitting height from the cockpit sole. There is an outside control station for use when fishing. The aft deck is wide enough to serve as wraparound seating for outdoor living. The deck is self-draining and broad enough to permit using deck chairs for lounging. The fixed scuttle over the aft entry door gives headroom for going below and provides rain shelter for opening the door for ventilation in inclement weather.

This is one of my favorite designs, I've enjoyed doing others like it in different sizes. The ***Solarium 44*** is a larger version. I've been working off and on, for the last several years, on some 65' variations for ourselves, both to live aboard and to contain our office, and these are also shown in this book.

The 38-footer was designed for a client who admired the 34 and wanted a bigger version with a bit more room all around. Her displacement is 28% larger, which is a good reflection of the amount of space gained. Power requirement is about 17 hp for hull speed, and an engine in the 20 to 30 hp range would be just about right.

There is room for an 11-foot dinghy on the aft housetop versus 9 feet on the 34. There is a large hold space under the pilothouse again, giving room to carry all sorts of gear and stores on board. She would make an excellent long range cruiser and liveaboard. Her styling would mark her as a classic and appeal to practical sailors for years to come, making her a good investment as well.

The second 38' version is another variation on our Solarium theme. The after end could be enclosed like on the 44, making a great living space. We've had interest in a Solarium 34, and this may be built soon too.

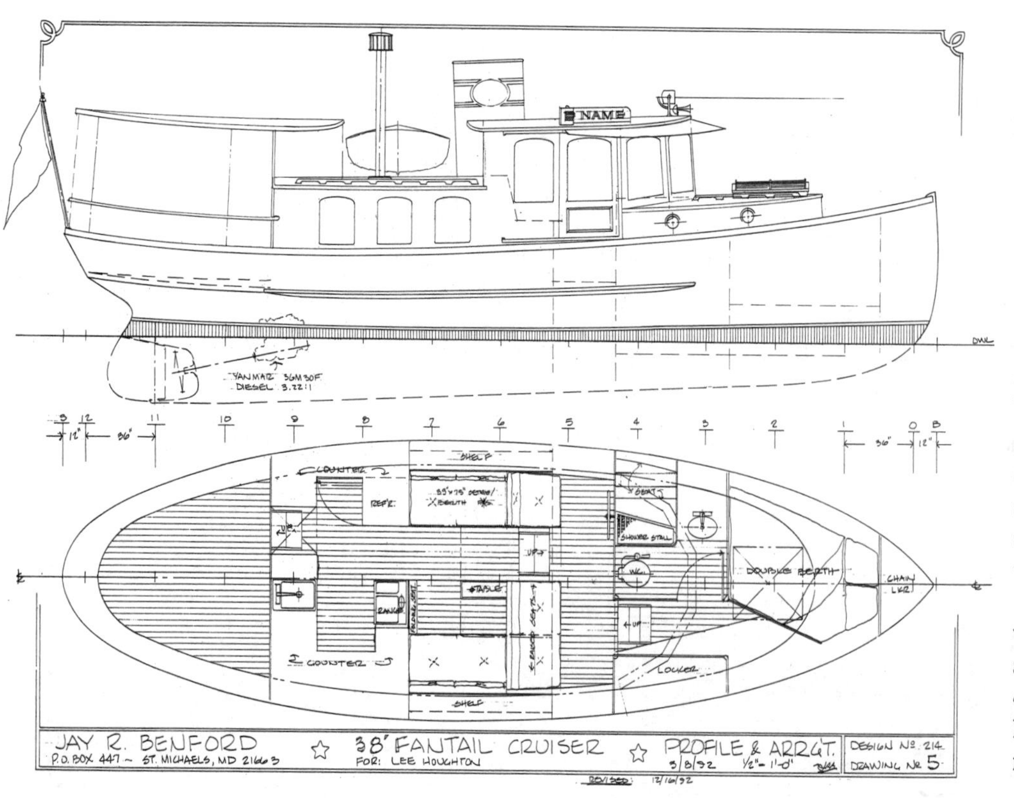

Solarium 44 Motor Yacht

Design Number 272
1980 & 1988

To Err Is Human...

Once upon a time, a couple of decades ago, I bought one of the most lovely yachts in the Pacific Northwest. Her name was (and still is) the ***Kiyi***. She was built in 1926, planked with Port Orford cedar on oak frames.

I found her one weekend in the spring of 1969 when I was wandering around one of the brokerage marinas in Seattle. I was just browsing, looking for interesting boats and only poking around because I was curious. Discovering the ***Kiyi*** in one of the slips was an experience not to be forgotten. I ended up spending hours poking over her and through her and the discussions with the salesman turned into making an offer on her.

She was only for sale because the prior owner had died. The widow agreed to terms and we went through the survey finding nothing worthy of comment. Soon, I was the proud new owner.

Cruising in her made apparent something that continued with my decade living on ***Sunrise***: anyone with a distinctive vessel has no trouble meeting people and making friends in new ports.

I progressed from using her for conventional cruising to making some modifications to the interior so I could live aboard and have my office aboard. Of course, the business was a lot more modest in those days, but I had two drawing boards, two file cabinets, a stack of plan files, and lots of bookcases aboard.

I lived aboard for about a year, living what seemed like an idyllic experience; cruising from port to port calling on clients and looking in on boatbuilding projects going on in the area.

However, this time coincided with a time of growth and expansion in my yacht design practice, and I began to have doubts about being able to carry on as I wanted on the boat. This led to one of the worst mistakes I can recollect making; I agreed to sell the ***Kiyi*** to a fellow who'd been inquiring about her for some time.

At the time, it seemed that buying a house ashore would be the solution to filling the needs of a growing design business. With 20-20 hindsight, the obvious thing that I should have done was just rent some larger office quarters. Instead, I went through a couple of years without a boat to cruise; an experience I don't wish to repeat.

In looking back on those times and thinking about how to do the same thing again, I've created some ideas which are shown here for similar styled fantail motor yachts. The 44-footer actually has more room in it than the ***Kiyi***, by virtue of a two foot wider beam and the enclosed solarium aft. There is a master stateroom forward with head and shower stall, raised seating in the pilothouse, a small second stateroom, an office with guest berthing in it, and the galley and dining area in the solarium in the stern.

The Solarium 44, as we've come to call her, provides an interesting study in how to adapt the spaces for living and working aboard. The solarium space can have screens fitted with the windows open in the summer and the windows closed and solar heating on sunny winter days. The weight of the office is centered in the boat, helping to keep the center of gravity down.

Duchess

The original Solarium 44 was started as a powerboat version using the hull of our 44' Cat Ketch design. Several years later, Katy Burke and Taz Waller were visiting us and saw the drawings for her. They soon decided that this would make a great liveaboard for them and set about building ***Duchess*** in a manner befitting her name. Some color photos of her are shown on the cover and in the color photo section in the front of this book. Take a look at what a nice job they've done in building this elegant yacht. This sort of nice detailing makes for great pride of ownership and is an encouragement to keep her in *yare* condition, which in turn means that she'll likely last for generations, giving pleasure to all who cruise on her and see her.

These fantail motor yachts are variations on the theme I've been developing since falling in love with the ***Kiyi***. Perhaps someday I'll have a chance to build one of them and live with it for a while. Maybe closer to ***Kiyi*** in size, say about 50 to 52' with a 13 to 14' beam...

Particulars:

Length overall	44'-0"
Length designed waterline	40'-0"
Beam	12'-0"
Draft	3'-0"
Freeboard:	
Forward	6'-0¾"
Least	3'-9¾"
Aft	4'-9¼"
Displacement, cruising trim*	20,950lbs.
Displacement-length ratio	146
Prismatic coefficient	.549
Pounds per inch immersion	1,450
Entrance half-angle	19°
Water tankage	200Gals.
Fuel tankage	250Gals.
Headroom	6'-5"

***CAUTION:** The displacement quoted here is for the boat in cruising trim. That is, with the fuel and water tanks filled, the crew on board, as well as the crews' gear and stores in the lockers. This should not be confused with the "shipping weight" often quoted as "displacement" by some manufacturers. This should be taken into account when comparing figure and ratios between this and other designs.

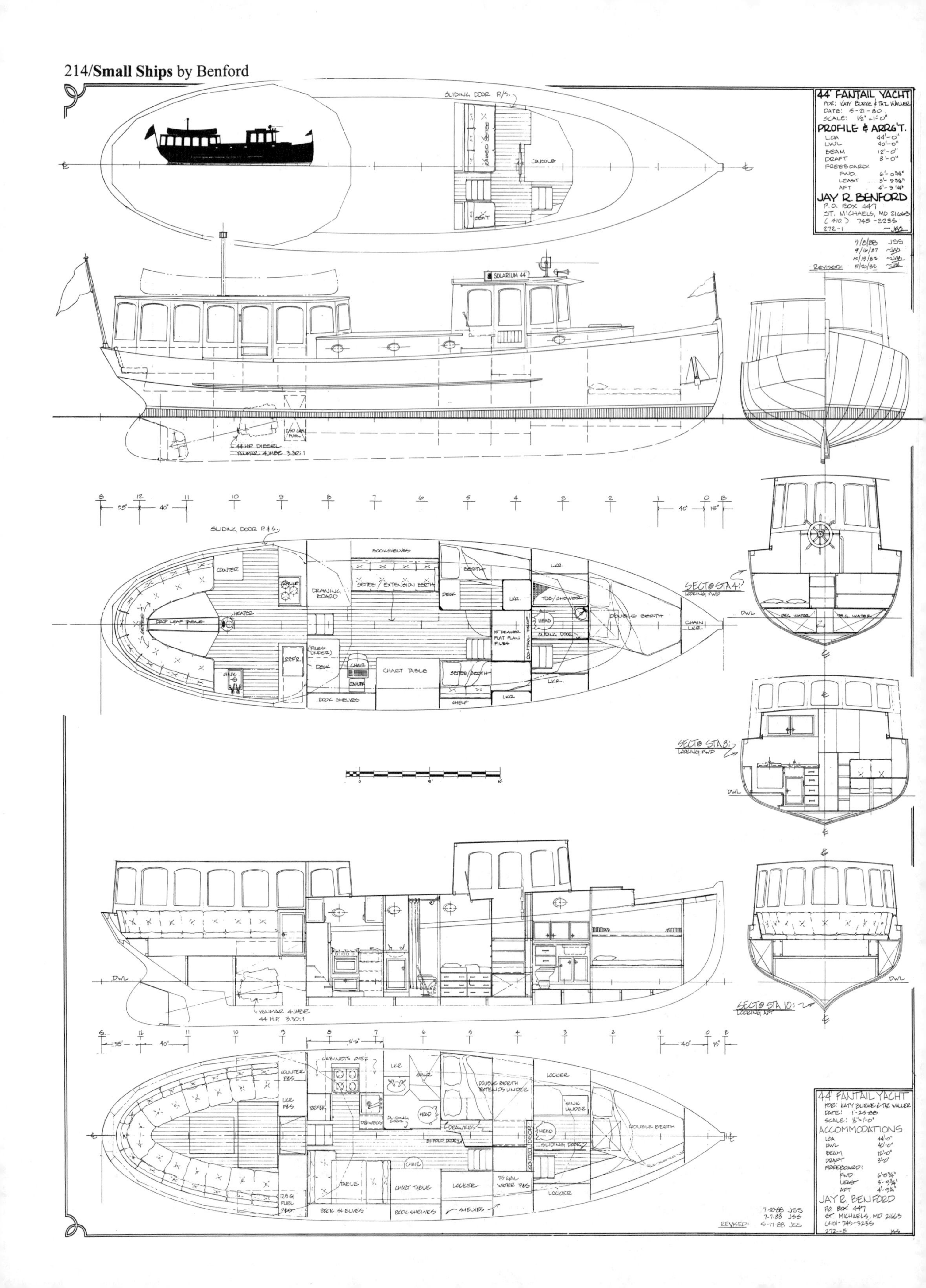
44' FANTAIL YACHT
FOR: KATY BURKE & TAZ WALLER
DATE: 5-21-80
SCALE: 1/2" = 1'-0"
PROFILE & ARRG'T.
LOA 44'-0"
LWL 40'-0"
BEAM 12'-0"
DRAFT 3'-0"
FREEBOARD:
FWD. 6'-0 3/8"
LEAST 3'-9 3/4"
AFT 4'-9 1/4"
JAY R. BENFORD
P.O. BOX 447
ST. MICHAELS, MD 21663
(410) 745-3235
272-1
SOLARIUM 44
SLIDING DOOR P/S.
CONSOLE
44 HP DIESEL
YANMAR 4JHBE 3.30:1
SLIDING DOOR P & S.
COUNTER
DROP LEAF TABLE
HEATER
SINK
REFR.
DRAWING BOARD
BOOKSHELVES
SETTEE / EXTENSION BERTH
CHART TABLE
BOOK SHELVES
DESK
BERTH
TUB/SHOWER
HEAD
SLIDING DOOR
DOUBLE BERTH
CHAIN LKR.
SETTEE/BERTH
SHELF
LKR.
SECT@ STA 4:
LOOKING FWD
SECT@ STA 8:
LOOKING FWD
SECT@ STA 10:
LOOKING AFT
DWL
YANMAR 4JHBE
44 H.P. 3.30:1
CABINETS OVER
COUNTER P&S
LKR P&S
REFER
DRAWERS
SLIDING DOOR
HEAD
DOUBLE BERTH EXTENDS UNDER
BI-FOLD DOORS
SLIDING DOORS
LOCKER
SINK UNDER
DOUBLE BERTH
TABLE
CHAIR
CHART TABLE
75 GAL WATER P&S
125 G FUEL P&S
BOOK SHELVES
SHELVES
44' FANTAIL YACHT
FOR: KATY BURKE & TAZ WALLER
DATE: 1-24-88
SCALE: 1/2" = 1'-0"
ACCOMMODATIONS
LOA 44'-0"
DWL 40'-0"
BEAM 12'-0"
DRAFT 3'-0"
FREEBOARD:
FWD 6'-0 3/8"
LEAST 3'-9 3/4"
AFT 4'-9 1/4"
JAY R. BENFORD
P.O. BOX 447
ST. MICHAELS, MD 21663
(410) 745-3235
272-6
REVISED:

Solarium 50

Design Number 290

1990

A recent design study for extending the ideas in the Solarium 44 to a larger size, this 50 has a raised flush deck section amidships instead of the trunk cabin on the 44.

There are two private double staterooms, each with its own head with shower stall. The after head also has a full-sized washer and dryer. The settee in the pilothouse could also serve for occasional guest berthing.

The solarium is also the dining and living room. The drop leaf table can seat a large gathering for meals and socializing in an area that has excellent ventilation and daylight.

Also shown are some ideas for doing some 44' x 14' variations (design number 236, 1985). The lines plans for this 44 and the 50 are some of the most refined ones we've done and will be the height of elegance afloat.

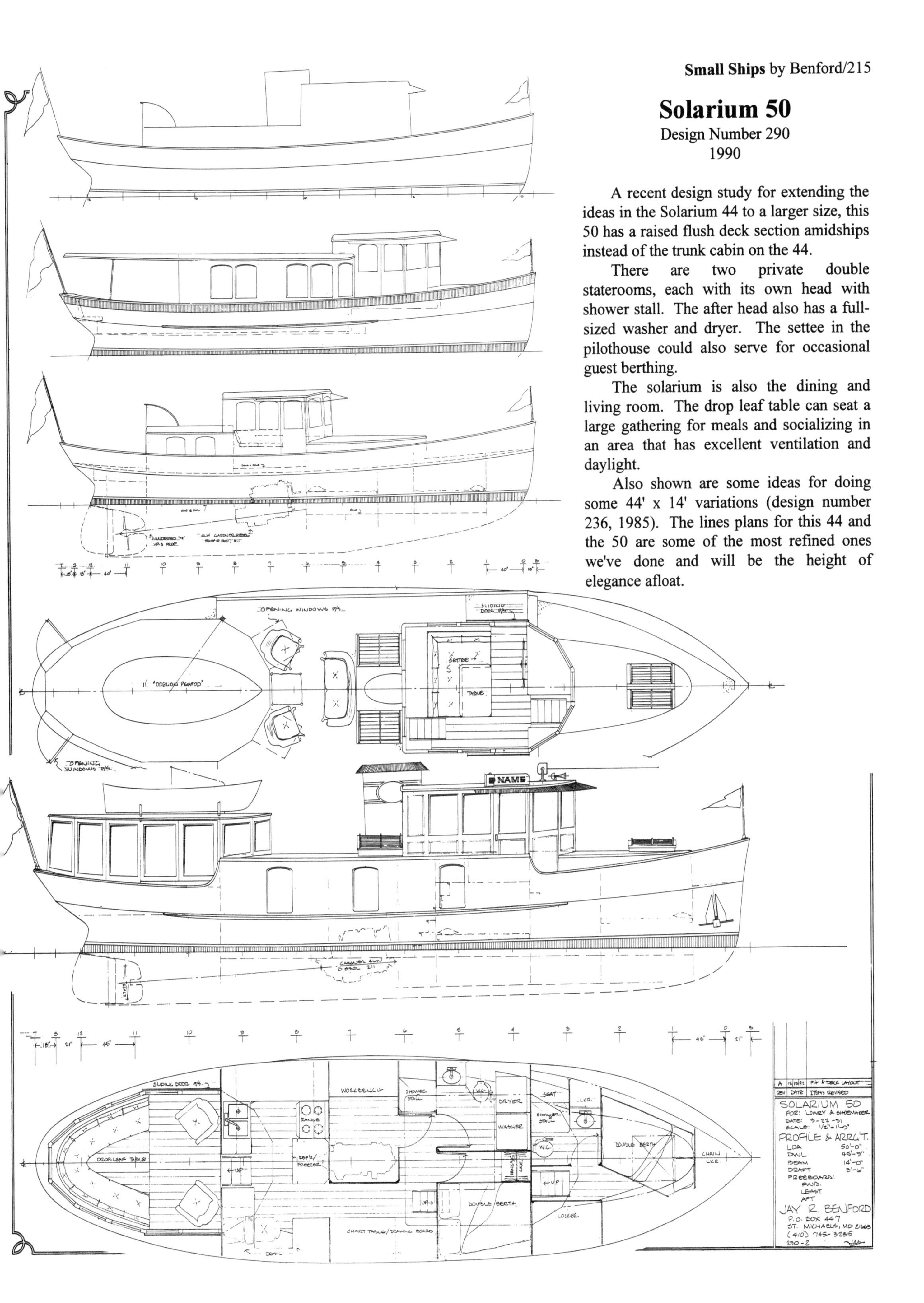

45' Fantail Yacht *Tiffany*

Design Number 296

1990

Our client for this boat is one of that select group that share my love for the old Mathis built-Trumpy designed fantail yachts from the 1920's. One of his businesses had been to make deluxe apartments in seemingly small spaces that had all the comforts of home and no feeling of being confined. He's turned those skills loose on the detailing of this boat and the result is a delightful and unique yacht.

The concept for the boat is a luxury traveling home for a couple. Their stateroom is in the bow, with a queen double on centerline, with drawers under the after end and a hanging locker on the aft bulkhead. Their head is in the main cabin, right behind the helm seat. If any guests do show up, there is the convertible sofa in the saloon.

The galley is on the passageway from the pilothouse to the saloon. It has all the essentials in a cleverly worked out layout. The saloon has a desk for writing projects to be done aboard, a sofa for relaxing, and a bookshelf for some reading and reference materials.

The deckhouse will have bright finished raised pa "like they used to make them" and a myriad of other detai make her an authentic small version of the classic fan From a distance we reckon it will be hard to tell her from of the original ones twice her size.

The drawing on this page shows one of the very preliminaries we did for this boat. It was 42' long at that You can see how the boat evolved from this to the version, with much of the details refined over time consultation with our client.

As the book goes to press we are expecting the signin the contract to build her to be executed, with her being over the next year or so. She should turn out to be a knock Stay tuned....

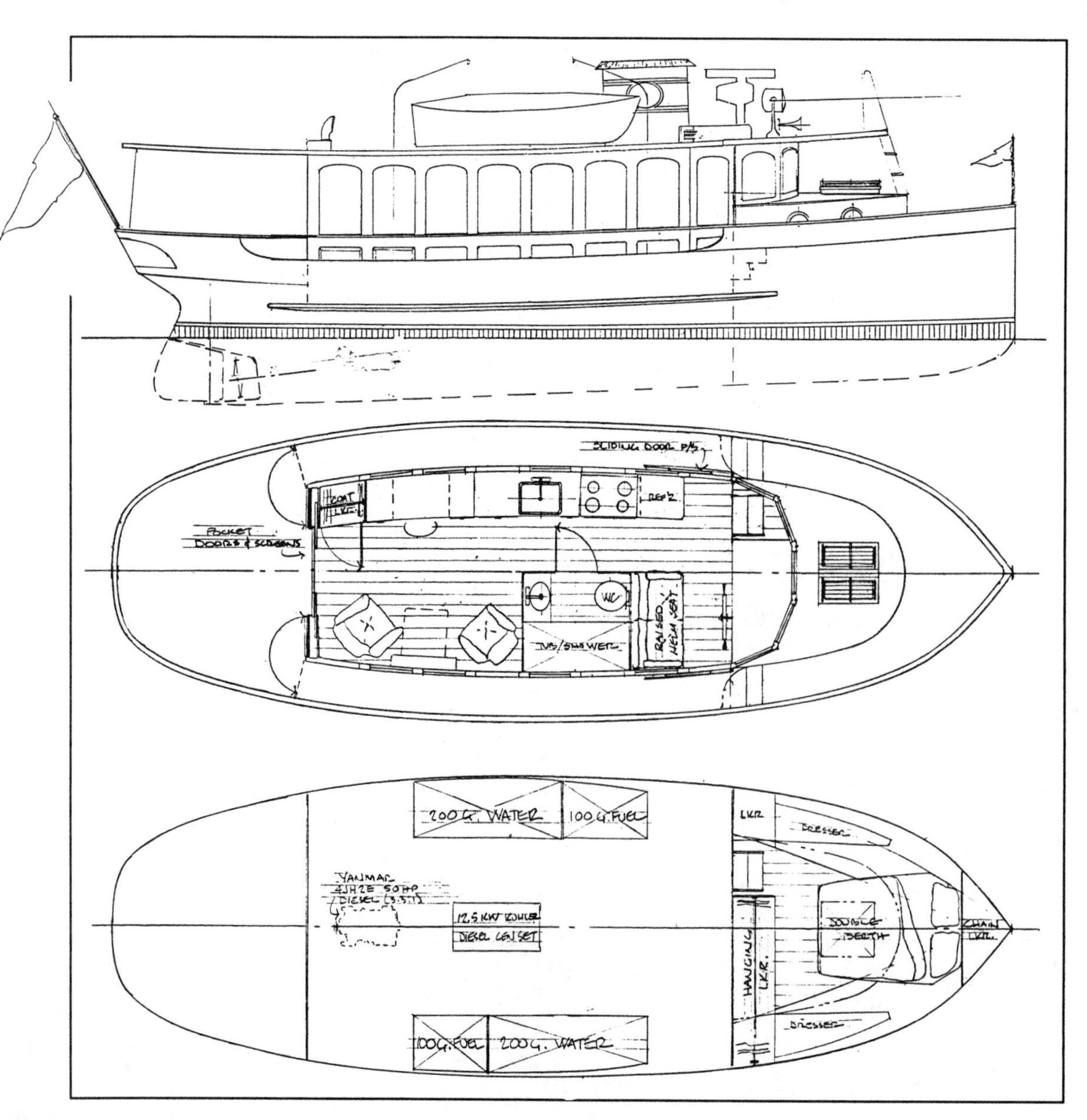

Particulars:

Length overall	45'-0"
Length DWL	42'-0"
Beam	14'-0"
Draft	3'-0"
Freeboard:	
Forward	5'-0"
Least	1'-6"
Aft	2'-0"
Displacement, cruising trim*	34,000
Displ.-length ratio	205
Lbs/inch imm.	2,225
Ballast	3,000
Water tankage	400 G
Fuel tankage	220 G
Headroom	6'-5±'
GM	5.6'

***CAUTION:** The displace quoted here is for the bo cruising trim. That is, with th and water tanks filled, the cre board, as well as the crews' and stores in the lockers. should not be confused wit "shipping weight" often quot "displacement" by manufacturers. This shoul taken into account when comp figures and ratios between thi other designs.

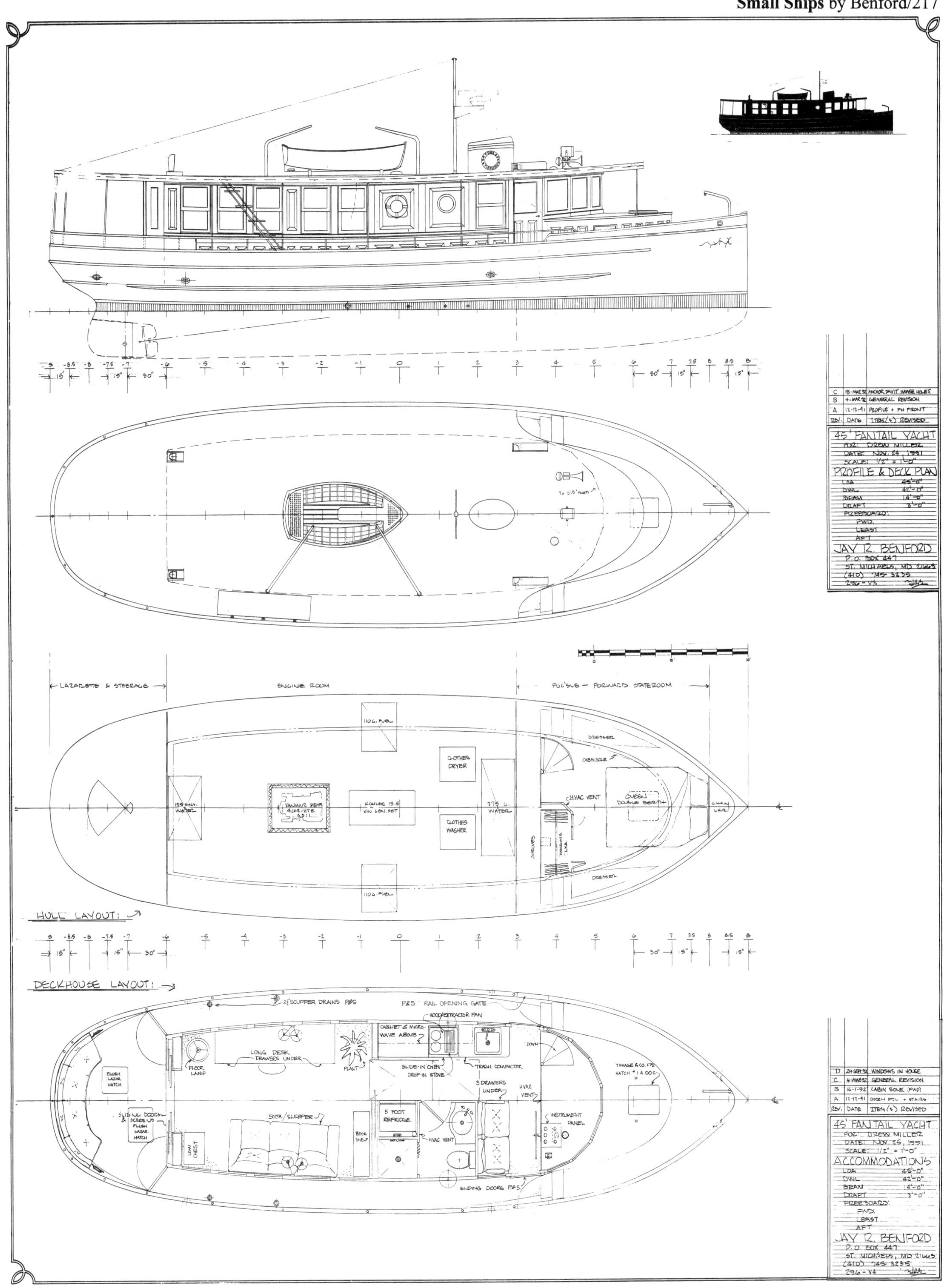
C 18-MAR 92 ANCHOR DAVIT HAWSE HOLES
B 4-MAR 92 GENERAL REVISION
A 12-12-91 PROFILE + PH FRONT
REV. DATE ITEM(S) REVISED
45' FANTAIL YACHT
FOR: DREW MILLER
DATE: NOV. 24, 1991
SCALE: 1/2" = 1'-0"
PROFILE & DECK PLAN
LOA 45'-0"
DWL 42'-0"
BEAM 14'-0"
DRAFT 3'-0"
FREEBOARD:
FWD.
LEAST
AFT
JAY R. BENFORD
P.O. BOX 447
ST. MICHAELS, MD 21663
(410) 745-3235
296-V3
LAZARETTE & STEERAGE
ENGINE ROOM
FOC'SLE — FORWARD STATEROOM
110 G. FUEL
CLOTHES DRYER
CLOTHES WASHER
KOHLER 13.5 KW GEN. SET
275 G. WATER
HVAC VENT
QUEEN DOUBLE BERTH
DRESSER
CABIN SOLE
SHELVES
HANGING LKR.
CHAIN LKR.
HULL LAYOUT:
DECKHOUSE LAYOUT:
2½" SCUPPER DRAINS P&S
P&S RAIL OPENING GATE
HOOD/EXTRACTOR FAN
CABINET & MICRO-WAVE ABOVE
LONG DESK DRAWERS UNDER
FLOOR LAMP
PLANT
SLIDE-IN OVEN DROP-IN STOVE
TRASH COMPACTER
3 DRAWERS UNDER
HVAC VENT
DOWN
FLUSH LAZAR. HATCH
SLIDING DOORS & SCREENS
SOFA/SLEEPER
5 FOOT REFRIDGE
BOOK SHELF
HVAC VENT
INSTRUMENT PANEL
SLIDING DOORS P&S
D 24 SEPT 92 WINDOWS IN HOUSE
C 4-MAR 92 GENERAL REVISION
B 16-1-92 CABIN SOLE (FWD)
A 12-12-91 QUEEN BTL. + STAIRS
REV. DATE ITEM(S) REVISED
45' FANTAIL YACHT
FOR: DREW MILLER
DATE: NOV. 25, 1991
SCALE: 1/2" = 1'-0"
ACCOMMODATIONS
LOA 45'-0"
DWL 42'-0"
BEAM 14'-0"
DRAFT 3'-0"
FREEBOARD:
FWD.
LEAST
AFT
JAY R. BENFORD
P.O. BOX 447
ST. MICHAELS, MD 21663
(410) 745-3235
296-V4

48 & 52' Fantail Houseyachts

Design Number 309
1989 & 1990

The first two fantail houseyachts shown are some evolutionary developments from our freighter yachts, with different styling on the sterns. They share a lot of similar features and the ideas shown here can be adapted into the freighters or we could redesign some of the freighters to have this sort of style.

In an attempt to hide the wide-body styling on the 48, we've shown the window panels inset from the tumblehome cabin sides, as they were built on the 45' Coaster ***Sails***. This sculpted relief creates a strong shadow line and gives vertical window panels for ease of interior outfitting.

The 48 has the interior at the mid-level maximized, with no waist decks and thus no entry at this level. We could sacrifice the pantry here and move it to the basement and use this space to pocket in an entry door on the starboard side. I have shown port and starboard entry gates onto the after deck as an alternative.

The 48's saloon has a freestanding fireplace, with the stack coming up through a corner of the master stateroom. This would allow radiant heat to both spaces. If the boat was to be lived aboard, particularly in an area like where I lived in the Pacific Northwest, a lot of fuel could be beachcombed. Like cutting your own firewood, it would warm you twice, once when you gathered it and once when you burned it.

The 52' Fantail Houseyacht was developed by pulling out the ends of the 48-footer by two feet each fore and aft. The change in appearance is dramatic, making for a much more elegant appearing boat. The four feet of additional structure would cost little more — it is only what changes might be made inside that would be a difference in cost between the two boats.

The 52 has good waist decks at about four feet off the water for boarding and side access to the boat. She has entries on both sides at the waist and onto the after cockpit. The cockpit has an elliptical seat built into that shape at the stern. This is a practical feature, since the overhang of the fantail shape means that the deck would not extend all the way to the bulwark rail.

With most of the additional length devoted to making the mid-level longer, the stateroom and head have been turned lengthwise and there are full length waist decks. The mid-level stateroom shows an office layout with settees that convert to make upper and lower berths.

The second 52-footer, with a 20' beam, has all her accommodations on two-levels. This is an idea in which I tried to combine some of the styling aspects of the older style fantail houseyachts in the next section with a lower deck opening directly off the saloon, like the Coasters.

The hull form is the beginning of an idea of how to make for an even beamier vessel than we've done on the Coasters. This would give more interior room down below and good upper walk-around decks, which still leave a good amount of interior room for the master stateroom. This ability to walk around the decks will be appreciated by the line handlers.

Her interior has a good sized office aboard. With the proliferation of cellular phones and mobile computing, it is becoming easier and more attractive to have a mobile office.

The fireplace is at the base of the stairs, which has a large opening above. This will let the heat rise up easily to warm the upper cabins.

With the lower cabins all on one level, there is not sufficient depth with the shallow draft hull form for a standing headroom shop and engine room. We've provided for a stairway down to the hull under the main stairs to ease the access. There is plenty of room for the machinery, tanks, stores, and long-term provisioning.

There is a full-size ("no-fold") chart table in the pilothouse. For those of us who still use full charts, in addition to the popular chart books, the ability to lay the charts out flat and store them flat is very much appreciated. I also found this makes a nice drawing board when anchored out in some quiet spot....

The displacement quoted for her is assuming plywood and epoxy glued and sealed construction. If she was built in steel, her displacement would be more like the other 52-footer, due mainly to the difference in structural weights.

We could also redesign the hulls of the Coasters to make 45x18 or 50x20 sizes with finer bows like this and sterns that are a bit fuller than shown here. This would produce hulls that take no more power to drive than the current ones, yet are roomier. With each new design we improve on the one that went before. The state-of-the-art is definitely a moving target!

Particulars:	**48'**	**52'**	**52'**
Length overall	48'-0"	52'-0"	52'-0"
Length designed waterline	43'-6"	49'-0"	50'-0"
Beam	17'-0"	17'-0"	20'-0"
Draft	3'-6"	3'-6"	3'-6"
Freeboard:			
Forward	9'-0"	9'-3"	8'-7½"
Waist	8'-9"	4'-1"	8'-6"
Aft	1'-9"	1'-9"	1'-9"
Displ., cruising trim*	96,000 lbs.	104,000	75,000
Displ.-length ratio	521	395	268
Water tankage	1,000 Gals.	1,000	1,200
Fuel tankage	1,000 Gals.	1,000	800
Headroom	6'-7"	6'-7"	6'-7"

***CAUTION:** The displacement quoted here is for the boat in cruising trim. That is, with the fuel and water tanks filled, the crew on board, as well as the crews' gear and stores in the lockers. This should not be confused with the "shipping weight" often quoted as "displacement" by some manufacturers. This should be taken into account when comparing figures and ratios between this and other designs.

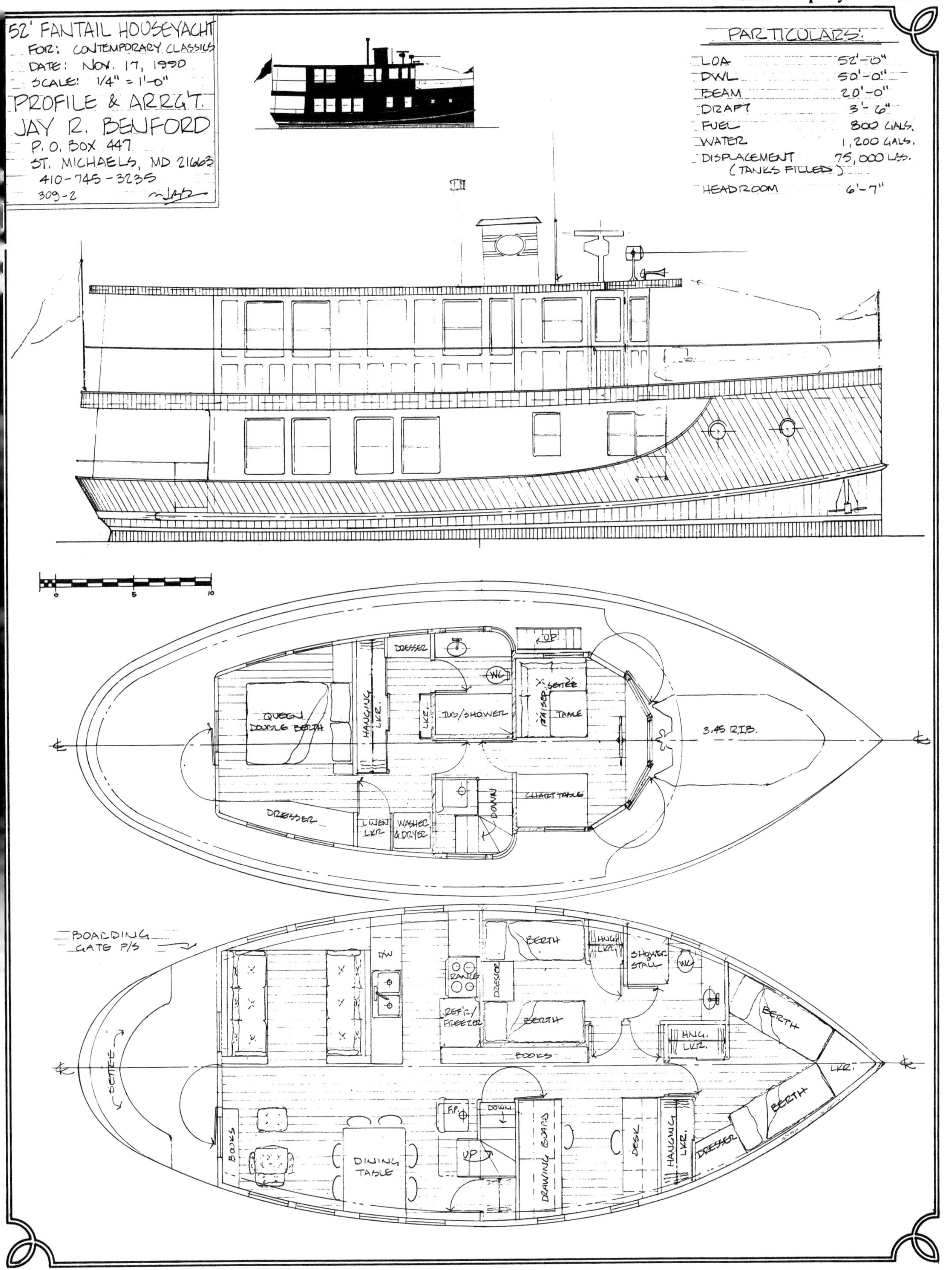

52' FANTAIL HOUSEYACHT
FOR: CONTEMPORARY CLASSICS
DATE: NOV. 17, 1990
SCALE: 1/4" = 1'-0"
PROFILE & ARRG'T.
JAY R. BENFORD
P. O. BOX 447
ST. MICHAELS, MD 21663
410-745-3235
309-2
PARTICULARS:
LOA 52'-0"
DWL 50'-0"
BEAM 20'-0"
DRAFT 3'-6"
FUEL 800 GALS.
WATER 1,200 GALS.
DISPLACEMENT 75,000 LBS.
(TANKS FILLED)
HEADROOM 6'-7"
0
5
10
UP
DRESSER
WC
SETTEE
QUEEN DOUBLE BERTH
HANGING LKR.
LKR.
TUB/SHOWER
RAISED TABLE
3.45 RIB.
CHART TABLE
DRESSER
LINEN LKR
WASHER & DRYER
DOWN
BOARDING GATE P/S
DW
BERTH
HNG. LKR.
SHOWER STALL
WC
RANGE
DRESSER
BERTH
REF'R/ FREEZER
BERTH
HNG. LKR.
SETTEE
BOOKS
LKR.
F.P.
DOWN
BERTH
BOOKS
DINING TABLE
UP
DRAWING BOARD
DESK
HANGING LKR.
DRESSER

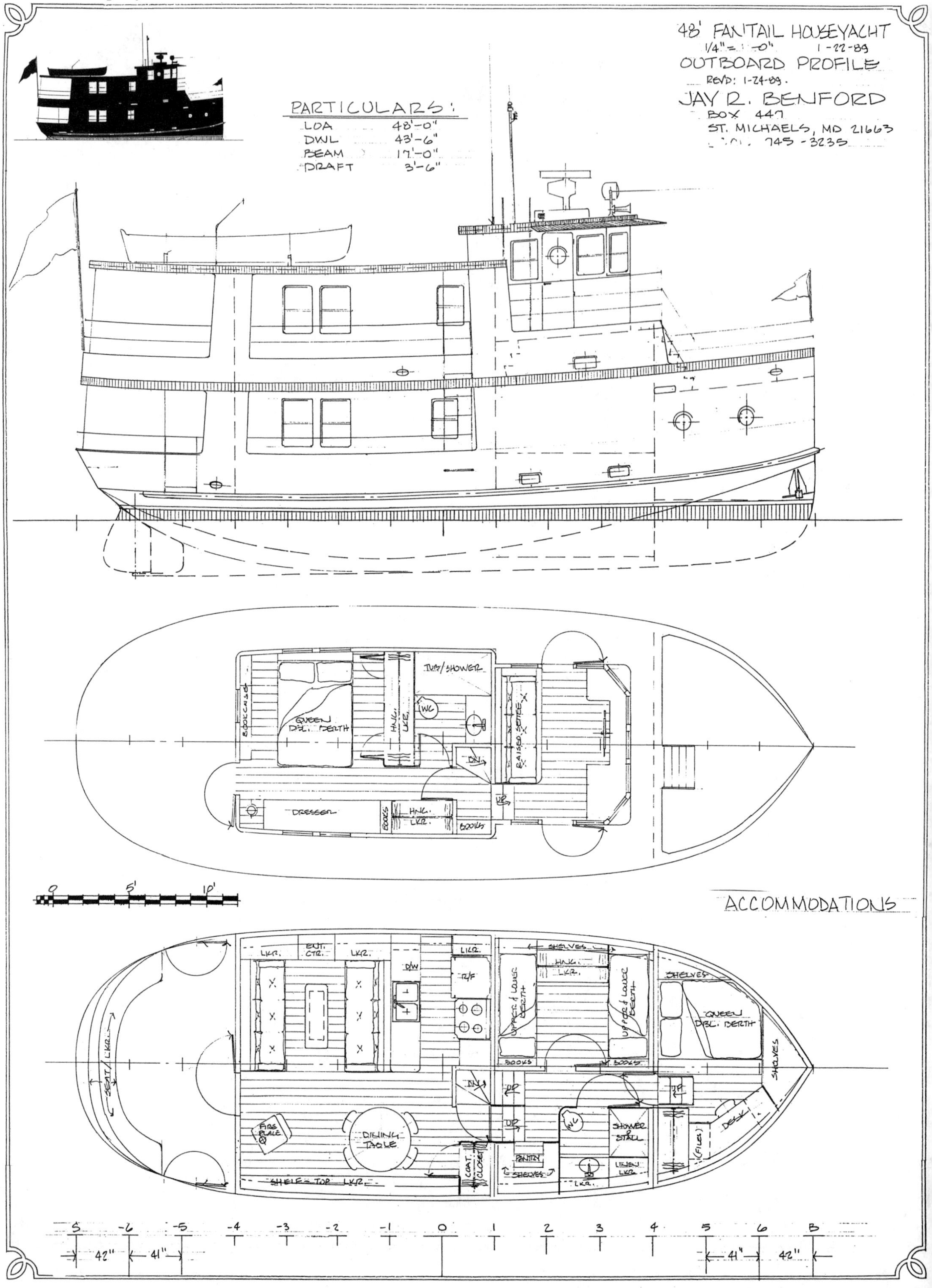

48' FANTAIL HOUSEYACHT
1/4" = 1'-0"   1-22-89
OUTBOARD PROFILE
REV'D: 1-24-89.

JAY R. BENFORD
BOX 447
ST. MICHAELS, MD 21663
745 -3235

PARTICULARS:

LOA	48'-0"
DWL	43'-6"
BEAM	17'-0"
DRAFT	3'-6"

ACCOMMODATIONS

TUB/SHOWER
WC
BOOKCASE
QUEEN DBL. BERTH
HNG. LKR.
D/W
RAISED SETTEE
UP
DRESSER
BOOKS
HNG. LKR.
BOOKS

LKR.
ENT. CTR.
LKR.
LKR.
D/W
R/F
SHELVES
HNG. LKR.
UPPER & LOWER BERTH
UPPER & LOWER BERTH
SHELVES
QUEEN DBL. BERTH
SHELVES
SEAT/LKR.
BOOKS
BOOKS
DN
UP
UP
WC
SHOWER STALL
DESK
FIRE PLACE
DINING TABLE
COAT CLOSET
PANTRY SHELVES
LINEN LKR.
LKR.
(FILES)
SHELF - TOP LKR.

0 5' 10'

S -6 -5 -4 -3 -2 -1 0 1 2 3 4 5 6 B
42" 41"
41" 42"

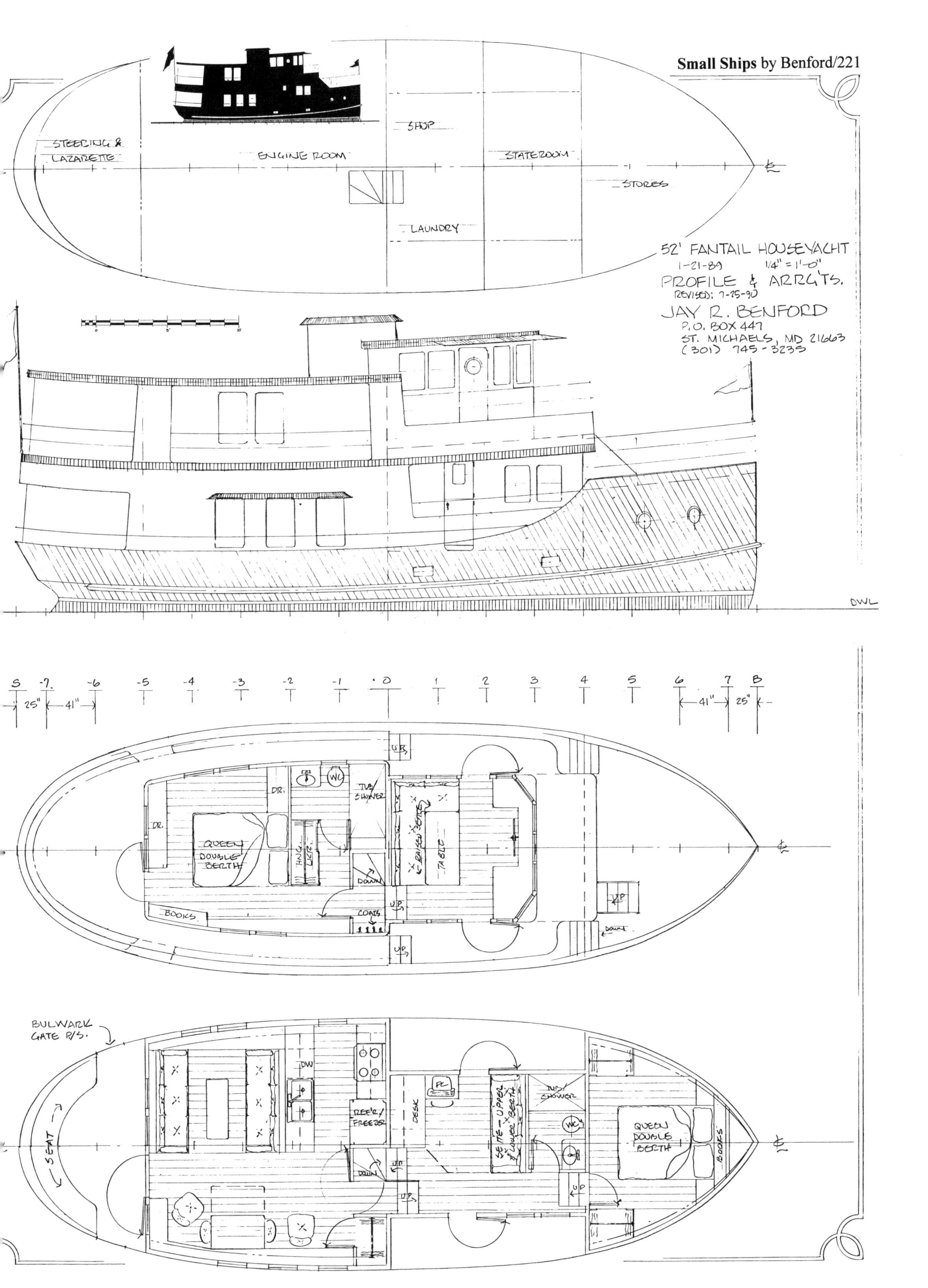
STEERING &
LAZARETTE
ENGINE ROOM
SHOP
STATEROOM
STORES
LAUNDRY
52' FANTAIL HOUSEYACHT
1-21-89 1/4" = 1'-0"
PROFILE & ARRG'TS.
REVISED: 7-25-90
JAY R. BENFORD
P.O. BOX 447
ST. MICHAELS, MD 21663
(301) 745-3235
DWL
QUEEN DOUBLE BERTH
DR.
WC
TUB/SHOWER
HANG. LKR.
RAISED SETTEE
TABLE
DOWN
UP
BOOKS
COATS
BULWARK GATE P/S.
SEAT
DW
REEF'R/FREEZER
DESK
SETTEE - UPPER & LOWER BERTH
QUEEN DOUBLE BERTH

KIYI

This 44-year-old "sailor's powerboat" is home and office for a naval architect who is making a name as designer of old-fashioned character boats. With his business split evenly between Seattle and Vancouver, he finds the venerable 50-footer also serves as essential transportation, offering many more pleasurable amenities than the freeway.

by Jay Benford

Like other boat nuts, I spend a fair amount of time wandering around boatyards, looking at boats – watching the way they grow into reality and the way they grow old, for there is much which can be learned by such observations and continuing visits over a period of time.

It was on one of these meandering trips last May that I found the KIYI for sale at a marina in Seattle. She was the personification of my idea of what a boat *could* be, and I wanted to possess her. After exploring the whole of her, from stem to stern, and determining what was aboard and more of the details about her, I made out an earnest money agreement with the brokers.

Benford at Kiyi's "flying bridge," which in deference to chilly wind and spray is fitted with a plastic windshield when boat isn't being photographed.

The next two weeks were spent putting my financial house in the proper order so that I could actually acquire her, in getting a haulout and survey (and doing the underwater painting at the same time), and getting sorted out with the finance company. It turned out that this was time well spent, for the whole of it was successful with the survey perfect, the finance company somehow being convinced that I was a good risk, and the ownership transferring into my name.

During this time and since then, I have had a number of interesting talks with various people about the *KIYI* and her history. She was built in 1926 by Schertzer Brothers in Seattle for the Philbricks. Schertzer Brothers had their shop on Lake Union in Seattle, right next door to where Vic Franck's is now, and their shop has been since used by the Edison Technical School for training young men in boatbuilding. At the time of her construction, the last owner, F.A. "Doc" Harvey had another boat under construction right alongside of the *KIYI*. He had known and loved her for many years before he had his chance to buy her some years ago. It was only after his death that she came on the market again and I had an opportunity to buy her.

The *KIYI* was designed by L.H. Coolidge, who has quite a number of rum runners, tugs, and other pleasure boats to his credit also. Another one of his well known pleasure boats is the *SINDBAD,* a black hulled 40' schooner in Seattle. She is now owned by Dick Wagner, proprietor of The Old Boathouse.

The construction of the *KIYI* appears to be first class throughout. Planking is one-and-one quarter-inch Port Orford Cedar over one-and-three-quarter-inch square bent oak frames on ten-inch centers. The backbone and other structural members are of equally good proportions, and she has maintained her shape and structural integrity quite well over the years.

On deck, she has an electric anchor winch with an anchor in the hawse pipe to starboard with the chain going down into the chain locker in the forepeak. The wildcat for the chain can be declutched and the niggerhead used for hauling in lock lines and the like. On top of the trunk cabin and overlapping onto the top of the pilothouse is a flying bridge that Doc

Reprinted courtesy of **Pacific Yachting** magazine from their May 1970 issue.

Harvey added, and it seems to blend in well. Just aft of that is the stack for the old dry exhaust, which now contains a fan for ventilation of the engine compartment or for drawing hot air off the overhead in the galley area, depending on how one works the controls inside.

The small boats are also carried on top of the trunk cabin. There is a small winch on the boom topping lift for lifting them on and off. The canoe I added to the boat, and it gets most of the use for going ashore and exploring. The 9' Penn Yan dinghy has a fitted cover, and makes a nice weatherproof locker for keeping gear out of the way. The dinghy doesn't get much use, and is soon to be replaced. The replacement is a most interesting boat. It is a hand crafted lapstrake, varnished 14' Whitehall pulling boat. These small boats are still being made by a young man in White Bear Lake, Minnesota, who is carrying on the tradition of producing fine small boats. Dick Wagner, at The Old Boathouse in Seattle, has some of these among his fleet of rowing and sailing rental boats, and is sales agent for them as well.

Aft, on the fantail, we have a large beach umbrella which is used for shade on overly sunny days and for shelter in the rain. There is a small table with a hole in the middle that fits under it and deck chairs for sitting around it. At the extreme aft end of the fantail, there is a built in seat with storage under one side and a 500 watt/110 volt generator under the other side. Inside the seatback a charcoal barbeque pit is stowed. This unit has a pipe leg that screws into the bottom of it, and sits in the flagpole socket over the stern when in use. Below the fantail there is the quadrant aft, with the emergency tiller and various clam guns and the like. Forward of this is a water tank on either side, about 50 gallons each, with the pressure water set and stowage between them. This area is accessible from inside the cabin or from on deck.

Going down the aft companionway, there is a hanging locker each side of the ladder. Then, to starboard is a convertible dinette with a bookshelf outboard of it under the side deck. Opposite is a settee with similar bookshelf and stowage behind and under the seat. Proceeding a bit further forward, we have the galley area, with range to starboard and sink and counter to port. Then, there is a box on centerline covering the engine, which serves as a buffet at times and presently is the home for the stereo set. To port of the engine box is the main head with a hot water shower – a great convenience since we live aboard. To starboard of the engine box is the electric refrigerator and

Her 143 HP Chrysler Royal pushes her at a majestic eight knots. Umbrella on stern is normally used only when moored, and the canoe gets more use than the dinghy.

more stowage. The passage along the starboard side of the engine box leads to the steps up to the pilothouse.

In the pilothouse, where a convertible settee used to reside, is my drawing board, with books and files and drawings stowed underneath. The forward end of the pilothouse has the controls for the boat, with a helmsmans seat to port. Just aft of this seat, there is a sliding door each side. Alongside the seat is a drop window port and starboard. Under the sole of the pilothouse there are fuel tanks and batteries each side, with stowage and access aft to the engine down the centerline.

Going forward from the pilothouse, one steps down into the focs'le. This is a private stateroom for two, with its own head and sink, and also a heater and the inverter for charging the batteries from shore power. There is a fair amount of storage under the berths, and some more bookshelves. In the forepeak, over the chain stowage, there is also some stowage space.

The original powerplant was a Hall-Scott 100 horsepower gas engine with a two to one reduction gear. The present engine is a Chrysler Royal, a flathead straight eight, rated at 143 horsepower. It has a 3.17:1 reduction gear and drives a 27 x 18-21 five bladed Coolidge propellor. The engine will turn this wheel up to 2300 rpm, at which point it is fully loaded, with quite a stern wave following. Normal cruising speed is 8 knots at 1700 rpm, at which point she is burning two and one quarter gallons per hour and has just over 10 inches of manifold vacuum. With a tank capacity of 200 gallons, this gives her a potential range of a bit over 700 miles.

With only 10 feet of beam on 50 feet of length (and 4' draft), there are some facets of her handling that are exaggerated from more conventionally proportioned vessels. Besides driving very easily when going ahead, because she carries practically her full draft up close to the plumb stem, the KIYI tracks quite well and tends to be reasonably easy to keep on course while cruising. However, she requires care in backing situations, for the large wheel torque wants to move the stern to port. This can be counteracted pretty well in most situations by taking it very slow and easy, but with cross winds or currents in the wrong direction it becomes rather tricky.

Another time in which the effect of the narrow beam becomes quite evident is in the manner in which she handles herself in various sea conditions. Waves coming at us from dead ahead back to practically abeam cause mostly an up and down pitching motion, usually of a quite gentle nature. Even in the kind of weather sometimes found crossing the Strait of Georgia or Juan de Fuca Strait, the KIYI will keep driving right on through. When the waves come at her from abeam to back on the quarters, some rolling motion is generated, and in windy weather it is more comfortable to set the course such that this rolling motion is minimized. The clinometer on board has registered a bit over twenty degrees in weather like this – which may not be that unusual, but certainly acts as a good test on the stowage practices of the crew. In a following sea a gentle surging motion is felt and heard in the engine.

The only way in which I can fault the boat would be a personal one. She is not a sailboat. But nevertheless, she is as close as can be to what I think of as a sailor's powerboat.

Contemporary Classics

52' & 60' Fantail Motoryachts

Designs Number 281 & 314

1989 & 1991

Once upon a time, elegant fantail motor yachts were found in harbors everywhere. They were widely appreciated for their beauty and performance. Now, the few remaining fantail yachts are large ones, mostly serving as charter boats.

In an effort to return some of this elegance to our harbors, the Benford Design Group has created these Contemporary Classic Fantail Motoryachts. They are designed drawing upon Benford's three decades of experience in the yacht design business, creating practical and functional boats which reflect a love of classic designs, combined with the use of the latest technology. The result are boats that share the classic character of the older vessels but have improved performance, stability, livability and reduced maintenance.

They have the seaworthiness for passage-making and the comforts to make living aboard a pleasure. Whether you cruise out of sight of land or on the thousands of miles of waterways and rivers, these vessels can take you there in comfort and style.

There's a galley with full sized kitchen appliances; side-by-side refrigerator and freezer, four burner stove with oven, deep double sinks, dishwasher, microwave, plus good cabinet space.

The dining table has chairs and the living room has full sofas, coffee table, book cases and plenty of windows for light and airy spaces.

The staterooms have beds that take standard sized sheets and have room for getting around them and making them. There's good closet space for clothes, dressers and plenty of storage. The bathrooms have elbow room for adults and there are separate shower stalls.

The roomy back porch (screening optional) has room for gatherings of friends, barbecues and enjoying living in touch with nature. The housetop extends to form a canopy over the side decks and after deck for shade from the sun and rain.

These fantail motor yachts are designed to be easily driven, achieving reasonable speeds with small engines. This means that the cost of the machinery will be low, reducing the first cost of buying the vessel. And, fuel consumption will be low, making the second cost of owning the vessel low too.

Actual speeds for the 60-footer are in the range of an honest 10 knots for top speed and 8 to 9 knots for comfortable cruising.

Is more speed necessary?

We tend to feel it is not. Once you are aboard your Contemporary Classic Fantail Motoryacht you will already have arrived at where you want to be. You can then move her to another location, but you'll still be aboard; just where you want to be.

Plus, in a boat that sips fuel, it's more affordable to go exploring and see a lot of the country. Traveling at reasonable speeds also means that you will get a chance to see all the scenery before you flash past it....

These vessels are available in either a single screw or twin screw version. A bow thruster is available as an option. The engine room is centrally located. It has standing headroom, like the rest of the boat. There is plenty of space to work on projects. The wet exhausts exit through the sides for short runs and easy servicing. Optionally, a dry stack can be fitted, with keel coolers on the engines instead of the heat exchangers.

Custom Steel Boats, near Oriental, North Carolina, has demonstrated their ability to create complex shapes in metal and they will build these fantail yachts with smoothly done steel work. Power away kits (or boats completed to what ever stage or level of finish you'd like to pay for) will be available.

Hull and decks are of welded steel, sandblasted and painted. The deckhouse can be painted steel or aluminum, or bright finished mahogany or teak. The latter is most in character with the style of the design and well worth the extra work for someone who appreciates the beauty of wood. Interiors are done with materials selected in collaboration with the owners, opening up a world of choices.

For those wanting the warmth of wood throughout the structure, we can provide a design for all wood versions. These would most appropriately be done with cold-molding the structure, so that it is all sealed up and there is no shrinking and swelling to start leaks. This would provide the wood boat lover with an authentic looking boat inside and out. Maintenance would be quite modest, mostly a matter of looking after the finishes on the structure and looking after the systems.

This 52-foot version of the 60-footer shares many of the same attributes. She's at about the lower limit for a vessel of this style to gracefully carry off the double decked layout. Her layout has many similarities to the 60, except the forward two staterooms share a head and the office space is slightly smaller.

If you would like to be able to spot your boat without having to read the name on the stern, if your taste in boats runs to boats that look like boats instead of imitating guided missiles, and if you're looking for a distinctive cruising yacht for living aboard and cruising in style and comfort, the Contemporary Classic Fantail Motoryachts offer practical and elegant solutions.

Particulars:	**52'**	**60'**
Length overall	52'-0"	60'-0"
Length designed waterline	49'-0"	56'-0"
Beam (17'-0" on one 52 vers.)	18'-0"	18'-0"
Draft	4'-0"	4'-0"
Freeboard:		
Forward	7'-7½"	8'-0"
Least	5'-7"	6'-0"
Aft	6'-5½"	6'-6"
Displacement, cruising trim*	81,000 lbs.	100,000 lbs.
Displacement-length ratio	298	254
Water tankage	1,000 Gals.	1,000 Gals.
Fuel tankage	750 Gals.	1,000 Gals.
Headroom	6'-7"	6'-7"

***CAUTION:** The displacement quoted here is for the boat in cruising trim. That is, with the fuel and water tanks filled, the crew on board, as well as the crews' gear and stores in the lockers. This should not be confused with the "shipping weight" often quoted as "displacement" by some manufacturers. This should be taken into account when comparing figures and ratios between this and other designs.

Shown below and on the last page of this section are some variations on a 52' x 17' beam size. These show some other ideas that we've considered and perhaps some aspect of them will show an alternative that would just suit how you think that yours should be done.

The stepped up pilothouse gives room for a wider cabin sole in the forward stateroom below and staterooms and head on the one on page 228. Both of these 52 x 17 versions have good workbench and shop spaces near the machinery. These can be used for maintenance and repair or for hobby projects.

All but the one below have some designated office space aboard for work or homework projects.

The one below shows a single screw version with triple rudders. Commercial ships have used this sort of system with great success, giving them increased and good maneuverability. I've shown them in the hard over position and you can see how they create a slot directing the flow in a manner that kicks the stern sideways. This combined with a bow thruster would provide good control similar to what one might have with a twin screw without having the second engine to maintain, but with the added cost of the bow thruster and rudders.

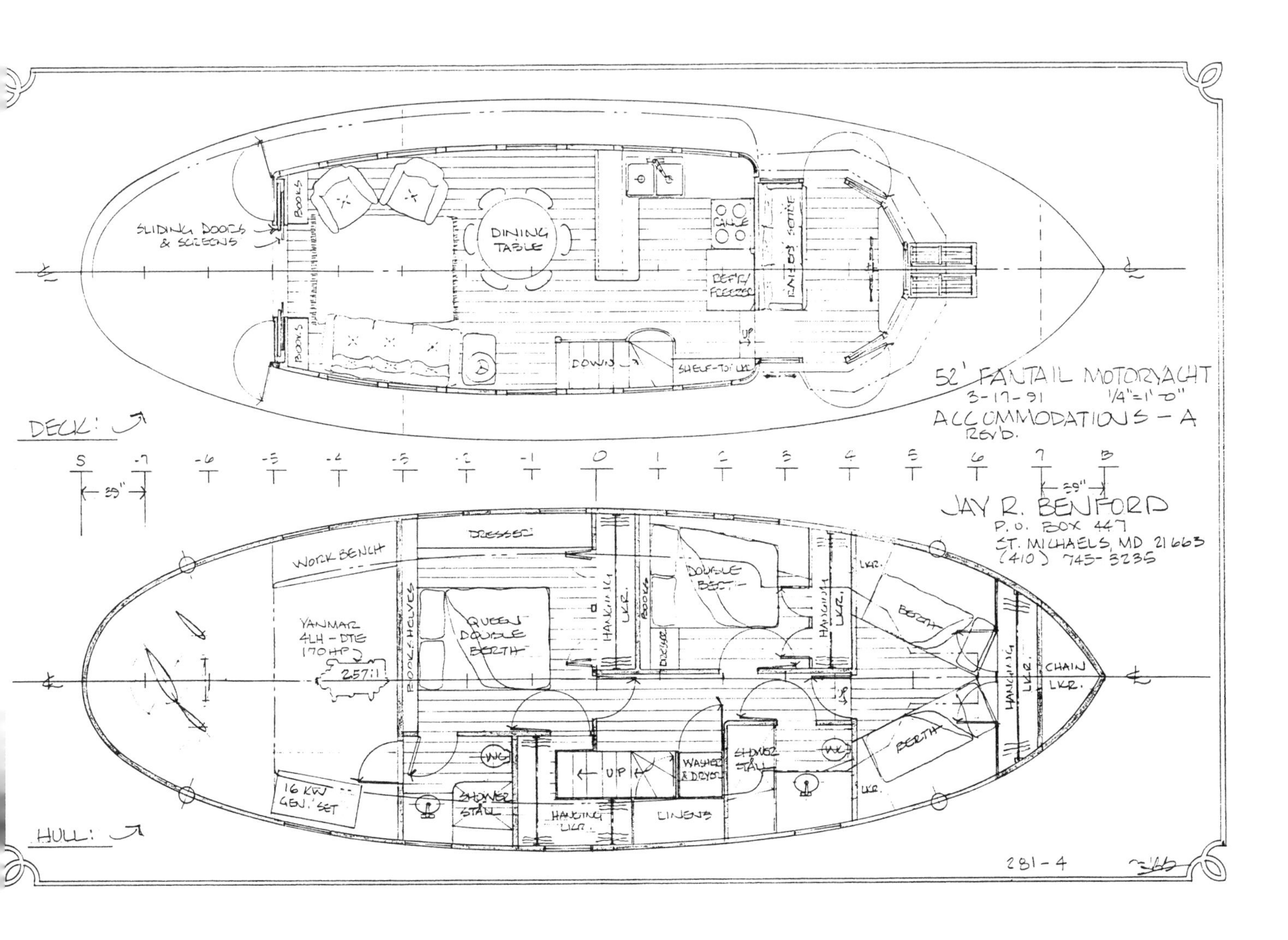

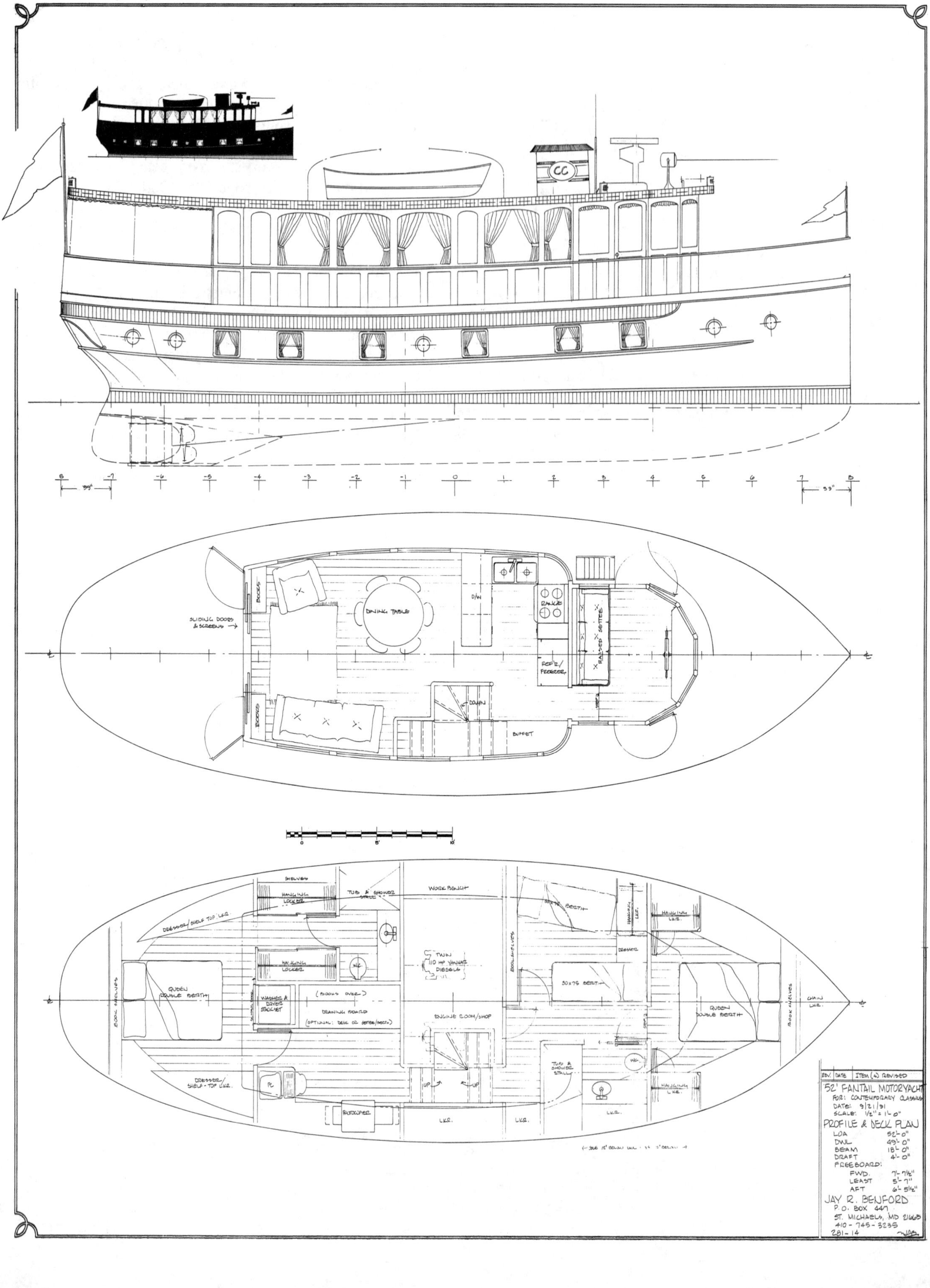

SLIDING DOORS & SCREENS →
BOOKS
DINING TABLE
D/W
RANGE
REFR/ FREEZER
RAISED SETTEE
DOWN
BUFFET
BOOKS

SHELVES
HANGING LOCKER
TUB & SHOWER STALL
WORK BENCH
BERTH
HANGING LKR.
DRESSER/SHELF-TOP LKR.
HANGING LOCKER
TWIN 110 HP YANMAR DIESELS
BOOKSHELVES
DRESSER
QUEEN DOUBLE BERTH
WASHER & DRYER STACK SET
(BOOKS OVER)
DRAWING BOARD
(OPTIONAL: DESK OR SETTEE/BERTH)
ENGINE ROOM/SHOP
30 x 75 BERTH
QUEEN DOUBLE BERTH
BOOK SHELVES
CHAIN LKR.
UP
UP
TUB & SHOWER STALL
DRESSER/SHELF-TOP LKR.
HANGING LKR.
LKR.
LKR.
LKR.

REV.	DATE	ITEM (S) REVISED

52' FANTAIL MOTORYACHT
FOR: CONTEMPORARY CLASSICS
DATE: 9/21/91
SCALE: 1/2" = 1'-0"

PROFILE & DECK PLAN

LOA	52'-0"
DWL	49'-0"
BEAM	18'-0"
DRAFT	4'-0"
FREEBOARD:	
FWD.	7'-7 1/2"
LEAST	5'-7"
AFT	6'-5 1/2"

JAY R. BENFORD
P.O. BOX 447
ST. MICHAELS, MD 21663
410-745-3235
281-14

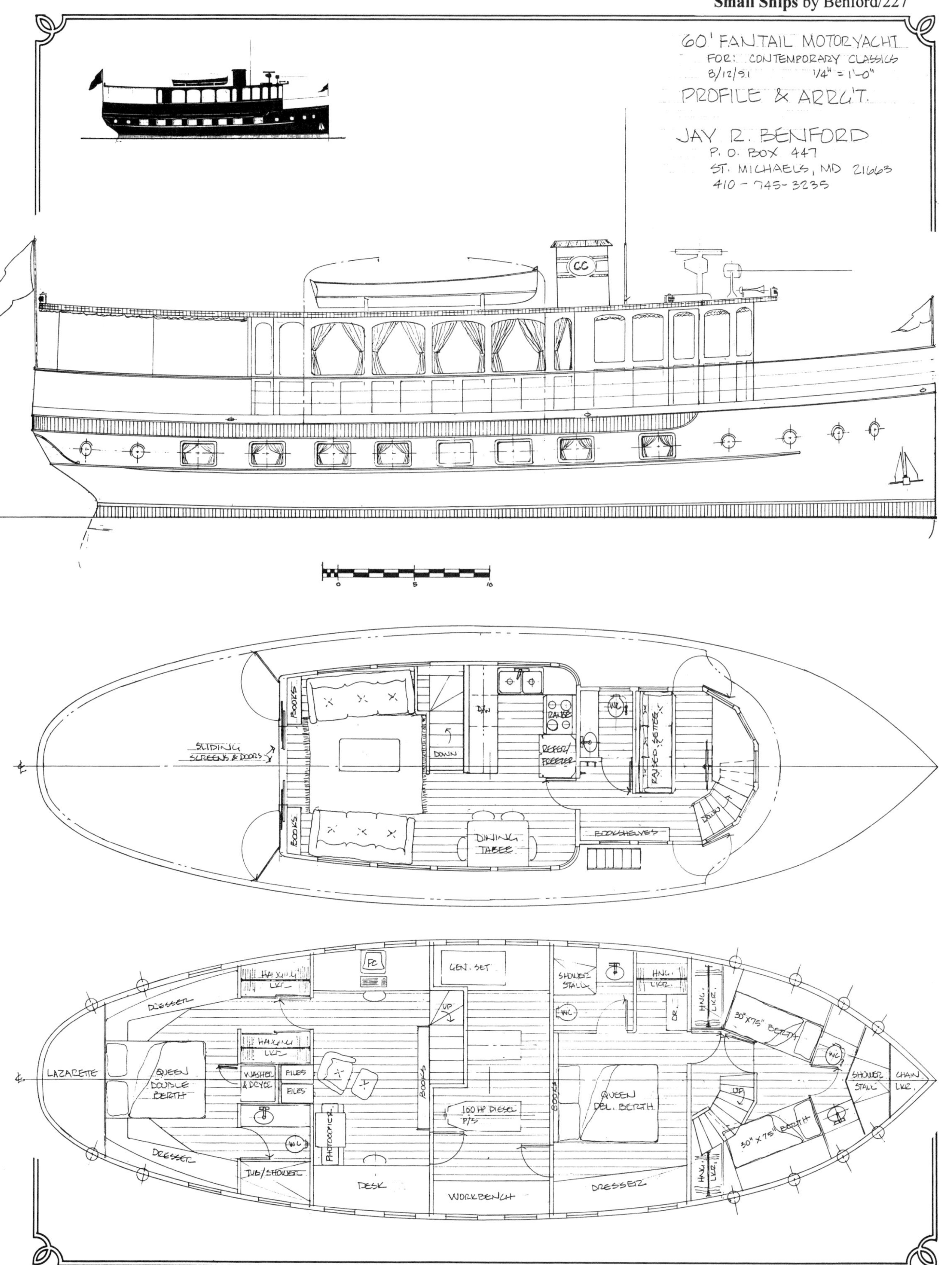

60' FANTAIL MOTORYACHT
FOR: CONTEMPORARY CLASSICS
8/12/91 1/4" = 1'-0"
PROFILE & ARRG'T.
JAY R. BENFORD
P. O. BOX 447
ST. MICHAELS, MD 21663
410-745-3235
CC
0 5 10
BOOKS
SLIDING SCREENS & DOORS
DOWN
D/W
RANGE
REFER/FREEZER
W.C.
RAISED SETTEE
BOOKSHELVES
DINING TABLE
DOWN
FC
GEN. SET
HANGING LKR.
SHOWER STALL
HNG. LKR.
DRESSER
UP
W.C.
DR.
30"X75" BERTH
LAZARETTE
QUEEN DOUBLE BERTH
WASHER & DRYER
FILES
FILES
BOOKS
QUEEN DBL. BERTH
SHOWER STALL
CHAIN LKR.
100 HP DIESEL P/S
PHOTOCOPIER
TUB/SHOWER
DESK
WORKBENCH
DRESSER
30"X75" BERTH
HNG. LKR.

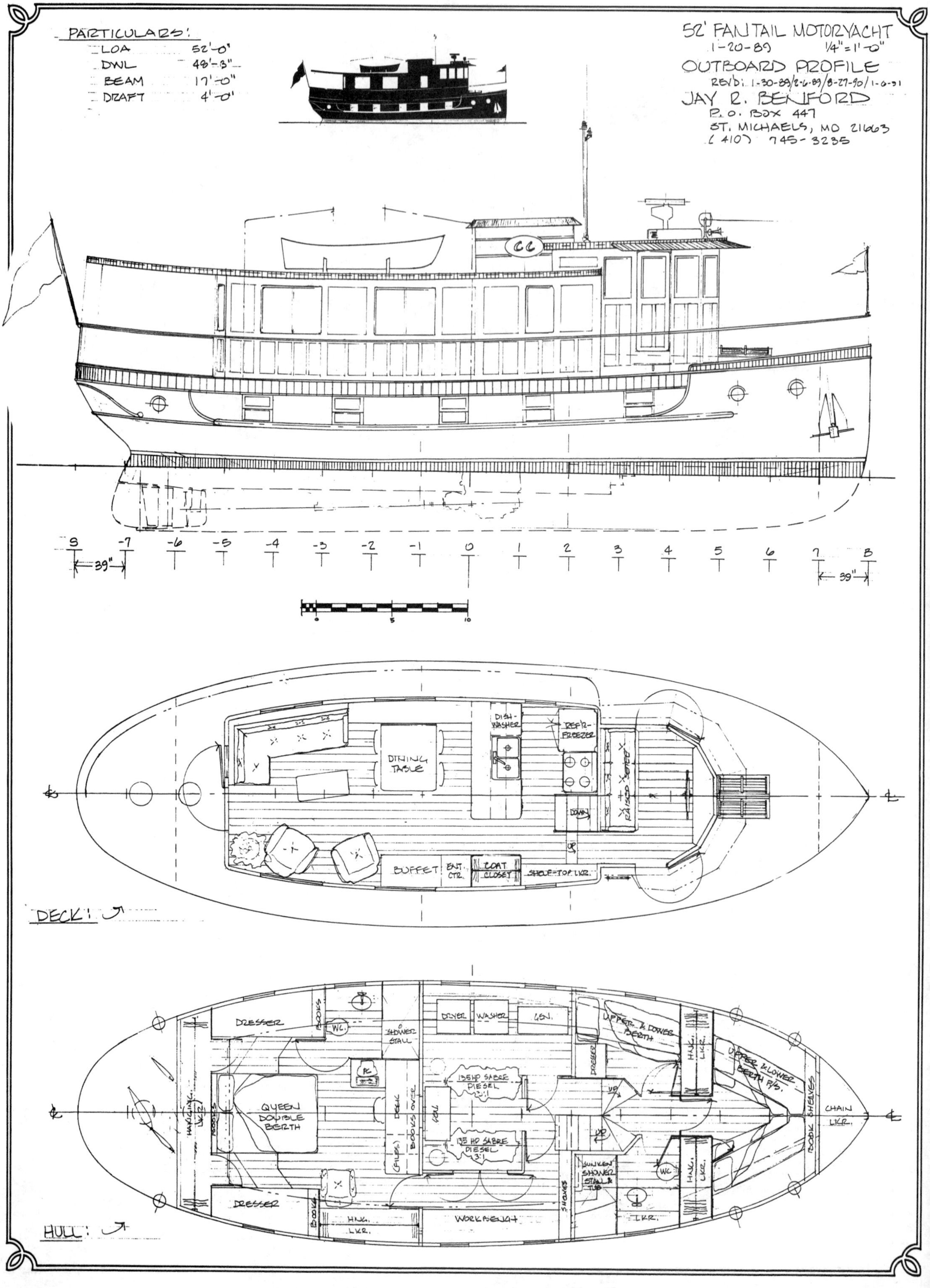

PARTICULARS:
LOA 52'-0"
DWL 48'-3"
BEAM 17'-0"
DRAFT 4'-0"

52' FANTAIL MOTORYACHT
1-20-89 1/4"=1'-0"
OUTBOARD PROFILE
REV'D: 1-30-89/2-6-89/8-27-90/1-6-91
JAY R. BENFORD
P. O. BOX 447
ST. MICHAELS, MD 21663
(410) 745-3235

S -7 -6 -5 -4 -3 -2 -1 0 1 2 3 4 5 6 7 8
39" 39"

0 5 10

DISH-WASHER
REF'R-FREEZER
DINING TABLE
RAISED SETTEE
DOWN
UP
BUFFET
ENT. CTR.
COAT CLOSET
SHELF-TOP LKR.

DECK:

DRESSER
BOOKS
W.C.
SHOWER STALL
DRYER
WASHER
GEN.
UPPER & LOWER BERTH
HNG. LKR.
UPPER & LOWER BERTH P/S.
DRESSER
135 HP SABRE DIESEL 3:1
135 HP SABRE DIESEL 3:1
HANGING LKR.
BOOKS
QUEEN DOUBLE BERTH
DESK
BOOKS OVER
(FILES)
UP
CHAIN LKR.
BOOK SHELVES
SUNKEN SHOWER STALL & TUB
W.C.
HNG. LKR.
DRESSER
BOOKS
HNG. LKR.
WORKBENCH
SHELVES
LKR.

HULL:

65' Fantail Yacht & Motorsailer/Office

Designs Number 182 & 310

1979 & 1990

These two 65-footers are both variations on a theme done just over a decade apart. Both are looks at the sort of motoryacht that I think would be a wonderful cruising home and office. The earlier one has more space devoted to the office and was done at a time when our business was smaller than it is today. The later one has more of a home-office layout which is shown in the optional layout drawing below.

The 65' Fantail Yacht has quarters forward that would be suitable for use with a paid crew, or can function as an additional guest cabin. I'd consider using it as the master stateroom with the office amidships and the kids in the aft stateroom, keeping the small aft double stateroom as the guest room. I like to sleep where I can hear what's happening with the anchor chain and the forward skylight would let me see what the weather is before I get up.

The deckhouse is laid out for convenient working and traffic flow. The galley is a couple steps from the dining table. The head is handy to the saloon and only a couple of steps aft from the pilothouse. The double sliding doors on the aft saloon bulkhead give a large opening to the spacious after deck. Sliding screens are also fitted so that fresh air is enjoyed even in insect weather.

Laid out as a single screw, we could fit a bow thruster or redesign her as a twin screw if maneuverability were a concern. My experience cruising ***Kiyi*** as a single screw with no thruster was that this was rarely a problem and a bit of planning ahead eliminated any potential problems.

The 65' Motor Yacht/Motorsailer has much more space devoted to the office. She would make having the office aboard possible with the size of our office condensed a bit. The saloon is the full width of the hull, with no side decks, and a bit shorter than the other 65's saloon. However, it does have most of the features, just arranged in a different manner. With the large, canopy-covered aft deck, there is still plenty of room for entertaining on deck and freely moving about. Tucked under the aft deck, with an octagonal lid over it, is a 6' hot tub with its own water tank underneath for storage underway. A nice bit of contemporary decadence for the crew to enjoy.

The motor sailing rig is an idea of what might be done to take advantage of the wind and ease the motion in beam seas — and ease fuel consumption. Although, with a calculated range of 4½ to 5 miles per gallon at 8 knots she's still quite economical.

Particulars:	**182**	**310**
Length overall	65'-0"	65'-0"
Length designed waterline	61'-0"	61'-0"
Beam	15'-0"	18'-6"
Draft	4'-0"	4'-6"
Freeboard:		
Forward	8'-0"	8'-6"
Least	4'-9"	6'-6"
Aft	5'-6"	7'-1"
Displacement, cruising trim*	49,500 lbs.	85,300 lbs.
Displacement-length ratio	97	168
Ballast	10,000 lbs.	
Ballast ratio	20%	
Sail area	1,020 sq. ft.	
Sail area-displacement ratio	12.11	
Prismatic coefficient	.568	.534
Pounds per inch immersion	2,700 lbs.	3,400 lbs.
Entrance half-angle	18°	19°
Water tankage	400 Gals.	1,000 Gals.
Fuel tankage	484 Gals.	800 Gals.
Headroom	6'-1" to 6'-10"	6'-3" to 6'-8"
GM		3.6'

***CAUTION:** The displacement quoted here is for the boat in cruising trim. That is, with the fuel and water tanks filled, the crew on board, as well as the crews' gear and stores in the lockers. This should not be confused with the "shipping weight" often quoted as "displacement" by some manufacturers. This should be taken into account when comparing figures and ratios between this and other designs.

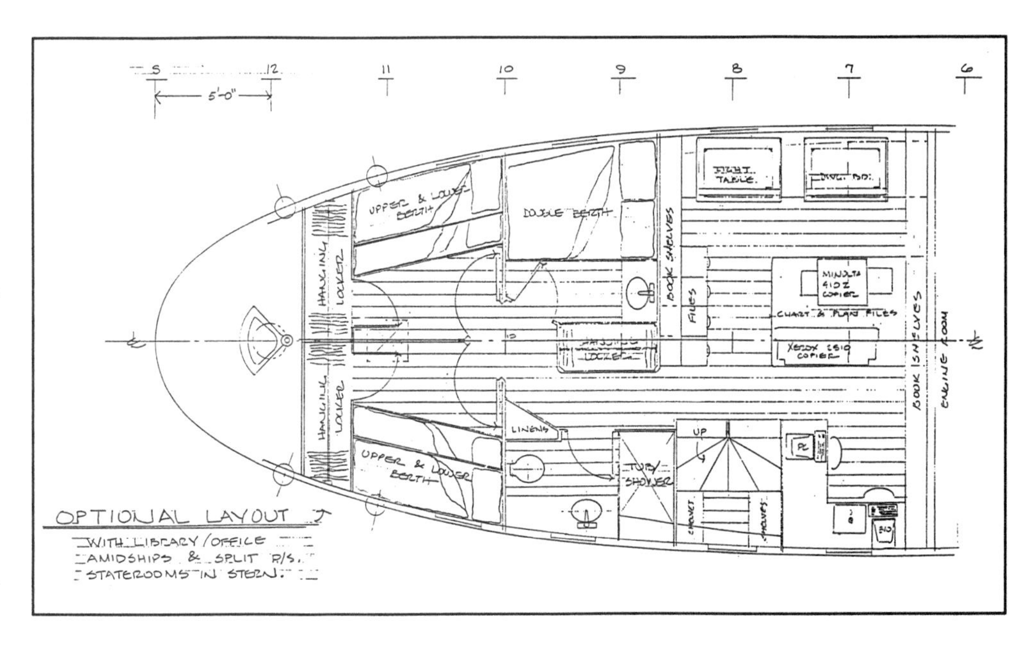

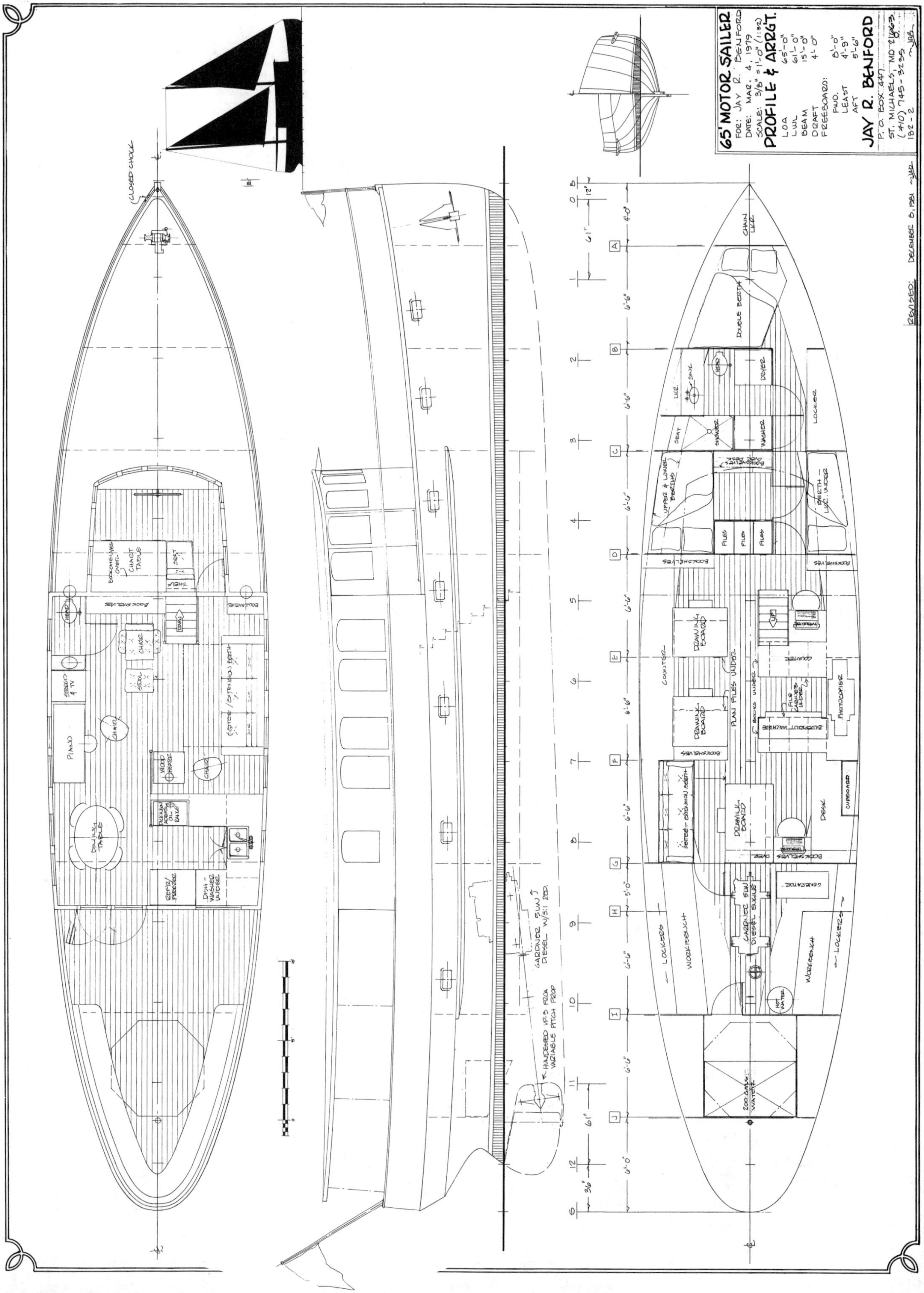

65' MOTOR SAILER
FOR: JAY R. BENFORD
DATE: MAR. 4, 1979
SCALE: 3/8" = 1'-0" (1:32)
PROFILE & ARRGT.
LOA 65'-0"
LWL 61'-0"
BEAM 15'-0"
DRAFT 4'-0"
FREEBOARD: FWD. 8'-0"
LEAST 4'-9"
AFT 5'-6"
JAY R. BENFORD
P.O. BOX 447
ST. MICHAELS, MD 21663
(410) 745-3235
182-2
REVISED: DECEMBER 8, 1981
CLOSED CHOCK
CHAIN LKR
DOUBLE BERTH
HEAD
SINK
LKR
DRYER
WASHER
SHOWER
SEAT
LOCKER
UPPER & LOWER BERTHS
BERTH - LKR. UNDER
FILES
BOOKSHELVES
DRAWING BOARD
COUNTER
UP
PLAN FILES UNDER
BOOKS UNDER
BLUEPRINT MACHINE
PHOTOCOPIER
SETTEE - EXTENSION BERTH
DESK
CUPBOARD
GENERATOR
GARDNER 5LW DIESEL ENGINE
WORKBENCH
LOCKERS
HOT WATER
200 GALS. WATER
CHART TABLE
BOOKSHELVES OVER
SEAT
DOWN
CHAIR
STOOL
STEREO & TV
PIANO
WOOD HEATER
SETTEE / EXTENSION BERTH
DINING TABLE
REFRIG. FREEZER
DISH-WASHER UNDER
GARDNER 5LW DIESEL W/3:1 RED.
HUNDESTED VP.3 FROM VARIABLE PITCH PROP

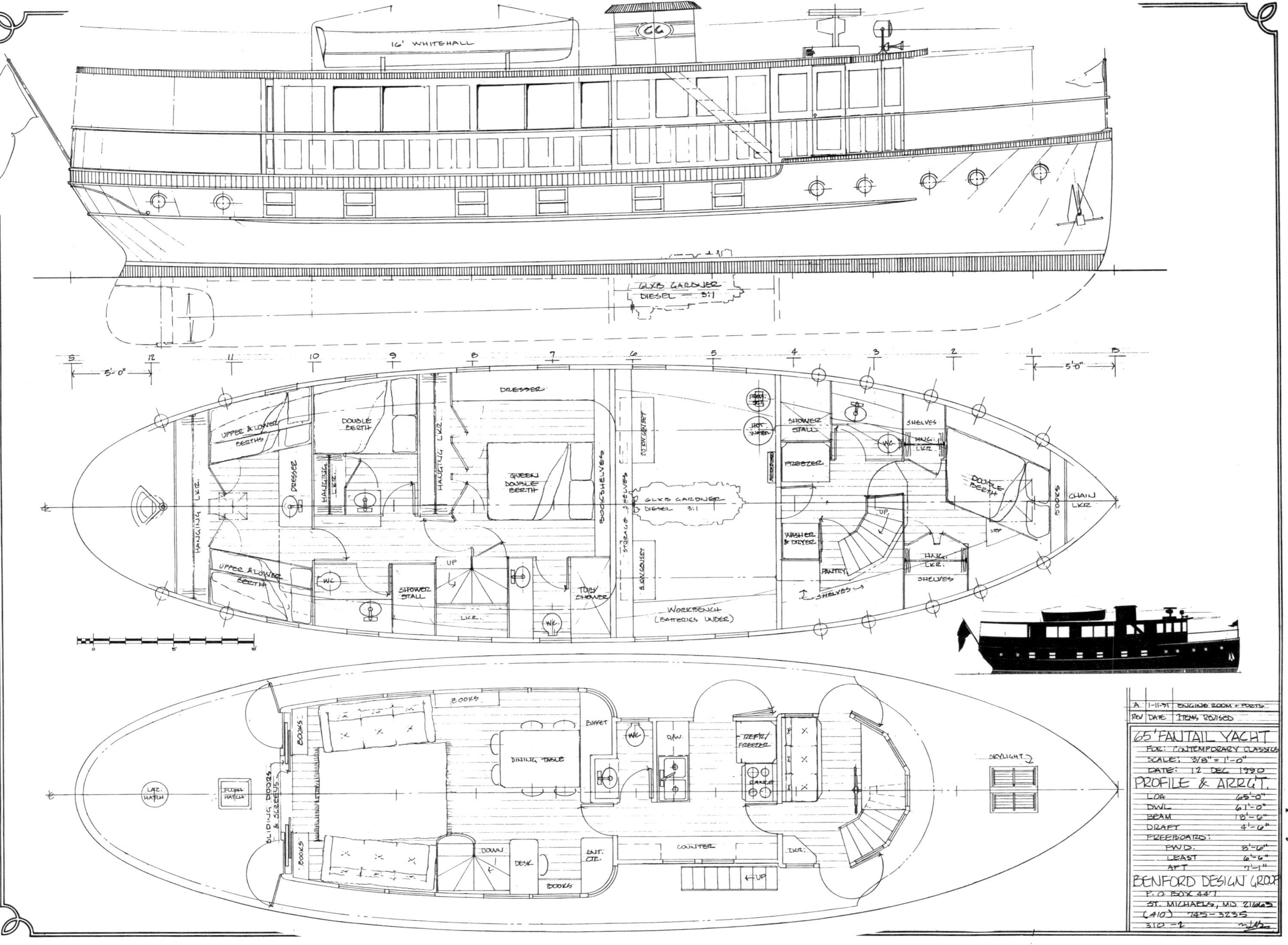
16' WHITEHALL
GLXB GARDNER DIESEL 3:1
5'-0"
DRESSER
UPPER & LOWER BERTHS
DOUBLE BERTH
HANGING LKR.
QUEEN DOUBLE BERTH
BOOKSHELVES
STORAGE SHELVES
20 KW GENSET
SHOWER STALL
FREEZER
WASHER & DRYER
PANTRY
SHELVES
HNG. LKR.
DOUBLE BERTH
BOOKS
CHAIN LKR.
UP
TOPY SHOWER
WORKBENCH (BATTERIES UNDER)
UPPER & LOWER BERTH
BOOKS
DINING TABLE
BUFFET
D/W.
REFR/FREEZER
RANGE
SKYLIGHT
LAZ. HATCH
FLUSH HATCH
SLIDING DOORS & SCREENS
DOWN
DESK
ENT. CTR.
COUNTER
LKR.
A 1-11-91 ENGINE ROOM + PORTS
REV DATE ITEMS REVISED
65' FANTAIL YACHT
FOR: CONTEMPORARY CLASSICS
SCALE: 3/8" = 1'-0"
DATE: 12 DEC 1990
PROFILE & ARRG'T.
LOA 65'-0"
DWL 61'-0"
BEAM 18'-6"
DRAFT 4'-6"
FREEBOARD:
FWD. 8'-6"
LEAST 6'-6"
AFT 7'-1"
BENFORD DESIGN GROUP
P.O. BOX 447
ST. MICHAELS, MD 21663
(410) 745-3235
310-2

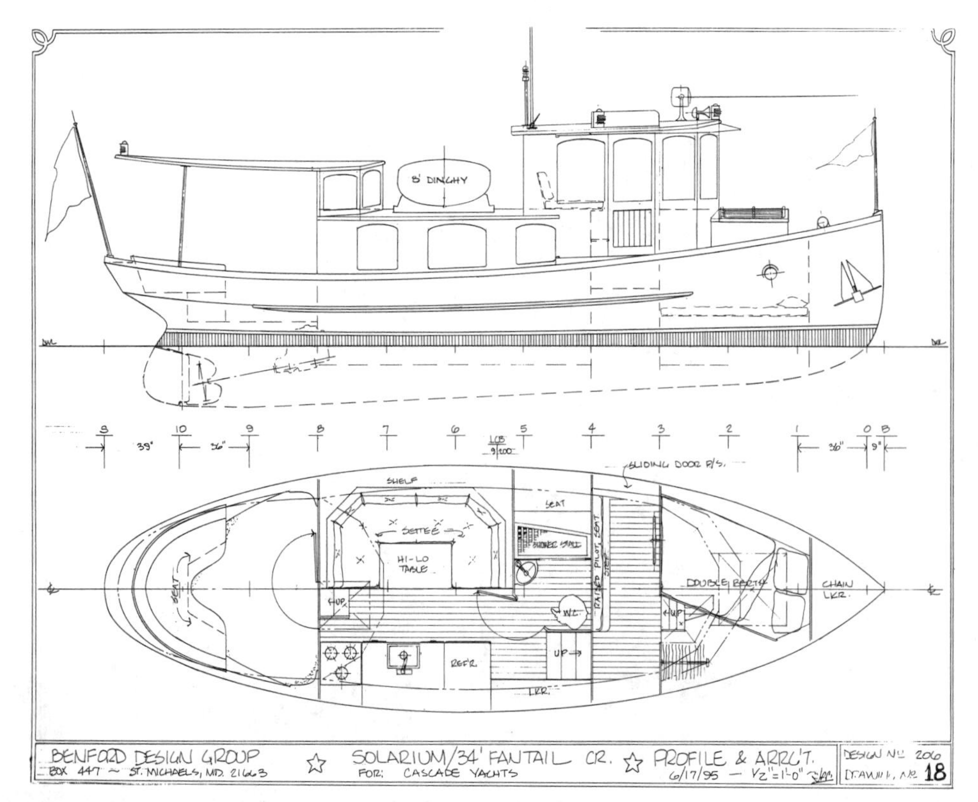

This page and the following one have some real visual treats. The Buckley Smith drawing above is the original 34' Fantail Cruiser. At right, we have the 100' Fantail Motoryacht above and the 65' Fantail Motoryacht below.

The one at left is the modified Solarium version of the 34. An alternate layout for the saloon could have port and starboard settee berths and the galley across the after end of the cabin.

In addition to the cold-molded plans for this boat, molded fiberglass hulls are available.

100' Fantail Yacht

Design Number 307

1990

I'm a member of that informal fraternity of boat lovers who have a special fondness for the old Mathis-built and Trumpy-designed fantail houseboats of the 1920's. Mathis only built this style for a few years in the latter part of that decade. Several of them are still afloat in various states of repair. Some do service as corporate charter and entertainment vessels, lending an air of elegance to such gatherings that is missing in the more modern yachts, which more often tend to look like guided missiles.

Back in the '20s this was the sort of boat that was called a houseboat, unlike the trailer-on-a-barge called houseboats today. It thus seemed criminal to call these designs houseboats, so we settled on fantail yachts as a better and more descriptive name for the type.

I searched for a long time to find an old one in anything resembling good structural condition, hoping to turn one into a cruising houseboat again, to function as a home for my family. Eventually I reluctantly came to the conclusion that it would be cheaper to build a new one than to do a proper rebuild and restoration of an old one.

Building from scratch would also let us make a few improvements in the designs, to add some needed stability and get a little more headroom. We could also make good use of more modern equipment, materials and building techniques. This would make the cost of owning the boat, known as the second cost, quite a bit less for the systems, fastenings, coatings, and equipment would all be new and not on the verge of breakdown.

The accommodations allow for crew quarters forward for up to eight. The owner's party, in the aft cabins, can be up to eight also. An alternate layout, that would be practical for how large yachts are actually used, would be to rearrange the owner's quarters to have port and starboard guest cabins all the way aft, each with their own head, and keep the master suite forward. In between, there would be space for a good sized office/library/study area at the base of the stairs.

The photos below show a couple of the originals and were taken a few years ago. The nice detailing on the curved saloon forward corner window and elegant fantail stern are evident. If you love this sort of boat as much as I do, I'd welcome an opportunity to discuss how we might create a contemporary one.

Particulars:

Length overall	100'-0"
Length designed waterline	95'-0"
Beam	20'-0"
Draft	4'-6"
Freeboard:	
Forward	10'-0"
Least	7'-0"
Aft	8'-0"
Displacement, cruising trim*	200,000 lbs. (est.)
Displacement-length ratio	104
Water tankage	2,500 Gals.
Fuel tankage	1,500 Gals.
Headroom	6'-6"

***CAUTION:** The displacement quoted here is for the boat in cruising trim. That is, with the fuel and water tanks filled, the crew on board, as well as the crews' gear and stores in the lockers. This should not be confused with the "shipping weight" often quoted as "displacement" by some manufacturers. This should be taken into account when comparing figures and ratios between this and other designs.

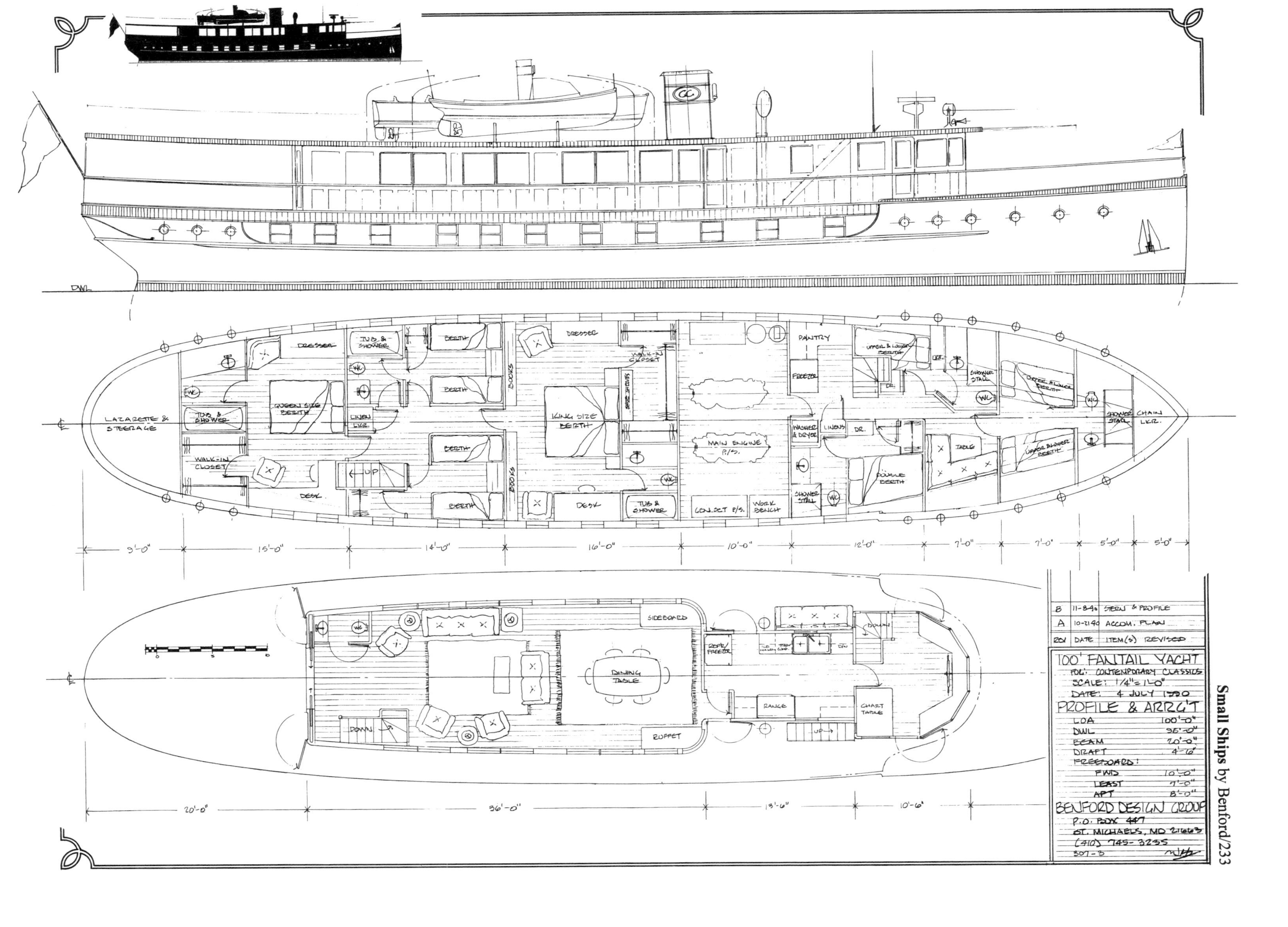
LAZARETTE & STEERAGE
TUB & SHOWER
WALK-IN CLOSET
DRESSER
QUEEN SIZE BERTH
DESK
LINEN LKR.
UP
BERTH
BOOKS
KING SIZE BERTH
WALK-IN CLOSET
TUB & SHOWER
MAIN ENGINE P/S.
CON. SET P/S.
WORK BENCH
PANTRY
FREEZER
WASHER & DRYER
LINENS
DR.
SHOWER STALL
UPPER & LOWER BERTH
DOUBLE BERTH
TABLE
CHAIN LKR.
9'-0"
15'-0"
14'-0"
16'-0"
10'-0"
12'-0"
7'-0"
7'-0"
5'-0"
5'-0"
DOWN
SIDEBOARD
DINING TABLE
BUFFET
REFR/FREEZER
RANGE
CHART TABLE
UP
20'-0"
36'-0"
18'-6"
10'-6"
B 11-8-90 STERN & PROFILE
A 10-21-90 ACCOM. PLAN
REV DATE ITEM(S) REVISED
100' FANTAIL YACHT
FOR: CONTEMPORARY CLASSICS
SCALE: 1/4"=1'-0"
DATE: 4 JULY 1990
PROFILE & ARR'G'T
LOA 100'-0"
DWL 95'-0"
BEAM 20'-0"
DRAFT 4'-6"
FREEBOARD:
FWD 10'-0"
LEAST 7'-0"
APT 8'-0"
BENFORD DESIGN GROUP
P.O. BOX 447
ST. MICHAELS, MD 21663
(410) 745-3235
307-3

The Custom Design Process

by Jay R. Benford

How to do business with a yacht designer and what is expected of the client.

Most people who have a boat custom designed have never had that experience before, and thus do not know what to expect of the naval architect and what is expected of them. In this section of the book, I will try to address these questions and define the way we, and most other yacht designers, do business. Much of the material herein is lifted straight out of our standard design contract forms with some elaboration for clarity. Since these forms were arrived at in collaboration with a number of other designers, they form a good basis for understanding the profession.

Accompanying this section are some of our preliminary designs that have yet to be completed. If one of them is either just what you like or suggests a variation that would be suitable, give us a call and we can tell you what is involved in completing the work. With finite time available to us, and the need to earn a living, we don't get to pursue all the ideas we generate.

Who Needs A Custom Design?

Fortunately, there's really no such thing as a "typical" custom design client. As all our clients are very much individuals, their differences and unique requirements are such that the most interesting and widely varying designs can be created. Perhaps some of the most pleasing clients to work with are those who have had a good deal of experience skippering a variety of boats. Such clients have selected aspects from each of their prior vessels to suit their particular, and often evolving, cruising, living, and working philosophy. In the process, they have also learned what they wish to avoid. In working with the individual custom design clients, however, the work flow most often follows a "typical" process, and the description of it here will provide a good idea of what can be expected.

The creation of a new yacht is similar in complexity of materials, equipment and form to the thousands of hours of design and engineering which go into the creation of a new airplane or automobile. A major difference arises, however, as the yacht designer and his client are able to afford only a few hundred hours in developing a new design. Thus the custom yacht design client must be prepared to approach his new design on a somewhat empirical and experimental basis. The yacht designer will bring forth his best efforts, together with his many successful past experiences, to create the finest new design possible to best fit the parameters set out by the client. A client with extensive past experience in boating is best prepared to venture into the realm of experimental ideas. For the new ideas to have the best chance of working successfully in filling the clients' needs, they must be based on tested past experiences.

Those without experience are best off starting with used boats first. This is a much more economical way of breaking into discovering what each cruise, with its varying conditions in weather, waters, locale and crew, will reveal, to help build up a store of ideas of just what their different experiences dictate as necessary for their next vessel.

Most of the inquiries we receive deal with having a new boat custom designed, or with modifying an existing design to meet some specific need and type of service.

How To Choose A Yacht Designer

The creation of a special boat to fill a unique set of requirements can best be accomplished by those with the specialized experience, day in and day out, in doing just this. A prospective client would do well to make an overview of the market with several thoughts in mind. Firstly, which of the many designers available has created yachts which you admire or indeed love? What publications with his designs are available? With what construction materials has he accumulated extensive experience? Next, a closer inspection of this newly narrowed field should include the various designers' experiences. How successful were their yachts? Were they merely handsome in appearance and poor performers in reality? Or were their yachts of high strung performance to the detriment of the comfort and enjoyment of the rest of the boat and/or crew? Did their yacht hold together or were there problems with the construction? If the latter, was it possible to determine whether it was a builder or designer problem? (Builders have been known to make changes to the plans without letting the designer know.)

Does the designer really care about his work, or is it just another job? Often a way to tell this is what the designer does with his spare time: does he seem to enjoy cruising and/or racing (thereby accumulating more knowledge to feed into future design work) or does he spend the majority of his time elsewhere, seemingly relieved to be away from the subject? The best designers truly love their work, see it not as a 9:00 to 5:00 job, but a something they would never wish to "retire" from. Those with such an approach to the art will naturally see their job as one in which they want to do their best effort. They know that future work comes from top present performance. This includes thoughtful consultation during all stages of the boat, from design through construction and into operation of the vessel. The result: a grand yacht and a happy client.

Custom Design Procedure

The custom design process usually begins with the designer receiving a letter or phone call from the client

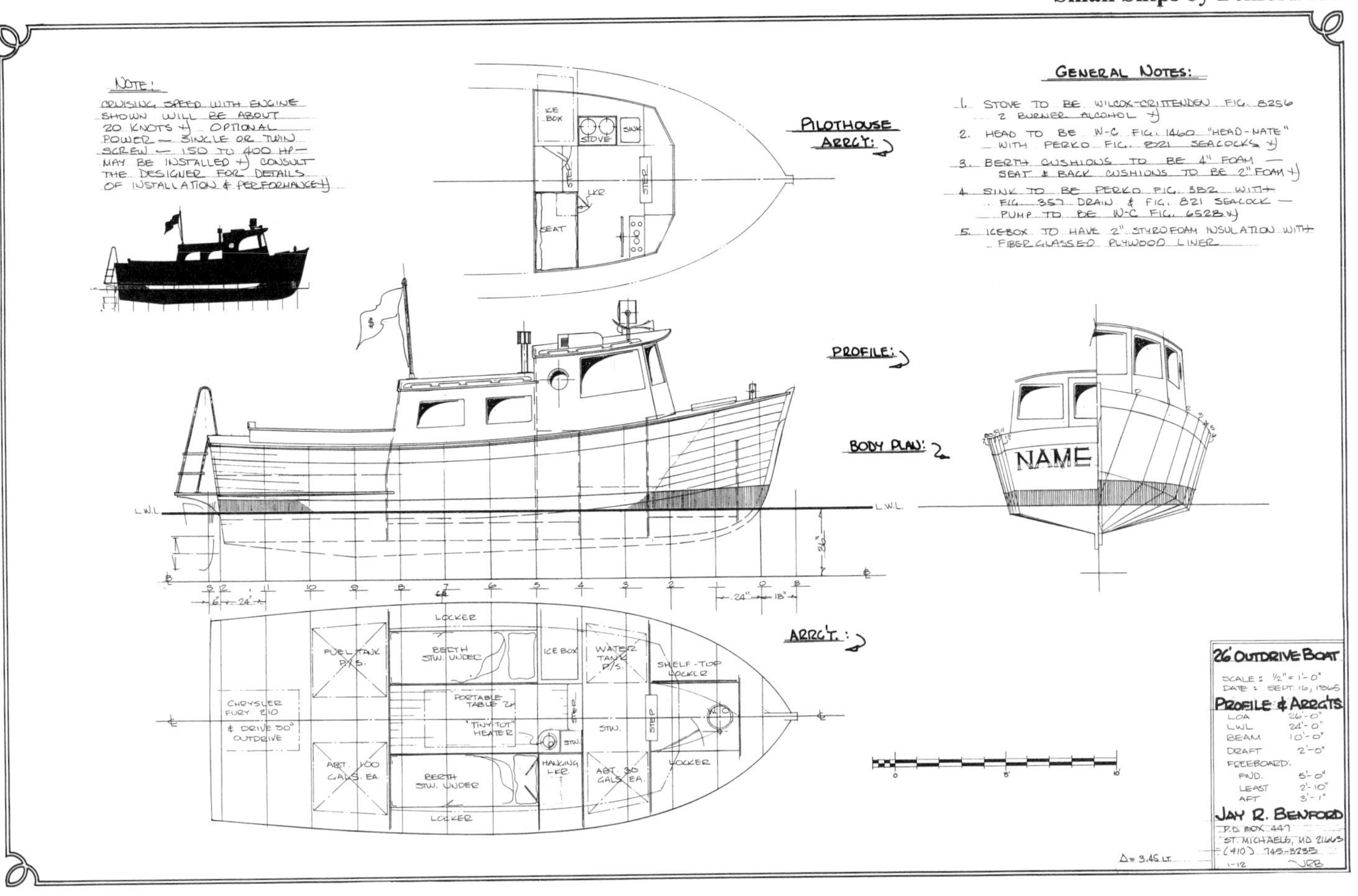
NOTE:
CRUISING SPEED WITH ENGINE SHOWN WILL BE ABOUT 20 KNOTS +) OPTIONAL POWER — SINGLE OR TWIN SCREW — 150 TO 400 HP. MAY BE INSTALLED +) CONSULT THE DESIGNER FOR DETAILS OF INSTALLATION & PERFORMANCE +)
PILOTHOUSE ARRG'T:
GENERAL NOTES:
1. STOVE TO BE WILCOX-CRITTENDEN FIG. 8256 2 BURNER ALCOHOL +)
2. HEAD TO BE W-C FIG. 1460 "HEAD-MATE" WITH PERKO FIG. 821 SEACOCKS +)
3. BERTH CUSHIONS TO BE 4" FOAM — SEAT & BACK CUSHIONS TO BE 2" FOAM +)
4. SINK TO BE PERKO FIG. 382 WITH FIG. 357 DRAIN & FIG. 821 SEACOCK — PUMP TO BE W-C FIG. 6528 +)
5. ICEBOX TO HAVE 2" STYROFOAM INSULATION WITH FIBERGLASSED PLYWOOD LINER
PROFILE:
BODY PLAN:
NAME
L.W.L.
ARRG'T.:
LOCKER
FUEL TANK P/S.
BERTH STW. UNDER
ICE BOX
WATER TANK P/S
SHELF-TOP LOCKER
CHRYSLER FURY 210 & DRIVE 50° OUTDRIVE
PORTABLE TABLE
"TINY TOT" HEATER
STEP
HANGING LKR
ABT. 100 GALS. EA.
ABT. 30 GALS. EA.
26' OUTDRIVE BOAT
SCALE: 1/2" = 1'-0"
DATE: SEPT. 16, 1965
PROFILE & ARRG'TS.
LOA 26'-0"
LWL 24'-0"
BEAM 10'-0"
DRAFT 2'-0"
FREEBOARD:
FWD. 5'-0"
LEAST 2'-10"
AFT 3'-1"
JAY R. BENFORD
P.O. BOX 447
ST. MICHAELS, MD 21663
(410) 745-3235
Δ = 3.46 LT.

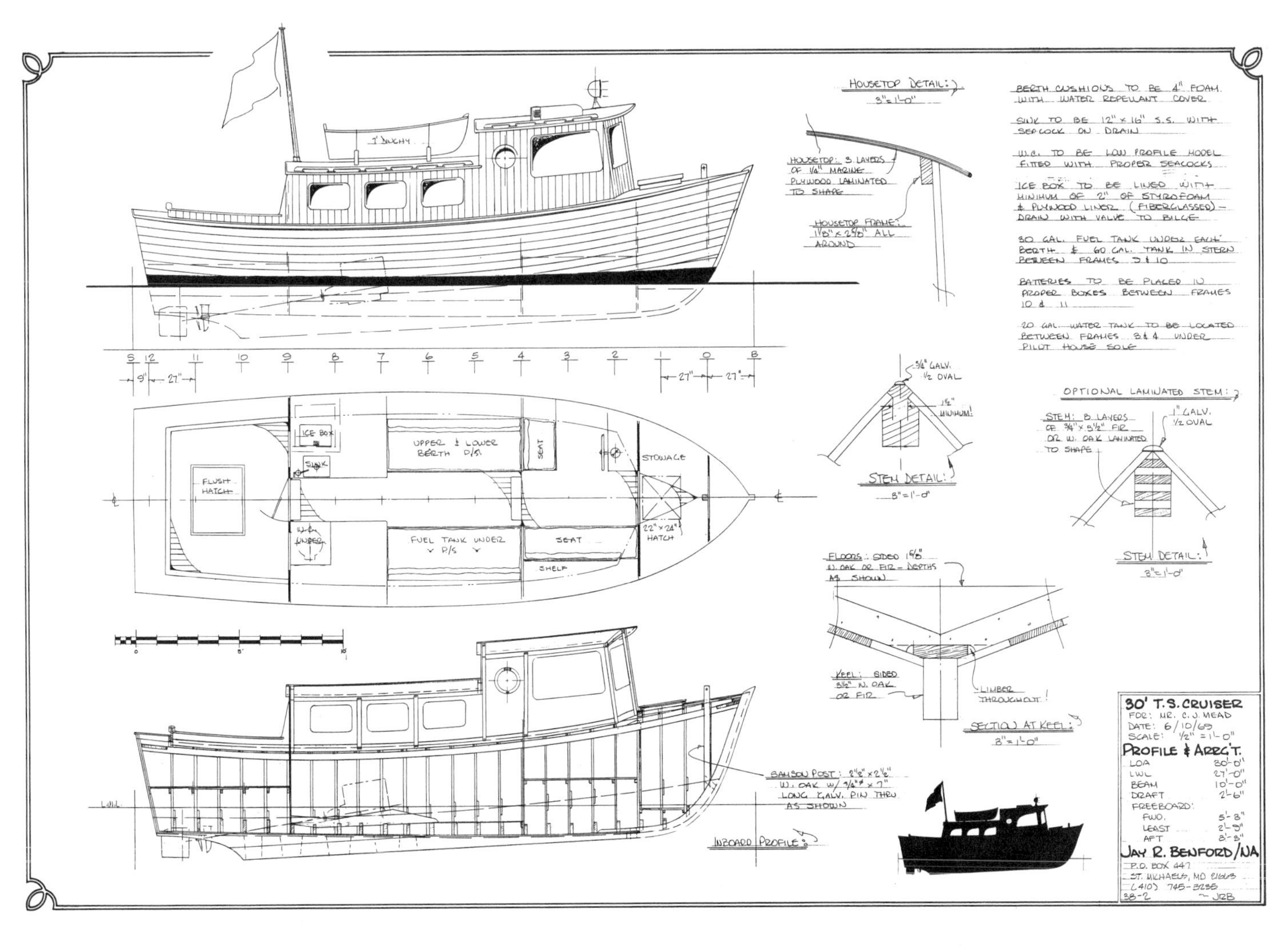
HOUSETOP DETAIL:
3" = 1'-0"
HOUSETOP: 3 LAYERS OF 1/4" MARINE PLYWOOD LAMINATED TO SHAPE
HOUSETOP FRAME: 1 1/8" x 2 3/8" ALL AROUND
BERTH CUSHIONS TO BE 4" FOAM WITH WATER REPELLANT COVER
SINK TO BE 12" x 16" S.S. WITH SEACOCK ON DRAIN
W.C. TO BE LOW PROFILE MODEL FITTED WITH PROPER SEACOCKS
ICE BOX TO BE LINED WITH MINIMUM OF 2" OF STYROFOAM & PLYWOOD LINER (FIBERGLASSED) — DRAIN WITH VALVE TO BILGE
30 GAL. FUEL TANK UNDER EACH BERTH & 60 GAL. TANK IN STERN BETWEEN FRAMES 9 & 10
BATTERIES TO BE PLACED IN PROPER BOXES BETWEEN FRAMES 10 & 11
20 GAL. WATER TANK TO BE LOCATED BETWEEN FRAMES 3 & 4 UNDER PILOT HOUSE SOLE
7' DINGHY
STEM DETAIL:
3" = 1'-0"
OPTIONAL LAMINATED STEM:
STEM: 8 LAYERS OF 3/4" x 5 1/2" FIR OR W. OAK LAMINATED TO SHAPE
1" GALV. 1/2 OVAL
ICE BOX
SINK
FLUSH HATCH
UPPER & LOWER BERTH P/S
SEAT
STOWAGE
FUEL TANK UNDER P/S
SHELF
22" x 24" HATCH
FLOORS: SIDED 1 5/8" W. OAK OR FIR — DEPTHS AS SHOWN
KEEL: SIDED 3 1/2" W. OAK OR FIR
LIMBER THROUGHOUT
SECTION AT KEEL:
3" = 1'-0"
SAMSON POST: 2 1/2" x 2 1/2" W. OAK W/ 3/8" x 7" LONG GALV. PIN THRU AS SHOWN
INBOARD PROFILE:
L.W.L.
30' T.S. CRUISER
FOR: MR. C.J. MEAD
DATE: 6/10/69
SCALE: 1/2" = 1'-0"
PROFILE & ARRG'T.
LOA 30'-0"
LWL 27'-0"
BEAM 10'-0"
DRAFT 2'-6"
FREEBOARD:
FWD. 5'-3"
LEAST 2'-3"
AFT 3'-3"
JAY R. BENFORD/NA
P.O. BOX 447
ST. MICHAELS, MD 21663
(410) 745-3235

inquiring into the possibility of creating a workable boat out of the ideas and thoughts he's collected over some period of time. From this point, letters are exchanged, or meetings arranged, to review the concept and its feasibility. The client often supplies photos and clippings of boats which appeal to him. Such details as the new vessel's intended use, desired range, speed, and rig, as well as the cruising and working philosophy of those who are to be involved in the new vessel are also discussed. Extent of design detail desired, and construction information required in the plans is also of pertinent interest. Once it's agreed that there is potential merit in the idea, the designer will explain the costs involved in the preliminary drawings.

With the payment of the retainer, the designer sets to work to give the idea some form on paper. When the designer is reasonably pleased with the drawing, two copies are sent to the client so that one can be marked with comments and questions and returned to the designer. The designer will answer the questions and give his opinions on the clients' comments, and then frequently revisions will be made to the drawings. These revised drawings will make the circuit again (sometimes more than once) until both the client and the architect are in agreement that the basic concept is well enough defined to proceed with working drawings.

Sometimes, however, the preliminaries are done as exploratory work towards seeing if an unusual idea will work, and if it is in the range of something near the projected budget. If it is not close enough to the mark, a second (or more) drawings may be made to see if the goals can be achieved by some other route. Occasionally, after the drawings are done and the quotes are in, the client decides that the boat is not what was in his mind or decides that he is unable to proceed beyond this stage. At this point, on bringing his account current for the preliminary design work, he can elect to stop the process and be under no further obligation to have the design completed.

Many times, the client will want to circulate the preliminaries for bids (also preliminary) to ascertain that the project will be within his budget. The designer usually assists in this, in recommending yards and in commenting on the bids as they are returned. Assuming the bids are reasonably close to what was expected, the next stage in the design process begins.

The evolution of the working (construction) plans and details involves perhaps the most hours of the total project, due to the extensive engineering, calculating, research, computer time, and drafting work which needs to be done. However, this part of the design usually runs more smoothly than the evolution of the preliminary conceptual work, because the gestation period has passed and the work which needs to be done — although innovative in a different way — is relatively straightforward.

However, because the design is an entirely new creation, and as such is as experimental as the previously-mentioned new plane or automobile design, there's always "fine-tuning" as the new ideas are tried and fitted into the total concept which is to be brought to life in the boat shop. The yacht designer, with fewer available hours and a much smaller staff, must depend partly on past experience, and even partly on intuition. However, the advances made in recent years in the computer industry have eliminated the need for an excessive amount of rule-of-thumb estimating. The result of the need for "fine-tuning" of a new design is that there is occasionally the need for changes while the boat is under construction. The client and the yard should be ready and able to accept this sometimes frustrating aspect of what is the natural result of working out the myriad problems inherent in bringing a brand new design to life for the first time. Naturally, the designer will make every effort to foresee as many of these problems as possible before they get to the shop, but often even the client does not realize what he has asked for, until he sees it in real life. Should he change his mind about any aspect of the design as it comes to life, this of course may affect many other parts of the vessel — all of which will need alteration accordingly, and which usually add time and cost to the project correspondingly.

As a general example, one area which can never be calculated with complete accuracy is the weights of new vessels. Even in the top production yards, which may be popping out hundreds of the same units annually, there is some variance in weights and trim between supposedly identical yachts. Each one requires slightly different trim ballasting after launching. To exercise very great accuracy in precalculating weights of a one-off custom yacht is thus, by comparison, almost a black art. Differences between construction practices and biases from one yard to the next are difficult to prejudge on the drawing board. When the designer does know which yard is to build the yacht, he can base his estimates on weights and trim of the vessel with somewhat greater accuracy, based on his knowledge of the particular practices and prejudices of that yard.

Custom Design Fees

The fee for a custom design will vary amongst designers, depending on the extent of detail put on the drawings, the time spent on the project, and the overhead costs of the office. Some designers will charge a percentage of the cost of the construction of the boat, often in the 7 percent to 15 percent range. However, this does not always provide the most incentive for the designer to keep the cost of the boat to the client as low as he can. An alternative method is for the designer to quote a flat fee for the work, after talking over the project with the client and after reviewing the cost records from prior design jobs of similar size and complexity. The advantage of this for the client is knowing in advance the extent of the design fees. The disadvantage is that there is a strong incentive for the designer to do it as quickly as possible, making the designer reluctant to encourage improvements in the design as they occur to the parties involved, for this would entail spending unbudgeted time on the project, even though the result might be a better boat. A third method is for the designer to bill for the time spent on this job at a previously quoted rate, or rates if several staff members will be working on the job. This method leaves the designer with a clear incentive to work towards the best

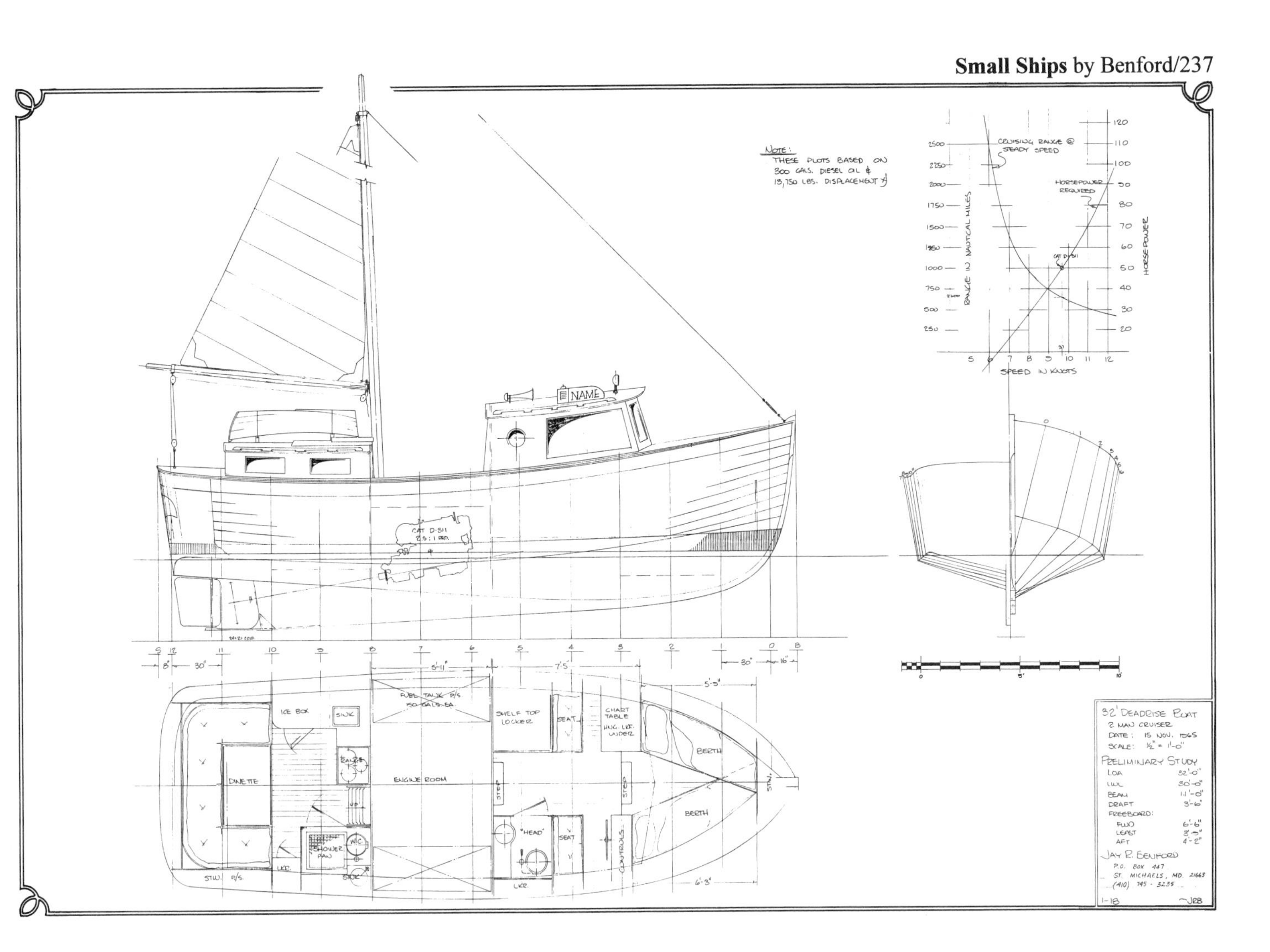
NOTE:
THESE PLOTS BASED ON 300 GALS. DIESEL OIL & 13,750 LBS. DISPLACEMENT
CRUISING RANGE @ STEADY SPEED
HORSEPOWER REQUIRED
RANGE IN NAUTICAL MILES
HORSEPOWER
SPEED IN KNOTS
NAME
CAT D-311
ICE BOX
SINK
FUEL TANK P/S
SHELF TOP LOCKER
SEAT
CHART TABLE
BERTH
DINETTE
ENGINE ROOM
STEP
"HEAD"
CONTROLS
SHOWER PAN
STW.
LKR.
32' DEADRISE BOAT
2 MAN CRUISER
DATE: 15 NOV. 1965
SCALE: ½" = 1'-0"
PRELIMINARY STUDY
LOA 32'-0"
LWL 30'-0"
BEAM 11'-0"
DRAFT 3'-6"
FREEBOARD:
FWD 6'-6"
LEAST 3'-3"
AFT 4'-2"
JAY R. BENFORD
P.O. BOX 447
ST. MICHAELS, MD. 21663
(410) 745-3235
1-18
JRB

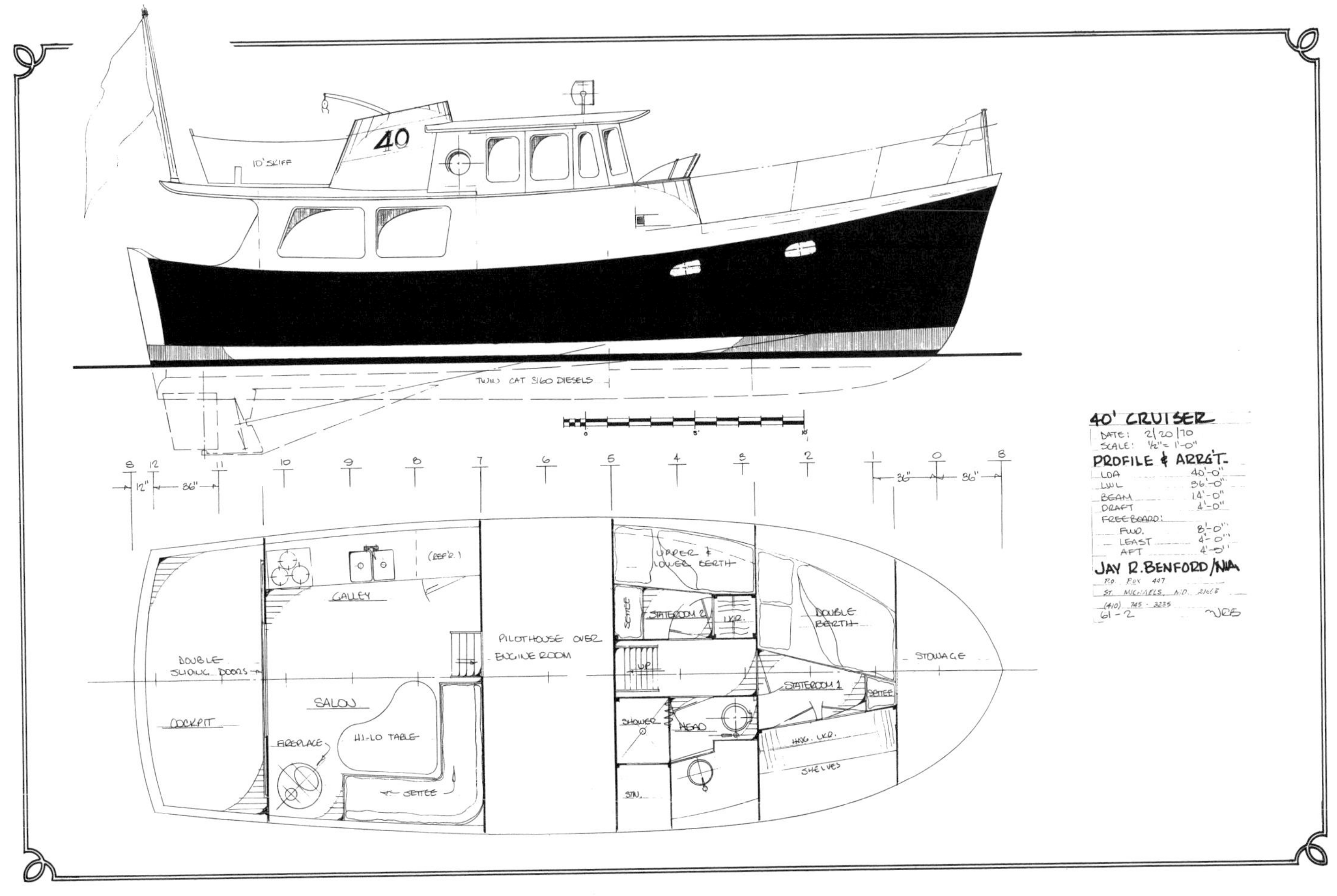
10' SKIFF
40
TWIN CAT 3160 DIESELS
40' CRUISER
DATE: 2/20/70
SCALE: ½" = 1'-0"
PROFILE & ARRG'T.
LOA 40'-0"
LWL 36'-0"
BEAM 14'-0"
DRAFT 4'-0"
FREEBOARD:
FWD. 8'-0"
LEAST 4'-0"
AFT 4'-9"
JAY R. BENFORD/NA
P.O. BOX 447
ST. MICHAELS, MD. 21663
(410) 745-3235
61-2
JRB
GALLEY
DOUBLE SLIDING DOORS
COCKPIT
SALON
FIREPLACE
HI-LO TABLE
SETTEE
PILOTHOUSE OVER ENGINE ROOM
UPPER & LOWER BERTH
STATEROOM 2
LKR.
UP
SHOWER
HEAD
STW.
DOUBLE BERTH
STATEROOM 1
HNG. LKR.
SHELVES
STOWAGE

possible boat, knowing that the creative time spent on it will be properly reimbursed. It also allows for varying the extent of the design detail required, depending on the needs of the client and the builder. This latter method is usually our preferred method of working and we've been quite flexible in the extent of the detailing supplied, once the basics are covered and we're assured that the result will be sufficient for a good boat to result. With unusual projects, or highly sophisticated concepts for very simple boats, it is often best to approach the job on an hourly rate basis, with frequent billings so that the client can keep track of the cost of the design work on a current basis.

Delivery Time For Custom Designs

Most small jobs and revision work can be expected to take a minimum of two to three months, and may be as long as six months or more, depending on the size of the yacht. If a specific delivery time is needed to fit a construction schedule, the designer appreciates being advised of this, so that schedules can be coordinated. Most designers have a number of projects "in process" at any one time, and just getting a position on the waiting list may take additional time.

Many custom design jobs, particularly on larger yachts, can be expected to take the best part of a year to gestate smoothly from the original concept into plans for the finished yacht. It is particularly important to allow plenty of planning time during the initial, conceptual work, so that the various new ideas can have the change to be absorbed, develop further, and finally evolve into the best end product. Once the preliminary conceptual work is over, the balance of the construction drawings flow reasonably smoothly.

Construction Bid Solicitation And Review

Construction bids can be solicited from recommended yards, if desired, and review and recommendations on them provided. If this service is desired, the designer needs to know of this in advance, so that it may be included in estimating of time and expense.

Material Lists

Most designers do not normally provide separate materials lists as a part of plans for a design, as materials and equipment are already specified throughout the drawings, and the builder can generally best familiarize himself with the design by compiling a materials list as required, upon studying the plans. Should separate materials lists be desired in the plans, the client should mention this in advance, so that it may be included also in estimating of design time and expense.

Inspection

We generally inspect construction progress at the builder's site in our local travel territory. The purpose of these inspection visits is to consult with the builder, to answer any question or problems which may arise, and to generally spot-check various aspects of the vessel to see that she is coming to life in the manner which we envisaged when we originally created the plans. Some visits also cover minor changes desired by the client and/or builder, and to approve such changes so that they will not drastically affect the successful outcome of the design. It is not the intention of the designer to accept liability for the execution of construction of the vessel in accordance with every detail of in the plans, on the basis of such inspection trips, for the available time and relatively few number of visits make it impossible to accurately determine this. Rather, we rely on the builder to construct the vessel in compliance with the specifications, tolerances and materials designated in the design, so that the designed performance of the vessel can be met, and we are happy to consult with the builder in an effort to best aid him in achieving this end.

We are pleased to inspect vessels which are built outside our generally traveled area for the cost of the travel and living expenses incurred in doing this. However, if it is a large yacht, and/or the time involved in the inspection process is considerable, then in addition, we charge for such consultation time. Quite naturally, such time subtracts from the designer's available time at the boards, but it also continues to give him/her important feedback for ongoing design work. Each designer's policies in this area is quite different from another's.

When The Boat Is Built . . .

After launching, we enjoy, whenever possible, partaking in the builder's sea trials. Our participation thus does not end either at the drawing board or the construction site, for we take the natural extension of our interest in our work which has stimulated this new custom design, by desiring to learn exactly how she reacts to her more demanding testing grounds at sea. Taking this sort of care, we believe that we are contributing to the sort of concern and attention to detail which we expect the builder will also have brought in seeing her come to life. The result is to enhance the yacht's worth, and her future resale value. The continuing success and longevity of our firm depends on our continuing ability to create fine designs. As custom yacht design is our first love and life's work, we want to ensure the continuance of our good name by taking the time and care necessary to see our designs successfully progress from the drawing board into the water.

Reuse Of Plans

Unless otherwise specifically prearranged, building rights for one boat only are included with the design fee or purchase of a set of existing plans. All rights of design ownership remain with the designer, and the designer should be consulted for arrangements for building rights for additional vessels. The drawings and prints supplied are instruments of service and remain the property of the designer. The designer may request their return on completion of the building of the subject vessel.

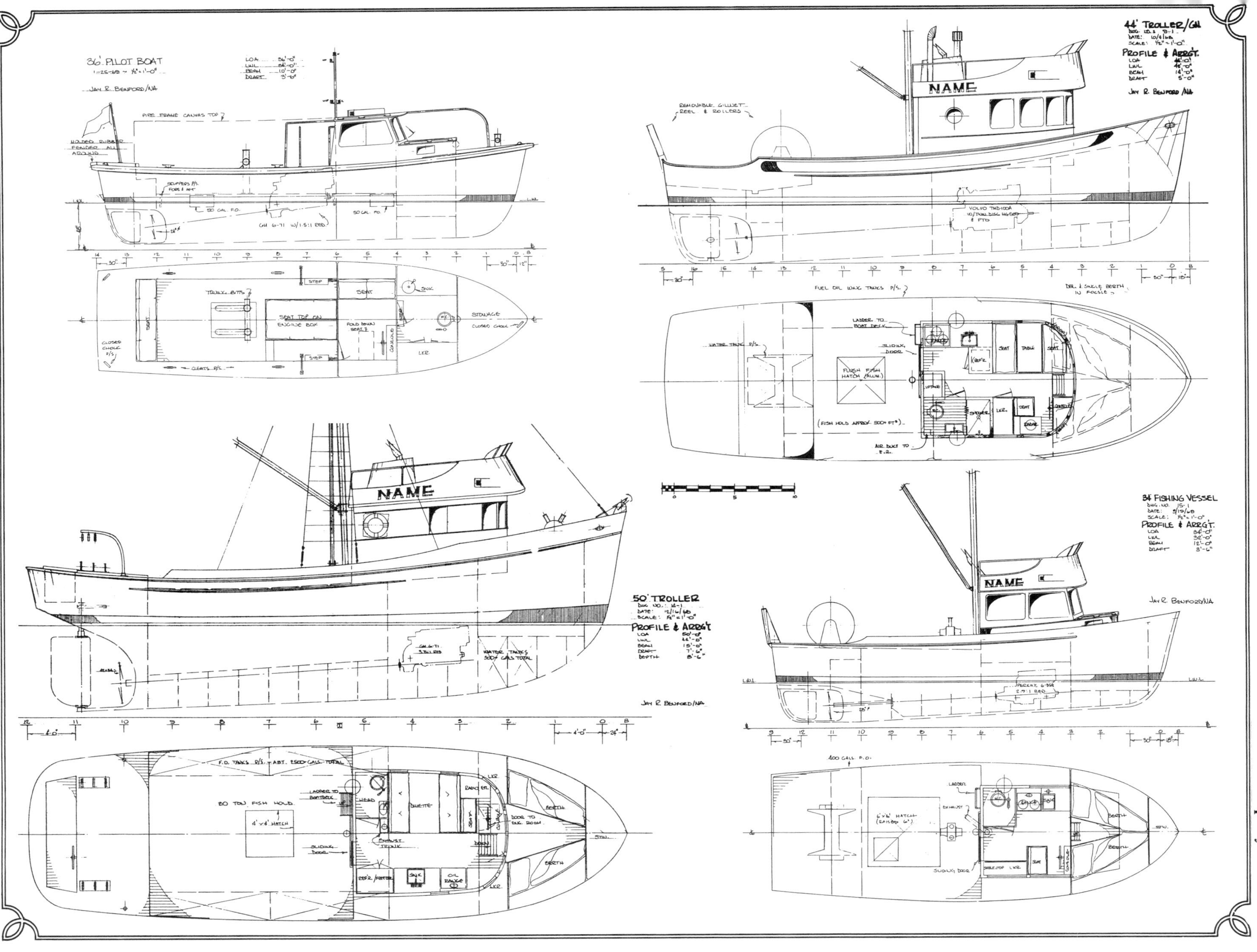
36' PILOT BOAT
JAY R. BENFORD/NA
PIPE FRAME CANVAS TOP
SEAT TOP ON ENGINE BOX
STOWAGE
44' TROLLER/GN
PROFILE & ARRG'T.
JAY R. BENFORD/NA
REMOVABLE GILLNET REEL & ROLLERS
FUEL OIL WING TANKS P/S.
FLUSH FISH HATCH (ALUM.)
NAME
NAME
NAME
50' TROLLER
PROFILE & ARRG'T
JAY R. BENFORD/NA
80 TON FISH HOLD
4'x4' HATCH
34' FISHING VESSEL
PROFILE & ARRGT.
JAY R. BENFORD/NA

Copyright

All designs, whether so marked or not, are protected by international copyright laws, and may be used and/or reproduced only with the permission of the designer. We have experienced copyright infringement in several forms. Some of these are the use of our drawings on some wrapping paper, on some business cards and stationery, on highway advertising signs, and in magazine advertisements. The great accessibility of the material makes the temptation to copy it without permission great. In most cases we would have been happy to have it used with a modest fee and/or a credit line as the source for the art. If you have the desire to use some of it just give us a call and let us come to an agreement in advance.

Expiration Of Quotes

When a designer quotes a fee for design work, quite naturally the quote would be valid for only a limited period of time. In our office, it is good for sixty days, unless otherwise stated. Thus, if a client desires to proceed with work on which a fee was quoted more than sixty days previously, it is advisable to contact us again to reaffirm the price.

Other Design Business Aspects

Beyond the creation of an entirely new design, other aspects of custom design work include modification to existing designs, consultation on other designs and/or vessels, and production design work.

Such modifications might include new interiors, new rigs, modified profiles, alternate engine-machinery, and/or alternate construction materials and methods. The basic considerations in doing such work are, to a small scale, quite similar to the creation of an entirely new design.

Consultation on such things as construction problems of any vessel, working out a new rig for an older vessel, or offering an opinion on construction method of another boat than our own design, is usually done on an hourly basis.

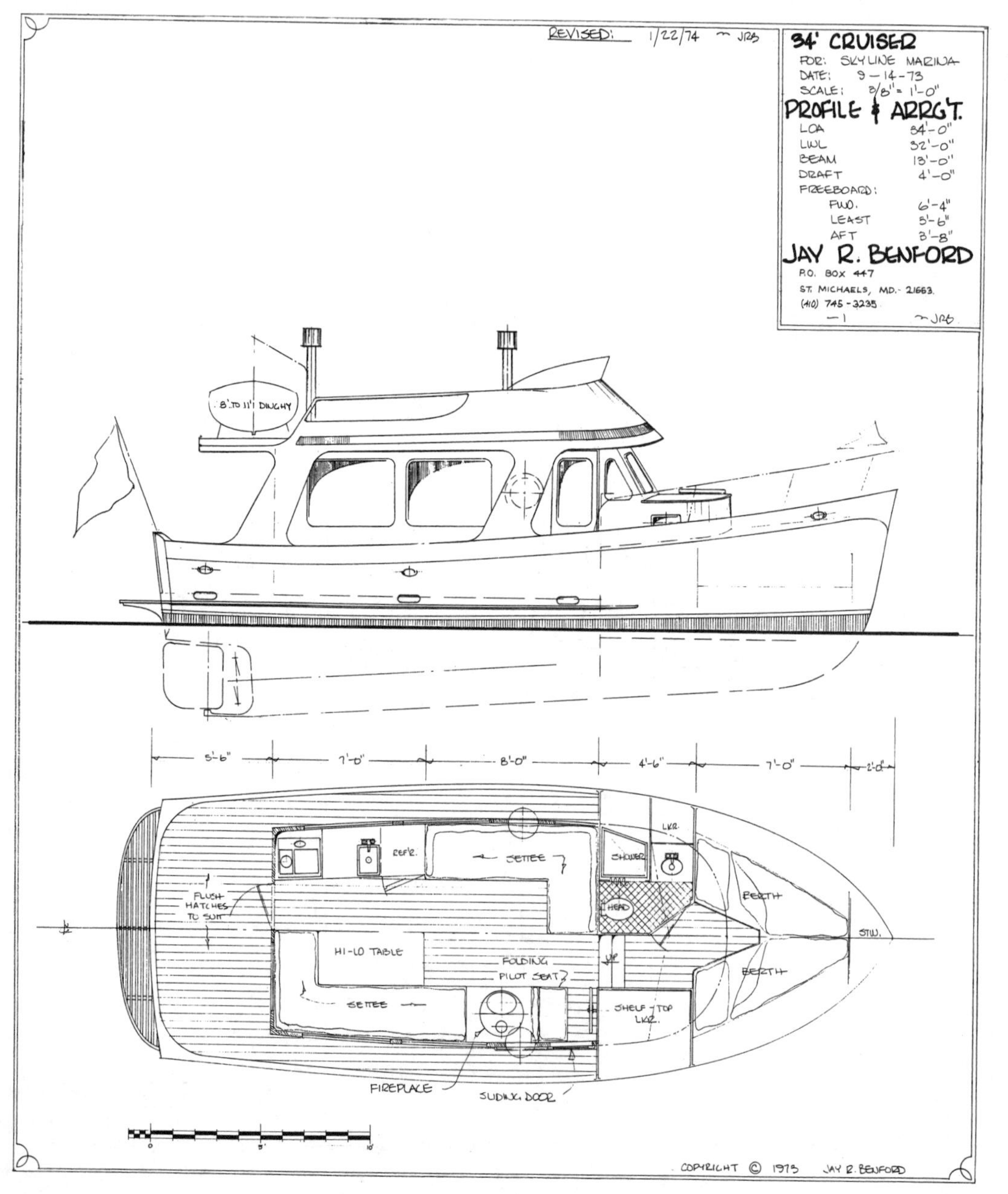

Production Designs

Designing for a boat which is to be built in production is another area of custom design. The basic approach is much the same as any other custom design. In addition to the process outlined earlier for proceeding with such work, the designer would also want such information as projected intentions for production quantities, construction schedules, location of construction site and intended or desired builder.

The Design Contract

The following pages have our design contract condensed without some specifics filled in. It gives an idea of what to expect when you proceed with a design. It can also be used by designers who didn't participate with the group that created it.

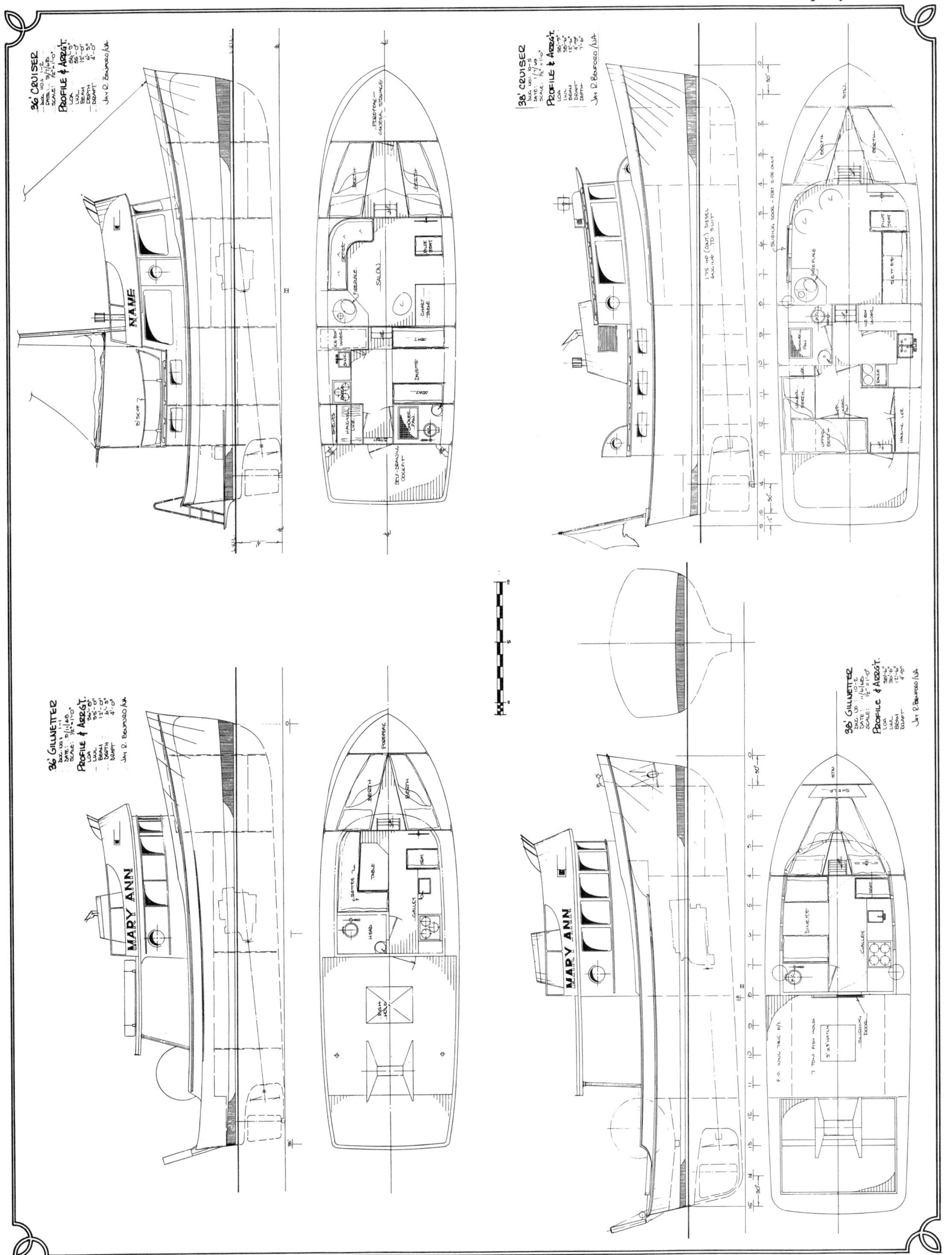
36' CRUISER
PROFILE & ARRGT.
NAME
38' CRUISER
PROFILE & ARRGT.
36' GILLNETTER
PROFILE & ARRGT.
MARY ANN
38' GILLNETTER
PROFILE & ARRGT.
MARY ANN

Benford Design Group

Naval Architects & Consultants
Specializing in Yacht Design

Mail: P. O. Box 447
Office: 605 Talbot Street
St. Michaels, MD 21663

Voice: 410-745-3235
Fax: 410-745-9743

(Note new area code above)

Latest revision: February 26, 1992

Design Contract

Agreement written as of the _________ day of _________ in the year of Nineteen Hundred and ______________________ between the **Owner** (or Client):

(Insert name and address)

and the **Architect** (or Designer):

Benford Design Group
P. O. Box 447
St. Michaels, MD 21663

For the following Project:

(Insert description of design to be created)

The Owner and the Architect agree as set forth below:

The Architect's Services consist of the elements listed below or of other elements listed as part of **Schedule B: Services & Practices.**

Article I: Preliminary Design Stage:

1) The Architect shall review the program furnished by the Owner to ascertain the requirements of the Project and shall review the understanding of such requirements with the Owner.

2) The Architect shall review, as applicable, with the Owner alternative approaches to design and construction of the Project.

3) Based on the mutually agreed upon program and Project budget requirements, the Architect shall prepare for approval by the Owner, Preliminary Design Documents including:

A) Outboard Profile sketch;

B) Accommodations sketch; and

C) Other documents as determined by the Architect to be required to illustrate the scope of the Project.

4) At the request of the Owner the Architect will submit copies of these documents to three (3) Builders of the Owner's choice for a preliminary bid estimate and assist the Owner in evaluating these estimates.

5) Compensation for the Preliminary Design Stage and any other services included as part of this fee under Article VI shall be paid as follows:

A) A fixed fee upon the execution of this agreement in the amount of ________________________________, or

B) A retainer of _______________ ______________ upon execution of this agreement, with charges per the Architect's hourly rates as shown in **Schedule A: Fee Schedule** for the balance owing for the work performed.

Article II: Construction Documents Design Stage:

1) Based on the approved Preliminary Design Documents and any further adjustments in the scope of the Project by the Owner, the Architect shall prepare, for approval by the Owner, Construction Documents containing information delineated in **Article IV** setting forth in detail the requirements for the construction of the Project.

2) Prints of each drawing shall be submitted to the Owner for approval. The Owner agrees to indicate approval or disapproval within ten (10) days of receipt. If disapproved, the Owner shall promptly advise the Architect in writing of the desired changes. The Architect will revise the plans accordingly and resubmit prints for approval. Failing such notification in writing, the Architect may deem the drawings approved by the Owner. Substantial changes requested to the drawings by the Owner may be billed by the Architect based on the terms for "Additional Services" as outlined in **Schedule B**.

3) The Architect shall be the sole determiner of whether or not changes to drawings are substantial.

4) At the request of the Owner the Architect will submit copies of these documents to three (3) Builders of the Owner's choice for a construction bid estimate, and assist the Owner in evaluating these estimates.

5) The Architect agrees to perform such theoretical calculations as are customary to insure the safety and seaworthiness of the design and reasonable speed underway. The Architect will furnish a carefully prepared estimate of the speed of the vessel, but does not guarantee such speed.

6) The Architect agrees to submit the lines of the vessel to a recognized agency for model testing at the Owner's request. Observation of the tank testing by the Architect will be at the Architect's discretion or as listed in **Schedule B: Additional Services**. The Architect's presence at such test shall not be interpreted as an endorsement of the test results.

7) Compensation for the Construction Documents Design Stage and any other services included as part of this fee under **Article IV** shall be:

A) A fixed fee of ___________________ ____, payable as follows: First payment of_____________ _______________on ____, second payment of ____ on

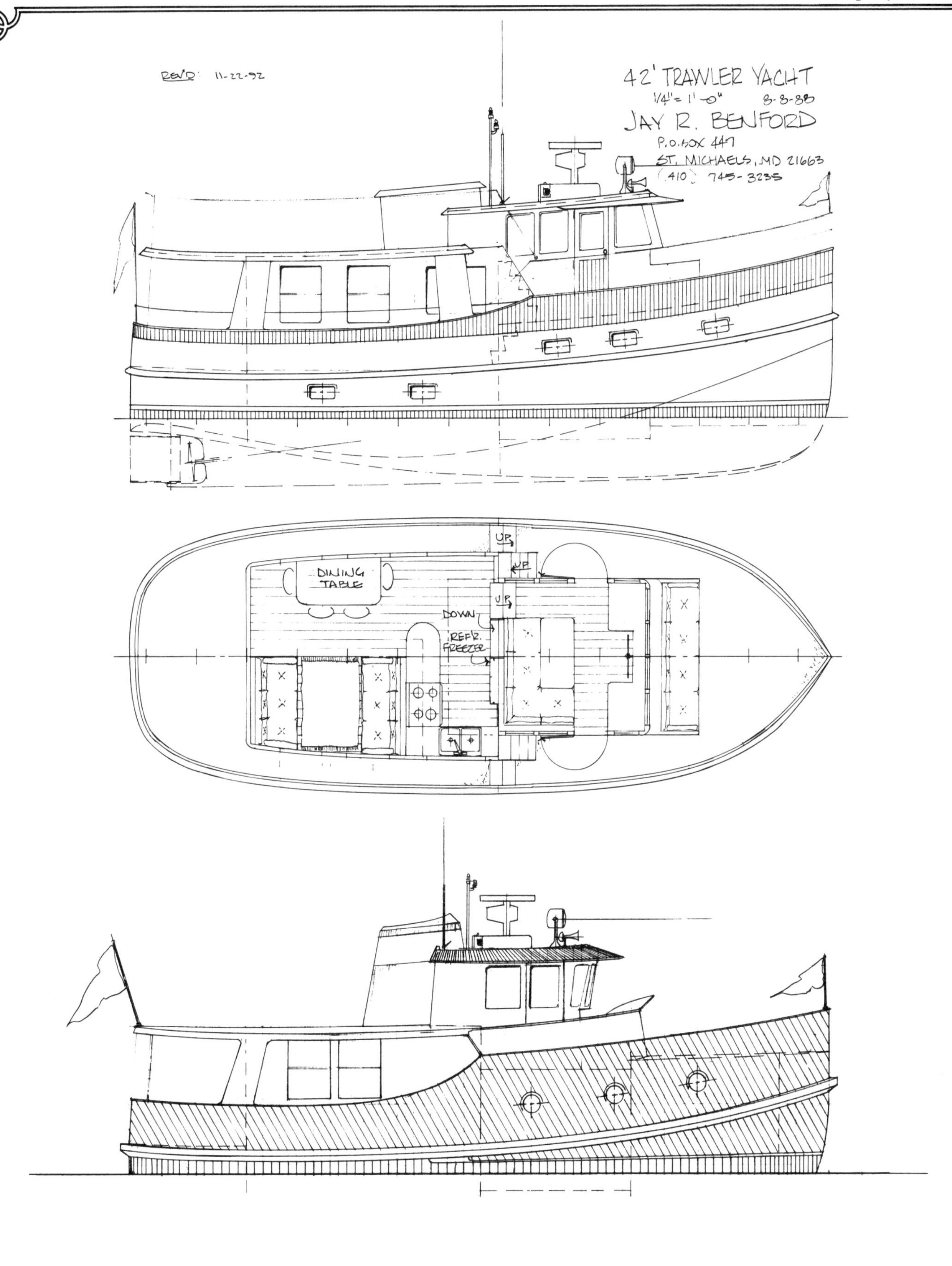
REV'D: 11-22-92
42' TRAWLER YACHT
1/4" = 1'-0"
8-8-88
JAY R. BENFORD
P.O. BOX 447
ST. MICHAELS, MD 21663
(410) 745-3235
DINING TABLE
UP
UP
UP
DOWN
REFR. FREEZER
42' MOTORYACHT
3/12/87
3/23/87
11/22/92
1/4" = 1'-0"
JAY R. BENFORD

________________, third payment of ____________________ on ____________________, fourth payment of ________________________________ on ________________, and a fifth payment of ________________________________ on completion of the drawing work. Or,

B) Hourly charges per **Schedule A: Fee Schedule** for the time spent on the Project. Billings will be submitted with each set of progress prints and the Owner will promptly pay these. (See **Schedule B** for more detail on Payments.)

Article III: Owner's Responsibilities:

1) The Owner shall provide full information regarding requirements for the Project including a program, which shall set forth the Owner's design objectives, constraints and criteria, including space requirements and relationships, vessel type and style, speed requirements, extent of equipment compliment anticipated, special equipment and systems requirements, and other general information and specifications as are pertinent to the design of the vessel.

2) If the Owner provides a budget for the Project, it shall include contingencies for bidding, changes in Work during construction, and other costs which are the responsibility of the Owner.

3) If the Owner observes or otherwise becomes aware of any fault or defect in the Project or non conformance with the Construction Documents, prompt written notice thereof shall be given by the Owner to the Architect.

4) The Owner shall furnish required information and service and shall render approvals and decisions as expeditiously as necessary for the orderly progress of the Architect's services and of the Work.

Article IV: Extent Of Agreement:

1) This Agreement represents the entire and integrated agreement between the Owner and the Architect and supersedes all prior negotiations, representations or agreements, either written or oral. This Agreement may be amended only by a written instrument signed by both Owner and Architect.

Article V: Architect Supplied Materials:

1) The information and materials to be supplied by the Architect during the Stages of the Design period are to include approximately the following lists:

The design work will be done in two Stages. The first Stage will be the Preliminary Design Phase, to define the concept of the Project. Drawings to be done include:

Preliminary Lines Plan,
Outboard Profile and/or Sail Plan,
Accommodation Plans,
Scantling Section, and
Initial Weight Study.

The second Phase, the Construction Documents Design Stage, will be to create the actual working drawings. These will include the final versions of the drawings and work done in the Preliminary Design Phase, plus:

Construction Plans and Profiles,
Inboard Profiles,
Sections of Construction Frames and Joiner and Outfitting,
Engine, Steering and Tanks, and
Deck and Rigging Plans.

Optionally, if desired or needed by the Builder, the Architect may provide:

Full Size Frame Templates on Mylar,
Plating Expansions,
Piping Schematics, and
Electrical Layout and One-line drawings.

2) The Architect may make necessary adjustments to the order in which materials are prepared or delivered based upon the Architect's determination of the most efficient design order.

3) This list is not meant to be all inclusive or to suggest that the Builder will not have to develop Shop Drawings of less significant details (which are to be submitted to the Architect for approval).

Article VI: Other Conditions Or Services:

1) It is understood that this vessel is not intended to be licensed by the U.S. Coast Guard for carrying passengers for hire.

2) (Add other specifics here)

Article VII: Attachments:

The Owner acknowledges receipt of **Schedules A, B,** and **C** which are expressly adopted and made part hereof.

(Add Owner & Architect signatures & dates)

Schedule A: Fee Schedule (effective January 1, 1992)

$/Hour **Service (Fill in current rates)**

Expert Witness
Consultation (not part of a design project)
Design Work by Jay R. Benford
Design Work by senior architect/engineer
Design Work by junior architect/engineer
Travel time by Jay R. Benford
Travel time by senior architect/engineer
Computer time running full-size frame templates from hulls designed on our computer facility

Computer Services to be charged as follows:

Performance Prediction of Benford design
Performance Predictions rerun with varied input

(Hydrostatics/stability studies quoted on request.)

Note: All work done under terms and conditions outlined in our standard **Design Contract** and **Schedule B: Services & Practices.** Work on all projects is started only after receipt of the retainer payment, and continued only while the account is current. Bills are due on receipt. 1.5% per month interest added to past due amounts.

Schedule B: Services & Practices

Additional Services

The following are among the Additional Services not included in the contract price unless so identified in the Design Contract. They will be provided if authorized or confirmed in writing by the Owner, and paid for by the Owner as provided in the Design Contract, in addition to the compensation for Basic Service.

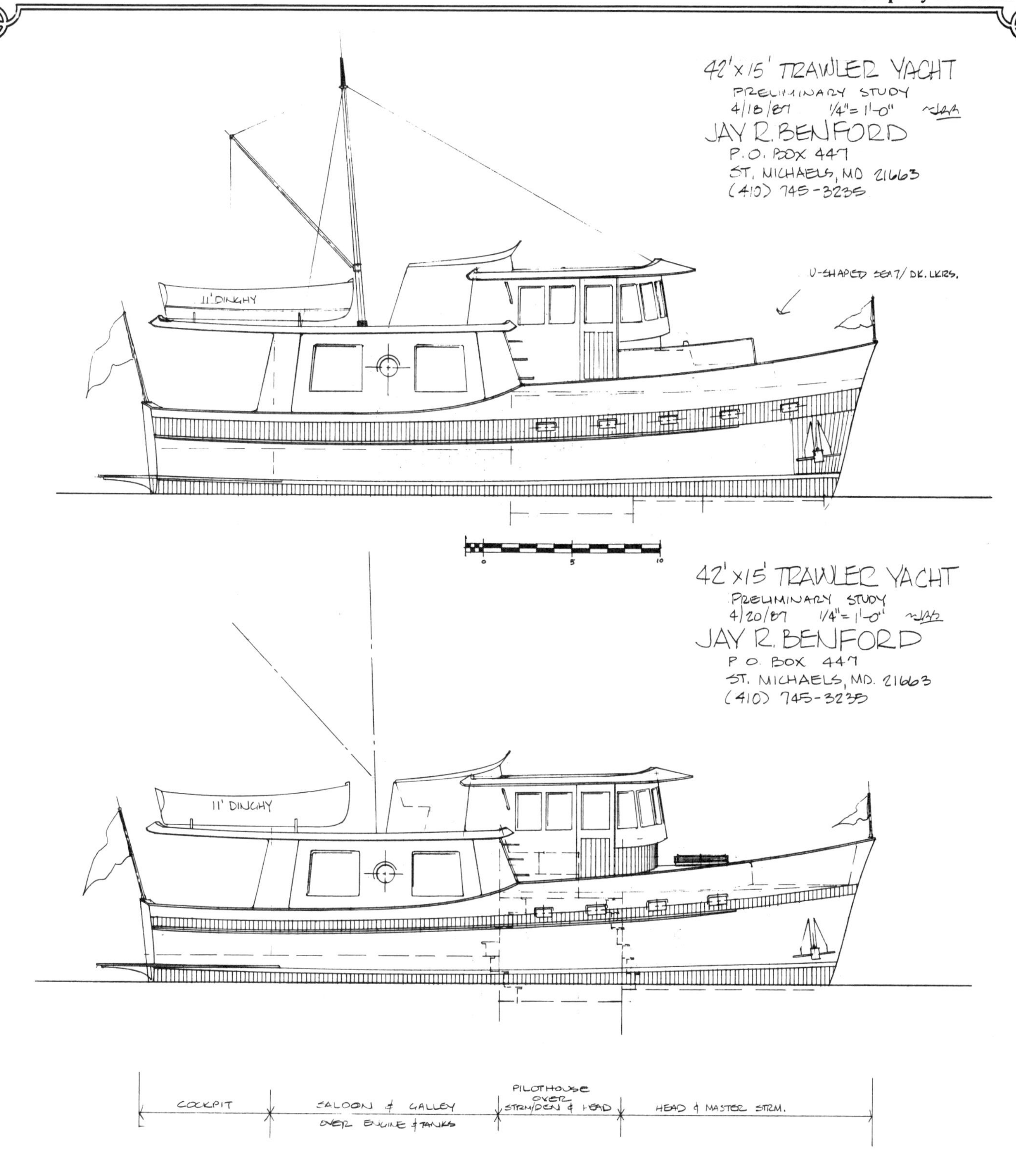

These two styling studies for a 42' Trawler Yacht show another approach to this type of boat.

Her dimensions are approximately 42' x 39' x 15' x 4'. Draft will vary with the power chosen.

The saloon is the full width of the vessel and as a result will have more room than is usually found there.

The lower profile has the freeboard reduced by six inches and an alternate forward trunk cabin.

These boats could be built in almost any medium, and with sufficient power and revised hull form, could go well beyond conventional hull speed.

Power could be either single or twin diesels. A bow thruster can be fitted.

1) Providing services to investigate existing conditions of a vessel or to make measured drawings thereof, or to verify the accuracy of drawings or other information furnished by the Owner.

2) To prepare or examine evaluations of the Owner's Project budget, Statements of Probable Construction Cost and Detailed Estimates of Construction Cost.

3) Submit Preliminary or Construction Documents to additional Builders, or revise Preliminary or Construction Documents to reflect changes necessary due to the Bid estimates received by the Owner.

4) Attend Tank Test, or other tests or meetings suggested by the Owner in the furtherance of the Design.

5) On-site observation visits requested by the Owner, or requested by the Builder and approved by the Owner.

6) Review of Builder's or Subcontractor's shop drawings, of Product Date, or Samples submitted by the Owner or Builder.

7) Provide additional detail drawings for the Builder, at the Owner's request.

8) Providing interior design and other similar services required for, or in connection with the selection, procurement or installation of furniture, furnishings, and other equipment, or the engineering of any unusual equipment or systems aboard. The Architect shall be the sole one to define what is regarded as unusual.

9) Making revisions in Drawings, Specifications or other documents when 1) such revisions are inconsistent with prior written authorizations, documents that have been deemed to be approved, or instructions previously given; 2) when such revisions are required by the enactment of codes, laws or regulations subsequent to the preparation of such documents, or 3) when such revisions are due to the Builder's preferences, or capabilities or other causes not solely within the control of the Architect.

10) Preparing Drawings, Specifications and supporting data and providing other services in connection with Change Orders provided such Change Orders are required by causes not solely within the control of the Architect.

11) Making investigations, surveys, valuations, inventories or detailed appraisals necessitated in connection with construction performed by the Owner.

12) Providing consultation concerning replacement of any Work damaged by fire or other cause during construction, and furnishing services as may be required in connection with the replacement of such Work.

13) Providing services made necessary by the default of the Builder, or by major defects or deficiencies in the Work of the Builder, or by failure of performance of either the Owner or Builder under the Contract for Construction.

14) Preparing a set of reproducible record drawings showing significant changes in the Work made during construction based on marked-up prints, drawings and other data furnished by the Owner and/or the Builder to the Architect.

15) Providing extensive assistance in the utilization of any equipment or system such as initial start-up or testing, adjusting and balancing, preparation of operation and maintenance manuals, training personnel for operation and maintenance, and consultation during operation.

16) Providing services to the Owner after the Architect's obligation to provide Basic Services under the Design Contract has terminated.

17) Providing any other services not otherwise included in the Design Contract or not customarily furnished in accordance with generally accepted design practice.

18) Compensation for Additional Services shall be in accordance with **Schedule A** for twelve (12) months after signing of the Design Contract. Thereafter, compensation for the Architect shall be at the then current revision to **Schedule A: Fee Schedule**.

Communications

The use of the telephone has proven to greatly speed up the design process. At the same time, we would suggest that the Client follow up all verbal authorizations or suggestions or ideas with written copies for our files. At any time, we may have a dozen or more projects in process, and while we try to keep good notes and sketches from such conversations, there is the possibility we may have missed something. Thus a fax or letter follow-up is the best insurance.

Computer Usage & Program Development

Computer rental time for equipment required beyond that owned by the Architect is billed at a rate permitting reasonable amortization of office programming expenses. Programming, including all documentation and work product is for the exclusive use of the Benford Design Group and shall be considered his property.

Construction Cost Estimating

The Benford Design Group may be asked by the Owner to make an evaluation of estimated construction costs of the completed vessel, or costs at any intermediate stage of completion. Evaluations of the Owner's project budget, statements of probable construction costs and detailed estimates of construction cost if prepared by the Architect represent the Architect's best judgment as a design professional familiar with the yacht construction industry. It is recognized, however, that neither the Architect nor the Owner has control over the cost of labor, materials or equipment, the builder's methods of determining bid prices, nor over competitive bidding, market or negotiating conditions. Accordingly, the Architect cannot and does not warrant or represent that bids or negotiated prices will not vary from the project budget proposed, established or approved by the Owner, if any, or from any statement of probable construction cost or other cost estimate or evaluation prepared by the Architect.

General Provisions

1) The Architect shall in dealings with the Builder be considered a representative of the Owner during the Vessel Construction Phase. As the Owner's agent, however, the Architect shall incur no pecuniary responsibility concerning the construction of the Project.

2) The Architect is to be furnished with written copies of all instructions supplied by the Owner to the Builder which could affect in any way the structural integrity, performance, or aesthetic value of the design.

3) On-site consultation by the Architect with the Owner and Builder may be arranged as outlined herein. The Architect shall at all times have reasonable access to the Work whenever it is in preparation or progress.

4) The Architect is to be presented with copies of all Change Orders issued which result in the modification of the vessel in any way from the original Construction Documents. No major changes, omissions, or other variations from the Construction Documents shall be made without the permission of the Architect. No such changes shall be considered approved by the Architect unless agreed to in writing by the Architect. The Architect may bill the Owner for time spent on such items as above.

5) It is understood by the Owner that additions to or deletions of equipment or parts of the vessel as designed by the Architect will affect the handling characteristics of the completed vessel. Any such additions or deletions by the Owner or Builder are at the sole risk of

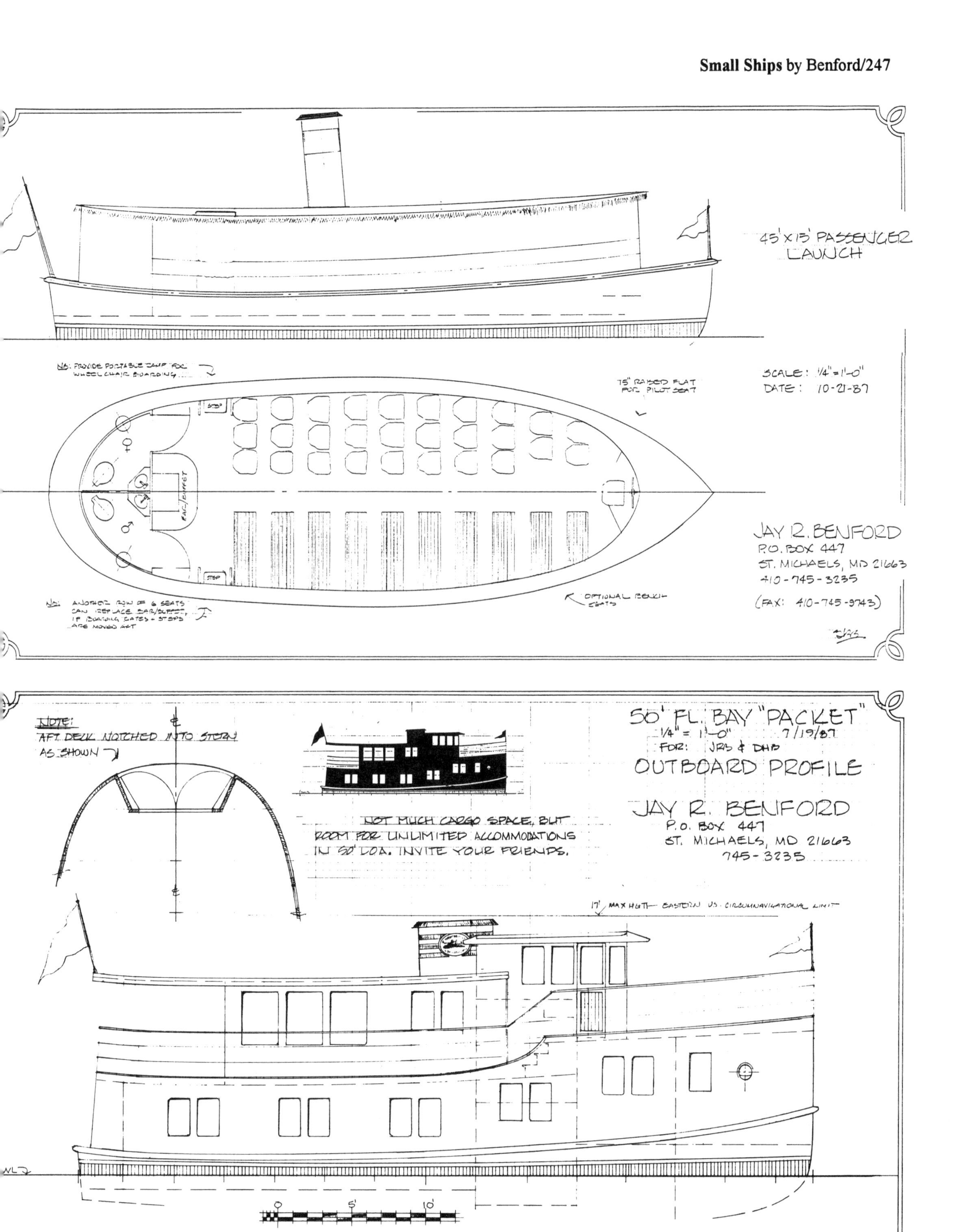
45' x 13' PASSENGER LAUNCH
NB: PROVIDE PORTABLE RAMP FOR WHEELCHAIR BOARDING
75" RAISED FLAT FOR PILOT SEAT
SCALE: 1/4" = 1'-0"
DATE: 10-21-87
JAY R. BENFORD
P.O. BOX 447
ST. MICHAELS, MD 21663
410-745-3235
(FAX: 410-745-9743)
NB: ANOTHER ROW OF 6 SEATS CAN REPLACE BAR/BUFFET, IF BOARDING GATES + STEPS ARE MOVED AFT
OPTIONAL BENCH SEATS
STEP
BAR/BUFFET
NOTE: AFT DECK NOTCHED INTO STERN AS SHOWN
50' F.L. BAY "PACKET"
1/4" = 1'-0"
7/19/87
FOR: JRB & DHB
OUTBOARD PROFILE
JAY R. BENFORD
P.O. BOX 447
ST. MICHAELS, MD 21663
745-3235
NOT MUCH CARGO SPACE, BUT ROOM FOR UNLIMITED ACCOMMODATIONS IN 50' LOA. INVITE YOUR FRIENDS.
17' MAX HGT - EASTERN US CIRCUMNAVIGATIONAL LIMIT
WL
0
5'
10'
-S
-6
-5
-4
-3
-2
-1
0
1
2
3
4
5
6
7
B
36"
41"
41"
31"
261-50
JRB

the Owner, and the Architect will not be responsible for the results thereof.

6) The Architect shall perform Basic and Additional Services as expeditiously as is consistent with professional skill and care and the orderly progress of the Work. This time allocation is to be in keeping with any time constraints or commitments outlined in **Schedule C: Current Work List**.

Guarantee

A yacht compares in complexity with a car or airplane which may be the result of many hundreds of thousands of hours of design and engineering development. This contrasts with the few hundreds of hours which the Architect and his Client can afford for the designing and engineering of a new yacht. In place of exhaustive scientific and engineering research, we must sometimes depend upon past experience, rule-of-thumb calculation, and even a degree of intuition. Naturally we use our best efforts in each new design calling heavily on our years of experience designing successful vessels. Because of these complexities, however, we are simply not in a position to be able to warrant or guarantee that a new yacht will be entirely free of defects, whether caused by our error or omission, or that of the Builder.

For example, it is difficult to predict with complete accuracy the weights incorporated in a new yacht, and to visualize in three dimensions all of the spaces in a new design or the integration of a complex mechanical or electrical system. In deciding to construct a new yacht, therefore, the Owner must be prepared for the possibility that some changes may be required as work progresses or even after the yacht is completed.

Thus, the Architect is not a guarantor with respect to the results of his design. However, the Architect does commit to perform his duties in accord with the appropriate standard of care.

Insurance

Many years ago it was practical for an architect to carry "errors and omissions" insurance to protect themselves against claims arising out of potential defects in yachts constructed to their designs. Today, however, even though our record in this regard has been excellent, the cost for coverage (if available at all) is so high that having coverage is no longer feasible. As a result it is now our policy (as with most other architects) to accept a design commission only if the client agrees to:

1) Accept the risk of all defects in the yacht whether caused by the Architect's error or omission or that of the Builder.

2) Bear the cost of "errors and omissions" insurance covering our work in connection with the yacht in question.

As a result, we must expect that the client is willing to agree to release Benford Design Group and its officers, employees, and subcontractors from any and all liability, suits or causes of action arising out of, or relating to the yacht being designed.

Legal Action

Unless otherwise specified, the Design Contract shall be governed by the law of the principal place of business of the Architect. Regarding all acts or failures to act by either party, any applicable statute of limitations shall commence to run and any alleged cause of action shall be deemed to have accrued in any and all events not later than the completion of the construction documents by the Architect.

The Architect and Owner agree that should legal action be pursued by either party, that the winning party shall be entitled to reasonable attorney's fees, in addition to other damages.

Liability

The Architect states and the Owner acknowledges that the Architect has no Professional Liability (Errors and Omissions) Insurance and is unable to reasonably obtain such insurance for claims arising out of the performance of or the failure to perform professional services, including but not limited to the preparation of reports, designs, drawings and specifications related to the design of this vessel. Accordingly, the Owner hereby agrees to bring no claim for negligence, breach of contract, or other cause of action, indemnity or otherwise against the Architect, his principals, employees, agents, sub-contractors and consultants if any claim in any way is related to the Architect's services for the design of this vessel.

The Owner further agrees to defend, indemnify and hold the Architect and his principals, employees, agents, sub-contractors and consultants harmless from any such claims that may be brought by third parties as a result of the services provided by the Architect for the Owner.

The Owner and the Architect waive all rights against each other and against the Builder, consultants, agents and employees of the other for damages covered by any liability insurance during construction. The Owner and the Architect shall require appropriate similar waivers from their Builder's consultants and agents.

Ownership of Plans

Unless otherwise agreed to by the Benford Design Group in writing the Owner is authorized to build only one vessel from the design contracted. Drawings and Specifications as instruments of service are owned by and shall remain the property of the Architect whether the Project for which they are made is executed or not. The design itself remains the exclusive property of the Benford Design Group. The Benford Design Group has the sole right to authorize subsequent use of such plans. The Owner, however, is entitled to receive complete sets of the plans developed for him.

The Owner shall be permitted to retain copies of Plans and Specifications for information and reference in connection with the Owner's use for maintenance, repair and operation of the completed vessel. Upon completion of the vessel all Plans and Specifications used by the Builder during construction shall be returned to the Architect.

(See section on **Sisterships** for information about reuse of plans and design fees due.)

Payments to the Architect

Payments for basic services, additional services, and reimbursable expenses shall be made promptly upon presentation of the Architect's statement of services rendered or expenses incurred.

Failure of the Owner to promptly pay the Architect the fees outlined in the design contract will be construed a breach of contract. In such case, the Architect has the option to continue work, suspend services until payment is rendered, or terminate the Project.

If the fees due the Architect are not paid within the time constraints permitted under the terms of the design contract, the Architect shall charge the maximum legal interest rate on all such sums from the due date until paid. If the Architect must engage a collector or an attorney to collect the fees due, the Owner shall pay reasonable collector's and/or attorney's fees whether or not legal action is instituted, and all of the Architect's collection expenses, including court costs if legal action is commenced.

If the scope of the Project or of the Architect's services is changed substantially the amounts of compensation shall be equitably adjusted.

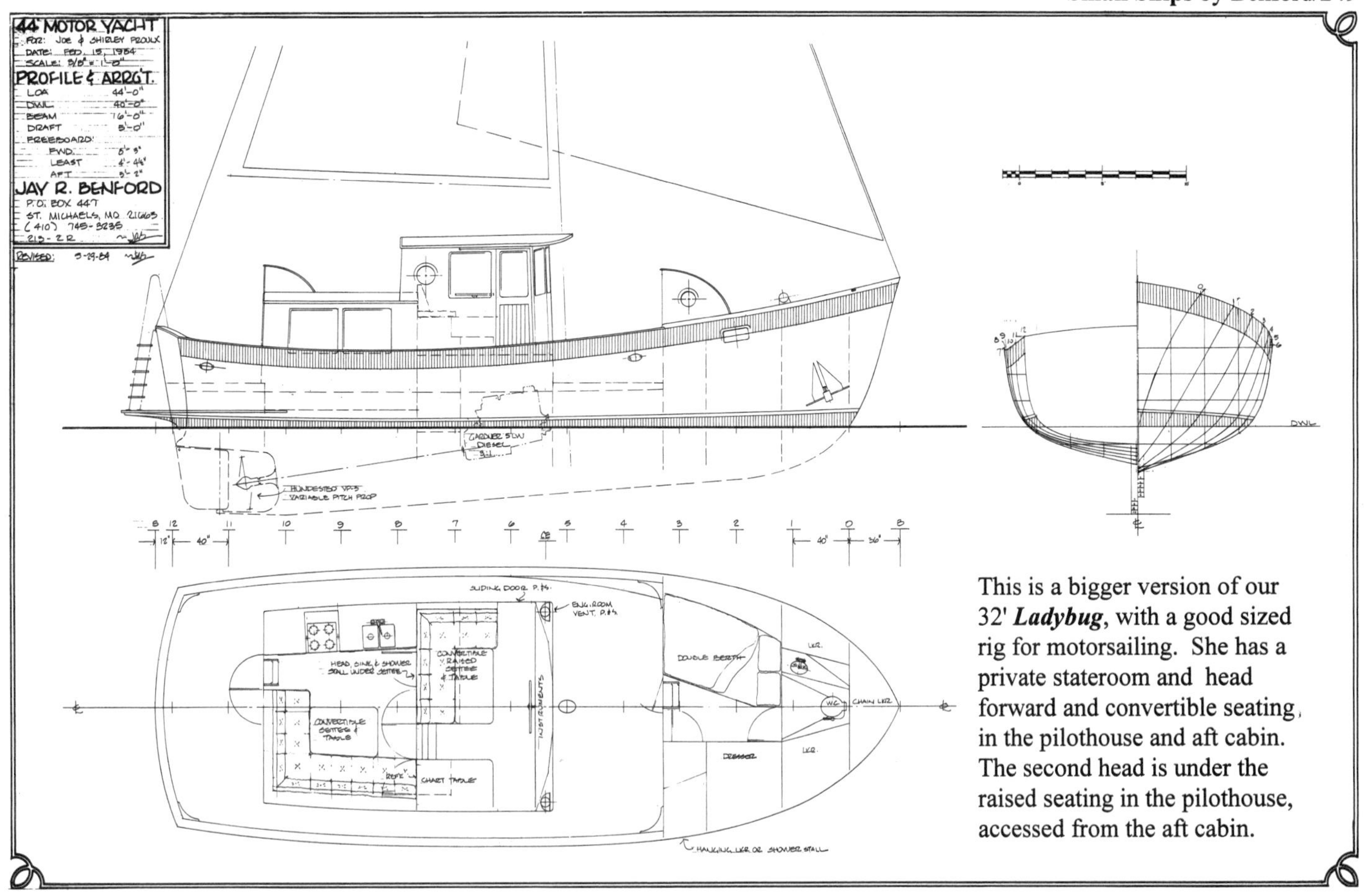

This is a bigger version of our 32' ***Ladybug***, with a good sized rig for motorsailing. She has a private stateroom and head forward and convertible seating in the pilothouse and aft cabin. The second head is under the raised seating in the pilothouse, accessed from the aft cabin.

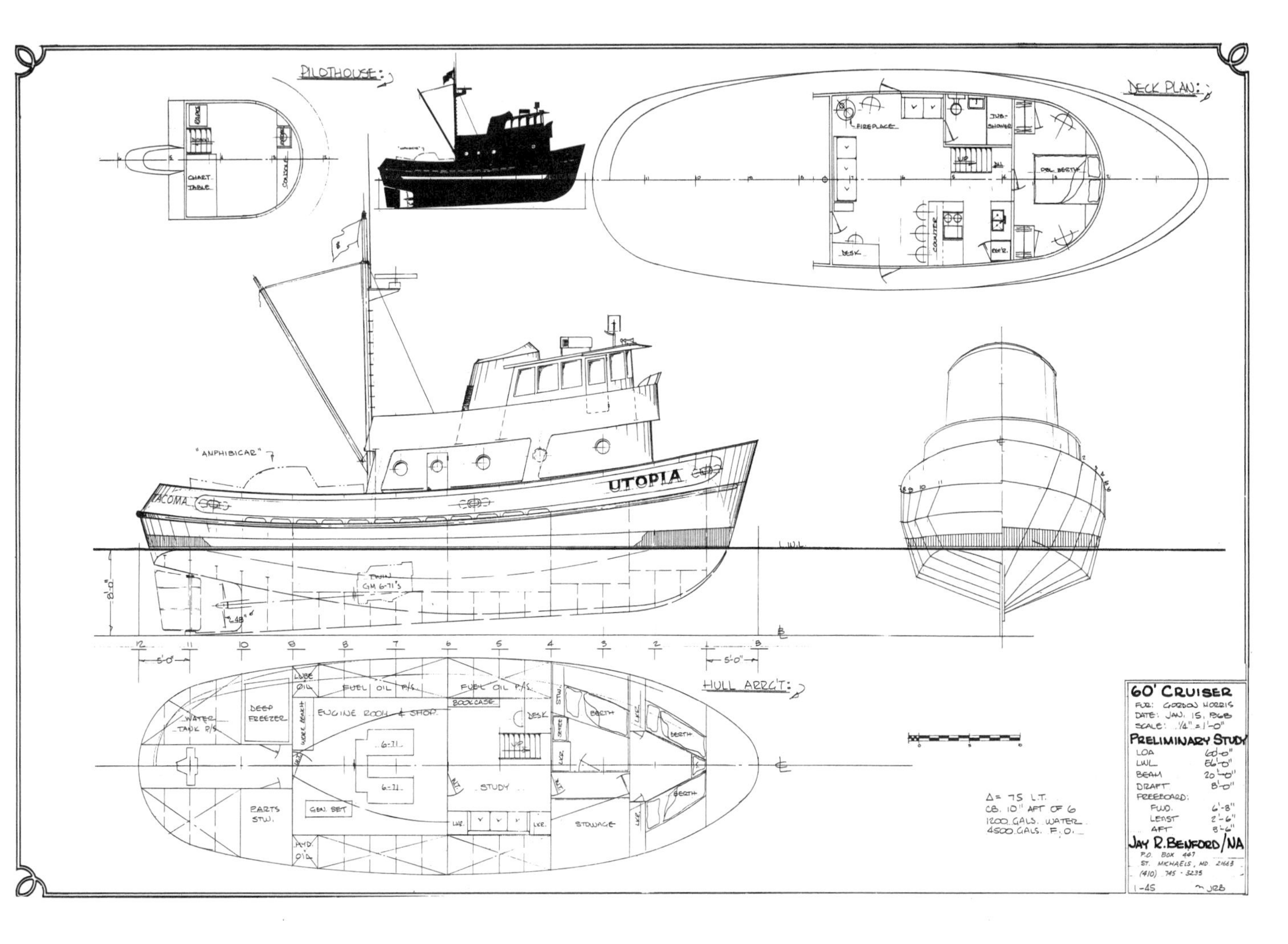

Publicity and Marketing

The Owner agrees that the name of the Architect shall appear in all articles, information, advertisements, and brochures prepared by the Owner unless otherwise requested by the Architect. Architectural drawings appearing in such documents are to originate from the Architect or are to be approved by the Architect in writing. The Architect maintains the right to use any test data, drawings, specifications, photographs etc. of the vessel for publicity, including advertising brochures, magazine articles, etc. The Owner's name, however, may not be used by the Architect in such publicity without the Owner's express consent.

Reimbursable Expenses

Reimbursable expenses are in addition to the compensation for basic and additional services and include actual expenditures made by the Architect and the Architect's employees and consultants in the interest of the Project for the following expenses:

A) Expenses of transportation in connection with the Project (Airline tickets of over $200 will be prepaid by the Owner), living expenses in connection with out-of-town travel, and long distance telephone, fax and electronic communications.

B) Expenses of prints, reproductions, postage, small package delivery service fees and handling of drawings, specifications and other documents, excluding reproductions for the office use of the Architect and the Architect's Consultants.

C) Expenses of data processing, photographic reproduction techniques, and photographic prints.

D) Expenses of renderings, artwork, models and mock-ups, tank testing fees and expenses.

E) Expenses of overtime work requiring higher than regular rates, if authorized in advance by the Owner.

F) Expenses of any additional insurance coverage or limits, including professional liability (errors and omissions) insurance, if requested by the Owner.

Compensation for reimbursable expenses as described herein shall be computed as a multiple of one point two (1.2) times the amounts expended by the Architect and the Architect's employees and consultants in the interest of the Project.

Services

In addition to preparing preliminary plans, bidding or contract plans, and building plans, Benford Design Group may be contracted separately to perform the following functions:

A) Recommendation of bidders, circulation of requests for bids, analysis of returns, assistance of a non-legal nature to the Client or Builder, or their attorneys, in the preparation of the building contract.

B) Checking of Builder's detailed building plans, consultations with builder, preparation of addenda to specifications to embody changes approved by the Client and Builder under provisions of the construction contract, consultation on requests for extras or credits.

C) Inspection visits to the Builder's yard can be made by the Architect to observe the progress and quality of the work being performed and to note if it is proceeding in general accordance with the plans and specifications. On such visits we often check specific items and take measurements on a sampling basis. In doing so we try to determine if defects or variation from the plans or specifications exist. We do not however supervise construction.

D) In performing such inspection visits we do not accept financial responsibility either for the cost of correcting any defect or variation from the plans or for any consequential damages caused thereby. Because we inspect only at stages of work and because of the limitations of time and of the measuring and testing facilities available to us, it is not possible for us to determine that the plans and specifications have been followed in every particular, or that details of construction and arrangement are perfect. Consequently we must rely on the Builder (and expect the Owner to rely on the Builder) to see to it that such details are properly taken care of, that the plans and specifications are followed and that defects, where they occur, are corrected.

Sisterships

The design may be licensed for serial production, by notation in the Design Contract. For this permission, the Architect will be paid a design royalty for each vessel as specified in the Design Contract. Exclusive production rights require the guarantee of a certain minimum number of royalty payments per year for the exclusivity to be in effect.

Successors and Assigns

The Owner and the Architect, respectively, bind themselves, their partners, successors, assigns and legal representatives to the other party to this agreement and to the partners, successors, assigns and legal representatives of such other party with respect to all covenants to the Design Contract. Neither the Owner nor the Architect shall assign, sublet or transfer any interest in this Design Contract without the written consent of the other.

Suspension of Project

If payments due the Architect by the Owner are not made in a timely manner in accordance with the terms of the Design Contract the Architect may at his option choose to suspend all work on the project until such payment is made. The Architect cannot be held responsible for any delays or other consequences that arise out of suspension of work resulting from either non-payment of fees by the Owner, or suspension of work requested by the Owner.

Tank Tests

Tank tests will be arranged by the Architect at the request of the Owner. The cost of these tests are billed on a time and material basis and may be expected to include tank costs, model costs, and design and analysis costs. Clients are expected to provide advance deposits to cover estimated tank billings. Tank test results, including all data and reports and the models themselves are to be considered the Architect's property and for their exclusive use.

Schedule C: Current Work List

Hull No.	Type	Scope of Project

(Add list of current projects here.)

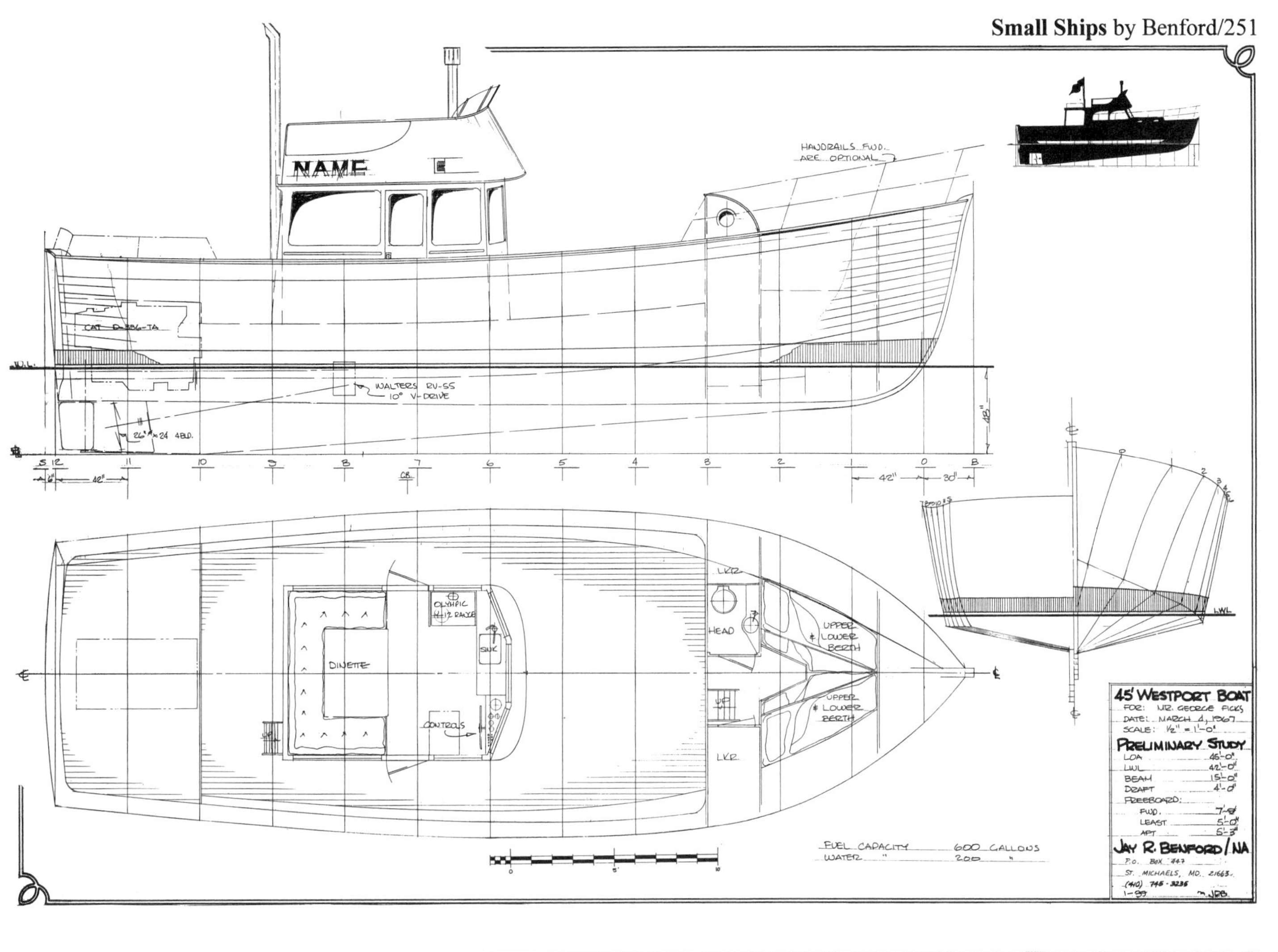
NAME
HANDRAILS FWD. ARE OPTIONAL
WALTERS RV-55 10° V-DRIVE
DINETTE
CONTROLS
SINK
HEAD
LKR
UPPER & LOWER BERTH
FUEL CAPACITY 600 GALLONS
WATER " 200 "
45' WESTPORT BOAT
FOR: MR. GEORGE FICKS
DATE: MARCH 4, 1967
SCALE: 1/2" = 1'-0"
PRELIMINARY STUDY
LOA 45'-0"
LWL 42'-0"
BEAM 15'-0"
DRAFT 4'-0"
FREEBOARD:
FWD. 7'-8"
LEAST 5'-0"
AFT 5'-3"
JAY R. BENFORD/NA
P.O. BOX 447
ST. MICHAELS, MD. 21663
(410) 745-3235

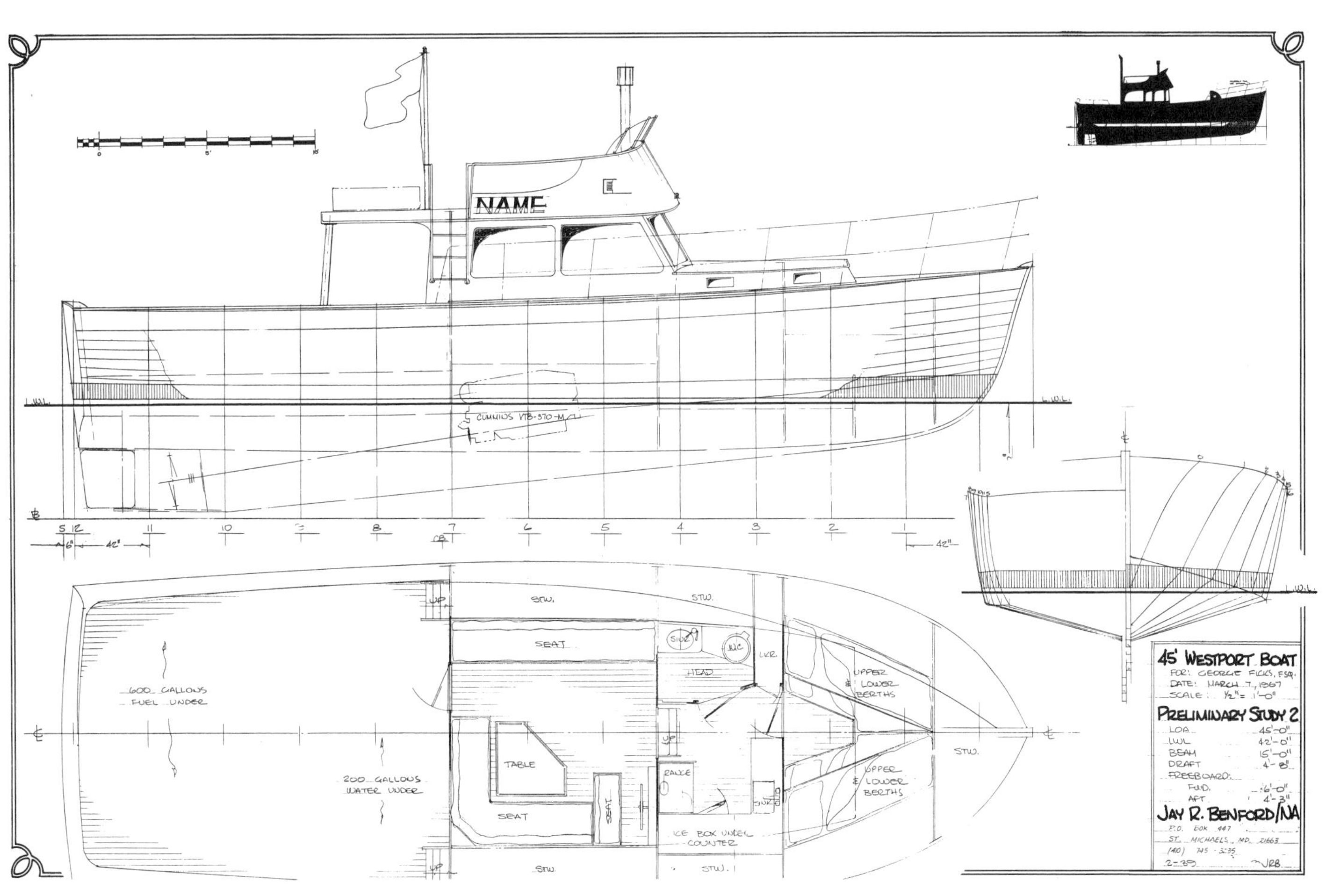
NAME
CUMMINS VT8-370-M
600 GALLONS FUEL UNDER
200 GALLONS WATER UNDER
SEAT
TABLE
HEAD
RANGE
ICE BOX UNDER COUNTER
UPPER & LOWER BERTHS
STW.
45' WESTPORT BOAT
FOR: GEORGE FICKS, ESQ.
DATE: MARCH 7, 1967
SCALE: 1/2" = 1'-0"
PRELIMINARY STUDY 2
LOA 45'-0"
LWL 42'-0"
BEAM 15'-0"
DRAFT 4'-8"
FREEBOARD:
FWD. 6'-0"
AFT 4'-3"
JAY R. BENFORD/NA
P.O. BOX 447
ST. MICHAELS, MD. 21663

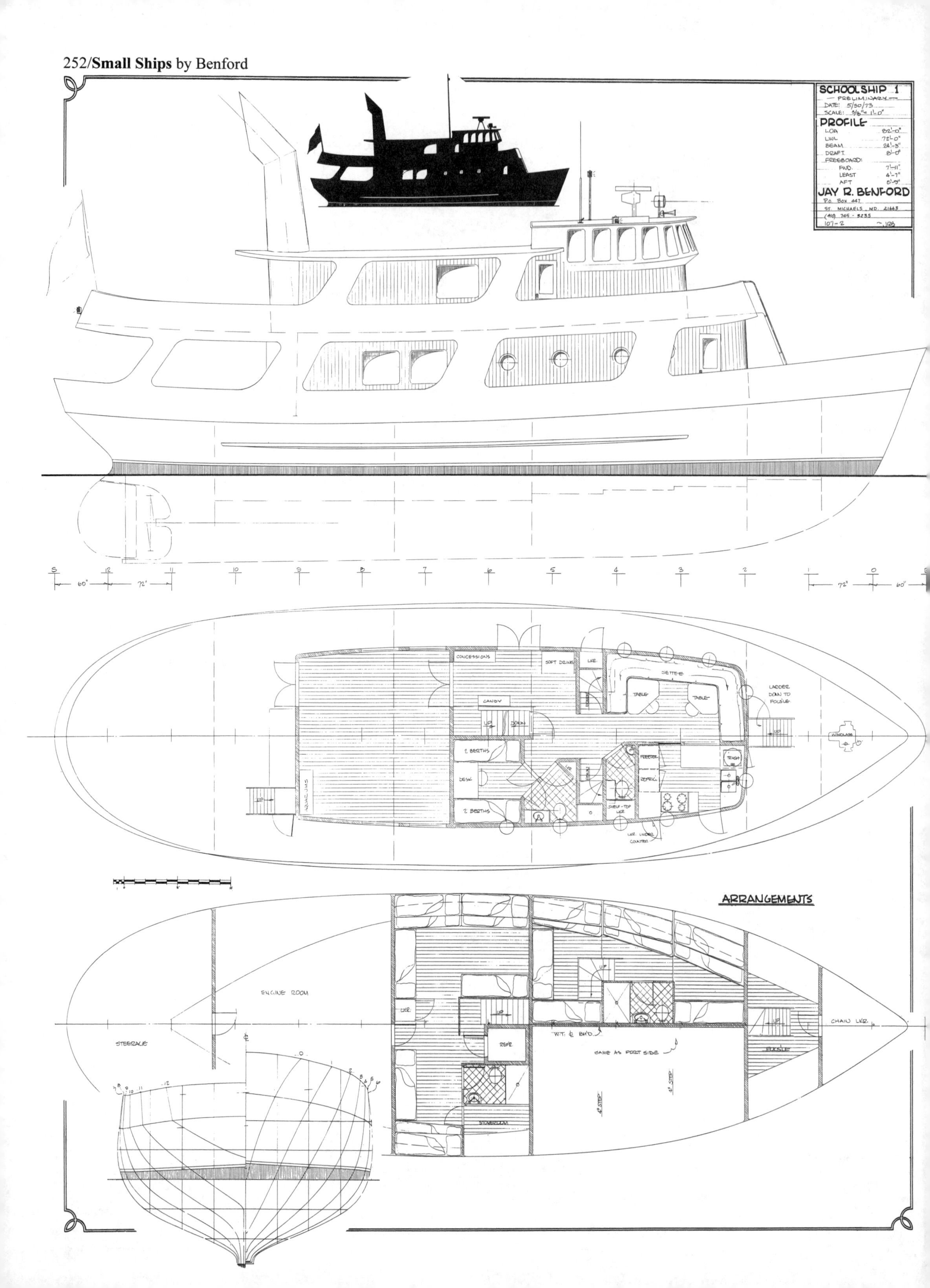
SCHOOLSHIP 1
— PRELIMINARY —
DATE: 5/30/73
SCALE: 3/8" = 1'-0"
PROFILE
LOA 82'-0"
LWL 72'-0"
BEAM 24'-3"
DRAFT 8'-0"
FREEBOARD:
FWD. 7'-11"
LEAST 4'-7"
AFT 5'-9"
JAY R. BENFORD
P.O. Box 447
ST. MICHAELS, MD. 21663
(410) 745-3235
107-2
CONCESSIONS
SOFT DRINKS
LKR.
SETTEE
TABLE
TABLE
LADDER DOWN TO FO'C'SLE
CANDY
UP
DOWN
2 BERTHS
DESK
2 BERTHS
FREEZER
REFRIG.
TRASH
SHELF-TOP LKR.
LKR. UNDER COUNTER
WINDLASS
ARRANGEMENTS
ENGINE ROOM
STEERAGE
LKR.
UP
REFR
W.T. & BH'D.
SAME AS PORT SIDE
CHAIN LKR.
FO'C'SLE
STATEROOM

Revisions & Design Modification Work

Another Design Service

As an adjunct to our custom design work, we are sometimes called on to help out with modifications to existing boats or revisions to a stock boat.

The drawings shown on this page and the following ones are examples of this sort of work. The boat below started life, as I recall, as a working gillnetter. The conversion work that we did on it involved making this conceptual drawing and providing some further measurements to the builder as to dimensions and scantlings.

The resulting cruiser worked out nicely, and the owners had a lot of fun with her. She was still going strong, last I heard.

The 26' motor whaleboat conversion to a cruising boat was a more involved job. The boat had been acquired as government surplus as an operational diesel powered launch. We tried to use as much of the original vessel as we could to make the modification as simple to build as possible, keeping in mind the owner's desires for styling and accommodations. We've since supplied these drawings to several others who've bought these hulls and they've been used in doing additional conversions.

Other work we've done in this vein are revised interiors for stock builder's hulls, like the 39' Trawler Yacht. This was done for a builder who wanted to modify a stock hull for another usage and we did the conceptual work and the scantlings for the new houses.

We've also done some modification work to reduce admeasured tonnage, repowering installations, and designing new rigs and keels for sailboats.

Like with the custom design work, it pays to get a designer whose work you like so that you're happy with the end results.

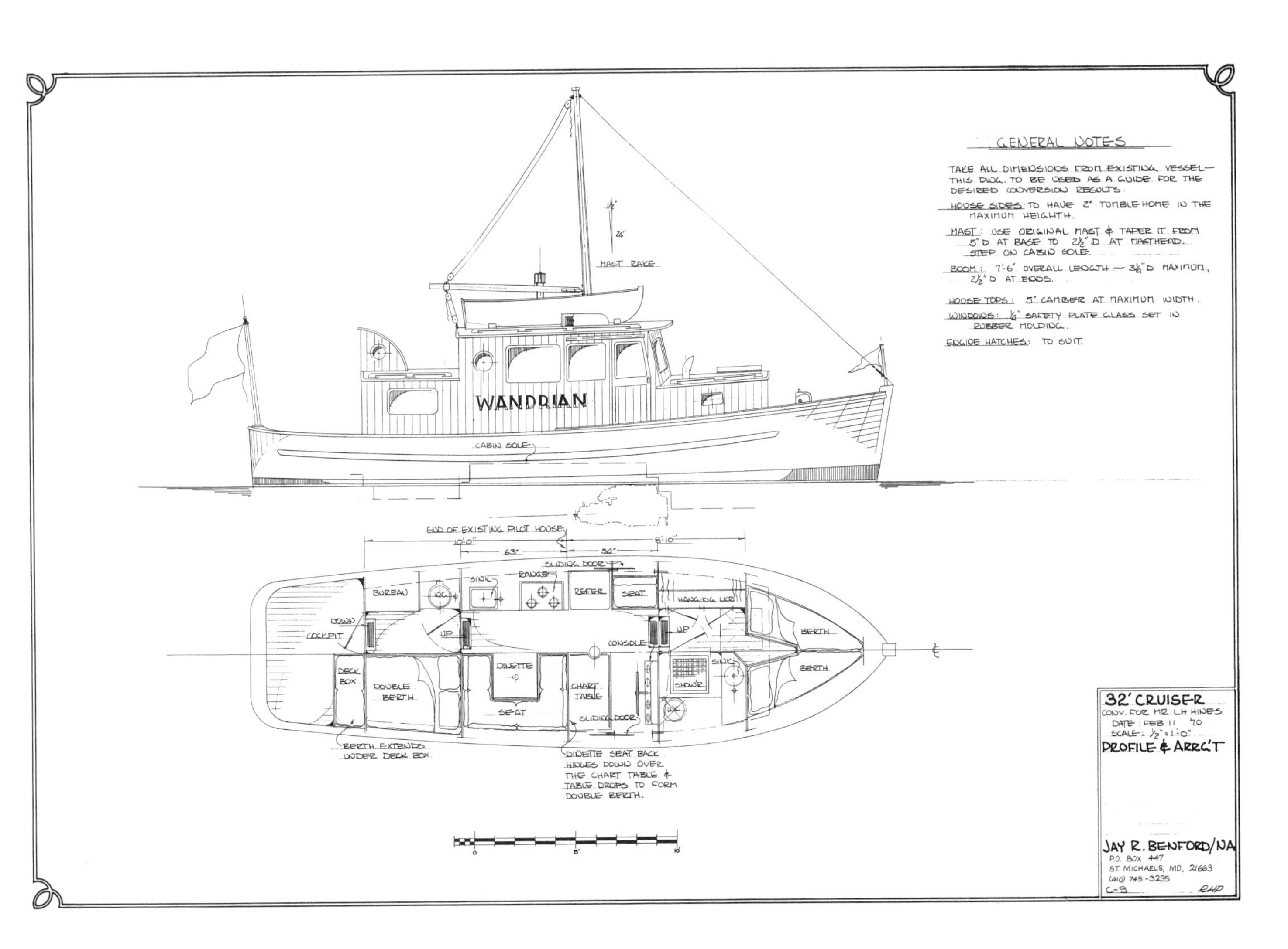

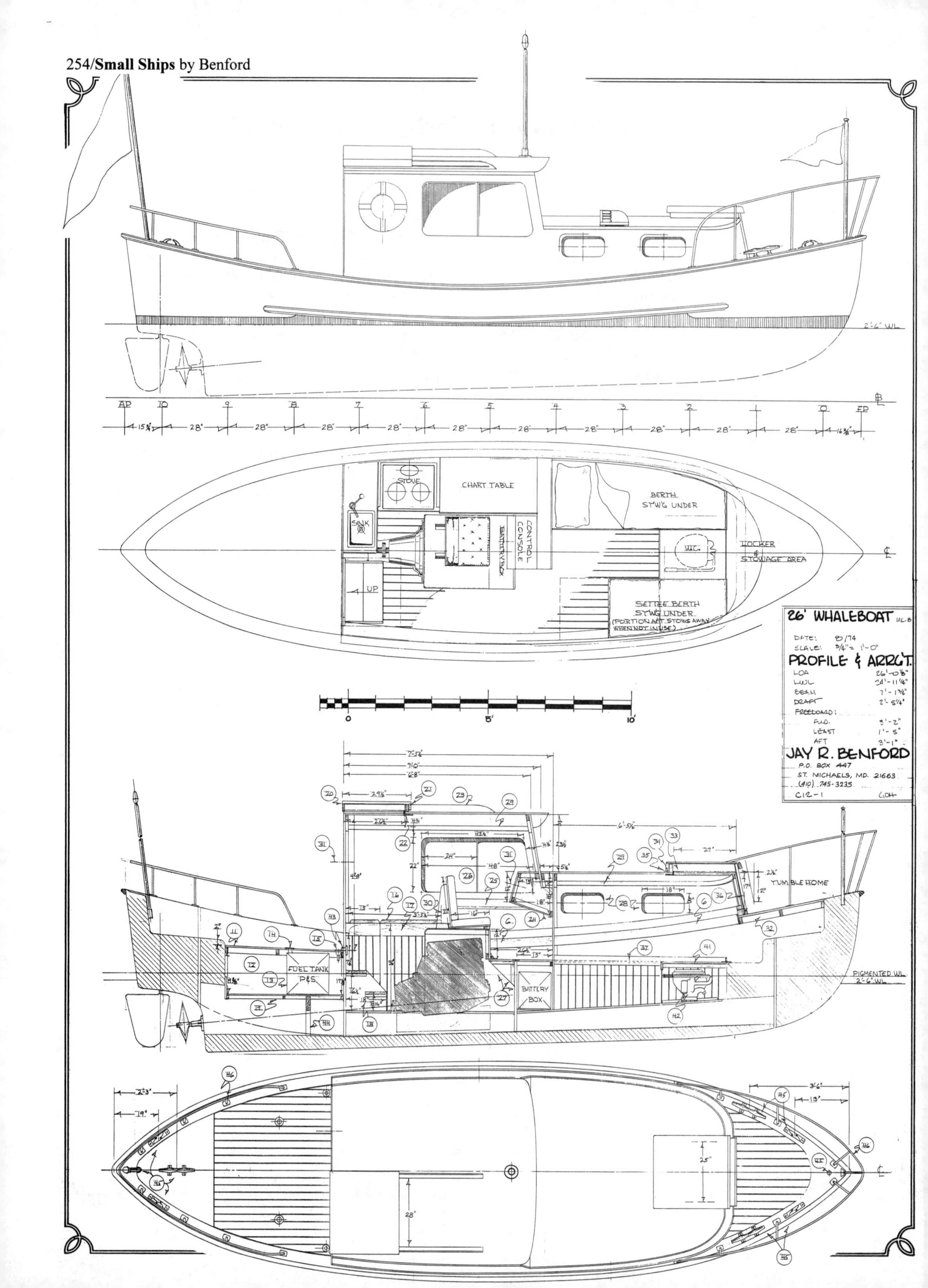
2'-6" WL
AP
10
9
8
7
6
5
4
3
2
1
0
FP
15⅜"
28"
16⅜"
STOVE
CHART TABLE
SINK
CONTROL CONSOLE
BATTERY BOX
UP
BERTH
STW'G UNDER
W.C.
LOCKER
&
STOWAGE AREA
SETTEE BERTH
STW'G UNDER
(PORTION AFT STOWS AWAY WHEN NOT IN USE)
26' WHALEBOAT MK. B
DATE: 8/74
SCALE: 3/4" = 1'-0"
PROFILE & ARRG'T.
LOA 26'-0⅛"
LWL 24'-11¼"
BEAM 7'-1¾"
DRAFT 2'-5¼"
FREEBOARD:
FWD. 3'-2"
LEAST 1'-5"
AFT 3'-1"
JAY R. BENFORD
P.O. BOX 447
ST. MICHAELS, MD. 21663
(410) 745-3235
C12-1
0
5'
10'
7'-5½"
7'-0"
6'-8"
TUMBLEHOME
FUEL TANK P&S
BATTERY BOX
PIGMENTED WL 2'-6" WL
2'-3"
19"
3'-6"
13"
25"
28"

NOTES

(1) 3/8" DIAM. BRONZE BOLT.

(2) 1" DIAM BRONZE WASHERS.

(3) MAKE UP PLATES FROM 1/8"x 1¼" BRONZE OR EVERDUR FLAT STOCK. CUT INTO 2" LENGTHS.

(4) HOUSE SIDES AND TOP ½" MARINE GRADE PLYWOOD.

(5) MOULDING MADE UP FROM ½"x 1¼" TEAK. SHAPE AS SHOWN.

(6) FACIA ½" X 3¾" TEAK. INSTAL ONLY WHERE ACCESS HOLES ARE VISIBLE.

(7) CUT 2"x 3" ACCESS HOLES TO ALLOW FOR INSTALLATION OF SILL BOLTS EVERY 10" O.C.

(8)(10) 2"x 2¾" FACIA SUPPORT BLOCKS HAVE KNIFE-GRADE THIOKOL APPLIED TO EDGES AS SHOWN PRIOR TO INSERTION AND ARE HELD IN PLACE BY A SCREW THROUGH A PIECE OF BAR STOCK 'TIL THE THIOKOL SETS UP.

(9)(10) SILL IS MADE UP FROM W. OAK OR CLEAR QUARTER SAWN FIR OR A. CEDAR AIR-DRIED AT LEAST TWO YEARS.

(INFO) ~~ACCESS TO~~ DECK PLAN, PROFILE AND SECTIONAL VIEWS MAY DIFFER. THESE DIFFERENCES ARE CONSTRUCTION OPTIONS OPEN TO THE BUILDER. IN ALL OPTIONS THE SCANTLINGS WILL REMAIN THE SAME WITH REGARD TO FRAMING.

(11) COCKPIT SOLE ½"x 3" TEAK STRAKES OVER ¼" MARINE GRADE PLYWOOD. BED STRAKES IN THIOKOL AND PAY SEAMS WITH THIOKOL. FRAMES AND BEAMS ARE 2¼"x 1¼" FIR, A. CEDAR OR SPRUCE.

(12) IF DESIRED A HATCH MAY BE LET INTO THE COCKPIT SOLE AND THE SPACE AFT THE FUEL TANKS USED FOR STOWAGE.

(13) REMOVE FUEL TANKS FROM ORIGINAL LOCATION AND POSITION AS SHOWN.

(14) FUEL TANK FILLER PIPES THROUGH COCKPIT SOLE AS SHOWN. USE PERKO OR W-C THROUGH DECK FITTINGS P.&S. NOT SHOWN ARE THE VENT TUBES. RUN THESE FROM JUST UNDER THE GUNWALES OUTB'D TO THE TOPS OF THE TANKS NEAR THE FILLER PIPES.

(15) INSTAL BALL-CHECK TYPE SCUPPERS AS SHOWN AT FWD OUTB'D CORNERS OF COCKPIT.

(16) 1½"x ½" TEAK FIDDLES INSTALLED ALONG FRONT OF COUNTER. NOTE: IF STOVE IS INSTALLED AS SHOWN IN THE ARRANGEMENT PLAN THEN THE FIDDLE WOULD BE PLACED ALONG THE BACK COUNTER ONLY.

(17)(39) FACE OF COUNTER MAY BE MADE UP FROM TEAK. FACIA AND STAKES AS SHOWN (FACIA ⅜"x 5", STRAKES ½"x 3") OR FROM ½" MARINE GRADE PLYWOOD. USE ¾" x ¾" FIR IN WAY OF FRAMING.

(18) LADDER 25½" WIDE. SIDES ARE ¾" MARINE GRADE PLYWOOD. STEPS ARE 9"x 1"x 24" W. OAK. STEP SUPPORTS ARE 1"x 1"x 9" W. OAK.

(19)(44) TANK AND STOWAGE DECK FROM ½" MARINE GRADE PLYWOOD. 1½"x ¾" Y. CEDAR IN WAY OF FRAMING. AFT HANGER SUPPORT FROM ½" MARINE GRADE PLYWOOD. SUPPORT UNDER TANKS IS MADE FROM 1½" Y. CEDAR AND IS SHAPED TO FIT THE INSIDE OF THE HULL. BED IN THIOKOL WHERE IT IS IN CONTACT WITH HULL. ALLOW 1" CLEARANCE AROUND SHAFT.

(20) HATCH COVER. SAME CAMBER AS HOUSE TOP. TOP IS ½" MARINE GRADE PLYWOOD. AFT BEAM 1"x 3" TEAK. FRONT BEAM 1"x 2½" TEAK. BRONZE SLIDERS ARE MADE UP FROM 1¼"x ⅛" FLAT STOCK EXTENDING THE LENGTH OF THE COVER.

(21) ALTERNATE HATCH COVER TOP IS FABRICATED FROM TWO PIECES OF ¼" MARINE GRADE PLYWOOD GLUED AND CLOUT NAILED AFTER ATTACHMENT TO BEAMS. THIS METHOD IS RECOMMENDED.

(22) STIFFENERS AROUND UNDER SIDE OF HATCH OPENING 3"x ⅝" W. OAK. FACIA AROUND INSIDE OF HATCH OPENING ½"x 2" TEAK.

(23) SLIDE RAILS AND CROSS PIECE AT FRONT END OF HATCH FROM 2"x 3" W. OAK. THE SLIDE RAILS ARE 5'-5" IN LENGTH, 2" AT THE BASE, AND 1¼" AT THE TOP. BED IN THIOKOL.

(24) CONTROL CONSOLE DIMENSION AS SHOWN. FRAME UP ½" MARINE GRADE PLYWOOD SIDES, FRONT AND BOTTOM WITH 1"x 1½" Y. CEDAR. THE CONTROLS (TYPE AND BRAND) ARE AT THE OPTION OF THE OWNER AND HE WILL PLACE THE CONTOL DROT. IN ACCORDANCE WITH HIS CHOICE OF CONTROLS.

(25) CHART TABLE IS ½"x 41½"x 21" PLYWOOD. 1¼"x ½" TEAK FACIA ALONG FRONT EDGE. 1"x 1" Y. CEDAR FRAMES 2"x 1" Y. CEDAR VERTICAL SUPPORTS AT FRONT. HEIGHT AND ANGLE TO BE DETERMINED BY THE OWNER.

(26) SLIDING WINDOWS PORT AND STARB'D. DIMENSION AS SHOWN.

(27) DIMENSION FRONT OF ENGINE COVER AS SHOWN. FRONT OF ENGINE COVER AND SOLE UNDER CONSOLE ¾" MARINE GRADE PLYWOOD. HEAVY DUTY HINGES AT FRONT END OF ENGINE COVER. LIGHT HINGES FOR BATTERY BOX COVER. FRAMING FOR TOP AND FRONT END OF ENGINE COVER ¾"x 1½" W. OAK. (2) TRIANGULAR SUPPORTS AT ENDS ¾" PLYWOOD BRACED WITH 1¼"x 1¼" W. OAK. FRAMING FOR SOLE OVER BATTERY BOX 1½"x ¾" Y. CEDAR.

(28) W-C OR PERKO PORT LIGHTS.

(29) FRAME 1¾"x 1" Y. CEDAR OR SPRUCE.

(30) DIMENSION SEAT AS SHOWN. BACK AND BOTTOM ½" PLYWOOD. MAKE UP BRACES AS SHOWN FROM ⅝ PLYWOOD. BRACE SUPPORT 1½"x 1" Y. CEDAR. ANGLE OF BOTTOM AND BACK OF SEAT TO SUIT OWNER.

(31) CENTERLINES OF DECK AND HOUSE WHEELS. SIZE OF WHEELS TO SUIT OWNER.

(32) FOREDECK CONSTRUCTION IS THE SAME AS FOR THE COCKPIT SOLE.

(33) FOREDECK HATCH SIDES 3¼"x 1⅜" W. OAK TAPER OUTB'D SIDE SUCH THAT THE SIDES ARE 1⅜" AT THE TOP. STIFFENERS AND FACIA SIZES ARE THE SAME AS FOR THE AFT HATCH.

(34) HATCH TOP IS MADE UP FROM ½" MARINE GRADE PLYWOOD AND 1⅜"x 2½" TEAK CONSTUCTED AS SHOWN.

(35) (2) HEAVY DUTY BRONZE HINGES.

(36) DOOR IS MADE FROM ¾" MARINE GRADE PLYW'D. THE DOOR IS TO OPEN TO PORT AND IS SUPPORTED BY TWO HEAVY DUTY BRONZE HINGES ON THAT SIDE. THE LATCH IS TO BE OF THE TYPE WITH HANDLES ON BOTH SIDES. A NEOPRENE GASKET IS TO BE PLACED AT ALL POINTS OF CONTACT. THE SILL IS MADE FROM 1½"x ¾" TEAK.

(37)(38) BERTHS - NOTE OPTIONS - FLATS ½" PLYWOOD. FRAMING ABOVE CABIN SOLE 1½"x 1" FIR, SPRUCE OR Y. CEDAR. AT SOLE 1"x 1" FIR, SPRUCE OR Y. CEDAR. FACIA ½"x 3½" TEAK. FRONT ½" PLYW'D OR TEAK STRAKES AS SHOWN.

(39) SEE NOTE #17.

(40) ALL CUT FIBERGLASS SUPPORTS (SUCH AS SEATS) ARE TO BE STRENGTHENED IN THE MANNER SHOWN.

(41) HEAD COVER ¾" PLYWOOD HINGED AT THE REAR. SUPPORT IN FRONT 1"x 2½" TEAK. EDGE SUPPORTS ATTACHED TO BERTHS 1"x 1½" FIR, SPRUCE OR Y. CEDAR.

(42) HEAD - "JUNIOR 51" SEACLO W-C FIG-1591.

(43) SILL 2"x 1½" W. OAK. FACIA 4"x ¾" TEAK.

(44) SEE NOTE #19.

(45) RELOCATE ORIGINAL EQUIPMENT CLEATS, CHOCKS AND MASTS AS SHOWN.

(46) MAKE UP RAIL AS SHOWN FROM ¾" STAINLESS STEEL TUBING.

(47) FACIA AROUND ENTRANCE 1¼"x ½" TEAK. CUT TO FIT. DO NOT ATTEMPT TO BEND TO SHAPE.

(48) BULKHEAD - ½" MAHOGANY PLYWOOD.

(INFO) ALL THROUGH-HULLS TO BE W-C OR PERKO BRONZE FITTINGS BEDDED IN THIOKOL AT THE TIME OF INSTALLATION. BE SURE THAT THERE IS EASY ACCESS TO ALL THROUGH-HULLS.

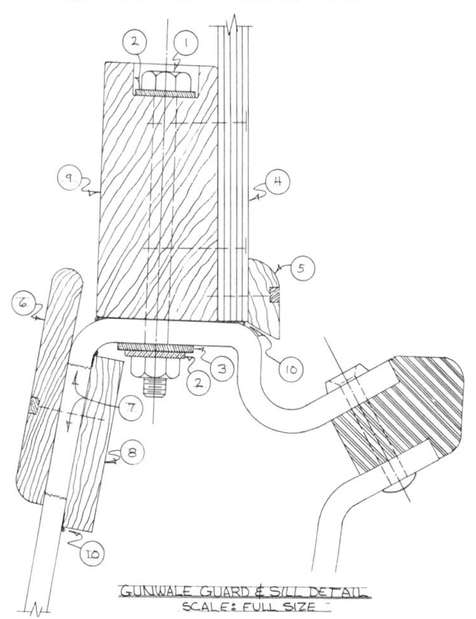

GUNWALE GUARD & SILL DETAIL
SCALE: FULL SIZE

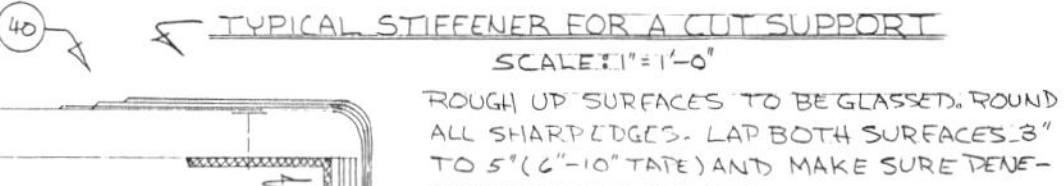

(40) TYPICAL STIFFENER FOR A CUT SUPPORT
SCALE: 1" = 1'-0"

ROUGH UP SURFACES TO BE GLASSED. ROUND ALL SHARP EDGES. LAP BOTH SURFACES 3" TO 5" (6"-10" TAPE) AND MAKE SURE PENETRATION IS COMPLETE.

HOLD IN PLACE WITH SELF-THREADING SCREWS

3"x 1/16" ALUMINUM SHEET METAL BENT TO FIT ANGLE BETWEEN THE PLYWOOD AND THE CUT F.R.P. SUPPORT

½" PLYWOOD

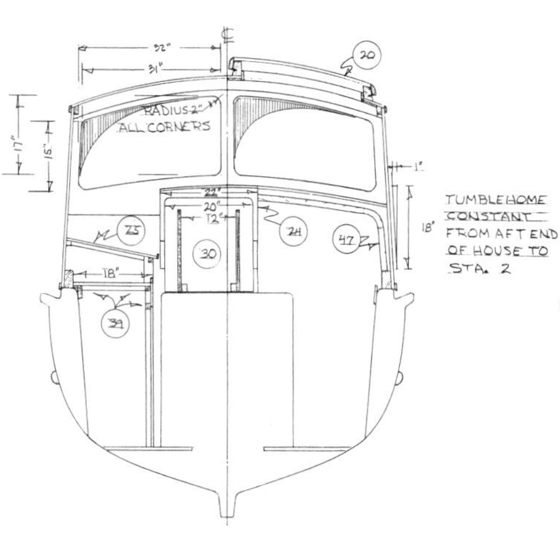

TUMBLEHOME CONSTANT FROM AFT END OF HOUSE TO STA. 2

SECTION AT STA. 7
LOOKING FWD.
SCALE: ¾" = 1'-0"

48
40
38

SECTION AT STA. 4
LOOKING AFT
SCALE: ¾" = 1'-0"

AFT-CABIN ROOF CAMBER
SCALE: ¾" = 1'-0"

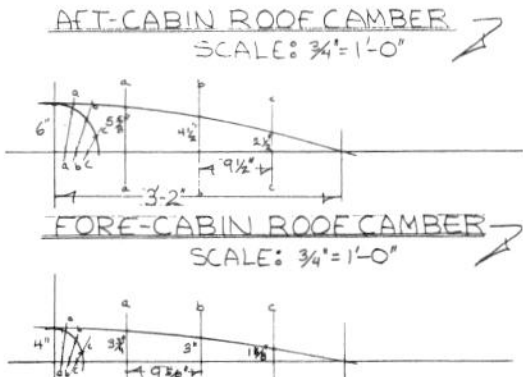

FORE-CABIN ROOF CAMBER
SCALE: ¾" = 1'-0"

0 5'

26' WHALEBOAT

DATE: 8/74
SCALE: AS NOTED

NOTES & DETAILS

LOA	26'-0⅛"
LWL	24'-11⅛"
BEAM	7'-1¾"
DRAFT	2'-5¼"
FREEBOARD:	
FWD.	3'-2"
LEAST	1'-5"
AFT	3'-1"

JAY R. BENFORD
P.O. BOX 447
ST. MICHAELS, MD. 21663
(410) 745-3235
C12-5 604

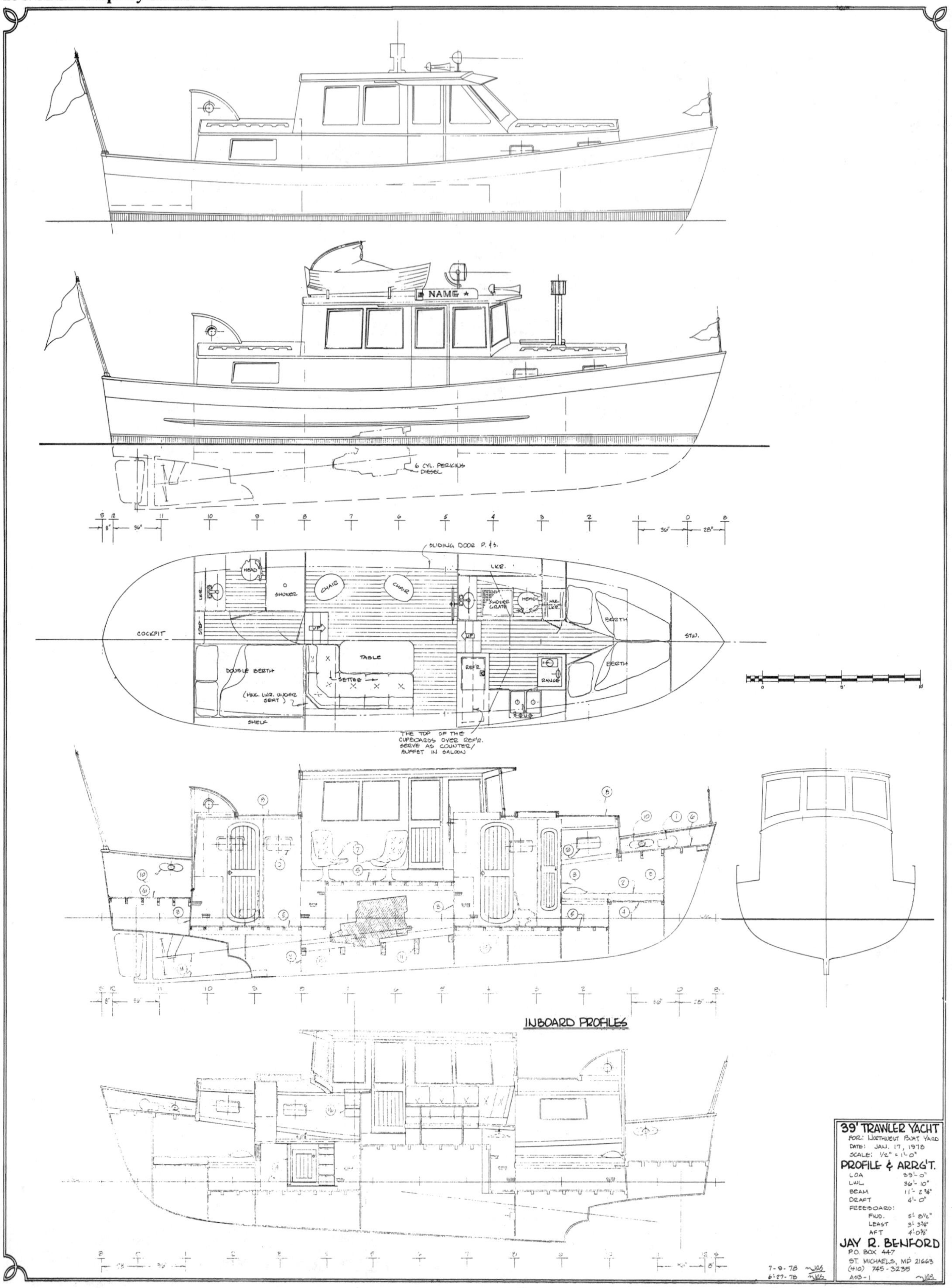

6 CYL. PERKINS DIESEL

SLIDING DOOR P. & S.

HEAD
SHOWER
CHAIR
CHAIR
LKR.
SHOWER GRATE
MEMO
HAN. LKR.
BERTH
COCKPIT
STEP
STW.
TABLE
DOUBLE BERTH
SETTEE
REF'R.
RANGE
BERTH
(HAN. LKR. UNDER SEAT)
SHELF

THE TOP OF THE CUPBOARDS OVER REF'R. SERVE AS COUNTER/ BUFFET IN SALOON

INBOARD PROFILES

39' TRAWLER YACHT
FOR: NORTHWEST BOAT YARD
DATE: JAN. 17, 1978
SCALE: 1/2" = 1'-0"

PROFILE & ARRG'T.

LOA	39'-0"
LWL	36'-10"
BEAM	11'-2 3/4"
DRAFT	4'-0"
FREEBOARD:	
FWD.	5'-8 1/2"
LEAST	3'-9 3/4"
AFT	4'-0 3/8"

JAY R. BENFORD
P.O. BOX 447
ST. MICHAELS, MD 21663
(410) 745-3235
A3B-1

7-8-78
6-27-78

Boatbuilding Materials

In this part of the book, we're taking a brief look at the various boatbuilding methods and materials. We're trying to cover each, giving the pros and cons and some opinions based on our experience. At the end there is a suggested reading list for those who want to delve further into the subjects and would like to have a starting point for their reading. Since the latest boatbuilding technology is always a moving target, some of this material will eventually be out of date. However, we like to think of our work as being mainly with "low-tech" materials and technology. We're certainly acquainted with the hi-tech world and use it for some of our work. But, use of hi-tech would not be appropriate for most of the boats in this book, not only for the expense but for the practicality of living with it and the costs of repair.

Aluminum Construction

Over the last several decades aluminum has gained in usage and acceptability. A significant portion of the custom building of larger yachts is done in aluminum, and some builders produce stock aluminum boats. This is due to the high strength to weight ratio of aluminum being attractive where speed is a prime consideration.

Great care in building is required to be sure that there are no abrupt areas in the structure. Much of this is taken into account in designing the framing and structure, but the plans must be carefully executed by the builder.

The builder must have specially trained welders to do a proper job of putting the materials together. Most of the work of cutting and drilling can be done with wood working tools.

The owner or operator must also exercise great care and caution to be sure that the boat's galvanic corrosion protection is maintained properly and that the electrical system remains isolated from the structure.

The 5000 series of aluminum alloys are normally used in marine work. They have good properties for this work and can often be left unpainted with the surface forming an oxide that protects itself from further deterioration. Of course there is no getting away from the need for antifouling paint. In this, there are specially formulated paints for use on aluminum to protect the aluminum from deterioration and slow the growth of barnacles.

If the boat is left unpainted above the waterline there will certainly be savings in maintenance costs on the coatings. However, most people paint yachts for esthetic reasons and then the coatings must be looked after. We've seen white painted topsides pick up enough yellowing at the bow in about five years that a cosmetic recoating is done. The basic paint was still in good order so the linear polyurethanes do work well.

Aluminum is a good thermal conductor. Thus all living spaces need to be well insulated to keep them dry and at comfortable temperatures. Since aluminum is quite a rigid material, care is needed to be sure that the machinery noises are well insulated and isolated from the living spaces. This means doing a good job on the isolation mounts for the machinery and sealing off any openings that can permit the transmission of airborne noise.

The majority of our past work with aluminum has been with larger sailboats that were operated with small crews. The lighter weight of the aluminum structure let them operate with smaller crews since the rigs were smaller than if the boats were built of steel. The trade-off is that the aluminum boats require more sophisticated repair facilities when and if they need work done. In powerboat design work, we would only recommend using aluminum on special use projects. For most of our cruising boats, it is hard to justify the extra expense and level of care required, both in building and in living with the boat.

We expect to see more use of aluminum in long-distance cruisers where they want to trade-off weight savings for additional fuel capacity. This is one area where it does make sense.

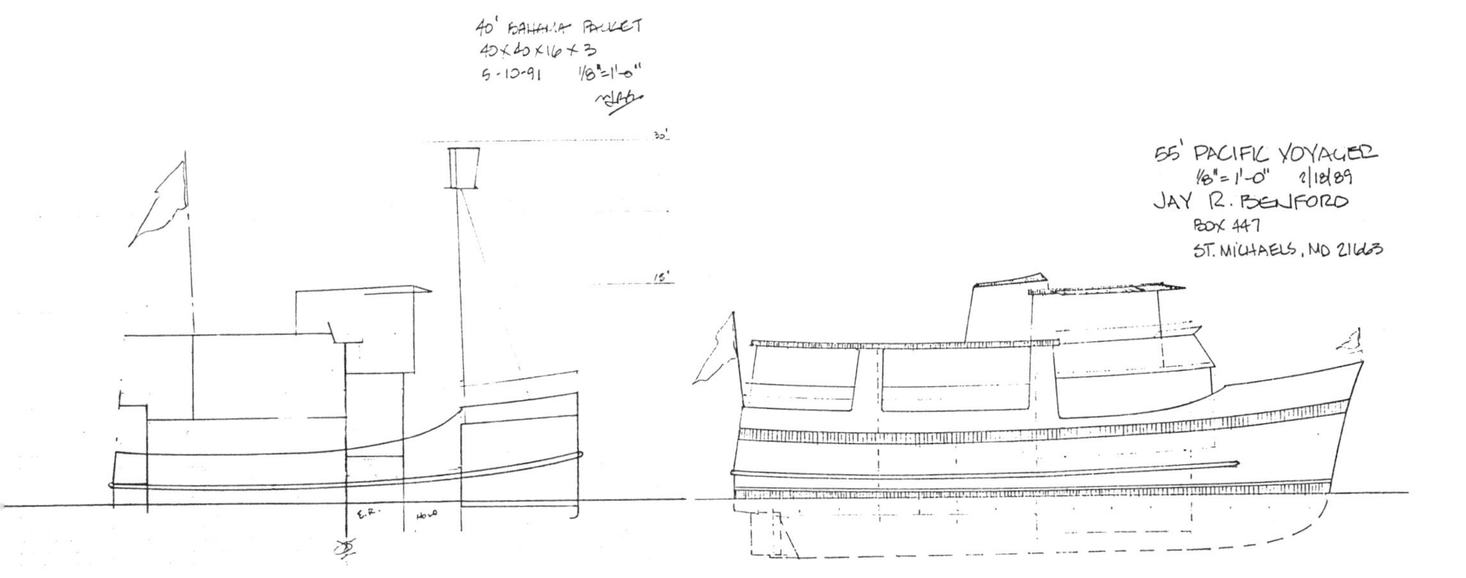

What About Ferro-Cement?

What Is Ferro-Cement?

For those who aren't familiar with Ferro Cement, it's a method of building in which the shell of the boat is built of a steel wire mesh and rod matrix which is permeated with a concrete mixture. The mortar is cured carefully to a minimum of 8,000 psi, to meet our specs. The resulting structure is a strong and rugged shell that will not burn, is not affected by ice, resistant to chemical attack and has a mortar that actually strengthens with age.

How Did It Get Started?

The first ferro-cement boats were built in the middle 1800's and are still in a museum in Europe. World War I saw the building of large concrete ships, some of which I've seen still afloat as breakwaters. During World War II, some experimental vessels were built which proved the material on some good sized working vessels and later on some yachts.

The building craze that began in the late 1960's was fueled mainly by promoters who spent more effort in selling franchises for their books and plans than in improving the technology. The few builders who built good quality boats often got overlooked with the vast numbers of amateur built boats sprouting up everywhere.

I had learned how to do it the right way while working for a licensee of the English Windboats Ltd. firm, who were the only ones building Lloyd's approved structures.

Following this experience, I did some additional work in the testing labs to develop a range of shell layups that would span the range of 12' to 90' designs we worked on. This knowledge gave us the technology to design much lighter and stronger boats than were commonly being done. It was an uphill battle to get people to recognize that chicken wire was not the best choice, though in the end almost everyone ended up using the same square welded mesh that we did.

Practical Ferro-Cement Boatbuilding

After a bit of this development and designing work, I teamed up and co-authored a book on the subject, titled **Practical Ferro-Cement Boatbuilding**. It was quite well received and widely distributed, going through 11,000 copies in four separate editions. This book covered the ideas, tools and techniques that we had developed to build lighter and better ferro-cement boats.

This book has been out-of-print for a long time now, though I still am able to occasionally pick up a copy in a used book store. Thus, we have a very few copies that we keep for those who are determined, usually against our advice, to build in ferro-cement.

What Ever Happened To Ferro-Cement?

Too many people who read Samson's ads saying they could build the hull and deck of a 45-footer for $2,000 didn't realize that was only the beginning of the money they would have to spend to make a complete boat. It usually took ten times that much to fully outfit the boat in those days.

Also, their publications stressed how easy and quick they were to build. Too many of the resulting boats looked like the builders took that advice literally. They were — and still are, for how do you dispose of cement? — an eyesore, enjoying very low, if any, resale value. Their being readily identifiable as ferro-cement has given a bad name to all ferro-cement boats. The good ones were always mistaken for wooden or custom fiberglass boats, and thus no credit was given to the medium of ferro-cement.

So, although it is still a viable way to build a tough and long-lasting boat, I have not been able to give it an unqualified recommendation for some time. It is very frustrating for me and much more so for the owners of the good boats to find that they can't get the same return on their invested time and money than if they had built in wood or fiberglass.

Do You Still Sell Plans For Ferro-Cement?

Yes, but only when we've given the caution above to the prospective buyers. Most of our few remaining ferro-cement sales are now going overseas where there is still a bit more viable market for the finished boats.

What About Buying A Used Ferro-Cement Boat?

With the negative attitudes prevailing about ferro-cement boats, their prices are usually quite low. The only problem in buying one is the difficulty in doing a proper survey. This is a two-part problem with the lack of many experienced surveyors and the difficulty of knowing what few clues to look for in the survey. If the seller has photos documenting the whole of the construction of the armature and the plastering, this is a big help. It they did mortar sampling and testing to assure the correct mortar strength this helps to prove the quality of the initial construction.

From there, it is a matter of looking at the level of finish and fairness of the structure, the quality of the detailing of how things are attached to the structure, and if there are any visible clues to maintenance work that has been overlooked. If you can find a good one that can be used as is and without any major investment in finishing it or adding equipment, then you might have some hope of reselling it later without loosing all *your* investment.

Fiberglass

Over the last several decades, fiberglass has become a sort of de facto standard for boatbuilding materials. Those of us old enough to remember its introduction recall when it was a novelty and no one knew how long it would last.

The most efficient use of fiberglass is building sister ships from a mold. There are a high number of man-hours required to get the plug, or form for the mold, truly faired and ready to lay-up the mold. Thus the best use of fiberglass is not in building one-off boats, for in building a one-off boat, it is hard to justify the expenditure of this amount of time, particularly if the boat is to be competitively priced.

Certainly the home builder can think of discounting the value of the time spent in fairing work, but the work is not pleasant and often tends to get short-changed to the detriment of the future resale value.

A one-off building technique that has been used with 'glass is the building of a large smooth table, using plastic sheet laminates (like Formica®) for the surface. On this are laid up large, flat, 'glass panels. These are then used to "plank" a hull, deck or house that has developed surfaces. The edges are trimmed and taped together and the outer gelcoated surface is repaired to make it look all alike. This method can give significant savings in labor when compared to having to manually fair all the exterior surfaces.

There are a number of core materials used with fiberglass. These serve to give insulation. They also add extra stiffness and rigidity by separating the inner and outer skins like the web of an I-beam separates the flanges for strength. Since the core is by definition sandwiched between two layers of fiberglass, it must be something that will be durable and not be subject to deterioration. Wooden cores, like end-grain balsa or plywood, can sometimes experience deterioration through water getting into them through breaks in the skin or careless installation of gear. If water does get into a wooden core, it can at best waterlog it and at worst have it all rot out and be very difficult and expensive to replace.

My preferred core materials are the closed-cell PVC foams, like Airex®, Divinycell®, Klegecell® or Termanto®, which are not subject to rotting if any water happens to get into the core. They also are good thermal insulation and can make for a surprising difference in eliminating condensation in lockers and living spaces. The techniques for installing the foam, whether in a mold or on a one-off, are well established and the material suppliers can give good guidance on how-to questions.

I have only specified closed-cell foams for core materials. These should not be confused with the cheaper foams that crumble and crush under ordinary use. Their closed-cell formulation means that water cannot migrate from one part of the foam to another and they are not subject to rotting.

The chemicals used in the resins and cleanup work are often unpleasant and need special care in handling. Some have explosive and/or toxic fumes, but there have been a lot of improvements of late and the are many more alternatives now.

The origin of the use of fiberglass in building boats is not officially known, and I would discount the stories about it being an attempt to dispose of overstocks of the foil-backed, household insulation materials....

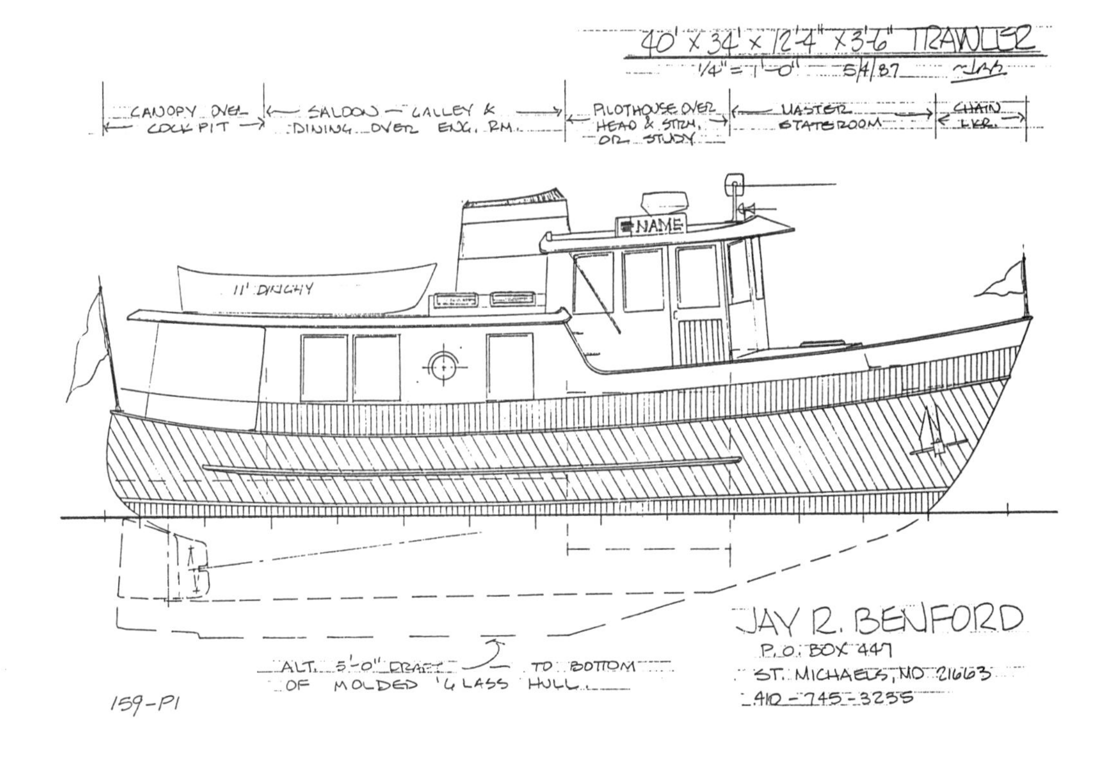

Boatbuilding In Steel

July 30, 1990
October 3, 1990

Why Steel?

In many ways, coastal cruising is harder on the boats than offshore voyaging. This is because, as anyone who's really been cruising knows, along the coast is where boats go aground and generally bump into things. To survive this gracefully, the boat must be built of a rugged material.

Steel is an excellent material for a coastal cruiser. It is rugged. And it is easily repaired almost anywhere. Too many of the modern cruising boats are built of high tech materials, and are not repaired easily in areas with only low tech facilities. Low tech repair services are what is usually found in the out of the way places that are interesting to cruise into and where the boat might have an adventure leading to needing a repair.

Steel is typically the heaviest material used to build cruising boats. This can be used to advantage by the designer in some types of boats. The added mass of the steel can be used to design a boat that gives the comfortable motion of a much larger boat. For anyone spending periods of time coastwise cruising, most of the time ends up spent not underway, but at anchor or tied up somewhere. At these times, the higher the stability the better. The way to do this with steel boats is to keep the waterplane wide with beamy and relatively shallow boats.

The additional mass of steel over other materials means the weight of one crew member is a smaller proportion of the total displacement. Thus, moving around on the boat has less effect on her heeling and trim and thus makes for a more comfortable boat.

Although steel is probably the heaviest material, it can also be one of the most affordable, if the design is done carefully. This is because steel can be used in large pieces and assembled quickly, if the design was done for developed surface construction.

Steel boatbuilding can be done by amateurs *if* they take the time to learn welding sequencing and take care about not distorting the plating. Having a design that is thought out for ease of assembly helps make for successful construction, whether by an amateur or professional.

If the designer uses more flatbar and cut plate framing instead of flanged shapes for framing, it will be easier for the builder to do a good job of sandblasting and painting. Then, using good quality finishes, with lots of zinc in the primer coats, should give good protection. Some of our boats have been built upside-down, with the sandblasting done before rolling over, so the sand will fall out easily. This is a time-saver for the builder and is reflected in the final costs, whether these are counted in hours or dollars.

We used to join aluminum superstructures to steel hulls by stainless steel bolts in insulated sleeves, having put neoprene gasketing in between the steel and aluminum. Now, we use the explosively bonded strips of steel and aluminum, which permits welding both metals to their respective sides of the strip.

However, our usual practice is to design the whole vessel in one metal. Usually the deckhouse has lots of holes in it for the windows and hatches, and thus is not a major weight item. If the allowance for it is made in the design phase, there is no reason to go to the added expense and complication of putting an aluminum deckhouse on a steel hull.

We've been designing our structures lately to make building and outfitting as easy as we can. Sometimes this makes for a little more steel weight, but the end result is a boat that goes together quicker and saves the builder and owner money on the first cost. Some of the things we do in this vein are making the engine beds and engineroom floors all one height, so the plating or sole just lays on them, arranging for as many duplicate pieces as we can in the structure, and laying out design of the hull form and deckhouse to make maximum use of standard size plates and framing member lengths.

Insulation

My first choice for insulation is for sprayed in place foam. This is put on over the zinc-rich primer, and eliminates any pockets in which condensation might occur. The condensation that can occur behind blocks of foam stuck on the plating is the primary candidate for rusting out the steel.

If it's not possible to have sprayed on foam, then an air space all around the plating, ventilated the way a traditional wooden boat should be done, is the second choice. If this is done, I would also put a fan in the ventilation system, to slightly pressurize the bilge spaces, forcing air flow through the whole structure. I've been aboard a Dutch built 45' ketch that was over 20 years old that was built with this sort of ventilation, and she was in excellent condition. True, she needed a lot more energy expended to heat her in cool weather, but you could look at the shell plating all over, readily ascertaining it's condition.

How Long Will It Last?

Life expectancy on a steel boat can be anywhere from less than a decade to several generations. The variable in this is how well the boat is designed, built, and most importantly, how it is looked after on a regular basis. Any indications of oxidation, like visible rust bleeds and pinhole rust spotting, calls for immediate action to stop the flow of oxygen to the steel and to re-protect the surface.

Why Not Wood?

Wood is the usually my first choice in boatbuilding methods. Both of the boats that I've lived aboard, to date, have been of traditional wooden construction. They were well built and have been well maintained over the decades. As a result, they both are looking quite well, in spite of being built in 1926 and 1973. **WoodenBoat** magazine did a survey a while ago and found that boatyard operators' records indicated that cold-molded wooden boats were the best investment when all costs of ownership were considered. The warmth and charm of living with a wood boat is hard to beat, even when putting a price on it.

Carvel Planked

Carvel planking is smooth planking applied over frames that are usually steam bent in place. Sometimes there are sawn frames in some working boats or heavier cruising boats. Both of my floating homes were built of plank-on-frame construction. The major advantage of this method is the ease of replacing any particular element of the structure, if need be, at a later date.

Carvel construction, to make it weather and watertight for comfort in living aboard, requires building of top quality materials. If the materials are not well chosen and well seasoned, there could be considerable movement in the structure as the wood shrinks and swells with moisture and humidity changes. This can lead to leaks, the most annoying of which happen through the deck or cabins. If the right materials are chosen, the structure will be stable, if it is well designed and constructed. If not, there will be the frustration and possible peril of the boat leaking.

A carvel planked boat often has an inner layer of planking inside the frames, called the ceiling. (The overhead is the underside of the housetop or deck.) This ceiling provides an air passage from the bilges up to the deck. The free flow of air through here keeps the boat from growing mold and rotting out. I often advise people who are going to leave their boats for a while to put on a solar powered ventilator to pressurize the bilges to force the air flow.

Professional boatbuilders often remark that carvel construction is the quickest for them to do, which can make it the most economical to purchase.

Lapstrake or Clinker

The planking method in which the bottom edge of one plank overlaps the top edge of the next one is called either lapstrake or clinker construction. This is most often seen in small craft or dinghies, where the lightness and stiffness gained in the overlapping makes for a sturdy but light boat. It has been used for some quite large boats too, like the Viking Longships.

The skills required for this method are similar to that for carvel except the close fits have to be made on the overlaps instead of the seams that butt to each other. A popular variation on this construction now is for epoxy gluing plywood planks, which can be done in conjunction with very little inside framing. Our own 11' Oregon Peapod dinghy was built by this technique and is light and sturdy.

Strip Planking

Strip planking usually is done with planks that are almost square in cross section. They are nailed and glued together to form a rounded shape over bulkheads and mold frames. If the spans are long enough, some additional bent or sawn frames may be fitted.

Strip planking will typically support itself over longer spans than carvel planking. However, it still needs something to give it strength across the grain such as the bent frames in a carvel planked structure. One approach that has been used is to rely on heavy cloth and resin sheathing, with the strips effectively becoming a core or spacer to hold the cloth skins apart. The downside to this is that some of the poorly applied coverings sheared off the cores from lack of bonding and the differing expansion or elasticity rates of the materials. Some of this can be forestalled by putting mechanical fasteners (staples, nails, or screws) through the skins into the core.

A better alternative is to cold-mold some veneers over the strips, applying them at say opposing sixty degree angles to the strips. These would have sufficient material to take the place of the ribs and help hold the strips together. This technique would usually be my recommendation for someone building their first round bilged boat. It allows for easy handling of smaller pieces and can be done in stages.

Cold-molding

Cold-molding refers to the cold bending, as opposed to hot or steam bending, and laminating of smaller pieces and layers of wood to form the shape of the structure. As currently practiced, this is usually done with epoxy glues and sealants to stabilize the moisture content of the wood and prevent shrinking and swelling. Combined with linear polyurethane paints this produces a structure that often needs less maintenance than fiberglass, since the gelcoats need buffing and waxing and eventually need painting.

The New Zealand method of cold-molding is to set up bulkheads and temporary mold frames. Over these are bent on substantial longitudinal framing, which is let into the bulkheads and permanently fastened to the bulkheads. Then, a minimum of three layers of planking are applied on opposing forty-five degree diagonals. These are glued and fastened to the longitudinals and the bulkheads.

The combination of strip-planking with cold-molded layers over it is more often used when weight savings are not a critical issue. This results in a smooth interior which is often easier to maintain and takes less work in making sure that the pieces added during outfitting do fit well. It also means that there are no collections of bilge water on the frames since it can freely flow to the lowest point on the hull. Adding limbers to the non-watertight bulkheads and floors will facilitate using a single pickup point.

Plywood

The popularity of good epoxy glues and sealers has given plywood boatbuilding a new life. With these, the plywood can be protected from water getting into the cores and setting up rot or deterioration there. The basic principle to remember is that whenever a hole is drilled in the ply be sure that it is done a little bit over size and all the exposed edges of the ply are sealed with epoxy before installing the new piece of hardware.

The quickest wooden boatbuilding is usually using plywood in sheet form over a developed surface form. (A developed surface is one on which the plywood will lay without any distortion, such as on a cone or cylinder.) We're using our Fast Yacht computer design software to speed up the work of developing the surfaces and "unwrapping" them to check the shape and size of the panels. From this, we can be sure that we've made them in a manner that will be an economical use of the materials.

Like most strip and cold-molded boats, plywood is usually sheathed with a layer of cloth to take the local impact loads and abrasion. This sheathing is not structural on most boats, but rather a protective skin to keep harm from coming to the wood underneath.

A Word Of Advice To Owner-Builders

What sort of boat is most appropriate for an owner-builder? Should he build a close copy (clone) of a stock or production boat?

No. Not if he wants to think about getting any sort of reasonable return on his time and money invested. He should instead think about building a boat this is not available as a stock boat. Something like this will set his boat apart and give it a unique position in the resale market.

If he really wants to own a Clone 40 then he should buy a Clone 40. Perhaps finding one that has been let go and doing the restoration and maintenance that it requires to bring it back to like-new condition. Or he could see if the Clone 40 builder sells kits and get the molded parts and pieces that the builder can do most efficiently. There is no sense in building a mold and a one-off with its very high time demands if one can get a molded hull that only takes a small number of hours and has an excellent finish built-in. It's somewhat like the question of whether one should reinvent the wheel.

But, if one wants a unique boat of a more classic or traditional type, the choices in the marketplace are much more limited. This is an area in which it does make sense to spend the time to do a one-off. There is a better chance of being able to sell it later on to someone who will pay a reasonable price for a boat that is different.

"But," you say, "I'm building this boat for myself. I don't care if I ever sell it."

That's all well and good if you don't care about what you do with your money. But, like buying insurance where you're betting on whether you get sick or die, the money put into a boat should have some insurance of preserving its value. The way to do that is to do the construction well, with finesse and skill, ***and*** build something that your heirs can sell and add some value to your estate.

So, we come back to my original point; build something that is not commercially available and something that will have a separate and distinct identity in the marketplace. Many of the plans in this book meet this criteria....

Further Reading

The Gougeon Brothers On Boatbuilding This is a first-rate building manual, covering using the Gougeon Brothers' WEST System™ epoxies for boatbuilding.

Boatbuilding by Chapelle. This books covers traditional wooden boatbuilding methods.

Boatbuilding Manual by Steward. More on wood building.

Ferrocement by Bingham. The best of what's still in print on ferro. Excellent outfitting chapters of use with any material.

Steelaway by Smith and Moir. A good introduction to and overview of steel boatbuilding.

Boatbuilding With Steel by Klingel. Includes section on aluminum by Colvin.

Steel Boatbuilding Colvin's 2 volume set covering steel construction and outfitting.

Just as the foregoing pages contain my opinions, prejudices and experiences, you will find the same is true in the pages of these other books. In the end, you will have to weigh all the evidence and make a decision on what materials to use based on your skills and what materials are available to you to use.

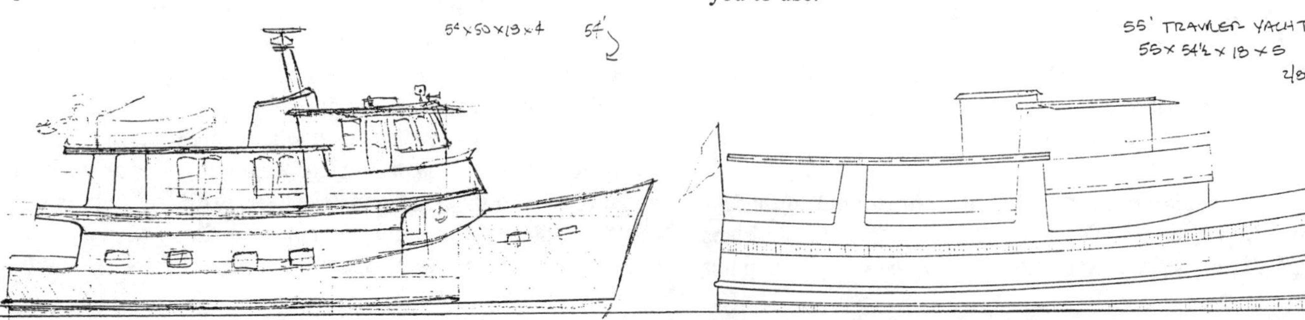

SEA INTERVIEWS JAY BENFORD

Tomorrow's Powerboat: Fuel economy with seaworthiness

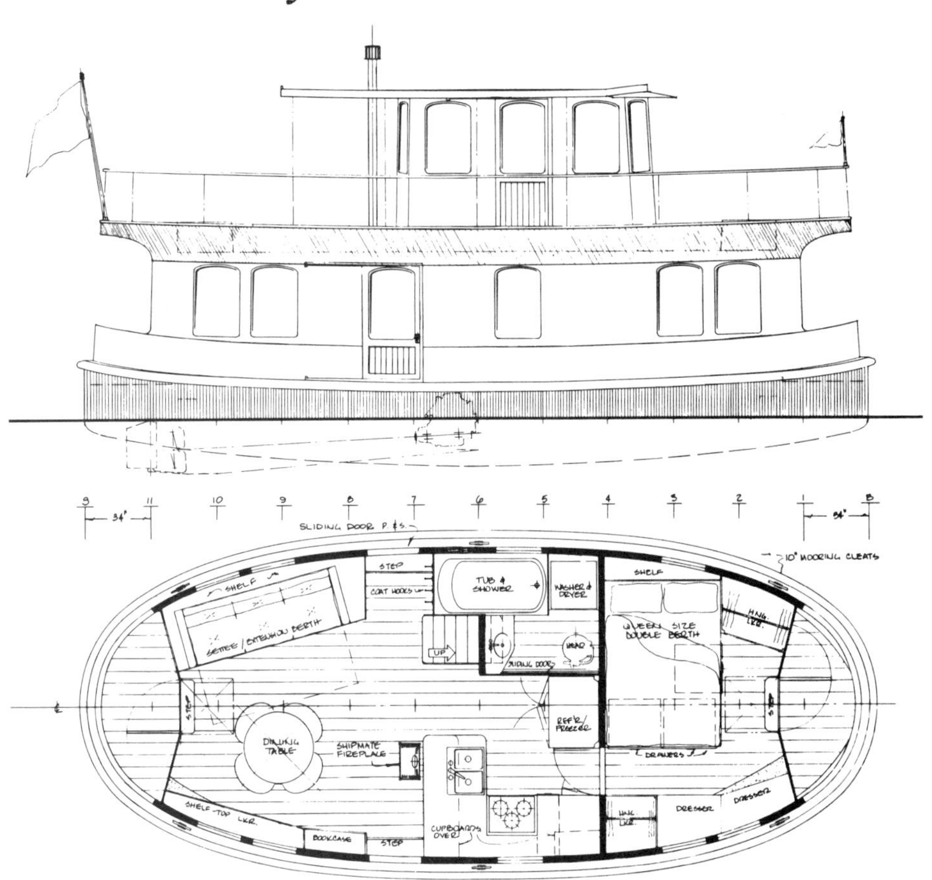

As the Washington State Ferry backed water and nosed into the dock at Friday Harbor on a crisp sunlit afternoon, the air was filled with the Dixieland beat of the Island City Jazz Band. Leaving the ferry, SEA's Bob Vollmer, in search of new yacht designs, found Jay Benford behind a drawing board in his office that overlooked the entire scene.

The firm of Jay R. Benford, Naval Architects and Consultants, has been specializing in yacht designs for many years, developing a highly respectable reputation in the Pacific Northwest and throughout the boating world. For 21 years, Benford has been professionally designing vessels for both commercial and pleasure boat use since his apprenticeship with John Atkin in Connecticut. This experience involved developing motorboat designs for commercial clients such as Foss Launch and Tug Co. in Seattle, followed by years of fishing boat design. Since then he has been creating primarily pleasure boats on a custom basis but with some production boat designs as well. One of Benford's more noteworthy clients was the author Ernest K. Gann, who commissioned the design and construction of *Strumpet*, a 35-foot trawler yacht.

It was this broad experience that prompted Vollmer to ask Benford to review the materials used in motor yacht construction.

Benford: The main design work that I did 20 years ago involved many motor yachts nearly all constructed of wood. Fiberglass is coming along now with many large boats in fiberglass production. Through the years we've done wooden yachts, fiberglass yachts, ferrocement yachts, steel and aluminum. There's been sort of a cycle going on where we're back doing a lot of wooden yachts again. The price of petrochemical materials and allied products has changed the economics of boatbuilding sufficiently to where wooden boatbuilding and, in some cases, even custom wooden boatbuilding is again viable.

We've found in a number of cases, when we're talking about a reasonable-size cruising yacht, a custom-built boat can be done in wood at practically the same price as a good quality fiberglass stock boat. This is because the custom builder, even though he has more man-hours involved, doesn't have to support all the marketing budget, the dealer network and the required mark-up. So, there is an opportunity here for many buyers to realize having an individual yacht as they want it without having to take what is available at a boat show.

SEA: That's a very interesting point.

Benford: The problem here is that they will have to be willing to wait for the boat to be built. Because of this, most of the major volume of boatbuilding today is in fiberglass. The last several years, however, have seen an increase in work for us in custom wooden yachts. Almost half of our work is in wooden yacht design. We also have seen an increase in requests for steel yachts and an occasional aluminum vessel. And, some clients are interested in developing production aluminum construction of larger yachts.

SEA: What would they be?

Benford: They are looking at a version of the design we call the Friday Harbor Ferry.

SEA: For pleasure boat use?

Benford: Yes. They want to set up a production line to build these boats in aluminum.

SEA: How large would they be?

Reprinted courtesy of Sea magazine from their November 1983 issue.

BENFORD

Benford: We're looking at various versions of the original Friday Harbor Ferry which was 34 feet. Under consideration are variations that run from 30 to 65 feet.

SEA: What about the use of the new core materials in hull construction and how will it affect performance?

Benford: A lot of the wood construction involves the use of epoxies that were not being used 20 years ago. Certainly the Gougeon Brothers, in their work promoting the West System, have had a lot to do with developing the use of epoxies. It's a fine way to build a boat and the end result is a low maintenance vessel similar to one constructed in fiberglass. This cold molded glass system is attractive for custom yacht building. One can build a custom one-off wooden yacht using this method and the cost is still pretty reasonable.

SEA: Looking forward to the next 10 years, Jay, what modifications in yacht design do you expect to see?

Benford: I think yacht design has always been an evolutionary combination of art and science. What we are experiencing now is the result of the continuing interest in more economical boats. This certainly will be an ongoing trend. There are basically two ways to achieve good fuel economy. First, and most obvious, is just to slow down. Decrease your power and you decrease consumption. The second is to construct lighter, more easily driven boats. We are working with ways to build lighter structures and outfit the boats with lighter gear.

One of the problems with trying to build light boats is that over the long haul very few owners have the mental discipline to keep them light. They keep loading them up with extra gear, accumulating spare parts, etc. every year. So, it is unrealistic to think that 10 years later the boat will have the same performance as it does when it's new.

I think we need more work in the refinement of hull form; future designs will have to be a combination of many things. A contributing situation, influencing yacht design for many years, is that moorage fees have been based strictly on length. It would make more sense to base the fees on the product of

The American Tug-Yacht 38, designed by naval architect Jay Benford for Cape-Bay Shipbuilders Corp. in Humarock, Massachusetts, closely follows traditional harbor tug lines with a fantail stern, wheelhouse wing doors, wide side decks and high bulwarks. "Heft" and "brawn" are accentuated by rugged rails and bumpers and oversize, heavy-duty hardware. An alternate deep keel configuration with a large prop aperture is available for commercial use.

Standard power is a four-cylinder, 70-hp diesel that gives a cruising speed of 7.5 knots with fuel consumption less than 2 gph. Optional power can be twin 50 to 70-hp diesels or a single diesel up to 250 hp. A brass wheel and binnacle and a brass air whistle on the stack operated from a pull chain in the wheelhouse are all provided as standard equipment.

The roomy enclosed wheelhouse includes a navigation station and an L-shaped settee and table with seating for four. The standard layout has two separate cabins with two berths in each and one enclosed head with sink and stall shower. The galley features a large sink, refrigerator, oil range and ample stowage lockers.

An alternative layout provides a private forward cabin with double berth, an enclosed head with vanity sink, and large stowage lockers. A second enclosed head with sink and stall shower is located beneath the wheelhouse settee. Aft there's a fully equipped galley with oil range and refrigerator along the port side, and opposite a convertible settee/extension berth, dinette table and two chairs. Alternate accommodation plans can be customized for individual owners.

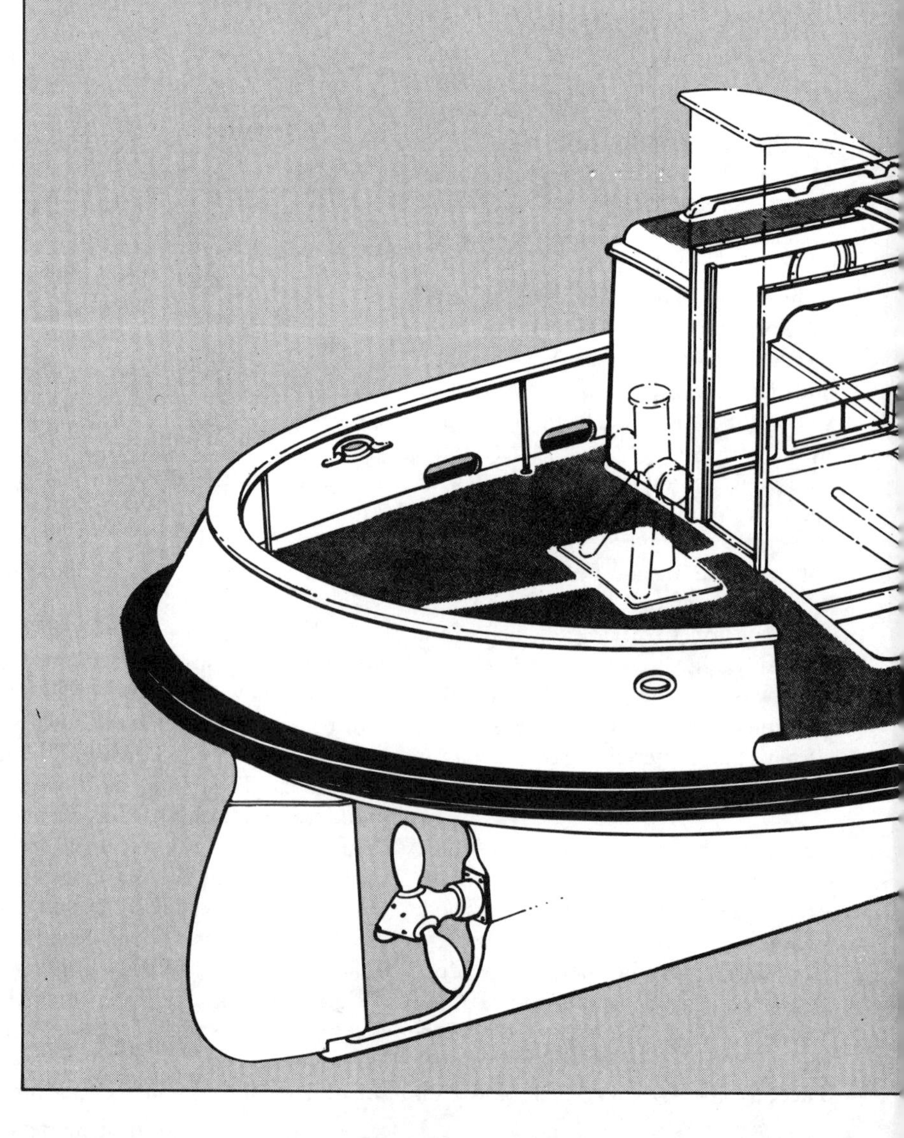

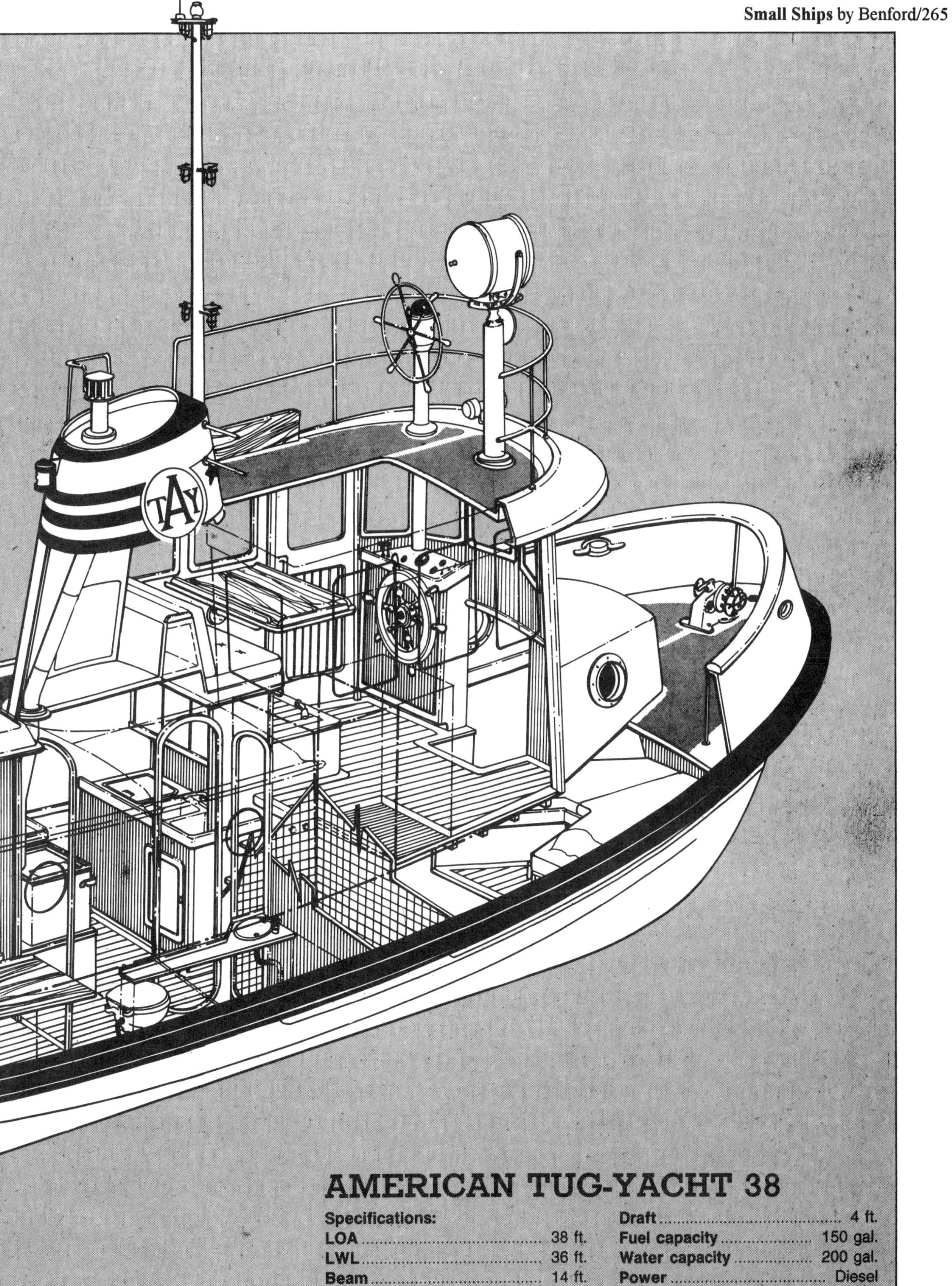

AMERICAN TUG-YACHT 38

Specifications:

LOA	38 ft.	**Draft**	4 ft.
LWL	36 ft.	**Fuel capacity**	150 gal.
Beam	14 ft.	**Water capacity**	200 gal.
		Power	Diesel

Cape-Bay Shipbuilders Corp., South River Yacht Yard, Humarock, MA 02047

BENFORD

the length-times-the-beam. The square footage of the berth space occupied by the boat would be a more rational basis for arriving at moorage fees. If this were done we would have clients who understood the results and accepted the idea of narrower boats similar to those designs popular at the turn of the century. A narrower boat is certainly more easily driven and subsequently more economical than a fatter boat.

SEA: What about the use of the new core materials in hull construction and how will it affect performance?

Benford: We've been using various cores for quite a few years now, and it is certainly one way to achieve lighter construction. I think there are limits that we keep testing to see how light it is reasonable to have a structure and still be able to hit a deadhead without holing the hull. Racing sailing yachts are willing to test the limits of ultralight construction more than a cruising yacht should. Cruising yachts, I feel, should be considerably more conservative in structure because it is approaching a lifetime investment, even if the original owner does not keep it that long. A cruising boat should be built to last a very long time.

SEA: Jay, do you have any graphs available for showing speed-to-horsepower ratios?

Benford: About four years ago I drew horsepower requirements for displacement and waterline level. For example, if we had a cruising boat that we estimated the load on board consisting of people, stores, fuel, water, machinery and structure to be 30,000 pounds with 40 feet on the waterline and we wanted to drive it at 8 knots, that would require approximately 36 hp. But, if we stretch the waterline to 50 feet, the power requirements would drop to about 21 hp and at a 60-foot waterline it would be under 15 hp. The benefits of making the boats longer and narrower are obvious. Of course, as you drop the power requirements your fuel consumption drops in direct proportion.

SEA: You feel that the present situation in the billing of berth space is, in fact, detrimental to the development of more fuel efficient vessels. What recommendations do you have to improve the situation?

Benford: To make this all happen, one of two things has to change. We have to either change the operating philosophy of how the moorages are built or let them put three boats in the space where there are currently two.

SEA: I wonder how practical that might be?

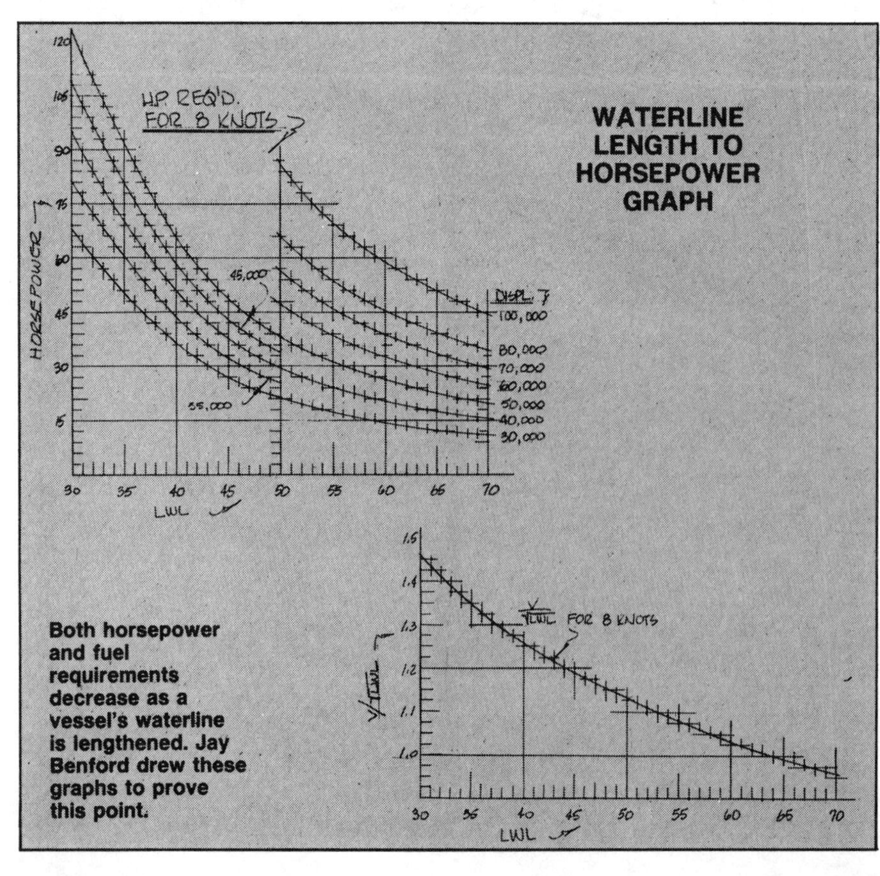

WATERLINE LENGTH TO HORSEPOWER GRAPH

Both horsepower and fuel requirements decrease as a vessel's waterline is lengthened. Jay Benford drew these graphs to prove this point.

Benford: Well, we could get back to the idea of mooring or anchoring out where boat length really doesn't matter and there's room to swing on an anchor.

SEA: In many harbors, don't we already have the problem of insufficient room to swing on an anchor?

Benford: I guess that depends on where your anchorage is. The northwest cruising area isn't quite as crowded as areas in California.

SEA: Jay, would you describe some of your present projects?

Benford: Currently in progress we have two that are strictly motor yachts and two motor sailers. The balance of volume changes from one month to the next. The two motor sailers are pilothouse auxiliaries with full power and full sailing capabilities. We achieved this by using a variable pitch prop so they will power well and the prop can be feathered to reduce drag allowing complete sailing ability. The two motor vessels are the ferry boat design that I mentioned earlier as well as a 38-foot tugboat yacht. My background in doing actual working tugs, 15 years ago, has carried over into this design and we did a tugboat that I think looks like a tugboat.

SEA: What area of pleasure powerboat design is most interesting to you and why?

Benford: For me it's the cruising boat that is also used as a liveaboard. I have been living aboard for years and enjoy it. I not only like my home aboard but would like my office aboard as well. We are looking at designs that can be used for housing as well as office space. One design under consideration is the Friday Harbor Ferry. I have designed an office version of that boat. Another long-term fantasy of mine is a larger motor yacht with motor sailing capabilities. We have a 65-foot design drawn up in a preliminary stage.

SEA: To both live aboard and work aboard is an ambitious project. You are aware, of course, of the slings and arrows of outrageous fortune that are put upon liveaboards along the West Coast?

Benford: Yes, quite aware.

SEA: How do you plan on getting around that problem?

Benford: The type of business I'm in lets me live in places where the problem isn't so severe. As long as I have mail service and a telephone hookup, I can carry on my business.

The Height Of Fuelishness

(This manuscript arrived, unannounced, in the mail one day. It caught our attention, being on a timely subject. Wanting to know more about it, we attempted to locate Ms. Wite, but were unsuccessful. We were able to reach Jay Benford. He confirmed the interview did take place, and said he thought Miss Wite was off on the ***Hatter****. Ed.)*

by Lilly Wite

One winter day recently, I boarded the Washington State ferry, ***Evergreen State***, at Anacortes for the two-hour ride through the San Juan Islands to Friday Harbor. The sky was sparkling blue, the air was cold, and I shivered under my blue, U-Vic float coat. I was on my way to meet and interview the reportedly reclusive and curmudgeonly yacht designer, Jay R. Benford. I had made the appointment only the day before.

Benford's salty and unique ketch, the ***Sunrise***, sat in her slip along side the Friday Harbor ferry landing. The Benford boat, I reflected, was something apart from most others I had seen, yet she was pretty and well-proportioned. I had a brisk walk off the ferry and through the hillside parking lot, where I turned left and flexed my calf muscles up another steep hill. A total of three and one-half blocks, following the directions given me, brought me to the old frame house that serves as the Benford office and design studio. It was painted off-white, and had a porch measuring about 7 feet by 10 feet. A large Herreshoff anchor leaned against the window post, and an old-fashioned set of spreaders seemed to grow out of the far corner opposite the door.

I knocked. Almost at once the door was flung open, and I faced a giant of a man, blond hair, red beard, broad shoulders and strong-looking arms and hands. He wore a blue chambray work shirt and heavy, whipcord trousers. Possibly in his early forties, he was rugged, handsome and competent-looking. "You're Miss Wite," he said. His blue eyes twinkled and his mouth broke into a broad, open grin. "I'm Frank Madd, a friend of Jay Benford's. I saw the ferry land. We've been expecting you."

I smiled, somewhat embarrassed at such a fast and warm welcome. He motioned me inside, took my coat, and asked me to wait while he checked to see if Benford was available. Then he dashed up a flight of stairs that led to the second floor, and was gone. I wondered if his parting look was questioning my attire. I had worn what has become almost a uniform for me — faded jeans and a **WoodenBoat** T-shirt. I enjoy being just me, although it seems to occasionally draw some attention.

Hull models, boat photos and drawings decorated the walls of the entry area. All the craft looked businesslike and sea-kindly, yet each had its own character and beauty. Sounds of rapid-fire typing came from an adjoining room, interspersed by the occasional shuffle of a file drawer closing. A sign beside the stairs read:

We are professionals, like doctors and lawyers, who sell our time and knowledge. Though we love to talk boats, we must give priority to our current design clients who have already paid for our services. Consultation is available by appointment at $120 per hour.

If Benford intended that notice to apply to this interview, he never mentioned it later.

Frank Madd came down the stairs and told me Jay Benford was free. He waited politely as I passed close by him, then followed me up to a large drafting room. One end of the room was partially closed to form a library and office, and its windows looked over the lovely harbor. Within, a man was working on some calculations, seemingly oblivious to my entry, but the next instant he rose, introduced himself as Jay Benford, and motioned me to a chair. Frank pulled another chair over close to me and sat down.

Benford was medium height, with a dark beard and dark rimmed glasses. In contrast to Frank, he seemed a quiet mover, contemplative, with a certain amused outlook from behind the neatly trimmed beard. For the huge number of designs he'd done, I was surprised when he told me he'd only started in the yacht design business in 1962. Frank Madd, he said, had been the technical and design editor of the old *Tiller* magazine, until it was bought by an international manufacturing and insurance conglomerate some time ago.

This very striking red-bearded man, I learned, was the only son of Clarence Madd, who originated the Madd Boat Works in Penobscot Bay, Maine. Their Hatter line of power cruisers remain collector's items today, and Frank owns and cruises the last one ever to come out of their yard.

Although I was becoming increasingly fascinated by this introduction, I felt compelled to begin on the stated purpose of my visit. I explained to Benford that I had come to learn about his concept of the ideal power cruiser for the Pacific Northwest.

Benford thought a moment, then said, "The proper power cruiser for this area ought to be a year-round boat. Did you notice there were almost no small boats on the water when you came over, yet the day is perfect for cruising? Too many people buy their boats in heated showrooms or at boat shows, where they are protected from the elements. They forget that the boat will be used in all sorts of weather. In this area, that usually means in dampness and cold."

"Thus, an ideal Northwest cruiser must have heat and good air circulation. Then it's pleasant to be aboard.

"The steering position must also be sheltered. It must not only be protected from the elements, it must provide a good view for the helmsman, with particular emphasis on ability to spot drift in the water. With heat, the whole area become comfortable, and others will join the helmsman, making the cruise more enjoyable for all. People feel inclined to repeat the experience, and thus do more cruising."

"I like the way you emphasize comfort and pleasure while cruising. That's always been my philosophy," I said, noticing Frank was smiling in agreement.

"Well, they are called *pleasure* boats," said Benford. "The whole exercise is soon abandoned if it's no longer pleasant. Witness the great numbers of boats that never seem to leave their slips. Here is first-rate evidence that something is amiss, either with the boat or with the owner's selection of a boat to satisfy his needs."

Benford paused here, looked out the window, then selected a well-used pipe from a rack on his desk. He picked up a humidor nearby, and from his overloaded shirt pockets produced a lighter and pipe nail. He gently tamped the tobacco in succeeding layers into the bowl, fired the lighter, and puffed thoughtfully several times. The flame alternately hovered, then disappeared into the bowl with each puff.

I asked what kind of boats were being designed today to satisfy this kind of cruising. His answer startled me.

"*Plus ça change, plus c'est la même chose.*"

"He means," Frank said, "the more things change, the more they stay the same. In other words, if we were better history students, we'd realize that most of the so-called 'innovations' and 'break-throughs' are just so much organic fertilizer."

I smiled at Frank's response. He obviously got fired up quickly, and I liked that in a man.

"Look at it this way," Frank continued. "Before about 1950, there were a lot of easily driven powerboats around. They had to be easily driven. Fuel costs, as a percentage of most people's incomes, were high. But as wages went up faster than prices, we were lulled into thinking we didn't need to be concerned with the cost of fuel. We ended up with floating condominiums with pointy ends driven by very thirsty engines. This trend was abetted by marinas and boatyards that charged solely on length, leading yachtsmen to ask designers for shorter and wider boats.

"Look at the boats my father built seventy-five years ago. The problems those boats solved should still be of concern to today's power cruisers. My father's boats were much narrower, and much more lightly loaded without all the gadgets and so-called conveniences people find so necessary today. Because they were narrow and light, they could be driven by modest sized engines with commendable economy.

"The Hatter line is a perfect example. My ***Hatter*** cruises effortlessly at ten knots. I have good visibility from the wheelhouse, and the diesel stove keeps the interior warm, even during this cold winter. In fact, starting this evening, I plan to go cruising for a few more days. I'll eat from the sea, drink homemade wine, and the fuel for the whole trip won't cost $10.00. What a great life!"

He looked steadily at me, and I hoped I wasn't reading too much into his gaze. I had to agree the cruise did sound like a lot of fun, and Frank looked competent to handle anything that came along.

Addressing myself again to Benford, who, I noticed, wore a gold wedding ring, I asked, "Where do you find boats like Frank's today?"

"Virtually all those boats are on the used boat market, if they're available at all," he replied. "Often they will have been loaded down with more gear and bigger engines over the years, but some are relatively unspoiled.

"There are also a number of powerboats aimed at the 12 to 16 knot range recently being marketed. One of these, with a smaller engine, would be more fuel-efficient."

"This brings me to what may be the most important factor in determining the extent to which boats are used," Benford continued. "That is fuel economy. Or, more correctly, the lack of fuel economy, which is usually the overriding reason why many cruises are never made. It simply costs too much to run the boat. Most powerboats aren't economical to operate, at least at the speeds people think they need for cruising."

"What is a proper cruising speed?"

"About ten knots, with the amount of drift in the water around here. This is fast enough to make reasonable passages in a matter of hours."

"What about the so-called Taiwan trawler yachts? How do they compare?" I asked.

"Most of them are heavy, and not shaped to be driven at ten knots," Benford answered. "If they were built to lighter scantlings, with modified shapes, they could cruise efficiently at ten knots. Taking them as they are. I would suggest getting one with a smaller engine, and driving it at a lower speed-length ratio for improved fuel economy."

I wanted to check my understanding of the term, 'speed-length ratio,' so I asked Mr. Benford to explain it for me.

"The speed-length ratio is the ratio of the boat's speed, divided by the square root of the waterline length. The hull speed of a displacement boat refers to the condition where the boat has a wave crest at the bow and a wave crest at the stern, with a trough in the middle. Usually, this occurs at a speed-length ratio of 1.33. For instance, a 36-foot waterline boat would generate this wave at 8 knots. Eight is 1.33 times 6, and 6 is the square root of 36."

I must have day-dreamed a bit, for thoughts of homemade wine and square roots were beginning to become thoroughly inter-meshed. Suddenly I was aware that Benford was talking again.

"....should not be thought of as an absolute limit," he was saying, "for there are too many cases of displacement boats exceeding 1.33. ***Strumpet***, a 35-foot displacement cruiser we did for Ernest K. Gann, ran right up to 1.52. She had a 32-foot waterline and made 8.57 knots in the trials. Since I'm advocating boats that cruise at 10 knots, either we must find ways of pushing smaller boats faster than 1.33, or design longer boats that make 10 knots within the 1.33 speed-length ratio parameters."

"How do you design smaller boats to move faster than 1.33?" I asked, getting my mind back on the interview.

Benford reached for one of several dozen notebooks that lined a section his library, and turned to this graph:

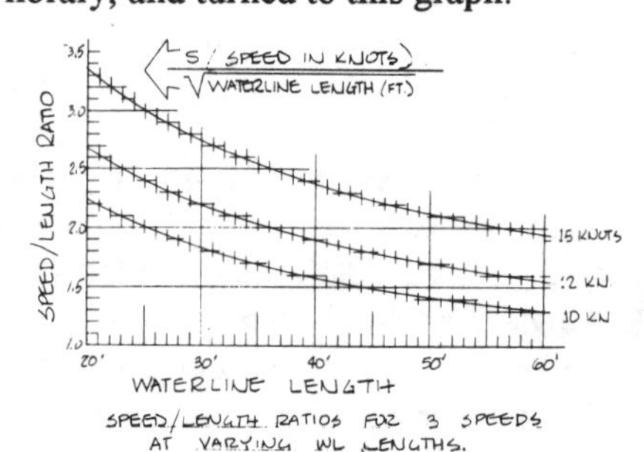

SPEED/LENGTH RATIOS FOR 3 SPEEDS AT VARYING WL LENGTHS.

"This graph shows the speed-length ratio at which a boat operates to achieve 10, 12, or 15 knots. A boat with a 25-foot waterline at 10 knot has a speed-length ratio of 2.0, while a boat with a 60-foot waterline at 10 knot has a speed-length ratio of only 1.3. The lower the speed-length ratio, within reason, the less power required, as this other graph shows.

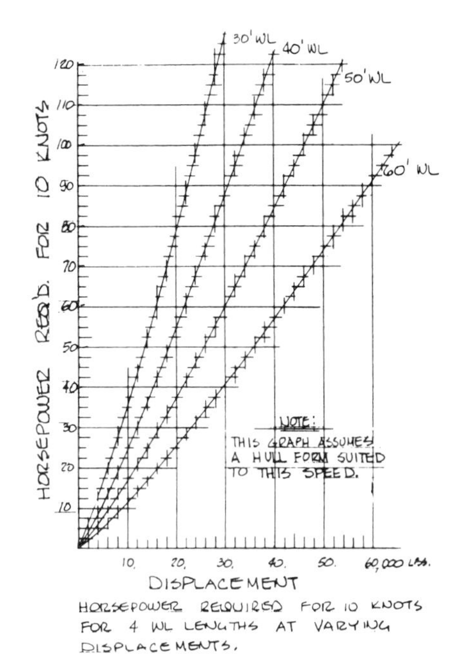

HORSEPOWER REQUIRED FOR 10 KNOTS FOR 4 WL LENGTHS AT VARYING DISPLACEMENTS.

"Let's say we've decided we need a 30,00 pound boat to carry us and our gear. Looking at these curves, if that boat were 30 feet on the waterline, we'd need about 127 horsepower for 10 knots. If we lengthen her to 40 feet, we'd be down to about 88 horsepower. At 50 feet, we'd need about 60 horsepower, while at 60 feet, we'd only need about 41 horsepower."

At this point, we were interrupted by a little black cat. She walked into the office, talking in almost human tones, hopped onto Benford's lap, then directly up to his shoulder. There she draped herself around his neck like a contented fur muff, purring and needling her claws in and out of his shirt. Mr. Benford gave her an encouraging little pat.

Benford resumed. "All this brings us to the specific hull shapes needed for increasing speed-length ratios. The most important thing is the shape of the stern. At higher speed-length ratios, we want to leave the aft crest of the hull speed wave further astern. Let me make a sketch of several sterns, to show how they affect performance."

As I turned from the window, he arose, cat still draped over his shoulders, and Frank looked in my direction. The four of us gathered around the drawing board. I was aware of Frank standing close to me.

"Stern A is much like the one we used on ***Strumpet***, and is seen on many sailing boats. Stern B is like that of the ***Kiyi***, which we lived aboard for a year, and is our favorite stern shape. Stern C is a transom stern, with the bottom just at or above the water at rest. All three of these sterns work well at speed-length ratios up to about 1.3 to 1.5. the curvature of the bottom as it rises towards the stern is too great for them to let the stern wave move aft far enough to raise their speed.

"For speeds in the 1.3 to 2.0 and up range, a stern like D would probably be best. Its transom is immersed even at rest, and the rise of the bottom toward the stern is very gentle. Stern E is most suitable for higher speeds, and is most often used on planing boats. There is no perceptible rise to the bottom on this type."

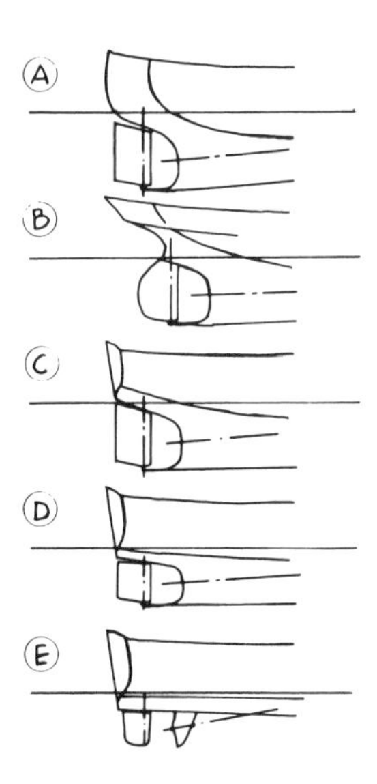

This business of comparing speed-length ratios and stern shapes to performance was starting to make sense to me. "So one of the things you're saying is that it's more efficient to make a long and narrow boat," I said. "But how can you look at an existing boat and tell how efficient it is?"

Before Benford could speak, Frank broke in. "If you don't have the actual fuel consumption data, just look at the height of the stern wave at cruising speed. If the wave's very tall, it would be the height of fuelishness to think she'd be very economical. Sorry, Jay, I couldn't let that one get by."

Benford groaned, trying to suppress a smile at Frank's pun. I liked it.

"Frank's quite right, Miss Wite. Those waves can only be generated by expending energy. The bigger the waves, the more energy wasted. When we lived on the ***Kiyi***, I loved watching her slip effortlessly through the water. There was almost no wake at cruising. The ***Kiyi*** is 50 feet long with a 10-foot beam, and an excellent example of an easily driven hull shape. Lately, Frank and I have been designing an economical cruiser, with the ***Kiyi***'s pretty stern. She's to be about 65 feet long, perhaps 60 feet on the water, with about nine and a half of beam and 3 feet of draft. Her displacement in average cruising trim should work out to approximately 15,000 pounds, giving her a displacement/length ratio of 31. We think she'd be able to go from Seattle to Southeastern Alaska and back on about 125 gallons of diesel oil, cruising at about 10 knots. That's $65 worth of fuel at current costs, or just over five cents per mile.

"By comparison, the typical planing boat of about half that length but with the same displacement would make the same round trip at about 20 knots and use about 1,250 gallons of gasoline. Its cost would be about $835, or 67 cents per mile. The savings potential is obvious."

I couldn't help but think how important this concept of fuel efficiency was. The ***Hatter*** was becoming more attractive by the minute, and it was nice to know there were ecological and scientific rationales for the choice, as well as economic. I asked what disadvantages there might be to the long and narrow boat.

"Well, the boat would only have the accommodations of a conventional 35-footer. She would also face higher charges when paying fees based solely on length. But the fuel savings on an actively cruised boat would more than offset this, plus you could put two of these in a slip that is usually used for only one today. You'd have to resist the temptation to load the boat with every toy offered at the boat shows, for weight is one of the detriments to fuel economy. Of course, the longer boat is less sensitive to loading, and would have more pleasant motion, so the crew would probably enjoy cruising more, which is the whole point, anyway."

By this time, the mid-winter sun was growing low, and Friday Harbor's trees were casting long, tranquil shadows over the water. A secretary carried in a handful of letters and laid them on Benford's desk for signature. I realized that Benford had quite succinctly explained his basic philosophy of the ideal Northwest power cruiser and our interview had come to an end. I thanked him, and looking away curiously from that little black cat, I turned to go.

Frank rose, his broad shoulders covering the doorway, and said, "Lilly, let me get your coat, and I'll take you to the ***Hatter***. I think you'll like what you see."

"I already do," I said, and followed him down the stairs....

An Update:

Since this article was first published in **Nor'Westing** magazine fourteen years ago a number of things have changed.

First is that in 1984 I relocated to the Chesapeake Bay area after eighteen years of living and cruising on Puget Sound. I certainly miss my friends and the great cruising in the Pacific Northwest. But, the business has benefited from the move and I've enjoyed seeing it grow even more back here. The temptation to move back to the Northwest is still great and who knows what may happen in the years down the road.

Second, bowing to the social niceties and hoping to live long enough to see many more of my boats built, I gave up the pipe a bit over a decade ago.

Third is that I think that even slower speeds may be the order of the day in these days of rising fuel costs and taxes. A lot of the boats in this book are more suited to cruising at speed-length ratios of about 1.1 to 1.3. This is even more economical and, for those of us who are moving from sailboats to powerboats, the speeds will still seem fast enough. For that 36' waterline boat this would be speeds in the range of 6.6 to 7.8 knots.

The graph below is another one of the tools we use in our preliminary design work. From it we can quickly read what sized engine to specify for a displacement boat, either power or sail. It is also useful in consulting on repowering old boats.

To use the graph, read the bottom axis to find the displacement of the boat. Then go up and find the horsepower required for it. For example, a yacht of about 23,000 pounds displacement would require about 35 horsepower.

Our 38' Tug Yacht operates at close to this displacement and owner feedback tells us that they run at a steady eight knots (a speed-length ratio of 1.33, or the so-called hull speed) while burning between two and two and a quarter gallons of diesel per hour. Converting this backwards on the engine curves tells us that they're actually using 35 to 40 horsepower.

Some of our slipperier sailboat hulls have done a little better for speed than would be indicated. Conversely, some of the very fat powerboats would require a bit more power than indicated.

This graph assumes a reasonably conventionally shaped and smooth hull form, driven to a speed-length ratio of 1.33. Bob Beebe's book, **Voyaging Under Power**, contains a good discussion of this subject too.

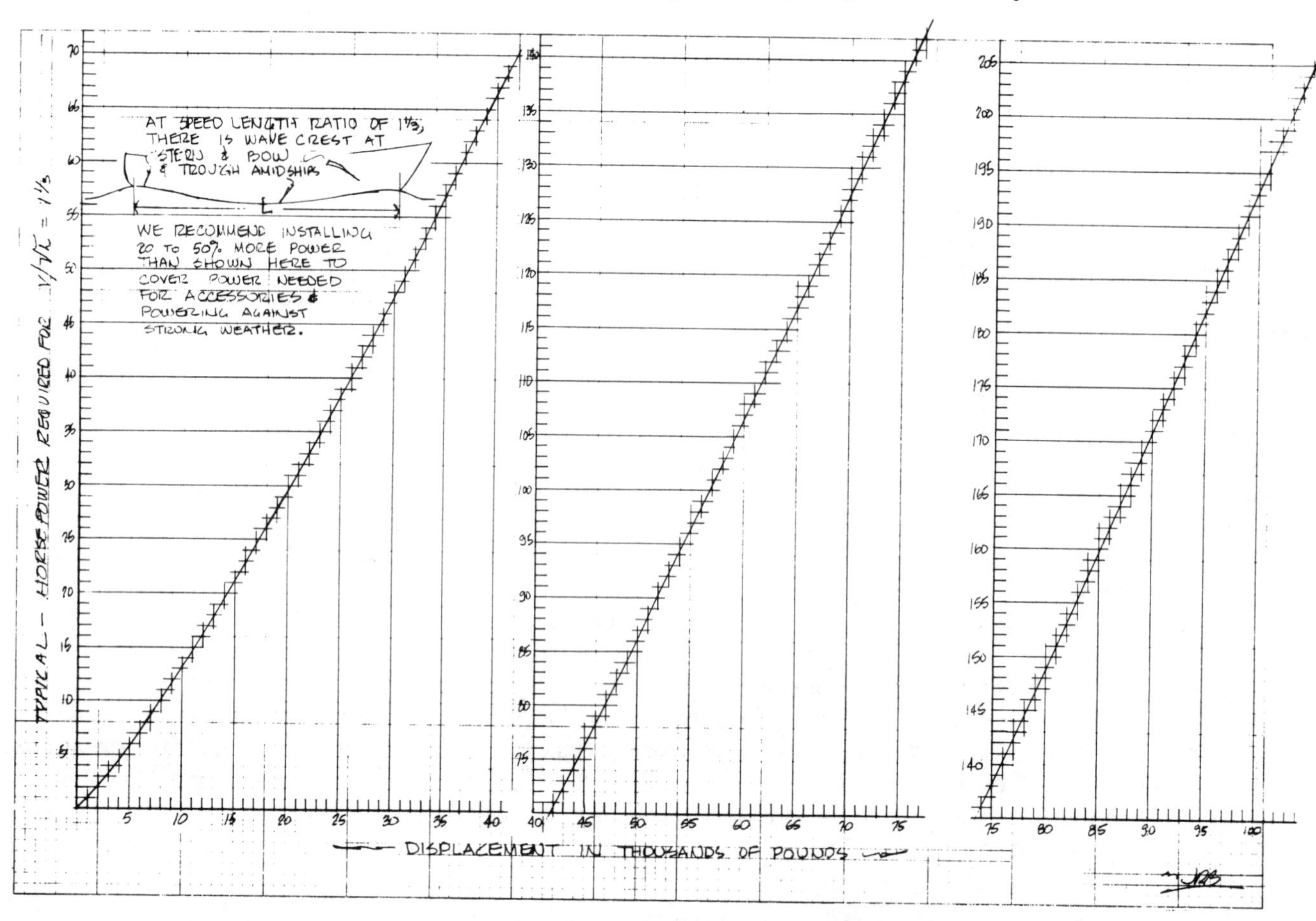

The State-Of-The-Art Yacht Design Office

Three decades ago, when I entered the profession of yacht design, high-technology was represented by the ability to use a slide rule and a planimeter. I'd gotten acquainted with the slide rule earlier and still use one for some things today. In a moment of shopping abandon, I splurged just over $40 and bought a planimeter. I quickly got onto using it and it still does faithful work for me today.

However, in the intervening decades electronics has taken the design world by storm. In the early '70s I got an HP-35 shortly after Hewlett-Packard brought out the first "shirt pocket calculator". It caused a revolution in the amount of work we could get out and the degree of accuracy we could attain.

About this time Bill Plice, a local computer programmer and sailboat enthusiast got into programming some yacht design programs and offering to run these at a reasonable cost. This got us into refining the offsets in the computer and coming up with more elaborate stability studies than we could otherwise afford the time to do manually. In the meantime, the price of computing power continued to drop and the availability of better design hardware and software expanded.

So, wanting to stay in the yacht design business for the long haul, I came to the conclusion that computerizing the design office was the best way to stay in business. This led to an investment over a thousand times as great as that planimeter a generation before. Fortunately, I'd been able to keep the business growing by investing back in the business. It's both a pleasure and honor to be among the very few who are able to make a living at doing something they love in the field of yacht design.

This major investment in the beginning of 1986 has led into another revolution in the way we do our design work. George Hazen's Fast Yacht software that we use is very quick and adept at hull form creation and modification. We start by creating a basic hull form on screen and then proceed to manipulate it into the size and detailed shape that we're seeking. Along the way we can take quick looks at the displacement, curve of areas, stability and coefficients.

If we want a more detailed check on the stability of the form, we can move into another section of the program and run full hydrostatics. If it's a sailboat, we can approximate the rig and run it through the performance prediction, generating polar plots of sailing performance in varying wind strengths. These can be continually refined throughout the design process so that we can meet whatever goals we set for ourselves.

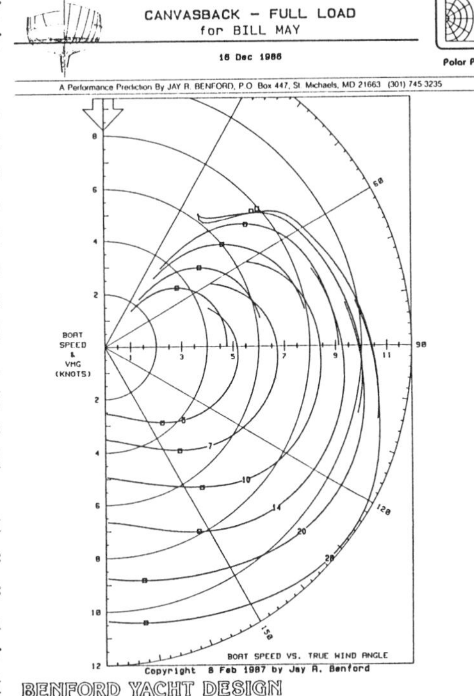

BENFORD YACHT DESIGN

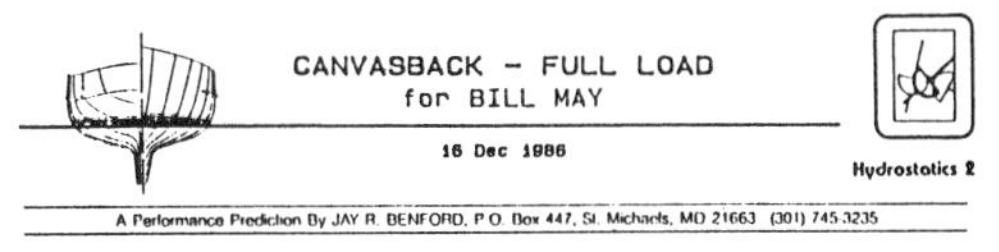

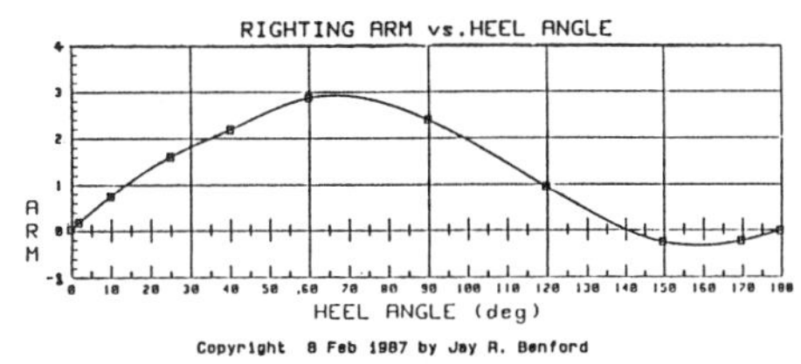

BENFORD YACHT DESIGN

We also use the performance prediction to check pre-computer designs and use them for a basis of comparison with the new boats being created, or just to see how well we did creating a good design. We're using this capability now to do some revisions and updates on some of our older designs, improving their performance and still retaining good stability and safety characteristics.

Fast Yacht also has the capability to design truly developable surface forms, making it ideal for doing steel, aluminum and plywood designs. When the hull form is created it is then possible to "unwrap" the surfaces, giving the flat plate layouts for the surfaces of the hull form. In years to come we'll see more builders taking advantage of this with computer driven numerical controlled cutters doing a lot of the work.

Spar and rigging design and analysis is another capability of the computer system. The computer can do a much more sophisticated design analysis than was practical with hand-held calculators and we can get this output as data that is usable by the client and/or builder.

Design of fin keels and the optimization of their shape and placement relative to the rig and hull is another program we can do. This lets us look at a lot of variables and optimize the keel. Then, we can turn the plotter loose and draw out the foils full size, if desired, for the builder.

Once we're done with the final version of the hull, we can turn to the plotter and generate full-sized templates of the new design. These have proven very accurate in the builders' shops and a great time-saver. If the builder still wants to check the lofting himself, we can provide offsets with the same high degree of accuracy.

Our Fast Yacht computer system is truly the state-of-the-art for contemporary yacht design work. We've also been able to use it to advantage with more traditional types and in updates to our older designs. I expect that we'll continue to expand the system and its capabilities over time so that we can keep providing the best yacht design work in a quick and timely manner.

Custom Yachts At Stock Boat Prices?

by Frank Madd

Custom yachts at stock boat prices? How can that be? Everyone "knows" that the production line boats are built more efficiently and thus they cost less — right? Is this a trick question? It sounds too good to be true — is it?

Maybe. A very definite maybe. ***If*** you're considering a quality cruising boat of, say, 35 to 40' and up. ***If*** you're willing to search out the good yards, out of the high rent districts. *If* you can wait while the builder finishes the jobs he has on hand and can then take on doing your boat. ***If*** you can find a designer who is clever enough to design a boat that is simple in its construction and concept, yet provides the features you want along with the seaworthiness necessary for how she'll be used.

Even though the stock boat builder may be more efficient, in the sense of using less man hours to build the boat, he has other costs the custom builder does not face. The production builder must amortize the cost of the tooling, molds, and patterns over each boat he builds; must carry on an expensive ad and promotional campaign (ad space does cost money, as do brochures and salesmen); and often must pay a dealer a commission for selling the boat to you.

Thus, the custom builder will save you the costs of the share of the tooling, and ad campaign costs, and the large sales markups (often 25 percent). All these savings can be put into the extra labor it will take to do the one-off boat's lofting and templating, and the interior detailing that you specifically want in your custom boat.

What types of construction are suited to producing reasonably priced custom yachts? The most economical are those using large sheet stock like plywood or steel, and having hull form developed for these large sheets to lay on easily. A properly designed hull of this type can be very handsome, with the chines becoming a part of the sculpture of the hull. Also there can be a topside chine which can provide a reinforcement point for a guard rail.

Paul Miller, who has his boat shop on Vancouver Island, has put together some pricing that indicates the Benford sailing dory designs can be built for even less than similar sized stock boats. His experience building a small one of these certainly confirms it.

Custom building also gives the owner a chance to become involved in the selection of almost everything that goes into the boat, from the materials for the hull to the hardware. The owner can also save some costs by doing the materials ordering-purchasing through the builder's accounts, if the builder is agreeable. This saves the builder's time, and this cost savings can be passed along to the owner. The owner can also save by acting as the go-fer, going for the supplies and hardware that the builder needs picked up. This gives the owner the chance to look at the things before they're installed on the boat, to be sure it is what he wanted.

The most persuasive advocate of this concept is John Guzzwell. His conviction is not lightly founded. He's been making a living as a professional boatbuilder for a score of years, and the quality of his work is recognized worldwide.

When John Guzzwell finished building the first Benford 37' Pilothouse Cutter ***Corcovado*** a bit over a decade ago, a review of the accounting revealed that she had indeed been created at the same price as a similar stock boat of the same size and type. In the process, a boat with several superior features had been created. She had a larger rig than the stock boat and sailed quite well, with some of the credit due to the use of the Hundested variable pitch prop that feathered under sail. She was a little lighter than the stock boat, with greater apparent stability. She also had better visibility from inside and a layout custom tailored to the owners.

Not every builder is capable of turning in this sort of performance. The builder must be located in a lower overhead area, have ready access to the supplies needed, and have good mental organizational skills in addition to being efficient in the use of his time. Each part of the job must be done in the proper sequence. Supplies must be ordered in advance so that they're on hand and there is no waiting for them to arrive. Thus, the builder won't have to work around an unfinished part of job, and lose some time coming back to finish it when the missing part arrives.

This sort of efficiency leads to keeping the building hours in line. Most builders work on a fee of so many dollars per hour. There is very little published information on how long it takes to build a boat. A decade ago, in an issue of *Cruising World* magazine, the Pardeys quoted some figures for building times. They noted that the consensus of the builders that they talked to took one hour to build two to two and a half pounds of displacement. Thus, a 20,000 pound boat would take 8,000 to 10,000 hours to build.

In talking with several builders in the Pacific Northwest, I found that they could normally build four to five pounds of displacement in one hour. This may have something to do with this being one of the few areas that still has a good deal of custom building activity.

The Pardeys' figure for hours is probably quite valid for the more elaborate forms of construction, or for operations without a lot of power tools and equipment. Also, a proper building in which to build the boats can save a lot of time fighting the elements.

For amateur builders, I would think that even more time should be allowed than the Pardeys indicate, unless the boat is of a simple construction and/or the builder is experienced in the work. Whichever sort of boat is being built, it's worth doing right the first time, though. The middle of a cruise is no time to find that the materials or workmanship aren't up to the job....

38' Packet

Design Number 334

1995

This design is another great liveaboard for a couple or a family. The accommodations have the master stateroom up and the guest stateroom in the bow. The head is up, adjoining the master, but with access from the passageway so it can be used by all the crew.

With two feet more beam than the 35' Packet, she can carry the raised saloon at the same level as the larger after deck. This also means the engine room can be under the saloon and the "basement" now can have a nice workshop.

The galley in the mid-level is close to both the pilothouse and the saloon. The saloon has a good-sized office in the forward port corner well suited to the tele-commuter.

Tankage is huge for a vessel of this size, permitting long periods away from shore-side facilities. With her great beam she has high stability needed for comfortable living aboard. While conceived as a coastwise cruiser, she could make Great Lakes crossings or transits to the Bahamas in reasonable weather.

Particulars:

Length over guards		38'-8"
Length-structural		38'-0"
Designed waterline		37'-6"
Beam over guards		18'-0"
Beam-structural		17'-4"
Draft, loaded		3'-6"
Freeboard:	Forward	7'-3"
	Waist	3'-9"
	Aft	1'-2"
Cruising Displ.*		60,000 lbs.
Displ.-length ratio		508
Prismatic coefficient		.585
Pounds per inch imm.		2,572
Tankage: Fuel		1,010 Gals.
Water		900 Gals.
Stability - GM		4' to 6.84'

***CAUTION:** The displacement quoted here is for the boat in cruising trim. That is, with the fuel and water tanks filled, the crew on board, as well as the crews' gear and stores in the lockers. This should not be confused with the "shipping weight" often quoted as "displacement" by some manufacturers. This should be taken into account when comparing figures and ratios between this and other designs.

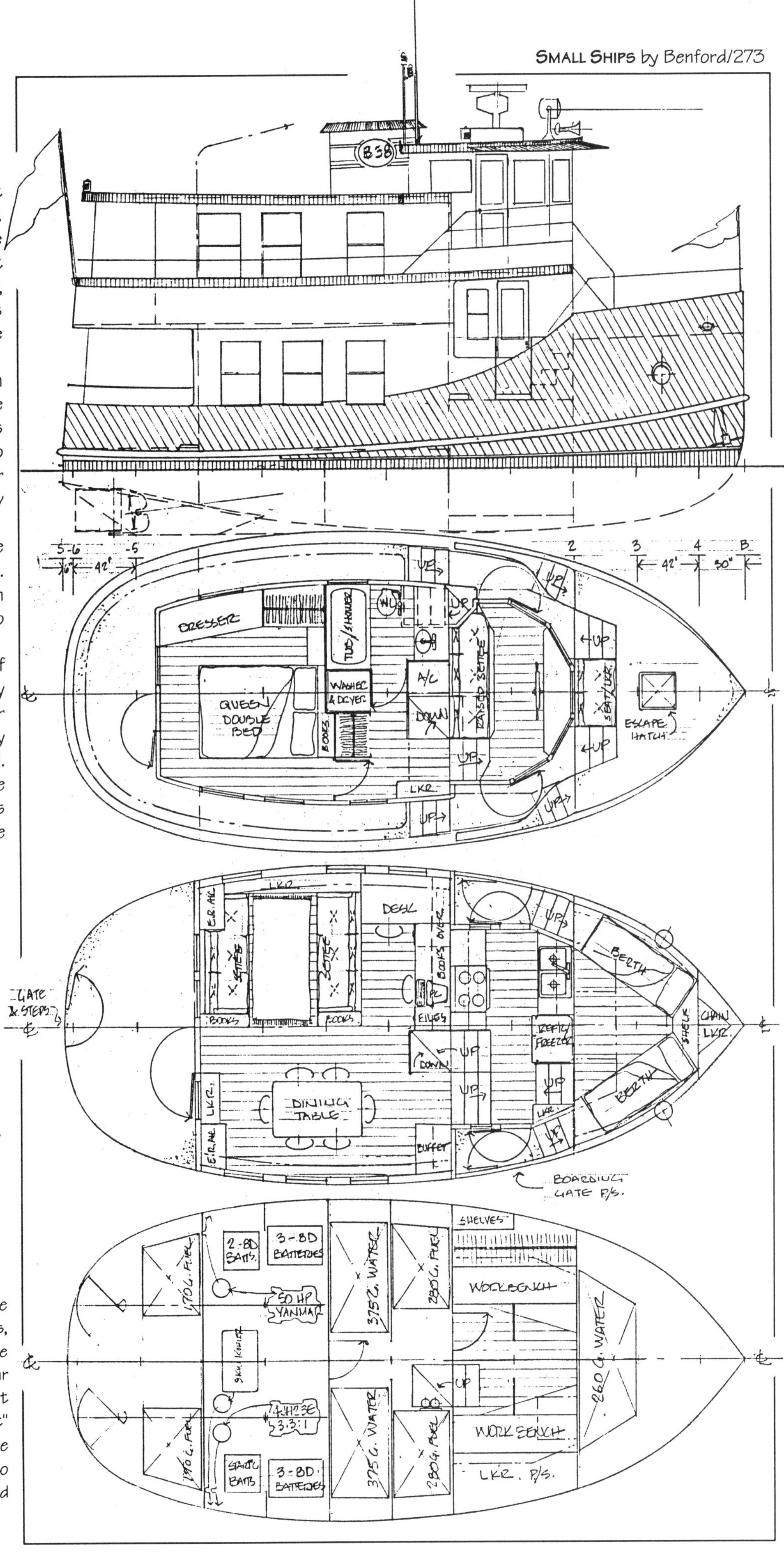

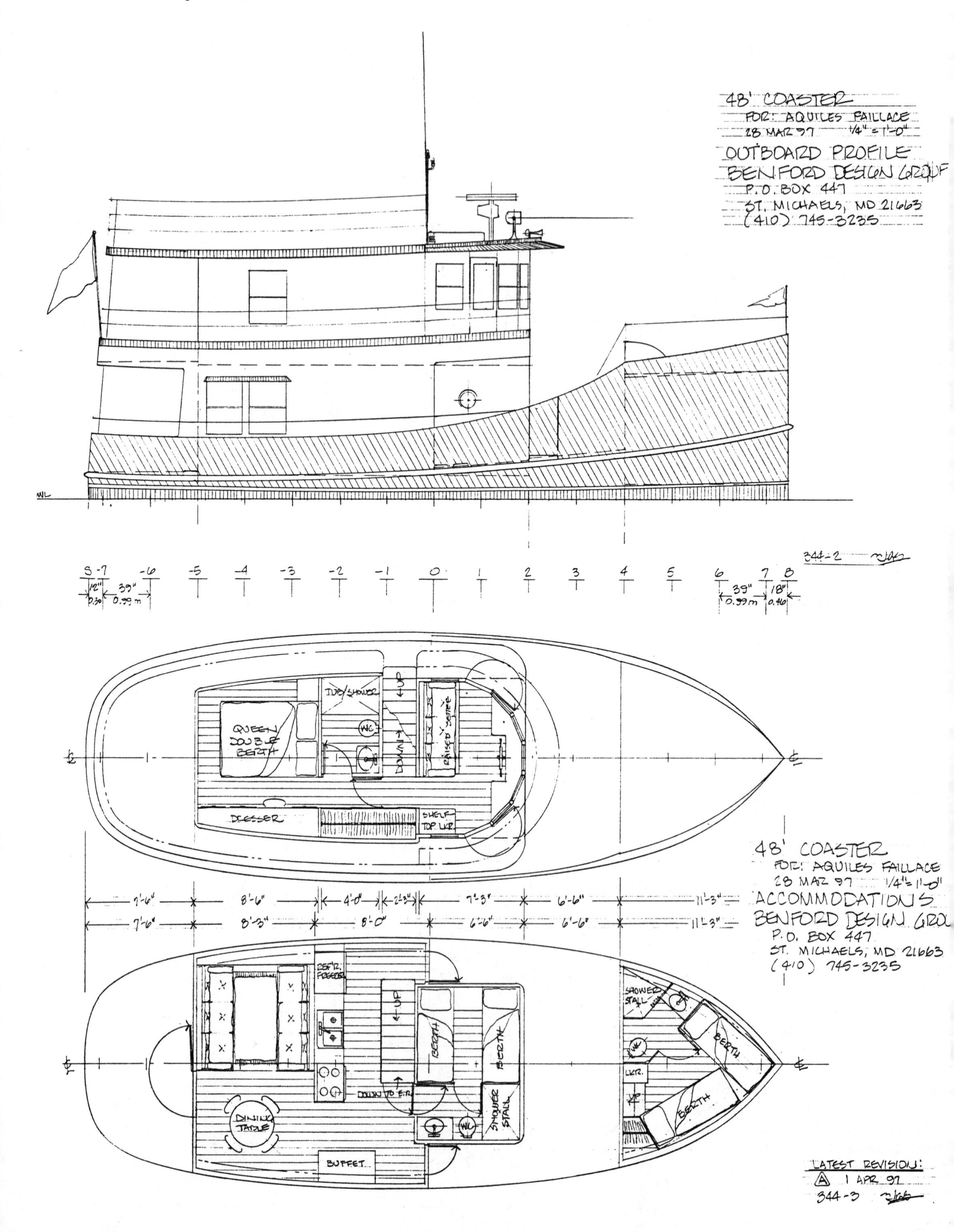
48' COASTER
FOR: AQUILES FAILLACE
28 MAR 97 1/4" = 1'-0"
OUTBOARD PROFILE
BENFORD DESIGN GROUP
P.O. BOX 447
ST. MICHAELS, MD 21663
(410) 745-3235
344-2
48' COASTER
FOR: AQUILES FAILLACE
28 MAR 97 1/4" = 1'-0"
ACCOMMODATIONS
BENFORD DESIGN GROUP
P.O. BOX 447
ST. MICHAELS, MD 21663
(410) 745-3235
TUB/SHOWER
UP
WC
QUEEN DOUBLE BERTH
DOWN
RAISED SETTEE
DRESSER
SHELF TOP LKR.
REFR. FREEZER
UP
BERTH
BERTH
DOWN TO E.R.
DINING TABLE
SHOWER STALL
BUFFET
SHOWER STALL
WC
LKR.
BERTH
BERTH
LATEST REVISION:
A 1 APR 97
344-3

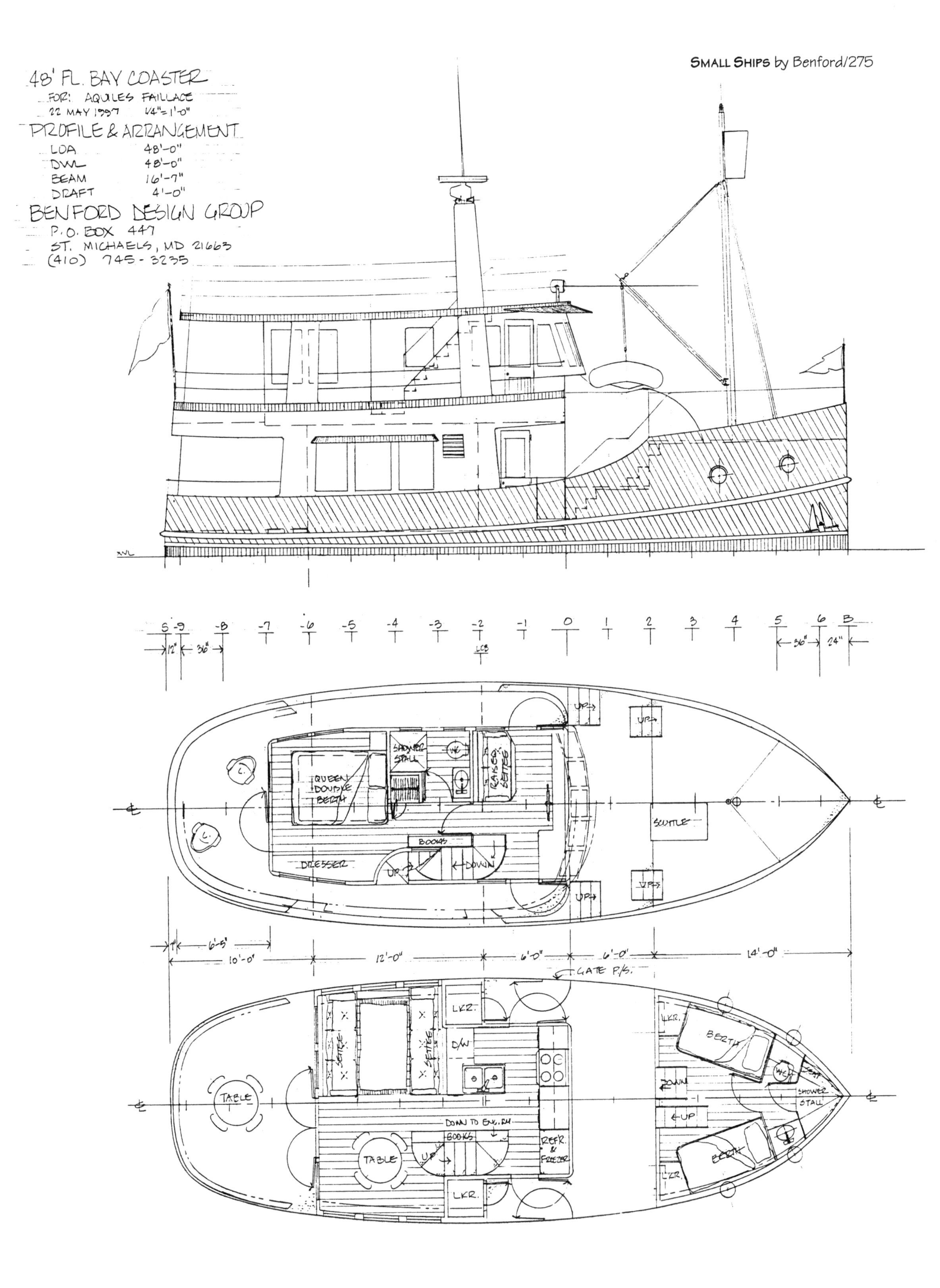

48' FL. BAY COASTER
FOR: AQUILES FAILLACE
22 MAY 1997 1/4"=1'-0"
PROFILE & ARRANGEMENT
LOA 48'-0"
DWL 48'-0"
BEAM 16'-7"
DRAFT 4'-0"
BENFORD DESIGN GROUP
P.O. BOX 447
ST. MICHAELS, MD 21663
(410) 745-3235
DWL
LCB
QUEEN DOUBLE BERTH
SHOWER STALL
W/C
RAISED SETTEE
BOOKS
DRESSER
UP
DOWN
SCUTTLE
10'-0"
12'-0"
6'-0"
6'-0"
14'-0"
6'-5"
GATE P/S.
LKR.
SETTEE
D/W
TABLE
DOWN TO ENG. RM.
REFR. & FREEZER
BERTH
SEAT
SHOWER STALL

48', 52', & 58' Coasters

Designs Number 344, 340, & 343
1997, 1996, 1997

These designs are more variations on our Florida Bay Coaster line. As this edition goes to press, the 58 is under construction the 48 is about to start, and The 52 is a proposal.

The hull form on all three is an evolutionary development of the more easily driven one used on the 65' ***Patriot.*** They are of all steel construction with wood used for the interior outfitting and some of the interior decks.

The 52 has enough of a well deck to carry a small car aboard. The 48 could carry a golf cart on its smaller well deck. The 58 has used the well deck space as additional interior living space and volume.

The 58 has three staterooms, plus room for crew and/or another stateroom below. The 48' versions show both two and three stateroom variations. The two stateroom one with the longer after deck is the one that is under construction.

The 48, as built, will probably have the vertical front windows shown on the three stateroom one instead of the sloped alternative shown. The sloped windows do show another alternative that can be considered when building one of them.

The 48 doesn't have the split level effect in the accommodations that the others have. She has the exterior stairs above the interior ones, giving good access to the boat of sun deck.

The 48 can have the well deck raised sufficiently to make for full headroom workshop underneath or for use as additional accommodations.

The stateroom under the pilothouse would make a good office, well suited to the tele-commuter.

All have generous tankage for a vessel of their size, permitting long periods away from shore-side facilities.

Particulars:		48'	52'	58'
Length over guards		48'-7½"	52'-7½"	58'-7½"
Length-structural		48'-0"	52'-0"	58'-0"
Designed waterline		48'-0"	52'-0"	58'-0"
Beam over guards		17'-2"	18'-7½"	18'-7½"
Beam-structural		16'-6½"	18'-0"	18'-0"
Draft, loaded		3'-9"	4'-0"	4'-6"
Freeboard:	Forward	7'-6"	7'-5"	8'-3"
	Waist	4'-3"	1'-10"	4'-6"
	Aft	1'-9"	1'-9"	1'-9"
Cruising Displ.* lbs.		87,000	113,000	131,000
Displ.-length ratio		351	389	300
Prismatic coefficient		.575	.567	.571
Pounds per inch imm.		3,276	3,881	4,348
Tankage: Fuel, Gals.		800	1,000	1,355
Water, Gals.		1,000	1,100	1,000
Stability - GM (Est.)		3.8'	3.9'	3.65'

***CAUTION:** The displacement quoted here is for the boat in cruising trim. That is, with the fuel and water tanks filled, the crew on board, as well as the crews' gear and stores in the lockers. This should not be confused with the "shipping weight" often quoted as "displacement" by some manufacturers. This should be taken into account when comparing figures and ratios between this and other designs.

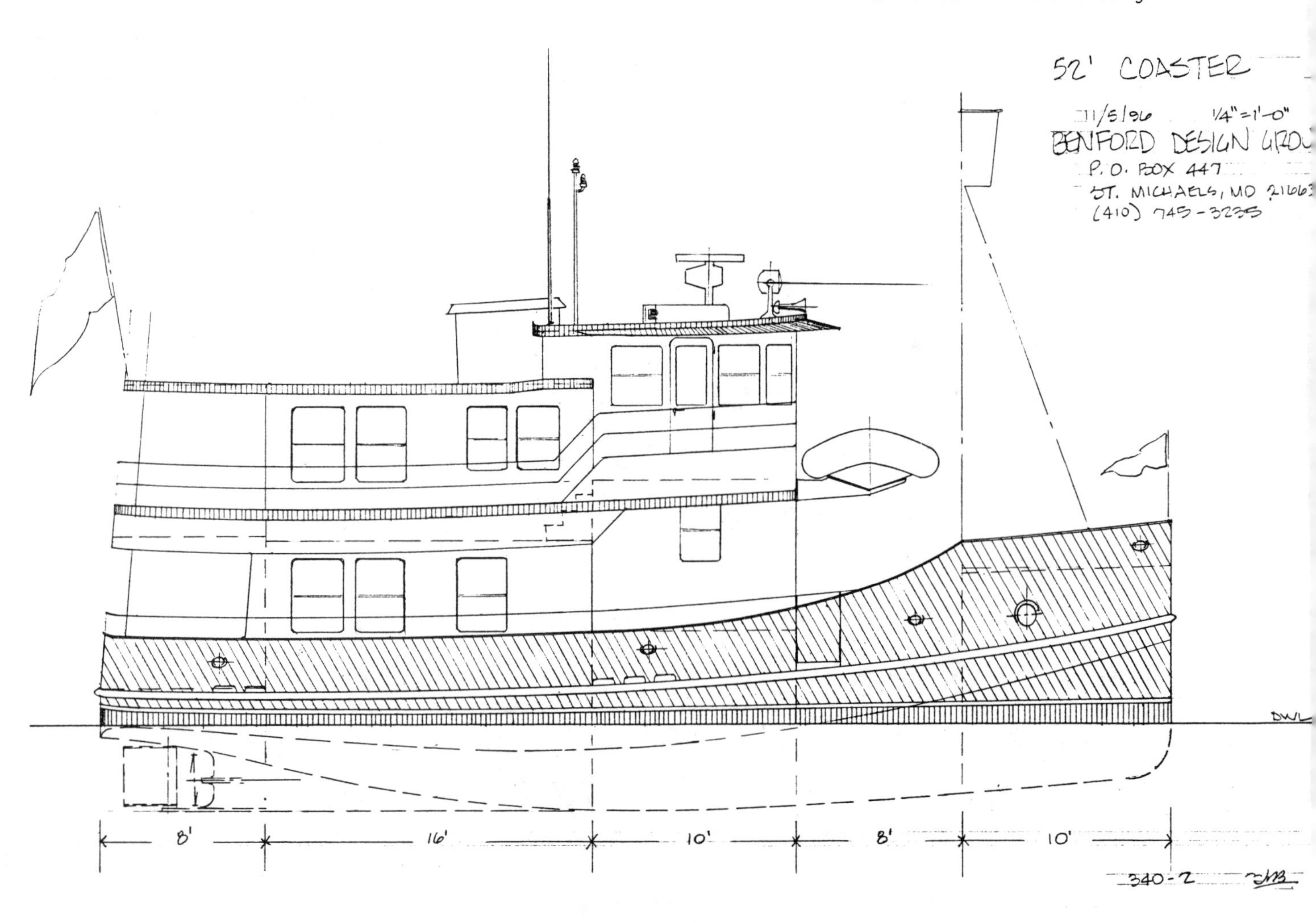

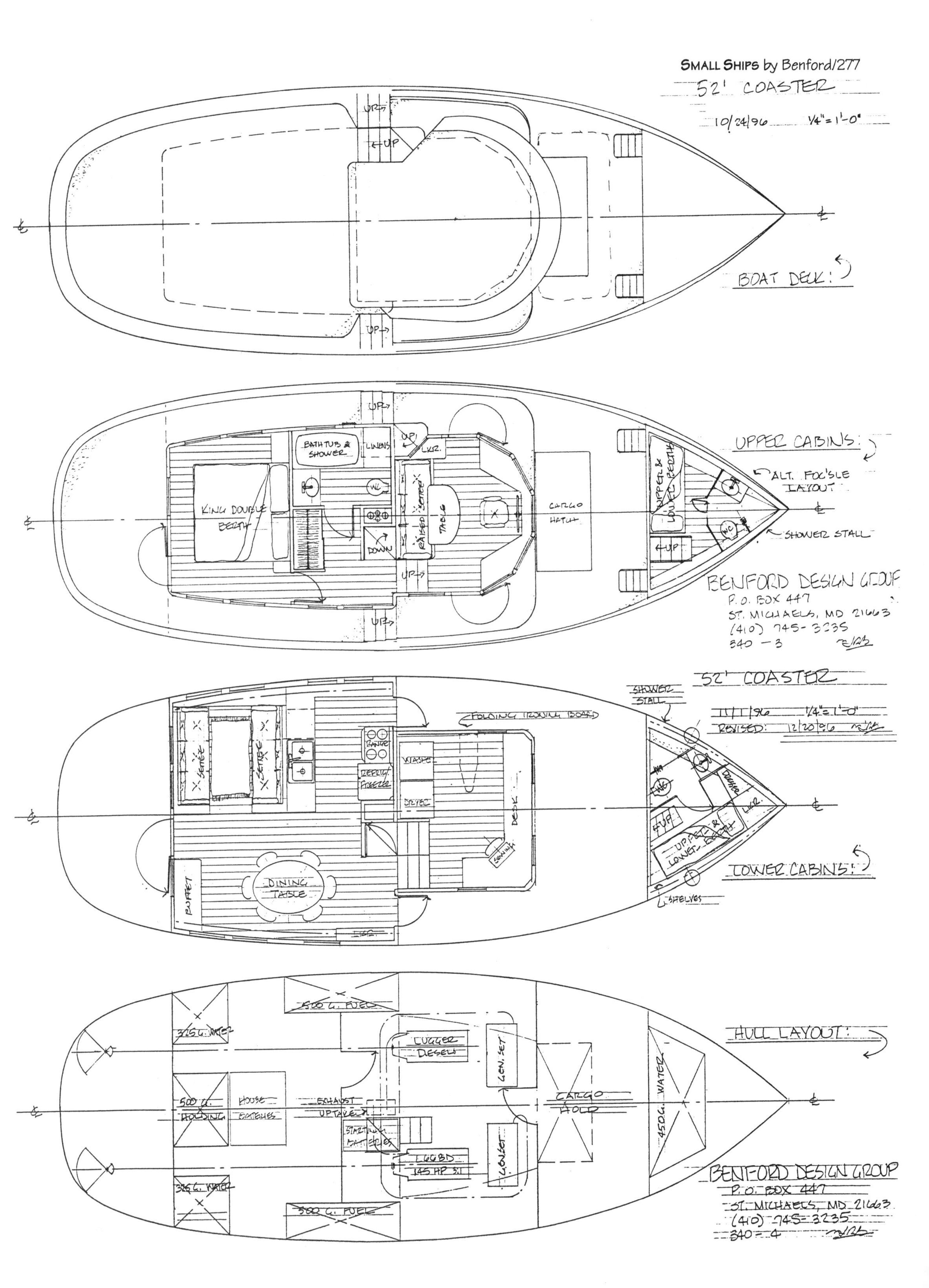
52' COASTER
10/24/96 1/4" = 1'-0"
UP
BOAT DECK
UPPER CABINS
BATHTUB & SHOWER
LINENS
LKR.
KING DOUBLE BERTH
WC
DOWN
RAISED SETTEE
TABLE
CARGO HATCH
UPPER & LOWER BERTHS
ALT. FOC'SLE LAYOUT
SHOWER STALL
BENFORD DESIGN GROUP
P. O. BOX 447
ST. MICHAELS, MD 21663
(410) 745-3235
340 - 3
52' COASTER
SHOWER STALL
FOLDING IRONING BOARD
11/1/96 1/4" = 1'-0"
REVISED: 12/20/96
SETTEE
RANGE
REFRIG./FREEZER
WASHER
DRYER
DESK
SEWING
DINING TABLE
BUFFET
LKR.
DRESSER
UPPER & LOWER BERTH
LOWER CABINS
SHELVES
HULL LAYOUT
500 G. FUEL
325 G. WATER
LUGGER DIESEL
GEN. SET
500 G. HOLDING
HOUSE BATTERIES
EXHAUST UPTAKE
STARTING BATTERIES
L668D 145 HP
CARGO HOLD
450 G. WATER
CLOSET
BENFORD DESIGN GROUP
P. O. BOX 447
ST. MICHAELS, MD 21663
(410) 745-3235
340 - 4

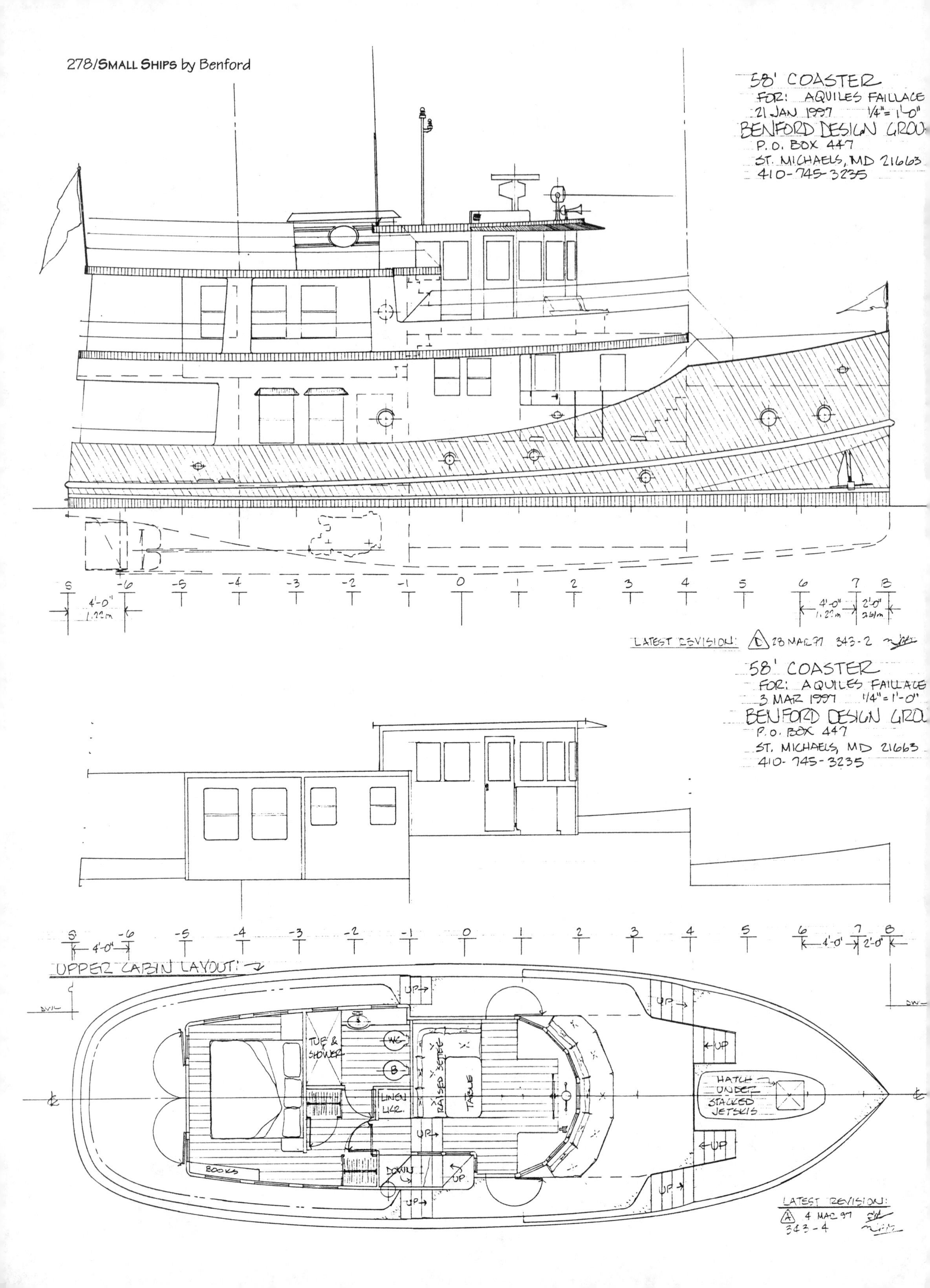
58' COASTER
FOR: AQUILES FAILLACE
21 JAN 1997 1/4"=1'-0"
BENFORD DESIGN GROU
P.O. BOX 447
ST. MICHAELS, MD 21663
410-745-3235
4'-0"
1.22m
2'-0"
.61m
LATEST REVISION: D 28 MAR 97 343-2
58' COASTER
FOR: AQUILES FAILLACE
3 MAR 1997 1/4"=1'-0"
BENFORD DESIGN GROU
P.O. BOX 447
ST. MICHAELS, MD 21663
410-745-3235
UPPER CABIN LAYOUT:
TUB & SHOWER
WC
B
RAISED SETTEE
TABLE
LINEN LKR.
BOOKS
DOWN
UP
HATCH UNDER
STACKED JETSKIS
LATEST REVISION:
A 4 MAR 97
343-4

DWL
5 -6 -5 -4 -3 -2 -1 0 1 2 3 4 5 6 7 8
4'-0" 1.22 m
4'-0" 1.22 m 2'-0" 0.61m
58' COASTER
FOR: AQUILES FAILLACE
24 FEB 1997 1/4"=1'-0"
BENFORD DESIGN GROUP
P.O. BOX 447
ST. MICHAELS, MD 21663
410-745-3235
LOWER CABINS:
GATE P/S.
D/W
UP
SHOWER STALL
BERTH
DR
REFR FREEZER
DINING TABLE
DOWN
UP
PANTRY
BOOKS
LKR.
CHAIN LKR.
LATEST REVISION:
3 MAR 97
343-5
DWL
500 GALS. HOLDING
3'6"/1.06m HEADROOM 5'/1.52m
6'-4"/2.06 m HEADROOM
58' COASTER
FOR: AQUILES FAILLACE
31 JAN 1997 1/4"=1'-0"
BENFORD DESIGN GROUP
P.O. BOX 447
ST. MICHAELS, MD 21663
410-745-3235
500 GALS. WATER EACH P/S.
HULL LAYOUT:
WORKBENCH
SHELVES
OPTIONAL DOUBLE BERTH
LUGGER L668T's
760 GALS FUEL
GEN SET
GEN SET
DESK
SHOWER STALL
575 GALS. FUEL
FREEZER
WASHER
175 HP 3:1
UP
DRYER
BERTH
SHELVES
LOCKERS OVER TANK
LKR.
LATEST REVISION:
4 MAR 97
343-6

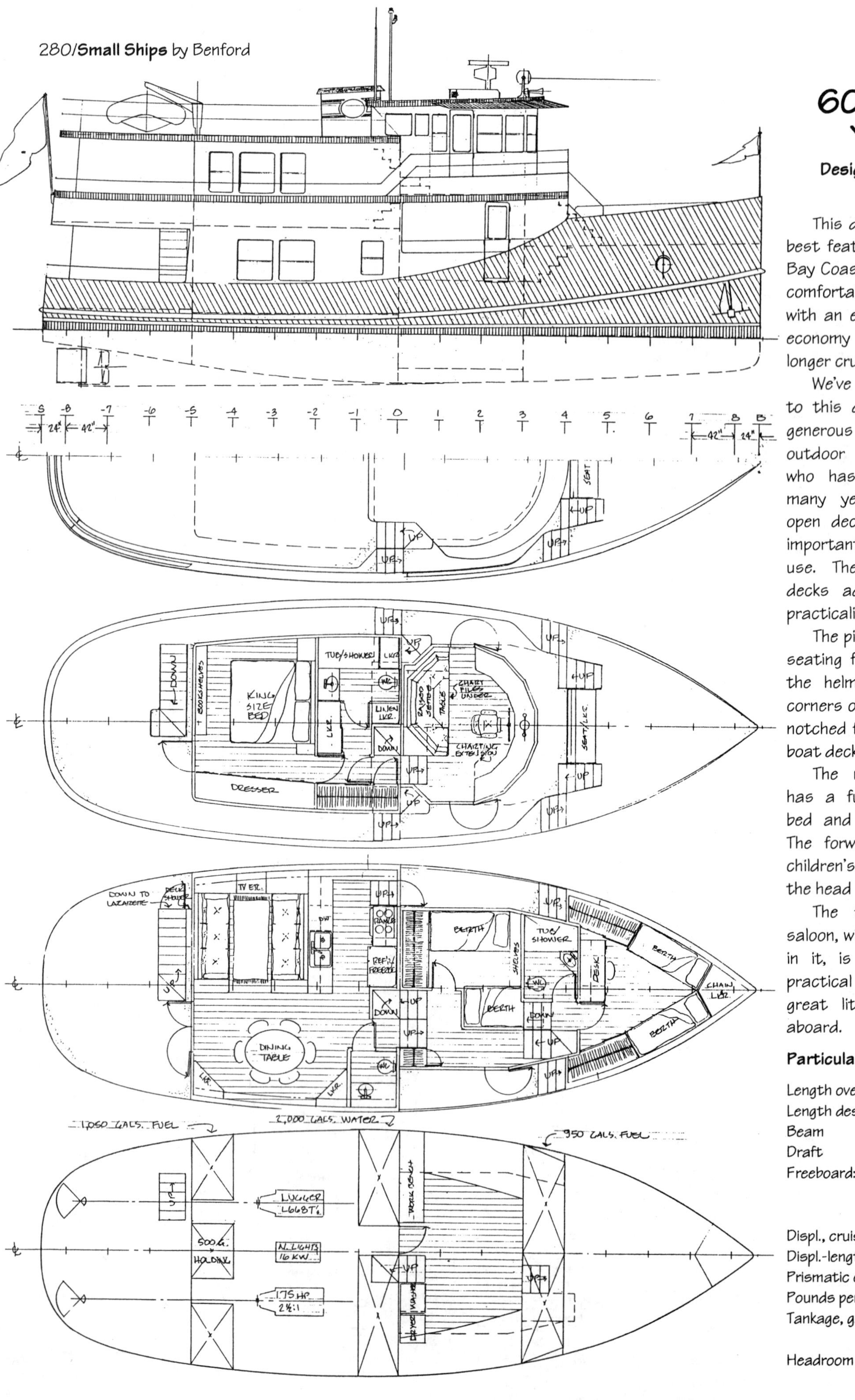

60' Motor Yacht

Design Number 337

1996

This design combines the best features of our Florida Bay Coaster designs (roomy, comfortable living quarters) with an easily driven hull for economy of operation on longer cruising.

We've added some length to this design to give very generous deck spaces for outdoor living. As anyone who has lived aboard for many years knows, these open deck spaces are very important and get a lot of use. The walk-around upper decks add greatly to her practicality & functionality.

The pilothouse has raised seating for the crew to join the helmsman. The after corners of the pilothouse are notched for stairs up to the boat deck.

The master stateroom has a full head, king-sized bed and generous storage. The forward two guest or children's staterooms share the head between them.

The family room style saloon, with galley and dining in it, is well proven as a practical solution. All in all, a great little ship for living aboard.

Particulars:

Length overall		60'-0
Length design waterline		60'-0"
Beam		20'-0"
Draft		4'-6"
Freeboard:	Forward	7'-9"
	Waist	4'-4"
	Stern	1'-9"
Displ., cruising. trim		154,000 lbs.
Displ.-length ratio		318
Prismatic coefficient		.589
Pounds per inch imm.		4,847
Tankage, gals.:	water,	2,000
	fuel	2,000
Headroom		6'-7"

INDEX

INDEX